Computational Finite Element Methods in

Nanotechnology

Computational Finite Element Methods in

Nanotechnology

Edited by
Sarhan M. Musa

CRC Press
Taylor & Francis Group
Boca Raton   London   New York

CRC Press is an imprint of the
Taylor & Francis Group, an **informa** business

MATLAB® is a trademark of The MathWorks, Inc. and is used with permission. The MathWorks does not warrant the accuracy of the text or exercises in this book. This book's use or discussion of MATLAB® software or related products does not constitute endorsement or sponsorship by The MathWorks of a particular pedagogical approach or particular use of the MATLAB® software.

CRC Press
Taylor & Francis Group
6000 Broken Sound Parkway NW, Suite 300
Boca Raton, FL 33487-2742

First issued in paperback 2017

© 2013 by Taylor & Francis Group, LLC
CRC Press is an imprint of Taylor & Francis Group, an Informa business

No claim to original U.S. Government works

Version Date: 20120615

ISBN 13: 978-1-138-07688-4 (pbk)
ISBN 13: 978-1-4398-9323-4 (hbk)

Library of Congress Cataloging-in-Publication Data

Computational finite element methods in nanotechnology / editor, Sarhan M. Musa.
    p. cm.
    Summary: "This book provides an introduction to the key concepts of computational finite element methods (FEMs) used in nanotechnology in a manner that is easily digestible to a new beginner in the field. It provides future applications of nanotechnology in technical industry. Also, it presents new developments and interdisciplinary research in engineering, science, and medicine. The book can be used as an overview of the key computational nanotechnologies using FEMs and describes the technologies with an emphasis on how they work and their key benefits, for novices and veterans in the area"-- Provided by publisher.
    Includes bibliographical references and index.
    ISBN 978-1-4398-9323-4 (hardback)
    1. Nanotechnology--Mathematics. 2. Finite element method. I. Musa, Sarhan M.

T174.7.C668 2012
620'.20151825--dc23
                                              2012018247

Visit the Taylor & Francis Web site at
http://www.taylorandfrancis.com

and the CRC Press Web site at
http://www.crcpress.com

Dedicated to my father Mahmoud, my mother Fatmeh, and my wife Lama.

Contents

Preface..ix

Acknowledgments xv

Editor..xvii

Contributors.. xix

1. Overview of Computational Methods in Nanotechnology............................ 1
   Orion Ciftja and Sarhan M. Musa

2. Finite Element Method for Nanotechnology Applications
   in Nano-/Microelectronics.. 19
   Jing Zhang

3. Modeling, Design, and Simulation of N/MEMS by Integrating
   Finite Element, Lumped Element, and System Level Analyses................ 41
   Jason Vaughn Clark, Prabhakar Marepalli, and Richa Bansal

4. Nanorobotic Applications of the Finite Element Method............................85
   S. Sadeghzadeh, M.H. Korayem, V. Rahneshin, A. Homayooni, and M. Moradi

5. Simulations of Dislocations and Coherent Nanostructures...................... 149
   Anandh Subramaniam and Arun Kumar

6. Continuum and Atomic-Scale Finite Element Modeling of Multilayer
   Self-Positioning Nanostructures.. 185
   Y. Nishidate and G.P. Nikishkov

7. Application of Finite Element Method for the Design of Nanocomposites........... 241
   Ufana Riaz and S.M. Ashraf

8. Finite Element Modeling of Carbon Nanotubes and Their Composites 291
   Mahmoud Nadim Nahas

9. Finite Element–Aided Electric Field Analysis of Needleless Electrospinning 311
   Haitao Niu, Xungai Wang, and Tong Lin

10. Molecular Dynamic Finite Element Method (MDFEM) 331
    Lutz Nasdala, Andreas Kempe, and Raimund Rolfes

11. Application of Biomaterials and Finite Element Analysis (FEA) in
    Nanomedicine and Nanodentistry..373
    Andy H. Choi, Jukka P. Matinlinna, Richard C. Conway, and Besim Ben-Nissan

12. **Application of Finite Element Analysis for Nanobiomedical Study** 401
   Viroj Wiwanitkit

13. **Finite Element Method for Micro- and Nano-Systems for Biotechnology** 423
   Jean Berthier

14. **Design of the Nanoinjection Detectors Using Finite Element Modeling** 477
   Omer G. Memis and Hooman Mohseni

15. **Finite Element Method (FEM) for Nanotechnology Application
   in Engineering: Integrated Use of Macro-, Micro-, and Nano-Systems** 505
   Radostina Petrova, P. Genova, and M. Tzoneva

16. **Modeling at the Nano Level: Application to Physical Processes** 559
   Serge Lefeuvre and Olga Gomonova

Appendix A: Material and Physical Constants 585

Appendix B: Symbols and Formulas 589

Index 607

Preface

Computational methods of nanotechnology continue to contribute to the progress of innovations in areas ranging from electronics, microcomputing, and biotechnology to medicine, consumer supplies, aerospace, environment, and energy production.

Nowadays, finite element methods (FEMs) are widely used in computational nanotechnology and nanoscience. *Computational Finite Element Methods in Nanotechnology* can be used for new developments and future interdisciplinary research by engineers, scientists, biologists, and information business managers. It also provides an introduction to key concepts in a manner that is easily digestible to a beginner in the field. This book may also serve as a single source of reference to the veteran in the field. It is intended for a broad audience working in the fields of physics, chemistry, biology, medicine, material science, quantum science, electrical and electronic engineering, medicine, optical science, computer science, mechanical engineering, chemical engineering, and aerospace engineering. The book has been written for professionals, researchers, and students who would like to discover the challenges and opportunities concerning the development of the next generation of nanoscale computational nanotechnology using FEMs. In addition, it emphasizes the importance of FEMs for computational tools in the development of efficient nanoscale systems.

This book has 16 chapters and 2 appendices. Indeed, the book also contains vast recent applications of microscale and nanoscale modeling systems with FEMs using Comsol Multiphysics and MATLAB®.

Chapter 1 presents an overview to computational methods in nanotechnology. It describes the latest developments in nanotechnology, which offer the possibility for revolutionary advances in fundamental sciences and engineering. It also focuses on nanoscale structures relevant to nanotechnology, modeling methods, and FEMs for nanoscale circuits.

Chapter 2 presents a novel finite element (FE)-based, thermo-electrical-mechanical-coupled model to study mechanical stress, temperature, and electric fields in nano/microelectronics. First, the governing equations of electricity, heat transfer, mechanical behavior, and piezoelectricity are provided. A rigorous framework of coupling these phenomena is then derived. Two case studies are presented that illustrate the application of the FE model in nano/microelectronics. The first case study describes thermal stress modeling in wafer-level three-dimensional (3D) integration. The second case study presents modeling of material degradation in heterostructure field-effect transistors.

Chapter 3 presents an integration of distributed element, lumped element, and system level methods for the design, modeling, and simulation of nano/microelectro mechanical systems (N/MEMS). The authors show that the benefits of lumped elements are computational efficiency and ease of parameterization; however, such lumped models must preexist. For models that do not preexist, or if the analysis of single component is necessary, distributed element methods are often beneficial. However, distributed elements are computationally expensive, especially for devices with a multitude of multiphysical components. Often, reduced order methods are used to minimize the degrees of freedom for a more efficient model at the expense of trading off some accuracy. The integration of both distributed and lumped element methods can be useful for modeling micro- and nanosystems. Although the authors conclude that not every N/MEMS can be represented using the methods discussed here, a large number of devices can be efficiently explored using these methods. The examples they provide include parameterized lumped carbon nanotubes (CNTs); novice-friendly design, simulation, and layout of MEMS; an automated lumped-to-distributed verification method; and the design exploration of a nanomechanical material property tester.

Chapter 4 provides some descriptions and challenges in nanorobotic science with a real view on the simulation of nanorobotic systems and macro-dimensions that is applicable to small-scale nanorobots. It presents different types of nanorobots and their applications. It focuses on macroscale nanorobots, where the FEM can be applied, and develops different methods of manipulation at the nanoscale using these nanorobots. It also describes manipulation at the nanoscale using macroscale nanorobots, called nanomanipulation. The authors introduce relatively comprehensive models for these devices and classify effective parameters in nanomanipulation. In addition, they provide an example of nanosized robots and study the linear and nonlinear behavior of electrical nanogenerators.

Chapter 5 presents FEMs to illustrate the simulation of structures and processes such as dislocations, growth of epitaxial films, and precipitation. Important parameters can be computed from these simulations, which include domain deformations and region of stability of dislocations in thin plates and critical thickness for the formation of misfit dislocations. Further, new structures/phenomena like "zero stiffness material structures" and "reversible plastic deformation due to elasticity" can also be discovered using simulations at the nanoscale. In all examples considered in the chapter, the assumptions involve limitations of the methodologies, and precautionary measures to be undertaken are stated in order to plan for better strategies.

Chapter 6 presents modeling of the self-positioning nanostructures that is performed by the continuum mechanics theory, the FEM, and the atomic-scale FEM with consideration of cubic crystal anisotropy. The authors derive the continuum mechanics solution for multilayer thin film structures subjected to initial strains under generalized plane strain conditions. The chapter applies FE modeling to estimate the curvature radius of self-positioning hinges. It also develops an atomic-scale FE procedure to model self-positioning nanostructures. The results are then compared through modeling of bilayer self-positioning nanostructures.

Chapter 7 presents FEMs applicable to the modeling and designing of nanocomposites, which can be used to predict the mechanical, thermal, optical, and morphological properties of nanocomposites.

Chapter 8 presents the use of FEMs to anticipate the characteristics of carbon nanotubes (CNTs) and their composites. The author models discrete molecular structures as an equivalent truss element by equating the molecular potential energy of nanostructured materials with the mechanical strain energy of the truss element model. This modeling method applies to a graphene sheet. The chapter shows that the FE model performs very well and gives good results. As the FE model comprises a small number of elements, it performs under minimal computational time. The model has also been used to investigate the properties of single-walled carbon nanotubes (SWCNTs) and multi-walled carbon nanotubes (MWCNTs). The obtained values of Young's modulus agree very well with the corresponding theoretical results, which demonstrate that the proposed FE model may provide a valuable tool for studying the mechanical behavior of CNTs and nanocomposites. The author shows that the obtained values of Young's modulus agree very well with the corresponding theoretical results and experimental measurements that are available in the literature. The results demonstrate that the proposed FE model may provide a valuable tool for studying the mechanical behavior of CNTs and nanocomposites.

Chapter 9 summarizes the recent progress made in using FEM to analyze the electric field formed in needleless electrospinning and the relationship between the electric field and needleless electrospinning performance. The authors show that the FEM is an effective method to calculate electric fields and intensity profiles in needleless electrospinning. The chapter provides a visualized electric field profile, which greatly assists in correlating with electrospinning experiments.

Chapter 10 presents how molecular dynamic (MD) simulations can be integrated into the FEM. A new type of FE is required for force fields that include multibody potentials. These elements take into account not only bond stretch but also bending, torsion, and inversion without using rotational degrees of freedom. Since natural lengths and angles are implemented as intrinsic material parameters, the developed molecular dynamic finite element method (MDFEM) starts with a conformational analysis. By means of CNTs and elastomeric material, the authors demonstrate that this pre-step is needed to find an equilibrium configuration before the structure can be deformed in a succeeding loading step. The chapter presents the theoretical background of the MDFEM as well as guidelines for its implementation and usage. The authors conclude that, apart from mesh generation techniques, which are not covered, all important aspects of MDFEM are discussed from a finite element analysis (FEA) software user's point of view: what time integration schemes are usually available and when to use which; what the difference is between natural and equilibrium bond lengths and angles; how to obtain an equilibrium configuration, or when inversion energy is important; and how it can be transformed to torsion energy. Two examples demonstrate the accuracy and efficiency of the introduced MDFEM elements. MDFEM provides a framework that performs more than simple MD simulations. Conventionally, MD programs are used in chemistry and physics to perform conformational studies based on force fields. The aim is to determine equilibrium states rather than to study the response of atomic structures under mechanical loading. The main benefit of MDFEM is that concurrent multiscale simulations, i.e., a combination of continuum and atomistic regions, are feasible. Complex models can be developed to predict the properties of composites containing nanoparticles, which determine the behavior of the macroscopic material. For such models, parametric studies in terms of computer-aided material design can be carried out to analyze the influence of changes in the atomic structure, namely, the particle size, distribution, or the particle–matrix interface. The results can then be used to identify the basic mechanisms that lead to the enhancement of characteristic values of such composites and subsequently exploited to improve the manufacturing processes. Chapter 10 also consists of two appendices (Appendix 10A—the Newton–Raphson method, and Appendix 10B—stiffness matrices of MDFEM elements), which provide additional information.

Chapter 11 presents a brief background of the current applications of FEA in nanomaterials and systems used in medicine and dentistry. The authors examine the processes used for the production of nanocoatings and further analysis related to the factors affecting the nanoindented biomaterials by FEA.

Chapter 12 presents the application of FEA and its usefulness in medicine, with a special focus on nanomedicine. The author provides examples of reports based on this technique. The chapter shows that FEM is mainly helpful to study the nanostructure of several organs in the human body in both physiological and pathological conditions and to study nanosystem drug delivery and targeting.

Chapter 13 presents a few examples of the potentialities of the FEM numerical approach for the design of microfluidics systems for biotechnology. The author concludes that microflows, chemical and biochemical reactions, and concentration transport can be modeled by using FEM methods. However, as pointed out in the introduction to the chapter, much remains to be done in the domain of multiphase microflows, where tracking interface motion and pinning is essential, and in the domain of the transport of large, deformable particles (e.g., vesicles and globules), where modeling the steric aspects and deformation associated to the local shear is a real challenge.

Chapter 14 concludes that, with a nontraditional geometry composed of nano-scale sensing and amplification nodes on thick absorption layers, the nano-injection detectors offer high-sensitivity photon detection and amplification. However, due to their nonplanar design, type-II band alignment, and the coupled detection/amplification mechanisms, the design and development of nano-injection devices require detailed, nonlinear, 3D FEM simulations. The authors provide the FEM simulation a multiphysics environment for stationary, parametric, and transient simulations, all based on the drift–diffusion equations in 2D and 3D. The model incorporates several nonlinearities such as the incomplete ionization of dopants, bimolecular recombination, Auger recombination, nonlinear mobility, impact ionization, thermionic emission, hot electron effects, surface recombination effects, and temperature effects. Furthermore, once the devices are optimized through simulations, the nano-injection devices are fabricated through micro- and nanofabrication. The measurements show amplification factors reaching 10,000 with low dark current densities and Fano noise suppression. The passivated nano-injection devices have bandwidths exceeding 3 GHz with a jitter of 15 ps. To form focal plane array infrared cameras, 320-by-256 pixel arrays of nano-injection detectors are hybridized. Also, with a pixel level responsivity exceeding 2500 A/W, the nano-injection focal plane arrays show a noise level of 28 electrons at a frame rate of 1950 fps. These imagers show two orders of magnitude improved signal-to-noise ratio compared to commercial SWIR imagers at thermoelectric cooling temperatures. All of these demonstrate the capabilities of nano-injection detectors and imagers and make them excellent candidates for demanding applications such as optical tomography, satellite imaging, nanodestructive material inspection, high-speed quantum computing and cryptography, night vision imaging, and machine vision for process control.

Chapter 15 presents the use of the spatial motion of the coupler from the traditional spatial four-bar linkage (SFL) as an output link (OL) and, in particular, as a carrier of the bioreactor chamber. Also, the authors provide the use of SFL with two spherical joints and two degrees of freedom (DoF), wherein the full rotation of the coupler or the bioreactor chamber, respectively, around its own axis is performed by a second actuator. The module of the bioreactor chamber with the actuator is connected to the coupler of the mechanism through a fixed connection.

Chapter 16 demonstrates that automatic meshing and its associated nano-objects is a convenient way to produce, in 2D and 3D as well, full spaces without any vacuum. As a consequence, it lets the boundaries display their specific behavior. This is compatible not only with all the geometrical shapes, circles, spheres, and so on but also with the shapes extracted from experimentation as shown with the bean in the chapter. Also, the authors conclude that automatic meshing can be seen as a way to be filled with nanoparticles and implement empty spaces between objects.

Two different applications are described all along the chapter: first, computation of mean values of physical constants of heterogeneous material, for instance, thermal and electric conductivity, permittivity, etc. with or without the grain joins influence; second, evaluation of capillary flows leading to an inside knowledge of permeability, and it letting out of the computation of the nanoparticles which acts only by their surfaces, but keeping them. On these geometries, it is possible to use all the classical FEMs used for drying, sintering, microwave heating, and so on. As a consequence, when a problem is solved in a small domain, it is possible to enlarge the solution to larger domains, just by matrix association. Comsol Multiphysics and MATLAB softwares are used (run on an HP Z800, 16 proc, 64 Gbits).

Finally, the book concludes with two appendices. Appendix A shows common material and physical constants, with the consideration that the material constants values varied from one published source to another due to many varieties of most materials and also because conductivity is sensitive to temperature, impurities, moisture content, as well as the dependence of relative permittivity and permeability on temperature and humidity and the like. Appendix B provides common symbols and useful mathematical formulas.

COMSOL and COMSOL Multiphysics are registered trademarks of COMSOL AB. For details visit www.comsol.com

MATLAB is a registered trademark of The MathWorks, Inc. For product information, please contact:

The MathWorks, Inc.
3 Apple Hill Drive
Natick, MA, 01760-2098 USA
Tel: 508-647-7000
Fax: 508-647-7001
E-mail: info@mathworks.com
Web: www.mathworks.com

Sarhan M. Musa

Acknowledgments

My sincere appreciation and gratitude to all the book's contributors. Thank you to Brian Gaskin and James Gaskin for their wonderful hearts and for being great American neighbors. It is my pleasure to acknowledge the outstanding help of the team at Taylor & Francis Group/CRC Press in preparing this book, especially from Nora Konopka, Michele Smith, Kari Budyk, and Joette Lynch. Thanks also to Dr. Vinithan Sethumadhavan from SPi Global for his outstanding suggestions. I would also like to thank Dr. Kendall Harris, my college dean, for his constant support.

Finally, the book would never have seen the light of day if not for the constant support, love, and patience of my family.

Editor

Sarhan M. Musa, PhD, is currently an associate professor in the Department of Engineering Technology at Prairie View A&M University, Texas. He has been director of Prairie View Networking Academy, Texas, since 2004. Dr. Musa has published more than 100 papers in peer-reviewed journals and conferences and is the editor of *Computational Nanotechnology Modeling and Applications with MATLAB®*. He currently serves on the editorial board of the *Journal of Modern Applied Science*. Dr. Musa is a senior member of the Institute of Electrical and Electronics Engineers (IEEE) and is also a 2010 Boeing Welliver Fellow.

Contributors

S.M. Ashraf (retired)
Materials Research Laboratory
Department of Chemistry
New Delhi, India

Richa Bansal
School of Mechanical Engineering
Purdue University
West Lafayette, Indiana

Besim Ben-Nissan
Faculty of Science
Department of Chemistry and Forensic
  Science
University of Technology
Sydney, New South Wales, Australia

Jean Berthier
Department of Biotechnology
CEA Leti
and
Department of Physics
University Joseph Fourier
Grenoble, France

Andy H. Choi
Faculty of Dentistry
Dental Materials Science
The University of Hong Kong
Pokfulam, Hong Kong, People's Republic
  of China

Orion Ciftja
Department of Physics
Prairie View A&M University
Prairie View, Texas

Jason Vaughn Clark
School of Electrical and Computer
  Engineering
and
School of Mechanical Engineering
Purdue University
West Lafayette, Indiana

Richard C. Conway
Faculty of Science
Department of Chemistry and Forensic
  Science
University of Technology
Sydney, New South Wales, Australia

P. Genova
Institute of Mechanics
Bulgarian Academy of Science
Sofia, Bulgaria

Olga Gomonova
Department of Higher Mathematics
Siberian State Aerospace University
Krasnoyarsk, Russia

A. Homayooni
Nanorobotic Group in Robotic Research
  Laboratory
College of Mechanical Engineering
Iran University of Science and Technology
Tehran, Iran

Andreas Kempe
Institute of Structural Analysis
Leibniz Universität at Hannover
Hannover, Germany

M.H. Korayem
Nanorobotic Group in Robotic Research
  Laboratory
College of Mechanical Engineering
Iran University of Science and Technology
Tehran, Iran

Arun Kumar
Department of Materials Science and
  Engineering
Indian Institute of Technology
Kanpur, India

Serge Lefeuvre
Eurl Creawave
Toulouse, France

Tong Lin
Australian Future Fibres Research and
    Innovation Centre
Deakin University
Geelong, Victoria, Australia

Prabhakar Marepalli
Mechanical Engineering Department
University of Texas
Austin, Texas

Jukka P. Matinlinna
Faculty of Dentistry
Dental Materials Science
The University of Hong Kong
Pokfulam, Hong Kong, People's Republic
    of China

Omer G. Memis
Bio-Inspired Sensors and Optoelectronics
    Laboratory
Department of Electrical Engineering and
    Computer Sciences
Northwestern University
Evanston, Illinois

Hooman Mohseni
Bio-Inspired Sensors and Optoelectronics
    Laboratory
Department of Electrical Engineering and
    Computer Sciences
Northwestern University
Evanston, Illinois

M. Moradi
College of Mechanical Science and
    Engineering
Semnan University
Semnan, Iran

Sarhan M. Musa
Department of Engineering Technology
Prairie View A&M University
Prairie View, Texas

Mahmoud Nadim Nahas
Department of Mechanical Engineering
King Abdulaziz University
Jeddah, Saudi Arabia

Lutz Nasdala
Faculty of Business Administration and
    Engineering
Offenburg University of Applied Sciences
Offenburg, Germany

G.P. Nikishkov
School of Computer Science and
    Engineering—Division of Information
    Systems
University of Aizu
Aizu-Wakamatsu, Fukushima, Japan

Y. Nishidate
School of Computer Science and
    Engineering—Division of Information
    Systems
University of Aizu
Aizu-Wakamatsu, Fukushima, Japan

Haitao Niu
Australian Future Fibres Research and
    Innovation Centre
Deakin University
Geelong, Victoria, Australia

Radostina Petrova
Faculty of Engineering and Education
Technical University of Sofia
Sofia, Bulgaria

V. Rahneshin
Nanorobotic Group in Robotic Research
    Laboratory
College of Mechanical Engineering
Iran University of Science and Technology
Tehran, Iran

Ufana Riaz
Materials Research Laboratory
Department of Chemistry
New Delhi, India

Raimund Rolfes
Institute of Structural Analysis
Leibniz Universität at Hannover
Hannover, Germany

S. Sadeghzadeh
Nanorobotic Group in Robotic Research
  Laboratory
College of Mechanical Engineering
Iran University of Science and Technology
Tehran, Iran

Anandh Subramaniam
Department of Materials Science and
  Engineering
Indian Institute of Technology
Kanpur, India

M. Tzoneva
Faculty of Engineering and Education
Technical University of Sofia
Sofia, Bulgaria

Xungai Wang
Australian Future Fibres Research and
  Innovation Centre
Deakin University
Geelong, Victoria, Australia

Viroj Wiwanitkit
Hainan Medical University
Haikou, People's Republic of China

and

Wiwanitkit House
Bangkok, Thailand

Jing Zhang
Department of Mechanical Engineering
Indiana University—Purdue University
  Indianapolis
Indianapolis, Indiana

1

Overview of Computational Methods in Nanotechnology

Orion Ciftja
Prairie View A&M University, Prairie View, Texas

Sarhan M. Musa
Prairie View A&M University, Prairie View, Texas

CONTENTS

1.1 Introduction 1
1.2 Nanoscale Structures Relevant to Nanotechnology 3
1.3 Modeling Methods 4
    1.3.1 Modeling of Carbon Nanotubes and Nanocomposites 5
    1.3.2 Modeling of Electronic Quantum Dots 6
    1.3.3 Modeling of Quantum Wires and Nano-MOSFET Devices 8
1.4 Finite Element Method for Capacitance Extraction of Interconnects in Microscale Circuits 10
    1.4.1 Microscale Single Interconnect Line on Si–SiO$_2$ Substrate 11
    1.4.2 Microscale Coupled Interconnect Lines on Si–SiO$_2$ Substrate 11
1.5 Conclusion 12
Acknowledgment 15
References 15

1.1 Introduction

If one considers which research areas in physics, chemistry, and engineering experienced the strongest growth in the last 10 years, then it is likely that material sciences and nanotechnology stand out as front runners. While it is fair to say that material sciences have always been important, it is also true to state that until very recently they were somehow limited in scope. Before the nanotechnology revolution, all material sciences research was basically dominated by physics and engineering. The major driving force behind such research were attempts in computer and information technologies to miniaturize transistors and electronic processors. Essentially, all was a top-down strategy: start with a macroscopic device and then try to make it smaller and smaller.

Nanotechnology introduced an absolute change in this mindset. Namely, the new nanoscale branch of modern material sciences is now more concerned with a bottom-up strategy. Nowadays, one wants to manipulate atoms/molecules in such a way as to form new artificial nanostructures with defined properties either by self-assembly or by self-organization. At present, nanotechnology reaches from nanoelectronics to biomedical applications, and the importance of the field can in no way be underestimated. Nanotechnology offers unlimited possibilities for advancement in many physical and engineering sciences. It also offers unprecedented possibilities for development of novel technologies, as well. A wide variety of nanomaterials are now used in engineering, pharmaceutical, biomedical products, as well as other industries. While nanoscale materials possess more novel and unique physical–chemical properties than bulk materials, they are not easy to study. Therefore, studies of nanomaterials have generated intense scientific curiosity, attracting much attention for the last few years.

Together with the experimental developments in nanotechnology, the fundamental techniques of theory and modeling have seen a revolution that parallels the advances on which the field of nanotechnology is based. The last two decades have seen the development of density functional theory (DFT), classical Monte Carlo (MC) techniques, quantum Monte Carlo (QMC) methods, molecular dynamics (MD) simulations, and fast multigrid algorithms. New insights have come from the application of these and other new theoretical and computational tools. Advances in computing and combination of new theoretical methods with high computer power have made possible the simulation of complex systems with million degrees of freedom.

Advances in nanotechnology have created a more pressing need for a better quantitative understanding of nanoscale systems. Absence of quantitative models and robust computational methods applied to newly observed nanoscale phenomena increasingly limits a quicker progress in the field. The use of the full potential of novel theoretical and modeling tools has the great beneficial effect to seriously accelerate widespread applications in many areas of nanotechnology. Realizing this potential, however, will require long-term fundamental research and expanded educational opportunities to train the next generation of scientists and engineers whose job is to overcome fundamental theoretical and computational challenges in nanotechnology. Although our ability to synthesize and fabricate various nanostructures such as quantum dots, quantum wires, carbon nanotubes, molecular magnets, etc., has constantly improved, we have not reached yet the phase of being able to incorporate them together in larger functional systems or devices.

From a theoretical and modeling perspective, it is not easy to study or model the properties of systems that span the whole range from macroscopic to microscopic length and time scales. It is also not easy to determine the transport mechanisms of various devices at the nanoscale. Studies of nanointerfaces generally are quite difficult, and it is not easy to describe with reasonable accuracy the response of nanoscale structures to external probes such as electric field, magnetic field, radiation, etc. Nevertheless, even though some of the challenges given earlier appear insurmountable, opportunities for research and discovery in nanotechnology outweigh the risks by far and large. New tools and techniques are giving us the ability to put atoms and molecules where we want them. Researchers are discovering new properties that emerge at nanometer length scales that are different from the properties of both individual atoms/molecules and bulk materials. Scientists and engineers have successfully synthesized and characterized a broad range of fundamental nanosystems with potentially useful properties. There is a convergence in length scales between inorganic nanostructures and biomolecules such as DNA and proteins. Nanostructures such as quantum dots are being used as biosensor assays. Overall, these are exciting times for the field of nanotechnology.

Latest nanotechnology developments offer the possibility for revolutionary advances in fundamental sciences and engineering. Nanotechnology raises research issues that are fundamental to a growing number of disciplines and the potential for applications of enormous economic and social significance. The basic theoretical approaches and modeling methods that go on par with such scientific developments have seen a revolution that parallels the experimental advances on which the field of nanotechnology is based. Advances in computing and combination of new theoretical methods with high computer power have shed more light on the properties of various nanoscale systems, but at the same time, these works indicate that current algorithms and numerical methods must be made more efficient and, perhaps, new ones should be invented. Clearly, the nanotechnology revolution has created an urgent need for more robust computational methods to understand the properties of matter at the nanoscale.

In this chapter, we will give a brief overview of some modeling challenges faced in the field of nanoscale research and will describe some of the most important computational methods already in use in various nanotechnology disciplines.

1.2 Nanoscale Structures Relevant to Nanotechnology

In recent years, interest in the area of nanotechnology has exploded worldwide including many institutes, laboratories, and universities where researchers from different disciplines have been working together in many aspects of nanotechnology. Nanotechnology represents a compelling case to bring groups of multidisciplinary scientists to work together on understanding phenomena at the nanoscale. Only this approach will allow us to have a share in the nanotechnology research, create strong educational programs, institute interdisciplinary research areas, spark additional collaborations between scientists, and at the end harvest all the expected benefits. This approach stimulates the formation of alliances and teams of scientists with diverse background to meet the challenge of developing a broad quantitative understanding of structure and dynamics at the nanoscale. Such cohesion between different disciplines is key to sparking additional collaboration across disciplinary boundaries and addressing some critical research issues in this fast-evolving field.

Given the rapid expansion of the field of nanotechnology, it is practically impossible to mention all the nanoscale structures or devices that are currently used or studied. Because of this, in this work, we are not even attempting to give a detailed description. On the contrary, we will focus our attention on few important nanoscale devices that, in our view, are relevant to the field of nanotechnology. We have placed in this category structures such as carbon nanotubes/composites, quantum dots, and quantum wires, just to mention a few. The key idea of our approach is to give a rather brief overview of the properties of such structures and then describe various computational methods used to study their properties. Well-characterized nanoscale elements like the ones mentioned earlier need to be quantitatively understood. Just as knowledge of the atom allows us to make and manipulate larger structures, knowledge of important nanoscale elements will allow us to reliably manufacture larger artificial structures with prescribed properties. Such nanoscale elements will be the centerpiece of new functional nanomechanical, nanoelectronic, and nanomagnetic devices. Thus, a quantitative understanding of the electronic, magnetic, transport, and mechanical properties of key nanoscale elements is crucial to building novel technological devices.

Without a coherent description of some of these key nanoscale elements, overall progress in the field of nanotechnology will be limited. In the following, we focus our attention basically on three groups of nanoscale structures relevant to nanotechnology:

1. Carbon nanotubes and composite systems
2. Electronic quantum dots
3. Quantum wires and nano-metal–oxide–semiconductor field-effect transistor (nano-MOSFET) devices

These important nanoscale structures are well defined by experiment and tractable using standard theoretical and computational tools. Moreover, they have been demonstrated to hold promise in future nanotechnologies. All computational methods devised to study the properties of these nanoscale structures attempt to solve accurately problems such as understanding the response of nano-building blocks and nanodevices to external probes, explore novel theories and models to predict behavior and reliability of nanosensors and devices, understand classical and quantum transport in nanostructures, and so on.

1.3 Modeling Methods

Studies of nanoscale structures offer great promise but require new theoretical approaches and computationally intensive studies. MD simulation methods can handle systems with tens of thousands of atoms; however, to fully exploit their power, algorithms need to be made scalable and fully parallelized. Such methods are especially useful in providing benchmarks for those systems, where experimental data are unreliable or hard to reproduce. Lack of clear prescriptions for obtaining reliable results that apply to nanostructures is another challenging problem for experiment and theory. While new experiments will need to be designed to ensure reproducibility and the validity of the measurements, the theoretical challenge is to construct new theories that would cross-check such conclusions.

DFT methods with standard exchange functionals are only partially satisfactory when calculating band structures of metals and, especially, semiconductors. Standard DFT fails even totally when describing van der Waals complexes (physisorption) or single-walled carbon nanotubes (SWCNTs), a behavior which is well known to many physicists. Within the DFT, the band structure calculations are generally routine task in solid-state physics; however, the accuracy of such calculations, especially with regard to band gaps, deteriorates when applied to nanoscale structures.

While great strides have been made in classical MC and QMC simulation methods, a number of fundamental issues remain. In particular, the diversity of time and length scales remains a great challenge at the nanoscale. Intrinsic quantum attributes like transport and charge transfer are difficult to incorporate into a classical description, and at best, MC simulation methods are effective and easy to implement only at zero or very low temperatures. Even though the QMC method and its variants, variational Monte Carlo (VMC), diffusion Monte Carlo (DMC), and Green's function Monte Carlo (GFMC), are currently the most accurate methods that can be extended to systems in the nanoscience range, major improvements are needed. Many nanosystems are driven more by entropy (temperature) than by energy. Modifications of current methods that are entropy (temperature) or free-energy friendly are badly needed. Current methodologies for free energy (temperature-dependent methods) such as the path integral Monte Carlo (PIMC) method are usually extremely complex and not friendly.

In contrast, continuum methods have had much success in the macroscale modeling and simulation of nanostructures. Finite element methods (FEM) are now the standard numerical analysis tool to study diverse problems. Therefore, the logical approach taken by many researchers in the desire to create truly multiple-scale simulations that exist at disparate length and time scales has been to couple various methods like MD, MC, and FEM in some manner and apply them to nanostructures. Unfortunately, the coupling of these methods is neither easy nor straightforward.

1.3.1 Modeling of Carbon Nanotubes and Nanocomposites

Carbon nanotubes are molecular-scale tubes of graphitic carbon with outstanding properties and unique characteristics. They are among the strongest fibers known and have remarkable electronic properties. For these reasons, they have attracted huge theoretical and experimental interest. Technological applications have also been forthcoming though at somehow slower pace given the relative high production costs of manufacturing high-quality nanotubes. The current huge interest in carbon nanotubes is a direct consequence of the synthesis of buckminsterfullerene, C_{60}, and other fullerenes, in 1985. The discovery that carbon could form stable, ordered structures other than graphite and diamond stimulated researchers worldwide to search for other new forms of carbon. The research was given new impetus when it was shown that C_{60} could be produced in a simple arc-evaporation apparatus readily available in virtually all laboratories. The fullerene-related carbon nanotubes generated this way contained at least two layers, often many more, and ranged in outer diameter from about 3–30 nm. The bonding in carbon nanotubes is sp^2, with each atom joined to three neighbors, as in graphite. The nanotubes can therefore be considered as rolled-up graphene sheets (graphene is an individual graphite layer). The strength of the sp^2 carbon–carbon bonds gives carbon nanotubes amazing mechanical properties such as very high stiffness, very high breaking strain, etc. These properties, coupled with the lightness of carbon nanotubes, give them great potential in applications such as aerospace.

The electronic properties of carbon nanotubes are also extraordinary. Especially notable is the fact that nanotubes can be metallic or semiconducting depending on their structure. Thus, some carbon nanotubes have conductivities higher than that of copper, while others behave more like silicon. There is great interest in the possibility of constructing nanoscale electronic devices from nanotubes, and some progress is being made in this direction. However, in order to construct a useful device, we would need to arrange many thousands of nanotubes in a defined pattern, and we do not yet have the degree of control necessary to achieve this task.

Thus, a lot of research is being done to design new nanoelectronic devices utilizing the extraordinary properties of SWCNTs. The design work comprises the exploration of a systematic functionalization concept for SWCNTs, the description of possible junctions between semiconducting and metallic parts of a nanotube, and the construction of more complex nanoelectronic devices.

In particular, a great deal of research has been focused to understand the electrical properties of carbon nanotubes and carbon-nanotube composites. Electrical properties of carbon nanotubes depend on aromatic structure and wrap style. Several studies [1,2] have indicated that nanotubes possess excellent electromagnetic properties. For example, one can use controlled dispersion of specifically designed carbon nanotubes into supporting polymer matrices. Experiments have shown that the percolation (onset of conductivity) for these nanocomposites is less than one-half of 1% by volume. The electrically conductive polymer nanocomposite materials, compared to conductive metal-filled systems, offer substantial weight savings, flexibility, durability, low-temperature processability, and tailored reproducible conductivity. The material's enhanced high-frequency absorption capability suits the cable shielding very well [3]. So far, the main challenge in such studies is to develop modeling methods capable of bridging the gap between nanoscale level structures and their micro-/macro-level counterparts.

A carbon nanotube consists of a large number of atoms. Thus, modeling of nanotubes requires sophisticated numerical tools, powerful computer processors, and three-dimensional visualization. The work generally begins with modeling of a single nanotube for its electrical conductivity and continues with modeling of the interactions of a nanotube with the surrounding polymer molecules in a polymer-based composite. Among the many available software for nanoscale modeling, we mention NWChem, ADF, ABINIT, VASP, and Spartan. Bridging the gap between nanoscale modeling and micro/macro modeling is a very challenging research task. The major challenge is to develop reliable methods for calculating effective properties that are passed on from nanotube level to nanocomposite level. Possible approaches include (1) accurate data management between two levels of computer codes, nanoscale computer code and microscale computer code, and (2) establishment and incorporation of constitutive relations into finite element software code to predict nanomaterial properties. Specific approaches of transferring modeling parameters (electrical conductivity) include the application of "exact scaling laws" [4] for electrical conductivity properties of polymer nanocomposites to bridge the orientation distribution of macromolecules of composites.

Finite element codes like ANSYS, COMSOL, and Nastran are available and have been applied to simulate electromagnetic field and a single electrode. Research work in finite element areas begins the modeling of a nanocomposite shell in electromagnetic field. Available electrical properties of nanocomposite materials derived from experimental data are then applied in conjunction with the finite element modeling. Results obtained from the finite element analysis are used as benchmarks for further studies. When computational modeling results become available, one starts the work of bridging the nanoscale to macroscale gap. Tasks may include data management, establishment of constitutive relations, probability theories, and other emerging methods. Studies of the influence of external factors such as electric field, magnetic field, and radiation effects on carbon-nanotube polymer-based composites and carbon-nanotube thin films are crucial to understand a variety of possible applications. This includes aerospace applications in which devices made from these materials would serve, for instance, as radiation shielding. While external factors will cause some degradation of given properties of these materials, it is also possible that they may enhance other desirable properties.

1.3.2 Modeling of Electronic Quantum Dots

The properties of electronic materials, such as single or coupled two-dimensional (2D) semiconductor quantum dots in presence of external factors such as electric field, magnetic field, or radiation, are not easy to investigate.

Semiconductor quantum dots are fabricated nanostructures in which charge carriers, such as electrons, are confined in a small spatial region usually in 2D. In heterostructures, such as GaAs/AlGaAs, the number, N, of electrons in the dot can be changed by adjusting the external electric potential of the metallic electrodes (gates). In many cases, the confinement potential of the electrons in the dot is quasi-parabolic and the number of trapped electrons varies from N = 1 or 2 to thousands of electrons [5–9]. The Bohr radius (size) of quantum dots is much larger than the Bohr radius of "real atoms." This implies that electromagnetic radiation with wavelength of the order of Bohr radius and ordinary magnetic fields (of the order of 1 T) controls the energy deposition and magneto-transport properties of quantum dots [10,11].

While the response of a quantum dot system to an electromagnetic field must be carefully investigated, the effects of a stationary magnetic field can be anticipated. From the theoretical point of view, a magnetic field applied perpendicular to the 2D quantum dot plane will change the nature of single-electron levels from those of the harmonic oscillator to Landau levels. In the presence of a perpendicular magnetic field, the electronic spectrum will consist of discrete energy levels and will strongly depend on the applied perpendicular magnetic field which also affects the electron's spin [12–14]. Exposure to radiation will qualitatively change the magneto-transport properties of a quantum dot. The simplest effect that arises naturally when radiation illuminates a quantum dot will be the tunneling of electrons from the parent quantum dot. First, the electron moves away from parent quantum dot, but if the radiation field is properly tuned, the electron can be driven back to his parent quantum dot, and phenomena such as quantum interference and tunneling may occur. It is useful to think of a "nano-machine" where the quantum dot provides the electron and the electromagnetic field accelerates the system. We can also think in terms of a "nano-interferometer" where the electronic wave function consists of two pieces as a result of tunneling from the parent quantum dot. One component of the electronic wave function is accelerated away from the quantum dot by the radiation; however, there is a second component of the wave function that remains bound to the quantum dot. When the two components superimpose, a "nano-interferometer" induced by a properly tuned external radiation may be created.

While the magneto-transport properties of isolated 2D quantum dots containing few (N = 1, 2, 3,...) electrons have recently been investigated both experimentally and theoretically [10–18], small arrays of quantum dots in which electrons in different quantum dots are weakly coupled to each other represent new challenges. From the modeling perspective, as a first step, one considers small (e.g., linear or other regular) arrays of few (N = 2, 3, ...) quantum dots each containing a small number (N = 1, 2, ...) of electrons. The inter-dot distance between nearest neighbor quantum dots can be experimentally adjusted. A given energy barrier separates the electrons of one quantum dot from tunneling to the other quantum dot. A two-well potential (with finite range) can be used to model the coupling of any two harmonic wells centered at a distance, d, apart. It is essential that electrons be allowed to tunnel between the dots. The tunneling properties of the electrons crucially depend on the energy barrier height or equivalently the inter-dot distance, d, and will certainly be affected by external factors.

Computational studies of the properties and the response to magnetic field and radiation of arrays of few weakly coupled quantum dots are very important. The study of weakly coupled arrays of quantum dots is a subject of great fundamental and practical interest since the magneto-transport properties of arrays of quantum dots can be dramatically and fundamentally different from the properties of the individual quantum dots that constitute them. While each individual quantum dot in the system still retains the set of well-defined charge states like in an atom, tunneling of electrons between two dots may fundamentally change the individual properties of each of the quantum dots including charging effects or Coulomb oscillations. Thus, one might reasonably expect novel magneto-transport features for the case of weakly coupled arrays of quantum dots. In principle, the interplay between quantum confinement, electronic correlation, inter-dot coupling, magnetic field, and external radiation should manifest itself in many interesting physical phenomena with possible technological applications. The simplest model able to capture the essential behavior of such a system should, at least, consider the following:

1. Arrays of $n = 2, 3$, quantum dots at equal distance, d, each containing $N = 1, 2, \ldots$ electrons

2. Inter-dot coupling between electrons much weaker than intra-dot coupling

3. Confined 2D electrons in each individual quantum dot

4. Magnetic field perpendicular to the 2D plane of the quantum dot

5. Parabolic confining potential with a finite range for the electrons in each of the individual quantum dots with a finite energy barrier separating the harmonic wells of one quantum dot from the other

6. Electrons in each of the individual quantum dots that interact with a Coulomb potential

7. External radiation (electromagnetic field, lasers, etc.)

Even though the model introduced earlier is quite idealized, one can immediately note the computational challenges faced when dealing with the case. At an elementary level, one might initially use standard quantum chemistry methods or molecular physics methods which rely on knowledge of individual quantum dots. Single-electron or reliable wave functions for electrons in an isolated quantum dot can be extracted from exact numerical diagonalization methods [15,16], Hartree–Fock methods [17], DFT methods [18], and QMC methods [19–21]. While the study of arrays of coupled quantum dots is a difficult problem, the analogy between real atoms and quantum dots ("artificial atoms") provides us with a powerful set of methods from molecular physics such as the valence bond Heitler–London (HL) approach or the linear combination of atomic orbitals (LCAO) approach [22]. For arrays or coupled systems of quantum dots, one can use prior results for isolated semiconductor quantum dots and their wave functions [23]. These wave functions or similar wave functions from literature can be used to construct the initial HL and/or LCAO wave functions, which would describe the arrays of coupled quantum dots.

   On the other hand, quantum calculations involve multivariable integrations; therefore, QMC simulations should be employed, as the next step. Among several computational methods, the QMC methods have the greatest advantage of all since they are not biased by approximations, are very accurate, and can be extended to larger systems in a straightforward manner. They are also known to give very reliable results for strongly correlated electronic systems of this nature.

1.3.3 Modeling of Quantum Wires and Nano-MOSFET Devices

Another challenging problem from the modeling perspective is the study of the properties of MOSFET devices at the ballistic quantum regime. In this regime, the interconnected wires of a nanoscale transistor are so thin that they can transport so few electrons that the whole system behaves as a quantum wire with electrons manifesting unusual transport properties. As transistor devices become smaller in size and hence faster in switching, the interconnect wires become narrower and hence more resistive and slower in transmitting signals. Since the advent of 250 nm technology at 1997, the signal transmission delay caused by interconnect wires has dominated the total delay and transistor miniaturizations. As we approach the next decade of the microelectronics revolution, microprocessors and related devices will not only get smaller and more powerful but also more difficult to understand and manufacture.

Numerical simulations are currently used to guide the development of analytical theories necessary for the understanding of ballistic field-effect transistors and the effects of external factors (magnetic field, electric field, radiation, etc.) on nano-MOSFET devices on the ballistic regime [24,25]. The developed analytical models provide insights into the performance of nano-MOSFETs near the scaling limits in the nano-regime and a unified framework for assessing and comparing a variety of novel transistors. As illustrated by recent engineering advances, it is evident that the MOSFET channel lengths continue to shrink rapidly toward the sub-10 nm dimensions called for by the International Technology Roadmap for Semiconductors [26,27]. Coupled with the use of high-mobility channel materials [28,29], nanoscale channel lengths open up the possibility of near-ballistic MOSFET operation with channel lengths essentially behaving as quantum wires.

For these reasons, accurate modeling studies of the ballistic operation in nano-MOSFETs are of paramount importance. The operation of a nano-MOSFET in the ballistic regime has recently been explored by simple, analytical models [30,31], as well as by detailed numerical simulations [32–36]. When considering a 10 nm sized transistor, we have to bear in mind that we are dealing with a small number of charge carriers. Control of charge and electrical current on a single-electron level will be required. Moreover, quantum phenomena will increasingly start to dominate the overall behavior of such structures. Finally, tinny structures have a large surface-to-volume ratio which is very challenging for conventional semiconductor devices. It is for certain that new theoretical concepts need to be developed and novel computational methods be introduced.

So far, modeling work in this direction has been focused on the following:

- Development of a ballistic, analytical model for nano-MOSFETs in the nanoscale regime incorporating external parameters (electric field, magnetic field, radiation, induced defects, etc.)

- Employment of ballistic models to further explore the physical properties of nano-MOSFETs in the quantum wire regime

- Quantification of the usefulness of the ballistic model in exploring new nanomaterials and nanostructures

- Quantification of the reliability implications for nano-MOSFETs in the ballistic nano-regime with specific emphasis on hot-electron injection

Analytical theories and quantum simulation methods (similar to those used for quantum dots) are used to investigate the properties of nano-MOSFETs in the ballistic regime. Semiclassical ballistic Boltzmann transport theory [34,37] and the nonequilibrium quantum transport Green's function formalism [38,39] are the main theoretical tools. A combined theoretical and modeling effort along these lines is the only alternative to capture the essential physics controlling the properties of nano-MOSFETs in the ballistic quantum regime.

1.4 Finite Element Method for Capacitance Extraction of Interconnects in Microscale Circuits

Due to the complexity of electromagnetic modeling, researchers and scientists always look for the development of accurate and fast methods to extract the parameters of electronic interconnects. In recent years, we have observed a magnificent application and development in the complexity, density, and speed of operations of integrated circuits (ICs), multichip modules (MCMs), and printed circuit board (PCB). For example, MCMs are extensively used to reduce interconnection delay and crosstalk effects in complex electronic systems. Multiconductor transmission lines embedded in multilayered dielectric media are known as the basic interconnection units in ICs and MCMs and have been characterized with the distributed circuit parameters such as capacitance C matrices under quasi-TEM conditions. Also, these distributed circuit parameters are very important factors in the electrical behavior and performance of other microwave integrated circuits (MICs) and very-large-scale integration (VLSI) chips.

In today's electronic circuits, there is very high density of devices in the circuit which requires an extremely large number of very-high-speed interconnects. Additionally, for advances in nanofabrication of high-speed ICs, it is essential to examine the limitations due to the parasitic coupled mechanisms present in micro-/nanoscale silicon-IC processes. To optimize electrical properties of micro-/nanoscale IC interconnects such as minimization of the length of the interconnection lines, an adequate attention must be given to the geometrical size of their transverse cross sections; the estimation of the transmission line parameters requires accuracy for system design. Furthermore, the determination of interconnect capacitance turns out to be a great challenge to the design engineers to obtain an optimum interconnect capacitance to reduce the resistance–capacitance delay time for micro-/nanoscale ICs.

Furthermore, multiconductor multilayered structures are essential for micro-/nanoscale ICs, MCMs, and PCB systems due to the important effects on the transmission characteristics of high-speed signals. Also, the transmission line effect on the micro-/nanoscale IC interconnects becomes extremely important for the transmission behavior of interconnect lines on a silicon–silicon dioxide (Si–SiO$_2$) semiconducting substrate. The conducting silicon substrate causes capacitive and inductive coupling effects in the structure. In this work, we will illustrate the power of the FEM through the design of micro-/nanoscale single and coupled interconnect lines on a Si–SiO$_2$ substrate. We will focus our attention on the calculation of the capacitance per unit matrices of microscale single and coupled interconnect lines on a Si–SiO$_2$ substrate, and we will determine the microscale quasi-static spectral for the potential distribution of the silicon-IC.

Many researchers have presented various kinds of methods for solving the problem in microscale. These approaches include equivalent source and measured equation of dielectric Green's function and boundary integral equation approach [40–43], CAD and quasi-static spectral domain [44–47], complex image method [48,49], quasi-stationary full-wave analysis and Fourier integral transformation [50], and conformal mapping methods [51]. We illustrate here that FEM with COMSOL is suitable and effective for modeling of inhomogeneous quasi-static microscale multiconductor interconnects in multilayered dielectric media [52]. The future extension of this work will be the modeling of inhomogeneous quasi-static nanoscale multiconductor interconnects in multilayered dielectric media.

The models are designed in 2D using electrostatic environment. In the boundary condition of the model's design, we use ground boundary which is zero potential ($V = 0$) for the shield. We use port condition for the conductors to force the potential or current to be one or zero depending on the setting. The microscale setup using FEM is suitable for the computation of electromagnetic fields in strongly inhomogeneous media, and it has high computation accuracy and fast computation speed.

1.4.1 Microscale Single Interconnect Line on Si–SiO$_2$ Substrate

Figure 1.1 shows the geometry of the model microscale single interconnect line on Si–SiO$_2$ substrate. The value of the single-line capacitance (C_s) using FEM, namely, the value of C_s, is 4.43×10^{-11} F/m.

Furthermore, Figure 1.2 shows the potential distribution of the model in microscale with their variations in spectra peaks and full-width half maximum (FWHM) from $(x, y) = (0, 500\,\mu m)$ to $(x, y) = (20\,\mu m, 502\,\mu m)$.

1.4.2 Microscale Coupled Interconnect Lines on Si–SiO$_2$ Substrate

Figure 1.3 shows the geometry of the model with the parameter values for microscale coupled interconnect lines on Si–SiO$_2$ substrate. The value of C_{11} and C_{22} which are the self-capacitance per unit length of line 1 and 2, respectively, is $C_{11} = C_{22} = 4.457 \times 10^{-11}$ F/m, and the value of the mutual capacitance C_{12} per unit length is $C_{12} = C_{21} = 1.734 \times 10^{-10}$ F/m.

Additionally, Figures 1.4 and 1.5 show the potential distribution of the model at the microscale design with their variations in spectra peaks and FWHM, from $(x, y) = (0, 502\,\mu m)$ to $(x, y) = (20\,\mu m, 502\,\mu m)$ and from $(x, y) = (0, 0)$ to $(x, y) = (20\,\mu m, 1500\,\mu m)$, respectively.

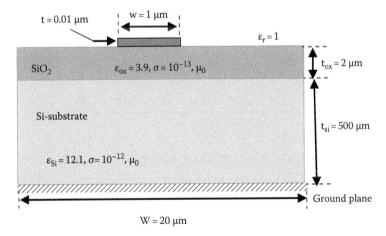

FIGURE 1.1
Cross section of microscale single interconnect line on Si–SiO$_2$ substrate. As seen above w is the width of the single conductor, t is the thickness of the single conductor, W is the width of the Si–SiO$_2$ substrate, t_{si} is the thickness of Si layer, and t_{ox} is the thickness of SiO$_2$ layer.

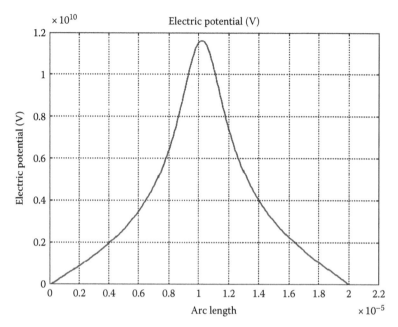

FIGURE 1.2
Potential distribution of microscale single interconnect line on Si–SiO$_2$ substrate from $(x, y) = (0, 500\,\mu m)$ to $(x, y) = (20\,\mu m, 502\,\mu m)$.

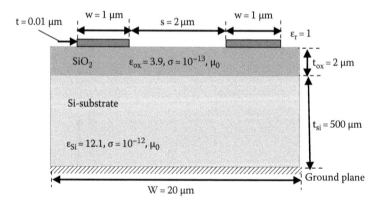

FIGURE 1.3
Cross section of microscale coupled interconnect lines on Si–SiO$_2$ substrate.

1.5 Conclusion

Nanoscale systems, though infinitesimal, are made up of thousands, even hundreds of thousands, of atoms. Thus, describing their properties requires significant theoretical skill and much computer power. Currently, scientists frequently employ combinations

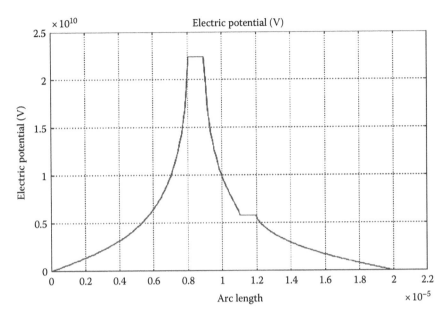

FIGURE 1.4
Potential distribution of coupled interconnect lines on microscale Si–SiO$_2$ substrate from $(x, y) = (0, 502\,\mu m)$ to $(x, y) = (20\,\mu m, 502\,\mu m)$.

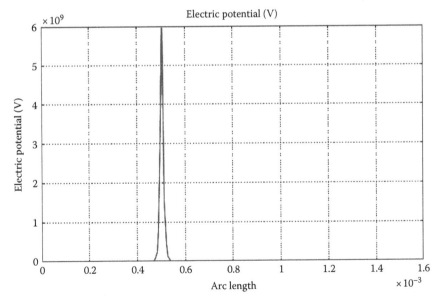

FIGURE 1.5
Potential distribution of coupled interconnect lines on microscale Si–SiO$_2$ substrate from $(x, y) = (0, 0)$ to $(x, y) = (20\,\mu m, 1500\,\mu m)$.

of theoretical strategies with precise modeling techniques as computer power grows. As several disciplines including physics, chemistry, biology, engineering, and medicine are affected by the nanotechnology progress, scientists must be aware of the important role played by computational methods applied to the field. With this recognition, we better prepare ourselves for the future challenges.

Nanotechnology offers great promise, but at the same time, the problems encountered while studying nanostructures indicate that current algorithms and numerical methods must be made more efficient and, perhaps, new ones should be invented. Thus, there is an urgent need for more robust computational methods to understand the properties of matter at the nanoscale. Without the use of the full potential of novel modeling tools, many opportunities in nanotechnology will be missed or delayed. In this chapter, we provide a brief overview of some of the main modeling challenges faced in the field of nanoscale research and describe some basic nanoscale structures that have great potential for future device applications.

Accurate modeling methods able to provide a quantitative understanding of transport, electronic, magnetic, and mechanical properties of these nanostructures are an integral part of the whole field of nanotechnology. Therefore, it is clear that there is a rich research agenda associated with nanoscale science and engineering that will provide us with many opportunities for advancement. The impetus for this interdisciplinary scientific work stems not only from the great intellectual promise of the field of nanotechnology but also from the revolutionary practical significance of this research. For example, the creation of systems of coupled nanostructures poses important questions for the engineering disciplines such as can we create complete and functional systems by coupling together very small objects that are stochastic in their individual behavior? Pursuing this agenda is well beyond the capabilities of any one of current research done alone and will require drawing on the tools and techniques of multiple disciplines. Thus, the new breed of nanotechnology researchers must be unafraid to cross disciplinary boundaries to other fields.

Along these lines, it is self-evident that research in nanotechnology is essentially an interdisciplinary research field which combines research activities in physics, engineering, chemistry, biology, and medicine. Accordingly, research in these fields requires a broad scientific background and working practice in several disciplines of mathematics and computer methods. While nanotechnology offers a multitude of research avenues and promising leads, many of such leads may turn out as dead ends. Therefore, guidance of practical research by theoretical means is a key prerogative. The nature of the objects that nanotechnology is dealing with makes computational physics, quantum chemistry, and theoretical modeling extremely important.

This chapter has outlined some key aspects of computational methods in nanotechnology. It is hoped that this brief overview spells out some key challenges and opportunities encountered in computational modeling of nanosystems. Hopefully, more computer scientists who are keen to contribute their works to the field of nanotechnology will enter the field and contribute to solve some of these computational challenges. Thus, this brief overview is intended to promote collaboration between computer scientists and other disciplines in the nanotechnology field. For all those who are interested in the field of nanotechnology, even the idea of building a system that consists of a large number of particles automatically forming into a designed structure will result in countless interesting scientific problems to face and solve for the future years.

Acknowledgment

This material is partly based upon work supported by the National Science Foundation under Grant No. DMR-1104795 (Orion Ciftja). Any opinions, findings, and conclusions or recommendations expressed in this material are those of the author(s) and do not necessarily reflect the views of the National Science Foundation (NSF).

References

1. G.Y. Slepyan, S.A. Maksimenko, A. Lakhtakia, O.M. Yevtushenko, and A.V. Gusakov, Electronic and electromagnetic properties of nanotubes, *Phys. Rev. B* 57, 16 (1998).
2. P.O. Lehtinen, A.S. Foster, A. Ayuela, T.T. Vehvilainen, and R.M. Nieminen, Structure and magnetic properties of adatoms on carbon nanotubes, *Phys. Rev. B* 69, 155422 (2004).
3. M.D. Alexander, Jr., C.-S. Wang, and P. Meltzer, Jr., *Electrically Conductive Polymer Nanocomposite Materials*, The Air Force Research Laboratory's Materials and Manufacturing Directorate, Wright-Patterson AFB, OH (2005).
4. M.G. Forest, X. Zheng, R. Lipton, R. Zhou, and Q. Wang, Exact scaling laws for electrical conductivity properties of nematic polymer nanocomposite monodomains, *Adv. Funct. Mater.* 15, 4 (2005).
5. L. Jacak, P. Hawrylak, and A. Wojs, *Quantum Dots*, Nanoscale and Technology Series, Springer, Berlin, Germany (1998).
6. R.W. Knoss (Ed.), *Quantum Dots: Research, Technology and Applications*, Nova Science Publishers, New York (2008).
7. R.C. Ashoori, Electrons in artificial atoms, *Nature* 379, 413 (1996).
8. L.P. Kouwenhoven and C.M. Marcus, Single-electron transistors, *Phys. World* 11, 35 (September 1998).
9. D. Heitmann and J.P. Kotthaus, The spectroscopy of quantum dot arrays, *Phys. Today* 46, 56 (June 1993).
10. S. Tarucha, D.G. Austing, T. Honda, R.J. van der Hage, and L.P. Kouwenhoven, Shell filling and spin effects in a few electron quantum dot, *Phys. Rev. Lett.* 77, 3613 (1996).
11. R.C. Ashoori, H.L. Stormer, J.S. Weiner, L.N. Pfeiffer, K.W. Baldwin, and K.W. West, N-electron ground state energies a quantum dot in a magnetic field, *Phys. Rev. Lett.* 71, 613 (1993).
12. O. Ciftja, A Jastrow correlation factor for two-dimensional parabolic quantum dots, *Mod. Phys. Lett. B* 23, 3055 (2009).
13. A.H. MacDonald and M.D. Johnson, Magnetic oscillations of a fractional Hall dot, *Phys. Rev. Lett.* 70, 3107 (1993).
14. C. Yannouleas and U. Landman, Collective and independent-particle motion in two-electron artificial atoms, *Phys. Rev. Lett.* 85, 1726 (2000).
15. P.A. Maksym and T. Chakraborty, Quantum dots in a magnetic fields: Role of electron–electron interactions, *Phys. Rev. Lett.* 65, 108 (1990).
16. F. Bolton, Fixed-phase quantum Monte Carlo study method applied to interacting electrons in a quantum dot, *Phys. Rev. B* 54, 4780 (1996).
17. A. Harju, S. Siljamaki, and R.M. Nieminen, Wigner molecules in quantum dots: A quantum Monte Carlo study, *Phys. Rev. B* 65, 075309 (2002).
18. M. Ferconi and G. Vignale, Current-density-functional theory of quantum dots in a magnetic field, *Phys. Rev. B* 50, R14722 (1994).

19. F. Pederiva, C.J. Umrigar, and E. Lipparini, Diffusion Monte Carlo study of circular quantum dots, *Phys. Rev. B* 62, 8120 (2000).

20. C. Yannouleas and U. Landman, Two-dimensional quantum dots in high magnetic fields: Rotating-electron-molecule versus composite-fermion approach, *Phys. Rev. B* 68, 035326 (2003).

21. O. Ciftja and A. Anil Kumar, Ground state of two-dimensional quantum-dot helium in zero magnetic field: Perturbation, diagonalization and variational theory, *Phys. Rev. B* 70, 205326 (2004).

22. W. Heitler and F. London, Wechselwirkung neutraler Atome und homöopolare Bindung nach der Quantenmechanik, *Z. Phys.* 44, 455 (1927); L. Pauling and E.B. Wilson, *Introduction to Quantum Mechanics With Applications to Chemistry*, 2nd edn., Dover, New York (1963); H. Haken and H.C. Wolf, *The Physics of Atoms and Quanta*, 6th edn., Springer, Berlin, Germany (2000).

23. O. Ciftja and M.G. Faruk, Two-dimensional quantum dot helium in a magnetic field: Variational theory, *Phys. Rev. B* 72, 205334 (2005).

24. C.O. Chui, S. Ramanathan, B.B. Triplett, P.C. McIntyre, and K.C. Saraswat, Germanium MOS capacitors incorporating ultrathin high-k gate dielectric, *IEEE Electron. Dev. Lett.* 23(8), 473 (2002).

25. Y.-C. Yeo, V. Subramanian, J. Kedzierski, P. Xuan, T.-J. King, J. Bokor, and C. Hu, Design and fabrication of 50-nm thin-body p-MOSFETs with a SiGe heterostructure channel, *IEEE Trans. Electron. Dev.* 49(2), 279 (2002).

26. International Technology Roadmap for Semiconductors (ITRS), 2011 Edition. Available at http://www.itrs.net/Links/2011ITRS/Home2011.htm

27. G. Timp, J. Bude, K.K. Bourdelle, J. Garno, A. Ghetti, H. Gossmann, H. Green, M. Forsyth, G. Kim, Y. Kleiman, R. Klemens, F. Kornblit, A. Lochstampfor, C. Mansfield, W. Moccio, S. Sorsch, T. Tennant, D.M. Timp, and R.W. Tung, The ballistic nanotransistor, *IEDM Tech. Dig.*, p. 55 (December 1999).

28. C.W. Leitz, M.T. Currie, M.L. Lee, Z.-Y. Cheng, D.A. Antoniadis, and E.A. Fitzgerald, Hole mobility enhancements in strained $Si/Si_{1-y}Ge_y$ p-type metal-oxide-semiconductor field-effect transistors grown on relaxed $Si_{1-x}Ge_x$ (x<y) virtual substrates, *Appl. Phys. Lett.* 79(25), 4246 (2001).

29. M.L. Lee, C.W. Leitz, Z. Cheng, A.J. Pitera, T. Langdo, M.T. Currie, G. Taraschi, E.A. Fitzgerald and D.A. Antoniadis, strained Ge channel p-type metal–oxide–semiconductor field-effect transistors grown on $Si_{1-x}Ge_x$/Si virtual substrates, *Appl. Phys. Lett.* 79(20), 3344 (2001).

30. S. Datta, F. Assad, and M.S. Lundstrom, The Si MOSFET from a transmission viewpoint, *Superlattices Microstruct.* 23, 771 (1998).

31. F. Assad, Z. Ren, D. Vasileska, S. Datta, and M.S. Lundstrom, On the performance limits for Si MOSFET's: A theoretical study, *IEEE Trans. Electron. Dev.* 47, 232 (2000).

32. Y. Naveh and K.K. Likharev, Modeling of 10-nm-scale ballistic MOSFET's, *IEEE Electron Dev. Lett.* 21, 242 (2000).

33. Z. Ren, R. Venugopal, S. Datta, M.S. Lundstrom, D. Jovanovic, and J.G. Fossum, The ballistic nanotransistor: A simulation study, *IEDM Tech. Dig.*, p. 715 (December 2000).

34. J.-H. Rhew, Z. Ren, and M.S. Lundstrom, A numerical study of ballistic transport in a nanoscale MOSFET, *Solid State Electron.* 46(11), 1899 (2002).

35. J. Knoch, B. Lengeer, and J. Appenzeller, Quantum simulation of ultra-short channel single-gated n-MOSFET, *IEEE Trans. Electron. Dev.* 49, 1212 (2002).

36. Z. Ren, R. Venugopal, S. Goasguen, S. Datta, and M.S. Lundstrom, NanoMOS 2.0: A two-dimensional simulator for quantum transport in nanoscale MOSFETs, *IEEE Trans. Electron. Dev.* 50, 1914 (2003).

37. Z. Ren, R. Venugopal, S. Datta, and M.S. Lundstrom, Examination of design and manufacturing issues in a 10 nm double gate MOSFET using nonequilibrium Green's function simulation, *IDEM, Tech. Dig.*, p. 107 (December 2001).

38. S. Datta, *Electronic Transport in Mesoscopic Systems*, Cambridge University Press, Cambridge, U.K. (1997).

39. S. Datta, Nanoscale device modeling: The green's function method, *Superlattices Microst.* 28, 253 (2000).

40. H. Ymeri, B. Nauwelaers, and K. Maex, On the modeling of multiconductor multilayer systems for interconnect applications, *Microelectron. J.* 32, 351–355 (2001).
41. H. Ymeri, B. Nauwelaers, and K. Maex, On the frequency-dependent line admittance of VLSI interconnect lines on silicon-based semiconductor substrate, *Microelectron. J.* 33, 449–458 (2002).
42. H. Ymeri, B. Nauwelaers, and K. Maex, On the capacitance and conductance calculations of integrated-circuit interconnects with thick conductors, *Microw. Opt. Technol. Lett.* 30(5), 335–339 (2001).
43. W. Delbare and D. De Zutter, Accurate calculations of the capacitance matrix of a multiconductor transmission line in a multilayered dielectric medium, *IEEE Microw. Symp. Dig.* pp. 1013–1016, Long Beach, CA (1989).
44. J. Zhang, Y.-C. Hahm, V.K. Tripathi, and A. Weisshaar, CAD-oriented equipment-circuit modeling of on-chip interconnects on lossy silicon substrate, *IEEE Trans. Microw. Theory Techn.* 48(9), 1443–1451, (2000).
45. H. Ymeri, B. Nauwelaers, and K. Maex, Distributed inductance and resistance per-unit-length formulas for VLSI interconnects on silicon substrate, *Microw. Opt. Technol. Lett.* 30(5), 302–304 (2001).
46. H. Ymeri, B. Nauwelaers, K. Maex, S. Vandenberghe, and D.D. Roest, A CAD-oriented analytical model for frequency-dependent series resistance and inductance of microstrip on-chip interconnects on multilayer silicon substrates, *IEEE Trans. Adv. Packag.* 27(1), 126–134 (2004).
47. E. Groteluschen, L.S. Dutta, and S. Zaage, Full-wave analysis and analytical formulas for the line parameters of transmission lines on semiconductor substrates, *Integration, The VLSI J.* 16, 33–58 (1993).
48. C.-N. Chiu, Closed-form expressions for the line-coupling parameters of coupled on-chip interconnects on lossy silicon substrate, *Microw. Opt. Technol. Lett.* 43(6), 495–498 (2004).
49. J.J. Yang, G.E. Howard, and Y.L. Chow, Complex image method for analyzing multiconductor transmission lines in multilayered dielectric media, *Int. Symp. Dig. Antennas Propag.* pp. 862–865, London, Ontario, Canada (1991).
50. H. Ymeri, B. Nauwelaers, and K. Maex, Frequency-dependent mutual resistance and inductance formulas for coupled IC interconnects on an Si-SiO$_2$ substrate, *Integration, the VLSI J.* 30, 133–141 (2001).
51. E. Chen and S.Y. Chou, Characteristics of coplanar transmission lines multilayer substrates: Modeling and experiments, *IEEE Trans. Microw. Theory Techn.* 45(6), 939–945 (1997).
52. COMSOL, Inc., http://www.comsol.com

2

Finite Element Method for Nanotechnology Applications in Nano-/Microelectronics

Jing Zhang

Indiana University—Purdue University Indianapolis, Indianapolis, Indiana

CONTENTS

2.1 Introduction...19
2.2 Coupled Thermo-Electrical-Mechanical Finite Element Model 20
2.3 Case Study 1: Thermal Stress in 3D IC Structures..................................21
    2.3.1 Introduction..21
    2.3.2 Experimental Data...22
    2.3.3 Finite Element Model 24
    2.3.4 Thermal Stress in Real 3D IC Structures....................................27
2.4 Case Study 2: Degradation in Heterostructure Field-Effect Transistors 34
    2.4.1 Introduction..34
    2.4.2 Results and Discussion 35
2.5 Conclusions..37
Acknowledgments 38
References..38

2.1 Introduction

This chapter presents a novel finite element (FE)-based, thermo-electrical-mechanical coupled model to study temperature field, electric field, and mechanical stress in nano-/microelectronics.

The governing equations of electrical transport, heat transfer, mechanical behavior, and piezoelectricity property are provided. A rigorous framework of coupling these phenomena is derived. Then two case studies are presented that illustrate the application of the FE model in nano-/microelectronics. The first study is thermal stress modeling in wafer-level three-dimensional (3D) integration which offers improved performance and functionality over conventional planar integrated circuits (ICs). The second one is modeling of the material degradation in heterostructure field-effect transistor (HFET).

2.2 Coupled Thermo-Electrical-Mechanical Finite Element Model

There are three components, heat transfer, electricity, and mechanical behavior, in the coupled thermo-electrical-mechanical phenomena as shown in Figure 2.1. These three components are interactively related. For example, between heat transfer and electricity, electrical resistivity is a function of temperature. Joule heating from electric current flow causes increase in temperature.

Between mechanical behavior and electricity, mechanical stress arises as a result of piezo-electricity effect. Between heat transfer and mechanical behavior, temperature change can induce thermal stresses.

Mathematically, the governing equations of electricity transport, heat transfer, mechanical behavior, and piezoelectricity are given as follows:

1. Electrical field
   Electric field is governed by the Laplace equation:

$$\vec{\nabla}\left(\frac{1}{\rho}\vec{\nabla}\varphi\right)=0 \tag{2.1}$$

   where
   ρ is the resistivity (Ω m)
   φ is the electric potential (V)

2. Thermal field
   Thermal field is governed by the Fourier equation:

$$-\vec{\nabla}\left(k_T\vec{\nabla}T\right)-\frac{\left(\vec{\nabla}\varphi\right)^2}{\rho}=0 \tag{2.2}$$

   where
   k_T is the thermal conductivity (W/m-K)
   T is the temperature (K)

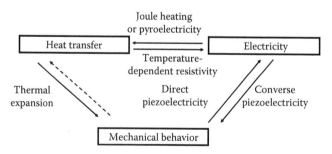

FIGURE 2.1
Diagram showing the coupling between heat transfer, electricity, and mechanical behavior.

3. Mechanical problem
   Mechanical problem is governed by the Navier equation:

$$(\mu + G)\frac{\partial \varepsilon}{\partial x_i} + G\Delta u_i - \frac{E}{1-2v}\left(\alpha_T \frac{\partial T}{\partial x_i}\right) = 0 \tag{2.3}$$

where
   α_T is the coefficient of thermal expansion (K^{-1})

4. Piezoelectric effect
   The piezoelectric effect is a transfer of electrical to mechanical energy or vice versa. The stress-charge form is

$$\text{Stress tensor: } \mathbf{T} = c_E \mathbf{S} - e^T \mathbf{E} \tag{2.4}$$

$$\text{Electric displacement vector: } \mathbf{D} = e\mathbf{S} + \varepsilon_s \mathbf{E} + \mathbf{P}_{sp} \tag{2.5}$$

where
   $\mathbf{S}$ is the second-order strain tensor
   $\mathbf{E}$ is the electric field tensor
   e is a third-order tensor of piezoelectric stress coefficients
   e^T is the transposed tensor of e
   c_E is the compliance
   ε_s is the permittivity tensor
   $\mathbf{P}_{sp}$ is the spontaneous polarization

2.3 Case Study 1: Thermal Stress in 3D IC Structures

2.3.1 Introduction

This section presents the application of the coupled model in 3D IC integration. Wafer-level 3D integration offers improved performance and functionality over conventional planar ICs. A primary driver for 3D ICs is the promise of reduced signal delay through shortened interconnects [1]. Moreover, monolithic wafer-level 3D integration promises increased functionality through the integration of diverse technologies, while maintaining the cost advantage of monolithically fabricated interconnects [2,3]. In wafer-level 3D integrations, a processed wafer (top) is bonded to another processed wafer (bottom) using benzocyclobutene (BCB). The top wafer is backside thinned to a few microns of Si. Cu inter-wafer interconnects are formed by interconnecting specified points in the multilevel metallization (MLM) layers of these two wafers, by etching, liner and Cu deposition, and chemical–mechanical polishing.

There are several design and/or processing concerns surrounding 3D chips. One primary concern for 3D ICs is the mechanical stability of the structures [4]. It is not feasible to predict total stresses in ICs, nor adhesion of the layers in ICs, and it is even difficult to relate structural failures to specific stress levels. Standardized tests are performed to evaluate the stability of ICs. Tests have been done to show that BCB bonding and thinning do not impact (planar) IC performance [5], and the wafers do not delaminate during standard reliability testing [6]. It is clear that the BCB bonded wafers are stable; however, the wafers in those studies did not have inter-wafer vias. One open question is the effect of thermally induced stresses, and whether they are significant during processing and during 3D IC operation, on inter-wafer Cu vias, as they can be for planar ICs [7,8]. Thermally induced stresses can be predicted using structural models that represent proposed structures reasonably well and the coefficients of thermal expansion (CTEs) of the materials present. In this chapter, we summarize the status of our FE-based modeling to address the question of thermally induced stresses in inter-wafer Cu vias to determine if they are a cause for concern for the stability of 3D ICs and whether more quantitative work is warranted. Computations using our FE model indicate that there is reason to be concerned about the reliability of inter-wafer Cu vias. However, these computations had no experiments against which to compare predicted results.

Thermal stresses in MLM structures in planar ICs have been studied both experimentally and by modeling. In XRD-based studies [9–11], stresses in the metal lines can be derived from the measured strain due to the change of lattice spacing. Wafer curvature methods [12,13] measure the changes in the curvature of the substrate on which a thin film is deposited. These experiments are useful for comparing with averages of stresses predicted by models. On the other hand, they are not used to identify localized failures, for example, individual failed vias. Direct imaging of the failures in carefully prepared samples using an SEM provides direct evidence for "qualitative comparison" with model predictions. FE-based analyses can provide predictions of detailed stress distributions and have been used to model the stresses and deformation in interconnect structures. Some models are based on the two-dimensional (2D) plane strain assumption [14–17], in which vias and/or metal lines extend infinitely in and out of the plane. Such models are useful for studying long metal lines. 3D models [16,18] have also been used to simulate thermal stress distributions in interconnect structures with complicated geometries. If model-predicted stresses exceed, or even approach, the yield strength of one or more materials, then there is cause for concern about the stability. Of course, it should be kept in mind that several assumptions routinely used in such analyses make quantitative comparison with experiment tenuous [16,18], and thermally induced stresses are only part of the total stress.

This chapter is organized as follows. Reliability data on 3D ICs are not available in the literature, so we first partially validate our modeling approach by comparing computed results with data on via chain test structures made with SiLK [7] or carbon-doped silicon oxide (SiCOH) [8] as the dielectric. These structures are used to test the reliability of MLM in planar ICs. The failure criterion we use is whether or not the computed von Mises stress exceeds the yield strength of materials in the structure [19–21]. Then we present modeling results from our study of thermal stresses in inter-wafer Cu vias that would form part of 3D ICs, as in the process developed at RPI, *that is*, bonded with BCB. Target values for design parameters, for example, inter-wafer via size, pitch, and the thickness of BCB, are estimated.

2.3.2 Experimental Data

In this section, we review two experiments from the recent literature that examine the reliability of planar IC MLM test structures using low-k dielectrics. Both Filippi et al. [7] and Edelstein et al. [8] performed thermal cycle tests on via chain structures. The geometry of

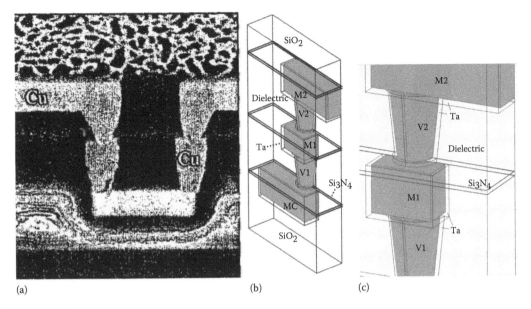

FIGURE 2.2
(a) Cross-sectional SEM image (with labels removed) (see Filippi et al. [7]) of the stacked via test structure using SiLK. (b) Schematic of test structure, labeling substructures as used in the text. Note that only one-quarter of the pictured structure on (a) is shown in the schematic in (b). (c) Localized details of the test structure.

the via chain test structure used in Ref. [7] is shown in Figure 2.2 (SEM on the left). The geometry of the test structures was not completely specified, and we were forced to choose reasonable values for some dimensions. The details of the structure used were not provided; in the following, we assume the structures are the same, with Filippi et al. [7] using SiLK as a dielectric and Edelstein et al. [8] using carbonized glass (SiCOH).

As described in Ref. [7], the repeated unit in the test structure consists of three metal levels (MC, M1, and M2) connected by two level of vias (V1 and V2). MC and M2 are local interconnects in the chain, and M1 is a landing pad in the stacked via structure. V1 connects MC and M1, and V2 connects M1 and M2. In the experiment, the via chain is 50 units long. Some of the following dimensions of the metal and barrier films were not reported and had to be assumed. Metal dimensions do not include liner thicknesses. Interconnects (M2 and MC) are taken to be 0.35 μm long, 0.31 μm wide lines, with an assumed height of 0.25 μm. M1 is a 0.31 μm by 0.31 μm square landing pad, with an assumed 0.25 μm height. The vias are taken to be tapered cylinders 0.22 μm (assumed) in diameter at the bottom and 0.3 μm in diameter at the top, and 0.35 μm tall. V1, V2, M1, and M2 are Cu embedded in the low-k dielectrics (SiLK or SiCOH). MC is tungsten embedded in silicon dioxide. Metal lines are passivated with an Si_3N_4 cap layer. Interconnects and vias are covered with Ta-based liners at their bottoms and side walls. Barriers and capping layers are taken to be of uniform 20 nm thickness.

Both groups [7,8] cycled the temperature between −65°C and 150°C at rates of 22°C per min (up) and −14°C per min (down). Filippi et al. observe statistically significant distributions of electrical failures of the SiLK-based system, with approximately 50% of samples failing by 1000 cycles (as estimated from Figure 5 of Ref. [7]). Figure 2.3a shows a failed V2 via, after the dielectric was removed by oxygen ashing [7]. The crack appears to coincide with a shear plane [7]. No failure was observed in the SiCOH system after 1000 cycles [8].

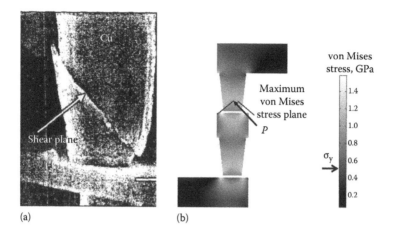

(a)   (b)

FIGURE 2.3

(a) SEM image from Filippi et al. [7] showing shearing and cracking along the bottom of a V2 via. (b) Computed von Mises stresses in Cu for SiLK-based via test structures. Note the stress concentrations near the top of the M1 landing pad and the bottom of the V2 via. Point P is 0.115 µm above the M1 along the centerline of the top via and is at or near the point of the maximum stress in V2. (From Filippi, R.G. et al., Thermal cycle reliability of stacked via structures with copper metallization and an organic low-k dielectric, in *2004 IEEE International Reliability Physics Symposium Proceedings*, April 25–29, 2004, Phoenix, AZ, IEEE.)

2.3.3 Finite Element Model

We constructed a model in COMSOL Multiphysics that is suitable for FE analyses of the test structure described earlier. As shown in Figure 2.2, our geometric model represents one-quarter of the via-chain repeat unit described earlier. We then used FEM-based thermoelastic analysis to compute stresses and strains due to temperature changes that are similar to those performed in the experiments [7,8]. The details of equations appropriate for thermoelastic models can be found in many places, for example, Ref. [22]. The materials are assumed to be isotropic and linear elastic with the material properties listed in Table 2.1. Linearity of deformation with temperature change has been confirmed for Cu by X-ray diffraction measurements over a large temperature range (25°C–400°C) [14]. The Young's moduli, CTEs, and Poisson ratios of Cu and other materials are considered to

TABLE 2.1

Materials Properties Used in via Chain Test Structure Simulations

Material	Thickness (µm)	CTE (ppm/°C)	Young's Modulus (GPa)	Poisson's Ratio
SiO$_2$	0.6	0.5	70	0.22
W	0.25	4.5	344.7	0.28
SiLK	1.2	66	2.5	0.40
BCB	1.2	52	2.9	0.34
SiCOH	1.2	12	16.2	0.30
Cu (V1,V2)	0.35	17	120.5	0.35
Cu (M1,M2,MC)	0.25	17	120.5	0.35
Ta	0.02	6.5	185	0.30
Si$_3$N$_4$	0.02	3.2	221	0.27

be temperature independent within the temperature range considered. Note that we use material properties at room temperature.

In order to evaluate if computed stresses indicate possible failure, a yield strength of Cu should be chosen. The mechanical properties of thin films can be significantly different from bulk samples of the same materials. In fact, there is a large variation of yield strength of Cu in the literature and has been reported to depend upon film thickness, grain size, and temperature. For example, the yield strength of thin film Cu was measured from 225 MPa for 3.015 μm film thickness to 300 MPa for 0.885 μm thick using a bulge test [23]. In another study, micromachined Cu thin-film beams were deflected until inelastic deformation was detected [24]. The yield strength was estimated to be 2.8–3.09 GPa.

In order to proceed in the presence of this diversity, we consider a range of yield strength. The yield strength of bulk materials is the lower limit, and one estimate for thin films is the upper limit. For bulk Cu, the yield strength of hard drawn Cu wires was measured between 414 and 483 MPa [25], assuming the yield strength is same as the tensile strength. In the following, we use 500 MPa, a round number, to be the lower limit of yield strength at room temperature for very small Cu films with small grains. For upper limit of yield strength, we assume the grain size is same as the via diameter in our planar IC MLM study, that is, 0.22 μm. Using a Hall–Petch formula [23], the upper limit of yield strength for Cu interconnects at room temperature is about 600 MPa. Although the lower limit is usually of most interest, for conservative estimates, the upper limit plays a role when comparing computed stresses with experimental results [7,8].

We use a static analysis, which assumes that the structure is in thermal equilibrium at the initial and final temperatures. This method has the advantage of being quickly applied, although it does not take into account transient aspects of the temperature trajectory. Simulations are carried out for a single temperature change. No creep/fatigue model related to multiple cycle tests is considered in our model.

We use linear, tetrahedral FEs within COMSOL Multiphysics [26] to obtain numerical solutions. We locally refine the mesh in some regions to resolve the stresses and strains in certain substructures, such as the barrier and capping layers, as shown in Figure 2.4. Mesh refinement adequacy is confirmed by the lack of significant change in local stresses upon varying the mesh density used to solve the model. A temperature change of 215°C was imposed on the structure, from an initial stress-free state at 150°C to −65°C, the same as the cyclic changes made in Refs. [7,8]. The periodic nature of the repeating via-chain unit is represented by applying symmetric boundary conditions on the sides of the simulation cells that cut transversely through the chain direction (the "transverse sides"), allowing points on these sides to remain in-plane. The simulation cell sides that cut vertically through the chain and are parallel to its direction (the "parallel sides") are also treated symmetrically. This set of symmetries means that the full line width of the interconnects and circular cross section of the vias are accounted for and that the model represents a large bank of via chains fabricated in parallel lines on a 0.6 μm pitch. The bottom surface of the model is considered to be bonded to a thick silicon substrate and constrained to move in-plane to remove components of rigid body motion of the cell from entering the FEM computation. The top surface is free to move, and the structure is taken to be at a stress-free reference state at its maximum temperature.

Figure 2.5 shows calculated von Mises stresses throughout both the SiLK-based and SiCOH-based structures, and Figure 2.3b shows the von Mises stresses in just the Cu on an expanded scale for the SiLK system. Points where the von Mises stress exceeds the yield stress of Cu indicate sites of potential failure. The thermoelastic model predicts that the stresses in the Cu vias in SiLK exceed the upper limit of yield strength of Cu (600 MPa). However, the stresses in Cu in SiCOH are below the lower limit of yield strength

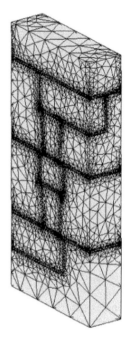

FIGURE 2.4
Tetrahedral FE mesh with local refinement to resolve the stresses in the barrier and capping layers.

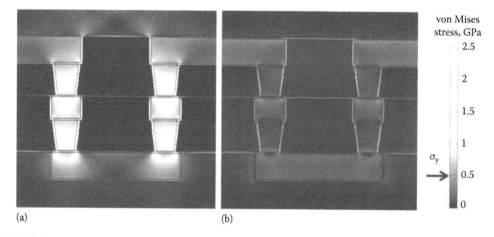

(a)   (b)

FIGURE 2.5
Computed von Mises stresses in SiLK-based system (a) and SiCOH-based system (b). Note that almost all of the V1, M1, and M2 Cu are above the top range of yield strength (600 MPa) in the SiLK system. None of the Cu is above the bottom of the range of yield strength (500 MPa) for the SiCOH system.

(500 MPa). The CTE of SiLK over temperature ranges common to BEOL processing is about 66 ppm/°C [27], which is considerably larger than that of Cu (approximately 17 ppm/°C [17]) in the same temperature range. From our results, it is not surprising that the strain induced in Cu metallization during temperature changes can lead to degradation and failure. Compared with SiLK, the CTE of SiCOH (12 ppm/°C [28]) is close to that of Cu. The thermal stresses in the Cu via are substantially lower.

Some parametric studies were performed to determine the sensitivity of our computed results to the model parameter values used; *that is*, we vary materials' properties within

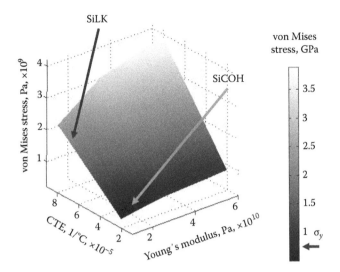

FIGURE 2.6
Calculated stress at point *P* from Figure 2.3, as a function of the Young's modulus and CTE of the surrounding dielectric. Poisson's ratio is taken to be 0.4.

reasonable ranges to identify the effects on thermal stresses. Figure 2.6 shows computed von Mises stresses at point P in Figure 2.3, which is 0.115 μm above M1 along the axis of V2. This point is at or near the point of the maximum stress in V2, and the stress here increases with increasing Young's modulus, CTE, and Poisson's ratio. This figure, in combination with yield strength, provides a failure criterion for Cu vias under different combinations of materials properties. As is reasonable, the figure also suggests that dielectric materials with lower Young's moduli and CTEs can reduce stresses in the Cu.

One concern in our simulation study is the treatment of the very thin barrier and capping films; for example, we assume they are of uniform thickness and ignore their granularity, that is, consistent with using idealized geometries for the vias and lines. However, small variations will have a larger effect on them than for the thicker materials. It is also not clear how the various material interfaces impact results. We assume ideal, perfectly adhesive films. We feel that it is important to gauge the effect of including barrier and capping films on the stresses calculated in our simulations. Simulations show that the stresses along the center of the top and bottom vias are about 10% higher when the volume occupied by barriers is replaced by Cu (Figure 2.7). It is higher because barrier films are stiffer than Cu, and the stresses are higher in the barrier films when they are included. As long as the barrier layers are thin, it does not matter much if they are included, or ignored, in the computations. They do not substantially change the stress distributions in the vias. It should also be noted that the stresses computed in the barriers surrounding the failed vias are larger than the yield strength of Ta (~350 MPa) [29] and the barriers could be expected to fail. In that case, ignoring them may provide a reasonable model for the local mechanical structure. The barrier and etch stop layers are not considered in the models of 3D IC structures discussed in the next section.

2.3.4 Thermal Stress in Real 3D IC Structures

We perform 3D IC simulations to calculate stress in a representative unit cell. The model consists of seven layers with embedded Cu interconnects (Figure 2.8). In order to derive design-related parameters and make the simulation tractable, we choose a repeated unit

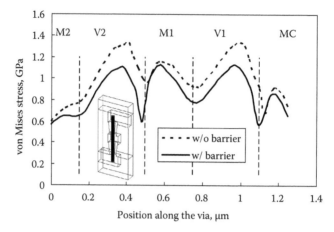

FIGURE 2.7
Stress distributions along the Cu via embedded in SiLK with and without barrier layers. The thick dark line in the inset shows the positions along the via where the stress values were taken.

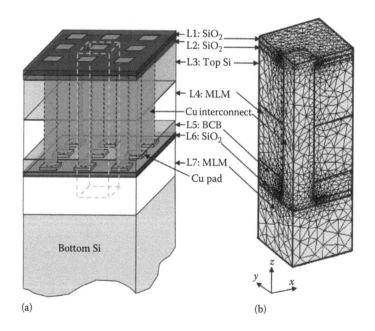

(a)   (b)

FIGURE 2.8
3D IC structure. (a) Schematic of several cylindrical via connecting square landing pads through the layered structure (not to scale). The dot-marked block is the unit cell. (b) A typical FE mesh. Taking advantage of symmetry, only one quarter of a via and pad has been included in the geometric model.

cell in the 3D IC structures (Figure 2.8). The top silicon wafer (L3) is thinned to 10 μm thick and bonded to the bottom wafer using a 2.6 μm thick BCB (L5). The bottom silicon wafer is not included in the unit cell model due to its large volume. Circular Cu vias, with an assumed 23.6 μm height, connect 10 μm thick MLMs (L4, L7) in the two wafers. Square landing pads are used at the end of Cu vias and embedded in the oxide layers (L1, L2, L7). Due to symmetry in the x–z and y–z planes (Figure 2.8), only one quarter of unit cell is used in the simulations. Barrier materials are not considered in the model due

TABLE 2.2

Material Properties Used in 3D IC Simulations

Material	Thickness (μm)	CTE (ppm/°C)	Young's Modulus (GPa)	Poisson's Ratio
SiO$_2$	1	0.5	70	0.22
Si	10	3.725	130.1	0.278
MLM	10	5	85	0.26
BCB	2.6	52	2.9	0.34
Cu (via)	23.6	17	120.5	0.35
Cu (pad)	1	17	120.5	0.35

to their small thicknesses, and as discussed earlier, their presence does not significantly affect computed stresses in our model of planar IC structures. The structure is assumed stress-free at 250°C (wafer bonding temperature). The material properties and dimensions of the model are given in Table 2.2. Young's modulus of BCB is found to decrease with temperature from 2.5 GPa at 25°C to 0.3 GPa at 180°C [30]. To be conservative, we use the value at 25°C, which will tend to overpredict the stresses in the Cu. We use the room temperature value of CTE for BCB, which wafer-curvature experiments [31] show to be largely temperature independent. For other material properties of pure materials, for example, Poisson's ratio of Cu, we use material properties widely reported at room temperature. The material properties of the MLM layer are determined by the volume average of each component (30% Cu and 70% SiO$_2$) [32]. For mechanical boundary conditions, the side walls of the unit cell are only allowed to shift within their respective planes due to symmetry, as is the bottom surface, which is assumed perfectly bonded to the thick silicon substrate at the bottom. The top surface is left free to move. As seen in the right half of Figure 2.8, the FE mesh is refined in areas where large stress gradients are expected.

Figure 2.9 shows the distribution of von Mises stress in the unit cell for via sizes from 1 to 6 μm in diameter and a pitch of 20 μm, for an initial temperature of 250°C and a final temperature of 25°C. The stresses in the via decrease as the via diameter increases.

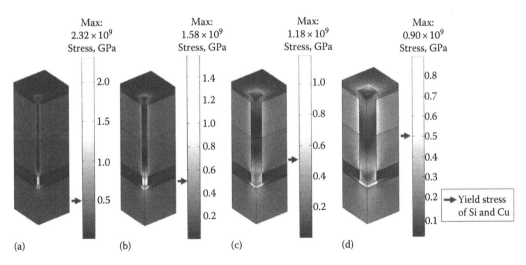

| Max: 2.32 × 10^9 Stress, GPa | Max: 1.58 × 10^9 Stress, GPa | Max: 1.18 × 10^9 Stress, GPa | Max: 0.90 × 10^9 Stress, GPa |

(a)　　　(b)　　　(c)　　　(d)

Yield stress of Si and Cu

FIGURE 2.9

Computed von Mises stresses in 3D IC structure with via size of (a) 1, (b) 2, (c) 4, and (d) 6 μm for 20 μm pitch.

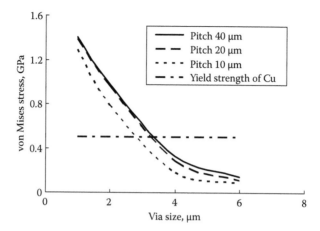

FIGURE 2.10

Computed maximum von Mises stress in Cu vias of different diameter and different pitch with a BCB thickness of 2.6 μm.

Figure 2.10 shows the von Mises stress at the center of the Cu vias half way up the BCB layer for vias from 1 to 6 μm in diameter and pitches of 10, 20, and 40 μm. The order of CTEs of the materials is BCB > Cu > MLM > Si > SiO$_2$, and the BCB layer is quite thin. A back-of-the-envelope calculation shows that the unconstrained thermal contraction of the Cu vias, were they not in the stack, is ~84 nm. This is significantly larger than the unconstrained thermal contraction of the stack of material in parallel with the Cu vias (~55 nm). Thus, in the 3D IC stack in which the vias are coupled to the other materials, the Cu vias act to constrain the vertical expansion of the entire structure. At each pitch, the maximum von Mises stress in the Cu increases as the via size decreases. This trend makes sense; as the cross section of the Cu decreases, the amount of stack it must constrain increases. The calculated von Mises stress decreases with decreasing pitch at constant via size, as the force per unit area is distributed to more vias. Our simulations indicate that plastic deformation of the vias is a concern. Combining yield strengths of Cu discussed in Section 2.3, we can estimate some target values for design parameters. The lower limit of yield strength (500 MPa) can be used for a conservative design. As shown in Figure 2.10, the via size should be larger than 3 μm at a pitch of 10 μm and 3.5 μm at a pitch of 40 μm.

Simulations were performed to evaluate the use of BCB bonding with small Cu vias at small pitches, such as might be used to bond logic and memory circuits. We assume a via diameter of 1 μm, vary the via pitch from 5 to 20 μm, and vary the BCB thickness from 0.5 to 3 μm. As shown in Figure 2.11, the thickness of BCB should be less than 1 μm, using yield strength of Cu 500 MPa, to avoid plastic yield of Cu vias for even the smallest pitch studied. This is due to the relatively high CTE of BCB compared to Cu. Although the "unconstrained stack" discussed earlier contracts approximately 55 nm, approximately 60% of that motion would come from the 2.6 μm of BCB. Over that same length, Cu contracts only 10 nm when unconstrained, so when they are coupled to each other, a significant compressive stress is induced in the lowest part of the via during cooling. By reducing the thickness of the BCB layer, this expansion can be reduced in magnitude, and the stress in the bottom of the via is reduced.

We also studied the effect of imposed temperature changes on the thermal stresses. Simulations of different initial stress-free temperatures, 150°C and 350°C, and cooling down to room temperature were performed to study thermal stress levels. The maximum

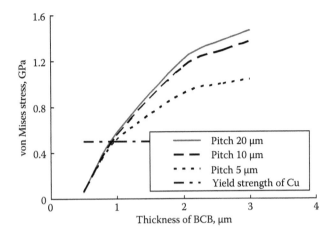

FIGURE 2.11
Computed maximum von Mises stress in Cu vias for different BCB thickness and pitch with vias 1μm in diameter.

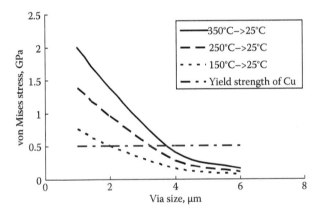

FIGURE 2.12
Computed maximum von Mises stress in Cu vias of different size cooling from 150°C, 250°C, and 350°C to 25°C.

von Mises stress in the Cu via is shown in Figure 2.12. As expected, the stress increases as the initial temperature increases. In Ref. [7], a threshold temperature range of 145.3°C (Table I in Ref. [7]) was found from statistical analysis of mean-cycle-to-failure, N_{50}, as a function of temperature changes ΔT (T_{max} was 150°C, ΔT varied from 150°C to 300°C) (Figure 8 in Ref. [7]). Below this threshold temperature, the test structures did not fail. Our simulations also show that the magnitude of temperature change is a major factor determining the stress level in the inter-wafer vias. If the structures cool down to room temperature, the maximum processing temperatures determine the thermal stress levels. The stress data are normalized by the temperature difference between initial and final temperatures and plotted in Figure 2.13. As expected for a linear analysis, the stresses at different initial temperature collapse into a single curve. Therefore, when thermally induced stress information is available for some temperature ranges, it is possible to extend this information into other temperature windows through this procedure.

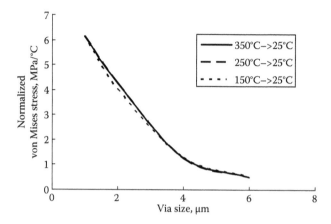

FIGURE 2.13
Normalized maximum von Mises stress by temperature change in Cu vias of different size, cooling from 150°C, 250°C, and 350°C to 25°C.

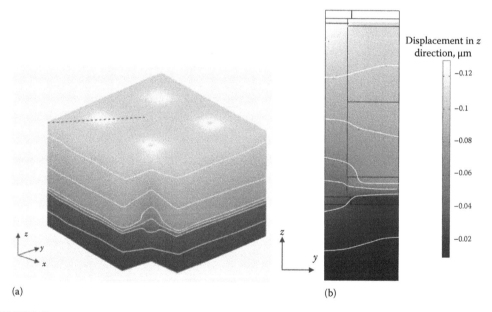

(a)   (b)

FIGURE 2.14
(a) Contour and isosurface (white curves) of vertical displacement (z) of the top surface of the 3D IC structures calculated for vias of 6 μm, a pitch of 20 μm, and a BCB thickness of 2.6 μm. The structure contains 15 unit cells (see Figure 2.8b). A unit cell is removed to reveal internal displacement. The black dashed line indicates the top of the slice shown on the right side of the figure. The black dashed line is also the location where the vertical displacements are measured in Figure 2.15. (b) Side view of vertical displacement (z) calculated with via of 6 μm. The white curves indicate isosurfaces of vertical displacements. The black lines delineate the undeformed structure.

Issues in addition to stresses in Cu can be of interest. As discussed earlier, Cu vias contract more than the surrounding materials, thus pulling down on these neighboring materials to which they are coupled. For example, Figure 2.14 shows that roughness is introduced into the top surface of the 3D IC structure; the vias and their surroundings are lower than areas away from the vias. This roughness, which can be on the order of

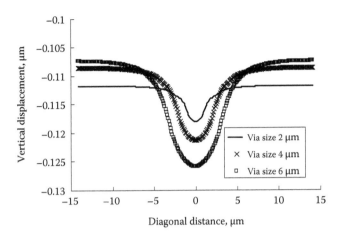

FIGURE 2.15
Vertical displacement along the diagonal direction (see Figure 2.14, pitch is 20 μm). Note the expanded vertical axis.

100 nm, may affect the downstream processes, and the adhesion of layers deposited after bonding and thinning. Figure 2.14 also shows the side view of displacement in the vertical direction. The same simulation was repeated with different via sizes, ranging from 2 to 6 μm. The resulting vertical displacements of the top surface are shown in Figure 2.15. The size of the depressed area increases with via size due to both the larger via diameters and the increased constraining forces transmitted by the via.

As indicated in Figure 2.9, at constant via pitch (20 μm), the maximum von Mises stress shifts from the Cu vias to the top Si substrate as the via size increases. At relatively low via densities (<6% area fraction), the maximum stress was observed in the Cu vias. Results suggest that Cu vias smaller than a critical size will yield plastically. As the via density increases (>10%), the maximum stress gradually shifts to the top, thinned, Si substrate. Potential yield or failure locations (e.g., failure in Cu versus Si) are also related to this critical via density. For brittle materials, such as Si, the critical fracture stress is a function of any flaw or existing damage to the specimen [33]. Such damage can be caused by wafer processing, such as grinding and etching. A carefully polished Si specimen has an average fracture stress as high as 800 MPa [34]. On the other hand, a backside processed wafer has the fracture stress of only 175 MPa [34]. In our simulations, the maximum stresses in Si occur in the vicinity of the interface between the Si and the Cu vias. As the via density reaches approximately 10%, the stress in Si exceeds the earlier estimate of the critical fracture stress. The shift in the location of the maximum stress can be understood by considering that the Cu via is in parallel with the other materials; *that is*, the Cu vias need to deal with the forces exerted by their surroundings. As via size increases or via density increases, the total Cu area increases. So, stresses are lower for the same forces.

Several avenues to improve our analysis have been identified. First, in this chapter, we partially validated our model using planar IC data [7] by comparison with experimental failure information—a qualitative response. The stress level at which vias fractured rather than just underwent plastic deformation was not studied in Ref. [7], and we are not aware of any such data in the literature. Previous studies [15,35,36] show that stresses predicted using FEM are higher than measured volume-averaged stresses. Though conservative computations are better than those that produce false assurances of stability, it would be better to explain the differences and refine the models. This effort will also be useful to evaluate assumptions in our model that the initial stress at high temperatures in the structure is zero. Thermal stress measurements using XRD of Cu lines passivated with TEOS oxide and methyl silsesquioxane dielectrics show that, depending on the processing conditions, initial stress at 400°C may be in the order of 100 MPa [36]. A careful evaluation of initial stresses is necessary for proper prediction of thermal stresses. Second, in our simulations, materials are assumed to be homogeneous, and no information on microstructure is explicitly included. It has been shown that stress levels increase with increasing average grain size for 0.35 μm Cu lines with passivation [37]. Grain boundaries are interfaces which serve as stress relaxation sites. Increasing the grain size reduces the number of grain boundaries and leads to higher stresses. A model linking stress and microstructure would be useful to our analysis. In addition, any gaps in the structure, which might occur during deposition in features, will decrease thermally induced stresses. Finally, the static analysis performed in this study does not account for any effects of creep or fatigue. von Mises stresses are used to predict plastic deformation, but not all structures destined to fail due to thermally induced stress do so with a simple temperature change, nor does plastic deformation always lead to electrical failure.

2.4 Case Study 2: Degradation in Heterostructure Field-Effect Transistors

2.4.1 Introduction

The second case study is the application of the FE model in wide bandgap (WBG) semiconductor materials. Electronic devices made of GaN and SiC have been commercialized and accepted by broad markets for their unique attributes. WBG semiconductor materials have been intensively studied in the area of growth, processing techniques, and device design and fabrication [38,39]. However, the severe operating conditions in terms of temperature, voltage, electric field, and frequency impose several design and/or processing concerns surrounding the WBG electronic devices. One primary concern is the stability of the structures caused by diffusion of electrically active impurities, void initiation and propagation, and interfacial instabilities. The performance of these devices is limited by the magnitude of the total voltage developed at the gate and drain before electronic breakdown occurs [38,39]. For example, Conway et al. studied accelerated RF life test on GaN HFETs at 10 GHz [40]. TEM was performed on devices stressed in air which showed feature on drain edge of gate in the AlGaN barrier layer [40] (see Figure 2.16). It is well known that the mechanisms of degradation and ultimately failure in WBG devices are very complex. Under operating conditions, the devices are usually subject to high electric fields, high stress/strain fields, high current densities, high temperatures, and high thermal gradients. Moreover, these phenomena are coupled together (Figure 2.1).

In order to fully understand the failure mechanisms in WBG devices, in this section, we develop a novel FE-based, thermo-electrical-mechanical coupled model to study induced stresses, temperature, and electric fields in WBG devices.

The 2D model of the transistor is shown in Figure 2.17. The FE mesh, materials, and critical dimensions are also given in the figure. The boundary conditions, as shown in Figure 2.3, are (1) electrical potential, (2) temperature, and (3) mechanical constraints. The transistor is operated under DC bias voltages V_{DS} up to 20 and −8 V at the gate. For the thermal boundary conditions, the bottom surface is set at constant 300 K. Other surfaces are thermally insulated. Also the initial temperature of the system is assumed as 300 K. We use

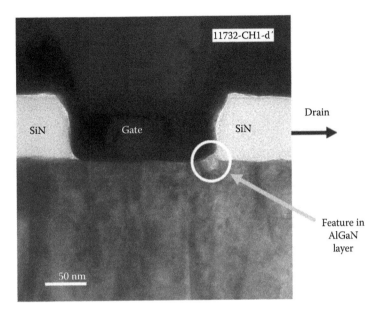

FIGURE 2.16
TEM image of damage feature developed in the AlGaN layer. (From Conway, A.M. et al., Failure mechanisms in GaN HFETs under accelerated RF stress, *CS MANTECH Conference*, May 14–17, 2007, Austin, TX, pp. 99–102, 2007.)

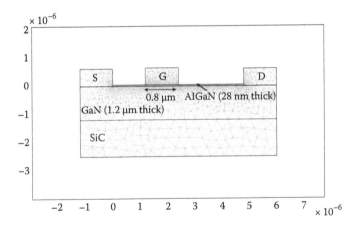

FIGURE 2.17
2D FE model of the transistor. The materials and critical dimensions are also given.

COMSOL Multiphysics to solve the governing equations listed in Section 2.2. Four modules in COMSOL Multiphysics are used for coupling simulations: (1) heat transfer by conduction module, (2) conductive media DC module, (3) plane strain module, and (4) piezoelectricity module. These four modules are capable of coupling all the governing equations.

2.4.2 Results and Discussion

Figure 2.18 shows the predicted electrical field distribution. Electrical field is maximal at the bottom corner of the gate near the drain side. This high electric field may cause the

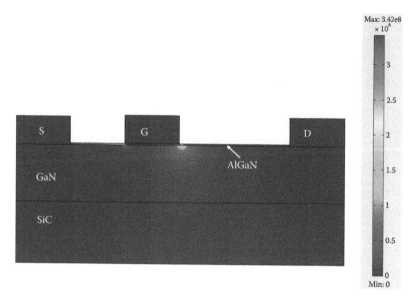

FIGURE 2.18
Predicted electrical field distribution.

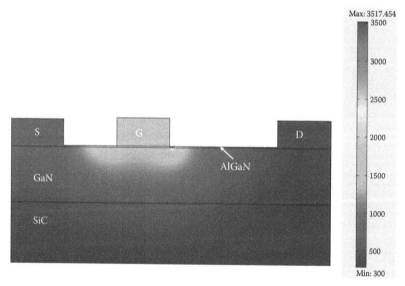

FIGURE 2.19
Predicted temperature distribution.

degradation of the transistor, which is in agreement with the experiments shown in Figure 2.16. Similar to the electric field, the maximum of temperature as shown in Figure 2.19 also occurs at the corner of the gate. In general, diffusion of atoms is accelerated at higher temperature. Therefore, there is a concern of material degradation under current operation conditions.

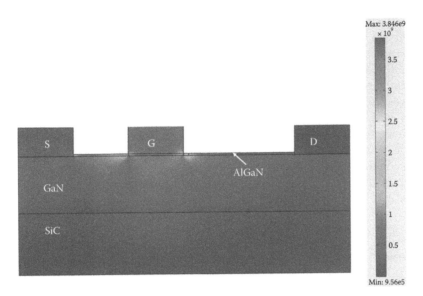

FIGURE 2.20
Predicted von Mises stress distributions.

The predicted stress is shown in Figure 2.20. The stress is caused by the combined thermal expansion and piezoelectric effects. The highest stress is at the bottom corner of the gate near the drain side, which is in agreement with experimental observations (Figure 2.16).

2.5 Conclusions

We presented a novel coupled thermo-electrical-mechanical FE-based model to study temperature field, electric field, and mechanical stress in nano-/microelectronics. The model was used to evaluate whether thermally induced stresses in Cu inter-wafer vias used in BCB-bonded (wafer level) 3D ICs are a reliability concern. Our modeling approach was first tested against data obtained from thermal cycling experiments on planar IC MLM via structures with Cu interconnects and low-k dielectrics. Computed von Mises stresses, assuming a single cooling step of the same magnitude as the experiments, are at a maximum near the bottom of vias in the test structure, which coincides with experimental observations of failure when SiLK was used. No failure in the via is predicted when SiCOH was used as the dielectric, which is consistent with experiments.

We then calculated thermal stresses in 3D IC structures bonded with BCB using a process developed at RPI due to changing temperature from 250°C to 25°C. Simulations show that the von Mises stresses in the Cu vias decrease with decreasing pitch at constant via size, increase with decreasing via size at constant pitch, and decrease with decreasing BCB thickness. We found that further study is indeed warranted; *that is*, the stresses in Cu vias either exceed or are close to the yield stress for temperature changes similar to those used in BEOL processing and inter-wafer via pitches and via sizes that may reasonably be used in 3D ICs. Our results also support the goal of using thin BCB. In general, thin BCB has been targeted in order to keep via aspect ratios down; however, it also helps keep the via stresses due to BCB contractions and expansions down. Target values for design parameters, for example, inter-wafer via size, pitch, and the thickness of BCB, were estimated. To improve upon the reliability of such design oriented computations, more quantitative simulation work is needed, *that is*, to compare computed stresses with measured stresses.

The model was also applied to conduct degradation analysis and optimization of the performance of electronic devices. Electrical field is maximal at the bottom corner of the gate near the drain side. This high electric field may cause the degradation of the transistor. Similar to the electric field, the maximum of temperature is also located at the corner of the gate. The simulated results are in agreement with experimental observations.

Acknowledgments

The work is partially supported by the United States Air Force Summer Faculty Fellowship Program managed by the American Society for Engineering Education (Contract number: FA9550-07-L-0052), and MARCO, DARPA, and NYSTAR through the Interconnect Focus Center. The author also thanks Don Dorsey, Max Bloomfield, Timothy Cale, James Lu, and Ron Gutmann for their assistance of the work.

References

1. Davis, J.A. et al., Interconnect limits on gigascale integration (GSI) in the 21st century. *Proceedings of the IEEE*, 2001. **89**(3): 305–324.
2. Guarini, K.W. et al., Electrical integrity of state-of-the-art 0.13 mu m SOI CMOS devices and circuits transferred for three-dimensional (3D) integrated circuit (IC) fabrication. In *IEEE International Electron Devices Meeting*, December 8–11, 2002. San Francisco, CA: IEEE.
3. Lu, J.-Q. et al., A wafer-scale 3D IC technology platform using dielectric bonding glues and copper damascene patterned inter-wafer interconnects. In *Proceedings of the IEEE 2002 International Interconnect Technology Conference*, June 3–5, 2002. Burlingame, CA: IEEE.
4. *The International Technology Roadmap for Semiconductors*. 2004. Available from http://www.itrs. net/Links/2004Update/2004_08_Interconnect.pdf
5. Gutmann, R.J. et al., A wafer-level 3D IC technology platform. In *Advanced Metallization Conference 2003 (AMC 2003)*. 2004. Montreal, Quebec, Canada: Materials Research Society.
6. Kwon, Y. et al., Dielectric glue wafer bonding for 3D ICs. In *Materials, Technology and Reliability for Advanced Interconnects and Low-k Dielectrics—2003*. San Francisco, CA: Materials Research Society.
7. Filippi, R.G. et al., Thermal cycle reliability of stacked via structures with copper metallization and an organic low-k dielectric. In *2004 IEEE International Reliability Physics Symposium Proceedings*, April 25–29, 2004. Phoenix, AZ: IEEE.
8. Edelstein, D. et al., Comprehensive reliability evaluation of a 90 nm CMOS technology with Cu/PECVD low-k BEOL. In *2004 IEEE International Reliability Physics Symposium Proceedings*, April 25–29, 2004. Phoenix, AZ: IEEE.

9. Greenebaum, B. et al., Stress in metal lines under passivation: Comparison of experiment with finite element calculations. *Applied Physics Letters*, 1991. **58**(17): 1845–1847.

10. Cornella, G. et al., Analysis technique for extraction of thin film stresses from x-ray data. *Applied Physics Letters*, 1997. **71**(20): 2949.

11. Flinn, P.A. and C. Chiang, X-ray diffraction determination of the effect of various passivations on stress in metal films and patterned lines. *Journal of Applied Physics*, 1990. **67**(6): 2927–2931.

12. Shen, Y.-L., S. Suresh, and I.A. Blech, Stresses, curvatures, and shape changes arising from patterned lines on silicon wafers. *Journal of Applied Physics*, 1996. **80**(3): 1388–1398.

13. Kobrinsky, M.J., C.V. Thompson, and M.E. Gross, Diffusional creep in damascene Cu lines. *Journal of Applied Physics*, 2001. **89**(1): 91–98.

14. Rhee, S.-H., Y. Du, and P.S. Ho, Characterization of thermal stresses of Cu/low-k submicron interconnect structures. In *Proceedings of the IEEE 2001 International Interconnect Technology Conference*, June 4–6, 2001. Burlingame, CA: IEEE.

15. Park, Y.-B. and I.-S. Jeon, Mechanical stress evolution in metal interconnects for various line aspect ratios and passivation dielectrics. *Microelectronic Engineering*, 2003. **69**(1): 26–36.

16. Paik, J.-M., H. Park, and Y.-C. Joo, Effect of low-k dielectric on stress and stress-induced damage in Cu interconnects. *Microelectronic Engineering*, 2004. **71**(3–4): 348–357.

17. Ege, E.S. and Y.-L. Shen, Thermomechanical response and stress analysis of copper interconnects. *Journal of Electronic Materials, Third Symposium on Materials and Processes for Submicron Technologies*, March 3, 2003. **32**(10): 1000–1011.

18. Sukharev, V., *Stress Modeling for Copper Interconnect Structures*. In *Materials for Information Technology*, E. Zschech, C. Whelan, and T. Micolajick, eds. 2005. Berlin, Germany, Springer-Verlag.

19. Herbert Goldstein, C.P.P. and J.L. Safko, *Classical Mechanics*, 3rd edn. 2002. Boston, MA: Addison Wesley.

20. Lee, S.-Y., J.-h. Jeong, and D. Kwon, Finite-element analysis of stress effect in metallization/passivation layers during thermal cycling. *Journal of the Korean Institute of Metals and Materials*, 2001. **39**(8): 949–955.

21. Swanson, J.A., S.M. Heinrich, and P.S. Lee, An elastoplastic beam model for column-grid-array solder interconnects. *Transactions of the ASME, Journal of Electronic Packaging*, 1999. **121**(4): 303–311.

22. Timoshenko, S., *Theory of Elasticity*, 3rd edn. 1970. New York: McGraw-Hill Companies.

23. Xiang, Y., X. Chen, and J.J. Vlassak, The mechanical properties of electroplated Cu thin films measured by means of the bulge test technique. In *Thin Films: Stresses and Mechanical Properties IX*, November 26–30, 2001. 2002. Boston, MA: Materials Research Society.

24. Gordon, M.H. et al., A simple technique for determining yield strength of thin films. *Experimental Mechanics*, 2002. **42**(3): 232–236.

25. Robert, C.W., ed., *CRC Handbook of Chemistry and Physics*, 46 edn. 1965. Boca Raton, FL: The Chemical Rubber Co.

26. *Comsol Multiphysics*, 2005, Comsol, Inc. Available at: http://www.comsol.com

27. SiLK D semiconductor dielectric, http://www.dow.com/PublishedLiterature/dh_004d/0901b8038004d89b.pdf?filepath=silk/pdfs/noreg/618-00317.pdf&fromPage=GetDoc

28. Grill, A., Plasma enhanced chemical vapor deposited SiCOH dielectrics: From low-k to extreme low-k interconnect materials. *Journal of Applied Physics*, 2003. **93**(3): 1785–1790.

29. Tantalum, manufacturing data. http://www.webelements.com/tantalum/physics.html

30. Zhao, J.-H., W.-J. Qi, and P.S. Ho, Thermomechanical property of diffusion barrier layer and its effect on the stress characteristics of copper submicron interconnect structures. *Microelectronics Reliability*, 2002. **42**(1): 27–34.

31. Hodge, T.C., S.A. Bidstrup Allen, and P.A. Kohl, In situ measurement of the thermal expansion behavior of benzocyclobutene films. *Journal of Polymer Science, Part B: Polymer Physics*, 1999. **37**(4): 311–321.

32. Stellbrink, K.K.U., *Micromechanics of Composites: Composite Properties of Fibre and Matrix Constituents*. 1996. Cincinnati, OH: Hanser Gardner Publications.

33. Chen, C.P. and M.H. Leipold, Fracture toughness of silicon. *American Ceramic Society Bulletin,* 1980. **59**(4): 469–472.

34. van Kessel, C.G.M., S.A. Gee, and J.J. Murphy, Quality of die-attachment and its relationships to stresses and vertical die-cracking. In *Proceedings—33rd Electronic Components Conference,* May 16–18, 1983. Orlando, FL: IEEE, 237–244.

35. Yeo, I.-S., P.S. Ho, and S.G.H. Anderson, Characteristics of thermal stresses in Al(Cu) fine lines. I. Unpassivated line structures. *Journal of Applied Physics,* 1995. **78**(2): 945–952.

36. Rhee, S.-H., Y. Du, and P.S. Ho, Thermal stress characteristics of Cu/oxide and Cu/low-k submicron interconnect structures. *Journal of Applied Physics,* 2003. **93**(7): 3926–3933.

37. Besser, P.R. et al., Microstructural characterization of inlaid copper interconnect lines. *Journal of Electronic Materials,* 2001. **30**(4): 320–330.

38. Winslow, T.A. and R.J. Trew, Principles of large-signal MESFET operation. *IEEE Transactions on Microwave Theory and Technology,* 1994. **42**: 935–942.

39. Trew, R.J. et al., RF breakdown and large-signal modeling of AlGaN/GaN HFET's. In *Microwave Symposium Digest,* June 11–16, 2006. IEEE MTT-S International, San Francisco, CA, pp. 643–646.

40. Conway, A.M. et al., Failure mechanisms in GaN HFETs under accelerated RF stress. *CS MANTECH Conference,* May 14–17, 2007. pp. 99–102.

3

Modeling, Design, and Simulation of N/MEMS by Integrating Finite Element, Lumped Element, and System Level Analyses

Jason Vaughn Clark
Purdue University, West Lafayette, Indiana

Prabhakar Marepalli
University of Texas, Austin, Texas

Richa Bansal
Purdue University, West Lafayette, Indiana

CONTENTS

3.1 Introduction..42
3.2 Lumped Modeling of Carbon Nanotubes...43
    3.2.1 Carbon Nanotubes..43
    3.2.2 Structural Mechanics Model...44
    3.2.3 Reduced-Order Modeling...47
3.3 Design and Simulation of Carbon Nanotubes.....................................48
    3.3.1 Sugar Design...48
    3.3.2 SugarCube Design and Simulation 50
    3.3.3 Applications..50
    3.3.4 Summary..53
3.4 Lumped Modeling of MEMS 53
    3.4.1 Sugar to SugarCube 54
    3.4.2 Librarian..57
    3.4.3 Parameterization..58
    3.4.4 Simulation...58
    3.4.5 Static Analysis 58
    3.4.6 Steady-State Analysis 60
    3.4.7 Sinusoidal Analysis 62
    3.4.8 Transient Analysis 62
    3.4.9 Optimization 63
    3.4.10 Layout..66
    3.4.11 Summary..67
3.5 Design and Simulation of N/MEMS...67
    3.5.1 Sugar Model..68
    3.5.2 SugarCube Model 69
    3.5.3 Carbon Nanotube Model in Sugar 70

3.5.4   First-Order Analysis of the Thermal Actuator 71
3.5.5   Thermo-Mechanical Response of the Device 73
3.5.6   Electro-Thermo-Actuator Model 75
3.5.7   Summary 76
3.6   Distributed Element Parametric Design of MEMS 76
3.6.1   Framework 77
3.6.2   Integration of Sugar with COMSOL 78
3.7   Integration of Sugar with Simulink® 79
3.7.1   Verification of Lumped Analysis 79
3.7.2   Integration of Sugar with Spice 80
3.7.3   Summary 80
3.8   Conclusion 82
References 83

3.1 Introduction

In this chapter, we discuss the integration of finite and lumped element methods for the design, modeling, and simulation of nano- and microelectromechanical systems (N/MEMSs). A reason to integrate finite and lumped methods is to achieve increased productivity and functionality in the design and analysis of N/MEMS. That is because N/MEMSs are multiphysical systems that can have some components that are best modeled by finite element (FE) methods (e.g., elements with nonlinear fields) and other components that are more efficiently modeled by lumped element methods (e.g., electronic circuit elements, carbon nanotubes [CNTs]). For commonly used components, it can often be advantageous to convert the FE model to a parameterized lumped model counterpart. There are also advances in the design of N/MEMS, where the design space can be initially quickly explored using lumped element methods, then subsequently importing the design into an FE tool for more refined analysis.

We limit the scope of our discussion of NEMS to that of CNTs subject to small deflection, and we limit our discussion of MEMS to that of thin flexures, comb drives, electro-thermo-mechanical actuators, and piezoelectric actuators. Such mechanisms have been used to create a great number of N/MEMS. Other types of transducers may be implemented using similar analysis methods. Much of this work is based on our recent efforts in the area of design, modeling, and simulation of N/MEMS for lumped and finite (or distributed) analyses [1–4].

The software tools we use are Sugar [5] and Spice [6] for lumped analysis, COMSOL [7] for distributed analysis, and Simulink® [8] for system level analysis. This set of tools is not unique for such applications, but they appear to be the least expensive. MATLAB® [9] and COMSOL are the only commercial software tools that we use.

Toward the goal of integrating of distributed and lumped element methods for the design, modeling, and simulation of N/MEMS, we present the following sections. In Section 3.2, we discuss lumped modeling of CNTs with respect to structural mechanic and reduced-order modeling. In reduced-order modeling, we reduce a 6N degree-of-freedom system to a 12N degree-of-freedom system, where N is the number of carbon atoms. In Section 3.3, we discuss how to design CNTs using our lumped model, which can be parameterized by chirality, diameter, and length. In Section 3.4, we introduce a novel tool for novice-friendly design and simulation of MEMS. We demonstrate its librarian, parameterization, simulation types, optimization, and novice-friendly layout for fabrication. In Section 3.5, we integrate the design of NEMS with MEMS by exemplifying the design of a nanomechanical material property tester. In Section 3.6, we describe a framework that integrates our lumped analysis tool with a commercial distributed element tool. In Section 3.7, we expand our previous discussion with the integration of lumped, distributed, and system level design, simulation, and lumped-to-distributed analysis verification. We summarize in Section 3.8.

3.2 Lumped Modeling of Carbon Nanotubes

Due to their excellent electrical and mechanical properties, CNTs have been used for broad applications ranging from nano-composites to NEMS. The extremely small size of CNT presents significant challenges in the evaluation of their mechanical properties. Design and analysis of CNT have been largely done with molecular dynamics tools, which limits access to molecular dynamics specialists and does not allow for exploration of CNTs as components of a system.

Many researchers have pursued the analysis of CNTs by theoretical modeling, generally classified into two categories: atomistic modeling and other techniques that include classical molecular dynamics, tight-binding molecular dynamics, and density function theory [10]. Structural mechanics–based models of CNT have also been developed by [11] and [12]. An online tool available on the nanoHUB.org by Ref. [13] simulates the pull-in behavior of CNT-based NEMS for different applied voltages. However, this analysis is restricted to cantilever and fixed–fixed boundary conditions. Although a computer program called CoNTub1.0 is available for construction of single- and multiwalled CNTs [14], it does not support evaluation of their response to applied loading. We have not found a preexisting tool for the design and simulation of M/NEMS with CNT components.

Our following CNT model is an extension of a structural mechanics model developed by [11] and later by [12]. Our model simulates the dynamic response of CNTs with constant stiffness using an assemblage of our linear flexure model that we previously described in Refs. [5,15]. To achieve a lumped CNT model, we reduce the number of degrees without reducing accuracy by using the matrix condensation approach. The resulting lumped CNT model has six degrees of freedom at each end-node terminal (x, y, and z, and rotations about x, y, and z). This facilitates the connection of CNTs to other elements to easily generate complex system configurations. A benefit of our reduced-order modeling is that it greatly reduces computation time and reduces the complexity of designing integrated nano–micro systems.

3.2.1 Carbon Nanotubes

CNTs are allotropes of carbon with a cylindrical nanostructure. The structure of an ideal nanotube is formed by rolling up a graphite sheet into a cylinder and is defined by the vector linking the two equivalent carbon positions that are matched together after rolling, called the chiral vector $\vec{c}$ (see Figure 3.1):

$$\vec{c} = n\vec{a}_1 + \vec{a}_2. \tag{3.1}$$

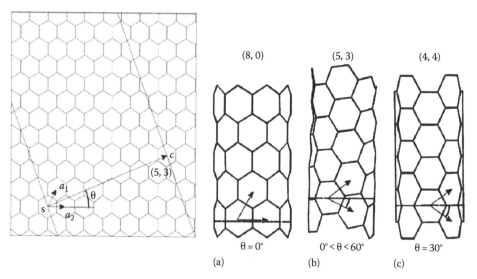

FIGURE 3.1
Nanotube geometry by means of graphitic plane rolling. CNT structures can be characterized by the chiral vector $\vec{c}$ and the chiral angle θ. They are usually classified as (a) zigzag $(n, 0)$; (b) chiral (n, m); or (c) armchair (n, n). (From Melchor, S. and Dobado, J.A., *J. Chem. Inform. Comp. Sci.*, 44, 1639, 2004.)

The angle θ formed between the chiral vector and a_1 defines the chirality of the CNT. Nanotubes that show symmetry in nature are called either armchair type $(n = m)$ or zigzag type $(m = 0)$. All remaining nanotubes with no symmetry are known as chiral. The graphitic sheet is made up of hexagonal lattices with a carbon to carbon, C–C, bond length of 1.415 Å. The diameter of a single-walled nanotube is given by $d = 0.0783(n^2 + m^2 + nm)^{1/2}$ nm. The mass of a single carbon atom is 1.9943×10^{-26} kg.

3.2.2 Structural Mechanics Model

We use the structural mechanics–based model suggested by [12] to develop the constant-stiffness model for CNT. This model is based on the notion that the carbon bonds in nanotubes may be represented as a geometrically framed structure, where the primary bonds between two nearest-neighboring atoms act like load-bearing members. That is, the carbon–carbon bonds are treated as solid rectangular flexure elements. The model neglects nonbond interactions caused by van der Waals forces and electrostatic forces. Table 3.1 summarizes the sectional properties that were used to develop the beam model in Sugar. Here, b, h, and l represent the width, height, and length of the beam element, respectively, E is Young's modulus, G is shear modulus, and ν is Poisson's ratio.

TABLE 3.1

Sectional Properties Used for CNT Model

CNT Mechanical Parameters		
$b = 0.127$ nm	$h = 0.086$ nm	$l = 0.1415$ nm
$G = 3260.32$ GPa	$E = 8476.84$ GPa	$\nu = 0.3$

Source: Li, C. and Chou, T.-W., *Int. J. Solids Struct.*, 40, 2487, 2003.

The CNT model is an extension of a linear flexure model existing in Sugar [15]. The beam model is modified with the properties given in Table 3.1 to represent CNT C–C bonds. We model the mass of the carbon atoms as lumped masses on the end of the Sugar flexures. Each lumped mass is a third of the mass of a carbon atom, because most often three such flexures (or bond types) form a carbon atom. The carbon atoms at the far ends of the CNT usually have two of these bonds.

The dynamics of the flexure can be described by a system of second-order ODEs of the form

$$M\ddot{q} + D\dot{q} + Kq = \sum F. \tag{3.2}$$

Here, M, D, and K represent the global mass, damping, and stiffness matrix, respectively. The global mass matrix is assembled from the elemental mass matrix [11]. By considering the atomistic feature of a CNT, the masses of electrons are neglected, and the mass of the carbon atom ($m_c = 1.9943 \times 10^{-26}$ kg) is considered to be concentrated at the flexure joints. Due to the extremely small radius of the carbon atom, the coefficients of mass matrix corresponding to torsional and flexural rotation are neglected. Only translatory displacement is considered. Hence the elemental mass matrix $[M]^e$ for CNT is given by

$$[M]^e = \begin{bmatrix} \dfrac{m_c}{3} & 0 & 0 & 0 & 0 & 0 \\ 0 & \dfrac{m_c}{3} & 0 & 0 & 0 & 0 \\ 0 & 0 & \dfrac{m_c}{3} & 0 & 0 & 0 \\ 0 & 0 & 0 & 0 & 0 & 0 \\ 0 & 0 & 0 & 0 & 0 & 0 \\ 0 & 0 & 0 & 0 & 0 & 0 \end{bmatrix}. \tag{3.3}$$

The elemental equilibrium equation can be written as follows:

$$Ku = f, \tag{3.4}$$

where

$$u = \begin{bmatrix} u_{xi} & u_{yi} & u_{zi} & \theta_{xi} & \theta_{yi} & \theta_{zi} & u_{xj} & u_{yj} & u_{zj} & \theta_{xj} & \theta_{yj} & \theta_{zj} \end{bmatrix}^T,$$

and

$$f = \begin{bmatrix} f_{xi} & f_{yi} & f_{zi} & m_{xi} & m_{yi} & m_{zi} & f_{xj} & f_{yj} & f_{zj} & m_{xj} & m_{yj} & m_{zj} \end{bmatrix}^T$$

are the nodal displacement vector and the nodal force vector of the element, respectively. The global stiffness matrix is assembled from the elemental stiffness matrix $[K]^e$, which is given by

$$[K]^e = \begin{bmatrix} K_{ii} & K_{ij} \\ K_{ij}^T & K_{jj} \end{bmatrix}, \tag{3.5}$$

where

$$
K_{ii} = \begin{bmatrix}
\dfrac{EA}{L} & 0 & 0 & 0 & 0 & 0 \\[2ex]
0 & \dfrac{12EI_x}{L^3} & 0 & 0 & 0 & \dfrac{6EI_x}{L^2} \\[2ex]
0 & 0 & \dfrac{12EI_y}{L^3} & 0 & \dfrac{-6EI_y}{L^2} & 0 \\[2ex]
0 & 0 & 0 & \dfrac{GJ}{L} & 0 & 0 \\[2ex]
0 & 0 & \dfrac{-6EI_y}{L^2} & 0 & \dfrac{4EI_y}{L} & 0 \\[2ex]
0 & \dfrac{6EI_x}{L^2} & 0 & 0 & 0 & \dfrac{4EI_x}{L}
\end{bmatrix},
$$

$$
K_{ij} = \begin{bmatrix}
\dfrac{-EA}{L} & 0 & 0 & 0 & 0 & 0 \\[2ex]
0 & \dfrac{-12EI_x}{L^3} & 0 & 0 & 0 & \dfrac{6EI_x}{L^2} \\[2ex]
0 & 0 & \dfrac{-12EI_y}{L^3} & 0 & \dfrac{-6EI_y}{L^2} & 0 \\[2ex]
0 & 0 & 0 & \dfrac{-GJ}{L} & 0 & 0 \\[2ex]
0 & 0 & \dfrac{6EI_y}{L^2} & 0 & \dfrac{2EI_y}{L} & 0 \\[2ex]
0 & \dfrac{-6EI_x}{L^2} & 0 & 0 & 0 & \dfrac{2EI_x}{L}
\end{bmatrix},
$$

$$
K_{jj} = \begin{bmatrix}
\dfrac{EA}{L} & 0 & 0 & 0 & 0 & 0 \\[2ex]
0 & \dfrac{12EI_x}{L^3} & 0 & 0 & 0 & \dfrac{-6EI_x}{L^2} \\[2ex]
0 & 0 & \dfrac{12EI_y}{L^3} & 0 & \dfrac{6EI_y}{L^2} & 0 \\[2ex]
0 & 0 & 0 & \dfrac{GJ}{L} & 0 & 0 \\[2ex]
0 & 0 & \dfrac{6EI_y}{L^2} & 0 & \dfrac{4EI_y}{L} & 0 \\[2ex]
0 & \dfrac{-6EI_x}{L^2} & 0 & 0 & 0 & \dfrac{4EI_x}{L}
\end{bmatrix}.
$$

Currently, our damping matrix (not shown) is proportional to the mass matrix. Its proportionality factor has yet to be rigorously determined. We refer to the aforementioned model as the "original CNT model."

3.2.3 Reduced-Order Modeling

Here, we describe a reduced-order model of the CNT. A nanotube is often shown to be a web-like structure comprised of simple bonds or flexure elements. However, the larger the number of elements or nodes used, the larger the size of the system stiffness and mass matrices by a factor of 6 (the degrees of freedom per node). For example, an armchair CNT model of chirality (4, 4) and length 10 nm in Sugar has 1769 elements. For such a seemingly small CNT, the stiffness and mass matrix have a size of 7212×7212. Thus, analysis of a large CNT structure, or a system of CNTs, can become computationally expensive.

Our model reduces the CNT structure of a given length, diameter, and chirality to an equivalent linear flexure structure, with only two 6 degree-of-freedom nodes (see Figure 3.2). The interfacial links are massless and only provide force, torque, and connectivity between each lumped CNT element. All information of the interior nodes is hidden in the reduced system matrix. The equivalent stiffness and mass matrix for the reduced system are generated using the matrix condensation technique [16]. A brief review of the matrix condensation technique is given in the following.

The first step is to partition the stiffness, displacement, and force matrices

$$\begin{bmatrix} K_{11} & K_{12} \\ K_{21} & K_{22} \end{bmatrix} \begin{bmatrix} x_1 \\ x_2 \end{bmatrix} = \begin{bmatrix} F_1 \\ F_2 \end{bmatrix}, \tag{3.6}$$

such that F_1 contains all the applied forces, F_2 contains only zeros, and x_1 refers to all the displacements we wish to retain. For our model, since we assume that forces are applied

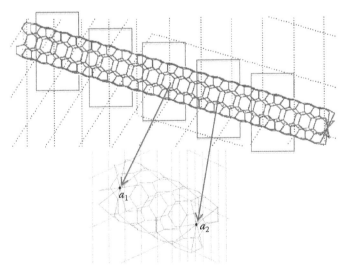

FIGURE 3.2
Lumped CNT section of a (4, 4) armchair CNT modeled in Sugar. In the previous figure, the complete length is obtained by a combination of six such reduced-order sections, linked at their terminal, or nodes (a_1 and a_2).

only at the CNT ends, these are the displacements of the first and the last node. Hence, the equation can be reduced to

$$[K_{11,reduced}][x_1] = [F_1],\qquad(3.7)$$

where

$$K_{11,reduced} = K_{11} - K_{12}K_{22}^{-1}K_{21}.$$

Similarly, the equivalent mass matrix M_c for the reduced degree-of-freedom system is given by

$$M_c = A_c^T M A_c,\qquad(3.8)$$

where

$$A_c = \begin{bmatrix} I \\ -K_{22}^{-1}K_{21} \end{bmatrix}.$$

The lumped model is particularly convenient when modeling CNTs as components of NEMS or integration with MEMS. The user can also make a long CNT as a combination of smaller CNT sections (as shown in Figure 3.2) to save computational cost. In the rest of this chapter, the model thus obtained is referred to as "reduced-order CNT model." Our reduced-order CNT model is just as accurate as the original CNT model for static deflections and low-order modes.

3.3 Design and Simulation of Carbon Nanotubes

3.3.1 Sugar Design

The CNT models of zigzag and armchair nanotubes are implemented in Sugar, a nodal analysis package for 3D MEMS simulation [5]. The user can simulate the structure and dynamic behavior of the CNT using either the original model or the reduced-order model. The original model is a hexagonal lattice structure comprising C–C bonds, which are modeled using CNT beam model described earlier. This model allows the user to analyze the deformation of geometry for the applied small forces or moments (see Figure 3.3).

However, as was mentioned earlier, using this model for simulating the geometry of large CNTs can be computationally expensive and time consuming. For such geometries, reduced-order models of symmetric CNTs can be used, the details of which were given earlier. In addition to the physics, graphics can also be computationally expensive. We have created a corresponding reduced-order display feature. Our display routine for our reduced-order model first generates a flat 2D image of the CNT with the specified properties and then maps this image onto a deformed 3D cylinder of required radius and length (see Figure 3.4).

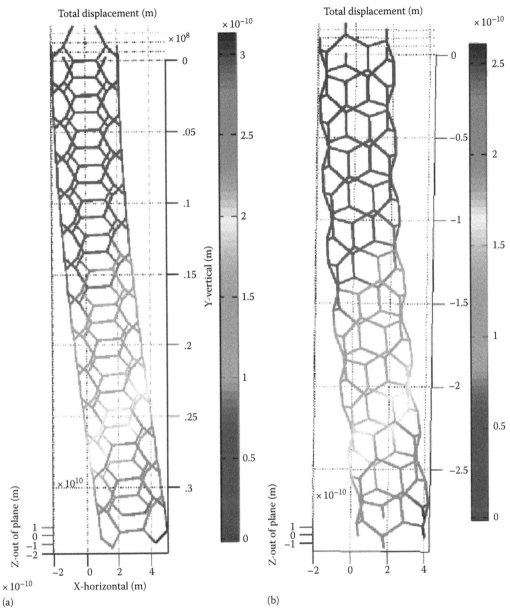

FIGURE 3.3
Deflected CNT models implemented in Sugar. Shown here are (a) armchair (3, 3) CNT and (b) zigzag (5, 0) CNT of length 3 nm each.

Our reduced-order model is highly efficient for predicting small linear displacements of the CNT. However, since a flat CNT image has been mapped onto its 3D surface for aesthetics, the image painted onto the surface of our reduced-order model may not be entirely accurate. Our lumped reduced-order model is much more computationally efficient than the original CNT model, both physically and graphically.

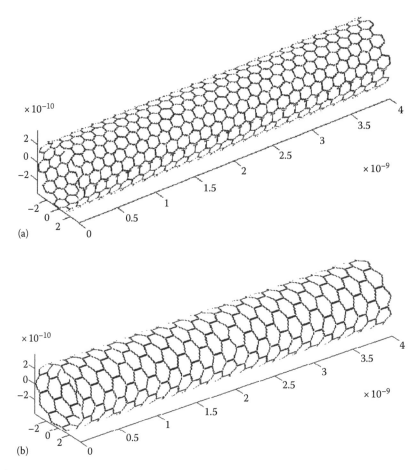

FIGURE 3.4
Reduced-order lumped CNT model implemented in Sugar. Shown here are (a) graphical display for armchair (10, 10) CNT and (b) zigzag (10, 0) CNT of length 4 nm each.

3.3.2 SugarCube Design and Simulation

The models for the two kinds of symmetrical CNTs are also available for use in SugarCube [14], which is a novice-friendly online tool for manipulating parameter values of ready-made N/MEMS that were initially configured using Sugar. The user can change various parameters like chirality, length and magnitude, and the direction of applied nodal forces. The model outputs the displaced structure and the displacements of the desired CNT node (Figure 3.5). Both original and reduced-order CNT models are available in SugarCube.

3.3.3 Applications

In this section, we discuss a few examples to demonstrate the application of our CNT model in N/MEMS. We exemplify our CNT model within a nanomaterial-testing device developed by [17], a nanomotor developed by [18], and our proposed NEMS comb-drive resonator.

Nanomaterial tester: As a first example, we demonstrate the use of the CNT model for performance optimization of the MEMS-based nanomaterial-testing device developed by [17].

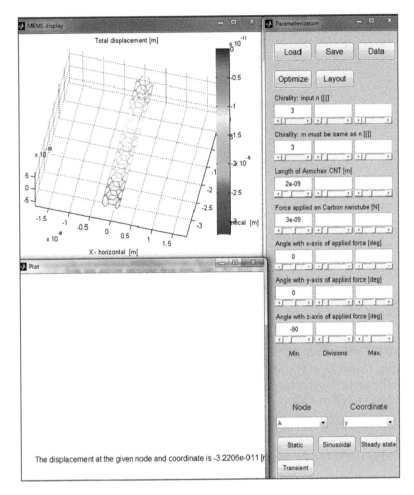

FIGURE 3.5
In SugarCube, the user explores performance using sliders. An armchair CNT (3, 3) of length 2 nm is simulated in SugarCube. An axial force of 3 nN yields a displacement of 3.2e–11 nm.

The chosen stiffness of the device's thermal actuator and load sensor is a strong function of the properties of the nanoscale specimen to be investigated. Hence, these properties need to be tailored for the prescribed nanostructure specimen in order to obtain sufficient or optimal performance. This essentially means that the same device cannot be used for all ranges of CNT specimen.

We have modeled the material-testing device in Sugar (see Figure 3.6). Using this model, the user is able to optimize the thermal actuator and electrostatic sensor of the MEMS-testing device to suit the expected properties of test specimens. Here, at node A is a single-walled CNT of diameter 0.78 nm and length 5 nm. For an average temperature rise of 150° of the V-beams, a specimen elongation of 172.8 nm was obtained using the Sugar CNT model. For the same specimen, the analytical model yielded an elongation of 149.6 nm [17].

The 13% discrepancy in the results may be attributed to the tester flexure geometry and our CNT model. The geometry of the folded beams in the tester device affects the stiffness of the load sensor and consequently the calculated specimen elongation. In the Sugar model, the folded beams are modeled as rectangular beam elements, instead of

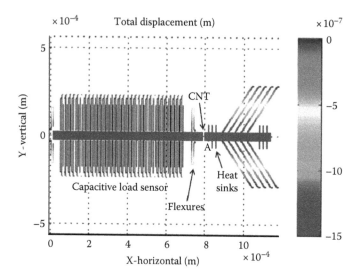

FIGURE 3.6
MEMS nanomaterial-testing device modeled in Sugar [1]. A CNT specimen of diameter of 0.78 nm and length of 5 nm (barely visible) is placed at node A.

arch-shaped structures. Also, our CNT model approximates the desired nanotube to the specified length, neglecting any incomplete hexagon structures in the lattice. This also contributes to the error in accuracy. Hence, these issues need to be addressed to increase the accuracy of our CNT model.

Nanomotor: Here, we exemplify the use of our CNT model in the design of a Zettl nanomotor [18]. It consists of an outer CNT sleeve surrounding a central CNT support. Figure 3.7 shows only the design of the nanomotor. Currently, small rotations are possible. However,

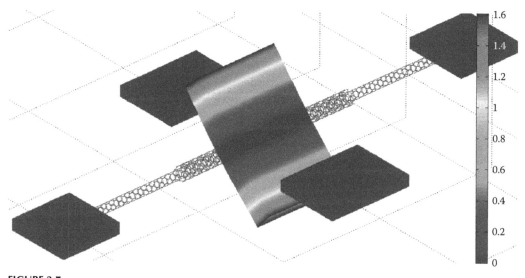

FIGURE 3.7
Zettl nanomotor concept. A multiwalled CNT is used as a rotational bearing for a proof mass. In practice, the proof mass rotates due to an applied electric field. (From Fennimore, A.M. et al., *Nature*, 424, 408, 2003.)

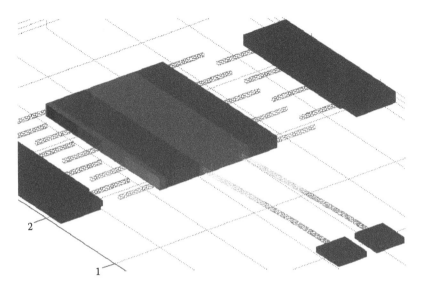

FIGURE 3.8
A proposed CNT comb-drive resonator. In this application, a pair of CNTs are used, flexures for a proof mass. CNTs protrude from proof mass to function as comb-drive fingers.

modeling of full rotation due to applied electric field analysis has yet to be implemented for complete analysis.

NEMS resonator: The CNT model can be used to explore the design of a NEMS comb-drive resonator. In Figure 3.8, we show a proposed resonator with flexures and comb fingers composed of CNTs. The CNT flexures have a diameter of 0.39 nm and length of 6 nm. The proof mass and anchors are polysilicon depositions over strategically positioned CNT flexures and comb fingers.

3.3.4 Summary

In this section, we introduced a new lumped mechanical model and online tool for simulating a system of CNTs. Our elemental and reduced-order CNT models are implemented in both Sugar and SugarCube. The models do not require any programming or extensive training to use. They allow both novice users to analyze the dynamic response of the CNT structure to small applied forces as well as experts to conveniently use CNT as components of N/MEMS. The user can either utilize the original CNT models for analyzing small structures or the more computationally efficient reduced-order model for simulating larger or a system of CNTs. Models for static analysis of symmetrical CNTs (zigzag and armchair) are currently available.

3.4 Lumped Modeling of MEMS

With the growing use of MEMS in education, research, and industry, there is a growing need for both novices and experts to easily and quickly predict the performance characteristics of MEMS. MEMSs allow computers to efficiently interact with physical phenomena.

There is an increasing presence of microscale subsystems that are comprised of components from a common set of electromechanical elements. For instance, consumer products that use MEMS gyroscopes include smartphones, Segway personal transporters, automotive stability control systems, video game controllers, etc. Although the applications for microgyroscopes may greatly differ—requiring various levels of precision, accuracy, or robustness—often the desired performance characteristics may be achieved by modifying the design parameters of basic microgyroscope systems (structure and electronics).

Computer-aided design (CAD) for MEMS tools is often used by experts to create unique MEMS or investigate higher-order behavior in preexisting MEMS. Such tools specialize in distributed analysis, for example, Refs. [7,19–21], or lumped analysis tools, for example, [5,22]. The types of MEMS that one investigates using lumped analysis tools are MEMS that can be configured with preexisting elements. Distributed analysis tools do not have such modeling constraints, but this comes at the expense of computational expediency and the required expertise in physics, CAD, and processing technology [23,24]. Such tools are sophisticated, often requiring high-end computers, technical support, and extensive training.

Conventional tools do not appear to accommodate novices who would like to explore variations in preexisting MEMS or professionals who would like to recycle or leverage from the MEMS knowledge base. These groups of individuals may potentially involve a much larger number of users than the typical number of advanced MEMS experts. To accommodate this larger group of individuals, key attributes might include a tool being readily accessible, easy to use, and requiring a minimal amount of time to yield results.

A preexisting tool that comes closest to these attributes is MEMSolver [25]. Although its developers had obviously set out to achieve an entirely different set of attributes, we compare MEMSolver to SugarCube because it is the closest-related tool. MEMSolver may be purchased and downloaded from Ref. [25]. It consists of a library of parameterizable analytical formulas of simple components of MEMS, such as cantilevers and diaphragms. However, the components cannot be combined into a complete system in MEMSolver. The library is not extensible by the user, and there are no practical bounds on parameter values. This latter issue may be a problem for novices that may not know the practical limits for fabrication or the limits for which a model is valid.

In contrast, entire systems may be parameterized in SugarCube. It is the only CAD tool that is able to display circuits with deflected structures in rotatable 3D. Much of SugarCube's library consists of MEMS designs found in the literature, and the library may be extended by its users, similar to a Wiki. The Wiki attribute is expected to help facilitate SugarCube's self-sustainment. SugarCube is also unique in using editable bounded sliders (discussed in the following). Not only are sliders quick and easy to use, but the bounded values provide practical design-rule limits, which may be overridden by the user by entering numerical values instead. Other key advances in SugarCube include its optimization features, its one-button layout array generator, and its wafer-level design [26,27].

3.4.1 Sugar to SugarCube

Our tool called Sugar [15] is what we use as the modeling and simulation engine for SugarCube. Sugar uses modified nodal analysis, similar to Spice [6], to mathematically represent a system (or network) of lumped elements. Spice is optimized for electrical elements, while Sugar accommodates electrical and mechanical elements for analysis and display, using a more versatile netlist language.

In Ref. [15], we experimentally validated Sugar's ability to fairly match experimental results with prediction using an advanced micromirror. This validation was a key

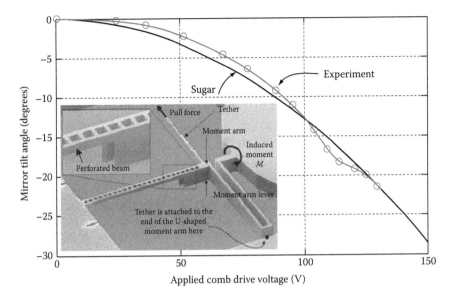

FIGURE 3.9

A micromirror and its Sugar representation. About a 1000 comb-drive fingers are supported by a pair of cosine-shaped flexures (out of view). The cosine-shaped flexures straighten once fully deflected, to stabilize the comb drive at large voltages. Upon actuation, the comb-drive array translates a pair of tethers that are attached to U-shaped moment arms. The moment arms convert translation to rotation, to rotate the circular mirror out of plane, which is supported by perforated torsional flexures (see inset in Figure 3.9). The Sugar representation is showing a view from underneath, showing the recessed plate for reducing its mass for a faster response time [5]. Validation of Sugar. Mirror angle versus voltage. Experimental measurements are shown as circles. The applied voltage in Sugar is parametrically swept from 0 to 150 V. The inset shows close-up view of the perforated torsional flexure that is modeled in Sugar.

milestone in the area of CAD for MEMS because the micromirror had such an intricate design that its performance is too difficult to predict using hand analysis, and it had such a large number of multiphysical components that it was too difficult to simulate using conventional distributed analysis tools on an average personal computer. For example, in Figure 3.8, we show scanning electron microscope images of an advanced micromirror and Sugar's representation of the micromirror. In Figure 3.9, we plot experimental data against simulation for the tilt angle as a function of applied voltage on ~1000 comb-drive fingers.

The skill to create a computer model of such a device is usually beyond the expertise of many individuals, which is where SugarCube becomes useful. Users that are familiar with Sugar may easily create a parameterized model of their MEMS for use by others as follows.

As an example, we show a simple netlist in Figure 3.10 and its corresponding image in Figure 3.11. By comparing node labels in the netlist to node labels in the figure, one is able to identify each line in the netlist with each element in the figure. For brevity, most of the netlist parameter values are not provided. However, the parameters that are provided in blue type are used to help demonstrate how to convert a Sugar netlist into SugarCube.

In Figure 3.12, we show netlist code that identifies which parameters from the netlist in Figure 3.3 will be accessible through the SugarCube GUI. Adding the code in Figure 3.12 to the original netlist code in Figure 3.10 completes the conversion from Sugar to SugarCube. The resulting netlist can still be used by Sugar, whereby the added SugarCube code has no effect.

```
anchor    p1    [A]    [l= w= h= oz= ]
etbeam    p1    [Ab]   [l=LHOT w=WHOT h= L1= oz1= L2= tec=ALPHA]
etbeam    p1    [bc]   [l=GAP w= h= oz= L1= L2= tec=ALPHA]
etbeam    p1    [cd]   [l= w= h= oz= L1= tec=ALPHA]
etbeam    p1    [Ed]   [l= w= h= L1= oz1= oz2= L2= tec=ALPHA]
anchor    p1    [E]    [l= w= h= oz= ]
resistor  cmos  [Af]   [l= R= oz= L1= L2= ]
voltage   cmos  [gh]   [l= V= oz= ]
short     cmos  [gE]   [l= R= oz= ]
ground    cmos  [g]    [l= V= oz= ]
```

FIGURE 3.10
Netlist of an electrothermal actuator. The corresponding image is shown in Figure 3.11. It consists of a thin hot arm and a wide cold arm. A difference in temperature between the two arms will show a lateral deflection once simulated in Sugar. For brevity, chosen parameterized quantities are given in blue type: length of the hot arm LHOT, width of the hot arm WHOT, gap between the hot and cold arms GAP, and the thermal expansion coefficient ALPHA (Figure 3.12).

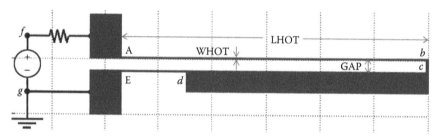

FIGURE 3.11
Sugar display of electrothermal actuator. Particular parameters and node names from the netlist given in Figure 3.10 are superimposed onto the image of the MEMS. Upon actuation, electrical current from the circuit flows from node A to b, causing the thin arm to get hot. Current flowing in the wide arm from node c to d is cooler because of less resistance from its wider width.

```
sugarcube * [ ] [LHOT = 'Hot arm length, m, 300e-6, 100e-6, 400e-6'
                 WHOT = 'Hot arm width, m, 0.8e-6, 0.5e-6, 6e-6'
                 GAP = 'Gap, m, 4e-6, 2e-6, 10e-6'
                 ALPHA = 'Thermal Expansion Coef, /C, 2.3e-6, 2.3e-7, 2.3e-5'
                 nodes = 'b, d']
```

FIGURE 3.12
Converting Sugar to SugarCube. Any Sugar netlist may be used for SugarCube by simply adding the netlist code that identifies which netlist parameters are to appear in the SugarCube GUI. The syntax is netlist parameter name, how the name appears in the GUI, unit, default, minimum, and maximum values. Finally, the nodes which are to be analyzed must be included. In this case, the SugarCube user may plot data from either node b or d of the structure (Figure 3.11).

We show SugarCube's framework in Figure 3.13, which illustrates how Sugar is integrated. The SugarCube user is not required to know how to use Sugar. The main algorithms in SugarCube include its librarian, parameterization, simulation, layout, and optimization. Due to the built-in automation, most of the algorithms in SugarCube are driven by simple mouse or pen commands requiring one or two button clicks.

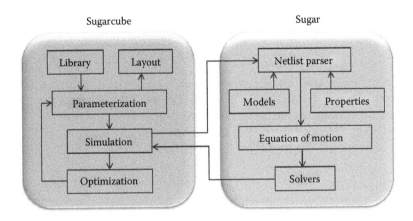

FIGURE 3.13
SugarCube Framework. SugarCube adds a simple and intuitive library, parameterization, optimization, layout features to the Sugar engine. The arrows indicate the direction of data flow. A library of MEMS netlists that is created in Sugar is uploaded into a hierarchical SugarCube library.

3.4.2 Librarian

SugarCube's librarian consists of a hierarchical collection of ready-made MEMS devices that are created in Sugar and imported into a SugarCube library directory folder. Many of the MEMS in the library are modeled after designs published in the literature. Each device in the SugarCube librarian may be accompanied with a preview image and a description since library file names alone may not be helpful to novices.

In Figure 3.14, we show an example of selecting a resonator from the resonators category in the SugarCube librarian. As shown in the figure, the image preview of the resonator

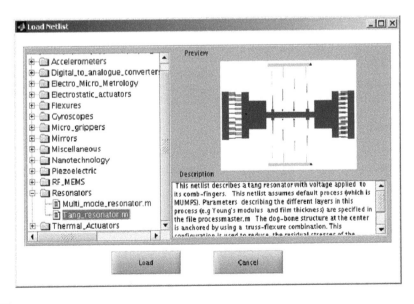

FIGURE 3.14
The SugarCube librarian. A hierarchical librarian is often the first step in using SugarCube. To assist with locating a desired MEMS design, an image and description of the designs are provided. The device's description may provide information such as what the device is, its common applications, and its reference(s).

and its description are provided for easy identification. Upon loading this device, it is displayed in the SugarCube GUI with modifiable design parameters.

Expert users may contribute to the library's content, which is expected to help with SugarCube's self-sustainment. In the previous section, we discussed how to convert a Sugar netlist to SugarCube. To allow SugarCube's librarian to show an image of the device when its file name is highlighted, the file's creator should save the image of the device using the same name as the netlist, with the proper file type extension. In addition, to allow SugarCube's librarian to show a text description of the device, the file's creator should write a commented description at the beginning of the netlist file.

3.4.3 Parameterization

Efficient parameterization of MEMS is one of the key features of SugarCube that allows users to quickly explore design spaces of ready-made MEMS. And experts might find that importing/exporting their unique designs to SugarCube might increase the productivity of their design cycle.

The parameters that SugarCube provides to users are defined by the model's creator. Common parameters might be geometry and material properties. However, properties such as chip acceleration or rotation, nanotube chirality, finite states, and whether to include sets of electrical or mechanical components may be parameterized as well.

The bounds of the sliders are meant to be realistic and reduce the chance of obtaining nonsensical results or an error. However, numerical values may be entered outside that exceed the suggested bounds. Single default values for each parameter are initially loaded from the librarian. It is up to the user to prescribe different single values or a range of values to sweep. If only single values are chosen, one data point usually results upon simulation. If one of the parameters is changed into a range of values, then usually a 2D curve results upon simulation. If two parameters prescribe a range of values, then usually a 3D manifold results upon simulation. The data can be saved and recalled for later use.

3.4.4 Simulation

We implement four types of solvers in SugarCube: static, steady-state, modal (or sinusoidal), and transient analyses.

3.4.5 Static Analysis

In static analysis, SugarCube determines a static equilibrium of the system due to constant efforts. For static analysis, the mathematical representation of a multidisciplinary system has the form

$$F(q) - K \quad q = 0, \tag{3.9}$$

where F is a multidisciplinary vector of applied efforts. The efforts may consist of voltages, currents, forces, torques, pressures, temperatures, etc. The vector q comprises of multidisciplinary displacements such as charge, translation, rotation, volume, entropy, etc. The constant of proportionality matrix K relates the displacements to the efforts. Since the effort is a function of displacement, (3.9) may be nonlinear. SugarCube uses our Newton–Raphson algorithm to solve (3.9). We show examples of static analysis in Figures 3.15 and 3.16. In Figure 3.15, we show the static analysis of an out-of-plane strain gradient of an

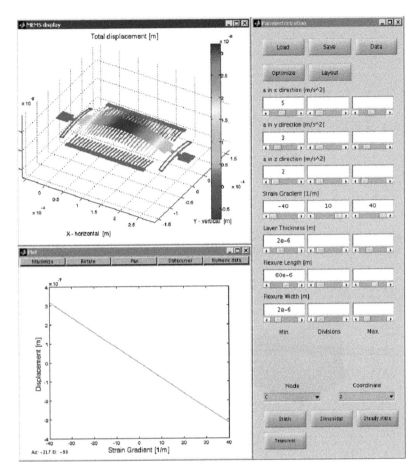

FIGURE 3.15
Strain gradient of ADXL accelerometer. In the BiCMOS process, a cantilever of length $L = 150\,\mu m$ may bend down by about $y = -0.3\,\mu m$ due to a negative strain gradient Γ. If only strain gradient is acting, the deflection of a cantilever is $y = \Gamma L^2/2$. Solving for strain gradient, we have $\Gamma = -26.7/m$. The default range of Γ is -40 to 40, which is applied to all structures. The deflection has been magnified to show the bowing effect.

analog devices ADXL-like accelerometer. Such a device is similar to the common air-bag deceleration (impact) sensor in most automobiles. Although the strain gradient is negative, which would cause a simple cantilever to bend down toward the substrate, in the ADXL, this negative strain gradient causes the backbone of the structure to bow upward out of plane due the compliance of its two support flexures. This strain gradient can also cause comb finger misalignment.

In Figure 3.16, we show the static analysis of an electrothermal actuator, sometimes referred to as the chevron or V-shaped actuator. Electric current flowing through its parallel flexures causes thermal expansion due to Joule heating. The flexures meet at a small angle to produce a magnified deflection. This type of deflection is sometimes mistakenly called buckling. In this model, average temperature is used because the amount of thermal expansion in a linear beam due to a distributed temperature along its length is the same as its average temperature. The amount of deflection for a given temperature highly depends on the initial angle of the flexures. Both the beam length and angle are swept in Figure 3.16. A plot of deflection shows that there is an optimal angle for which these types of actuator should be configured.

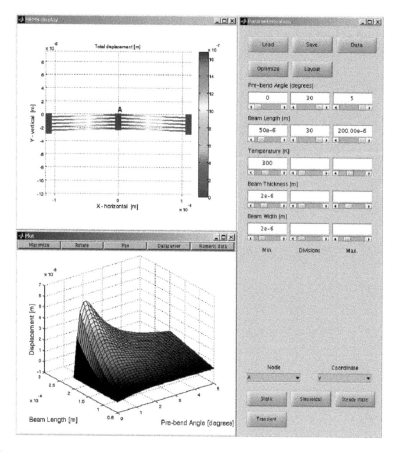

FIGURE 3.16
Static analysis of a thermal actuator. A 3D manifold shows the deflection of the actuator tip versus a change in width of the hot arm and a change in gap between hot and cold arms.

3.4.6 Steady-State Analysis

In steady-state analysis, SugarCube computes the frequency response of the system to an external excitation. This response is obtained by computing the transfer function of the equations of motion for the system

$$M\ddot{q} + D\dot{q} + Kq = Bu$$

$$y = L^T q$$

(3.10)

where
   M, D, and K are the multidisciplinary mass, damping, and stiffness matrices of the system, respectively
   The vector q describes the state of the system
   B is the input influence array to indicate the position of the excitation
   u is the input excitation
   y is the output vector
   L is the output influence array that is chosen to extract the components of interest of state vector q

Upon computation, the transfer function is obtained as

$$H(s) = L^T \left(s^2 M + sD + K\right)^{-1} B \qquad (3.11)$$

where s is a complex variable and takes a value of $s = j\omega$, with $\omega \geq 0$ being the excitation frequency.

In Figure 3.17, we show an example of the frequency response of an MEMS gyroscope. The response includes both magnitude and phase as a function of frequency. What else is unique about SugarCube is that its frequency response may also be parameterized as a manifold. For instance, in Figure 3.18, we show frequency response for a crab leg structure (inset) that is parameterized by the size of its proof mass.

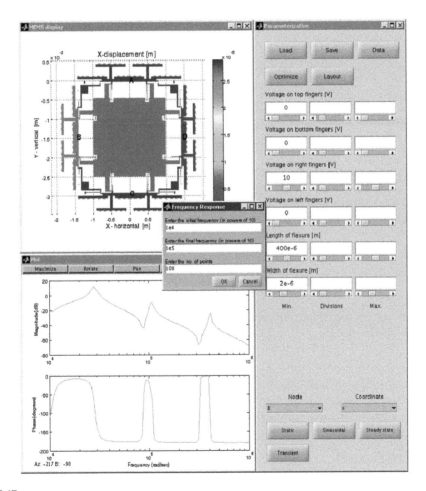

FIGURE 3.17

Frequency response of a gyroscope. This MEMS gyro consists of 2000 comb fingers and orthogonal movable-guided flexures. These flexures allow the proof mass to translate with 2 degrees of freedom and resist rotation. On the other hand, a set of fixed-guided flexures allows each comb drive only 1 degree of freedom. The magnitude and phase of the x-coordinate of node C are swept from $10\,k$ to $1\,Mrad/s$. The inset prompt shows the entrees for new frequency response of $10\,k$ to $100\,k$ with a 100 point resolution.

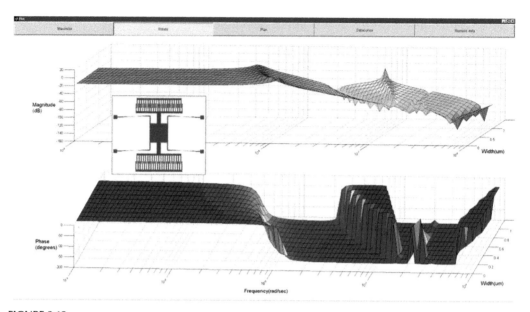

FIGURE 3.18
Parameterized frequency response of a crab leg resonator. The frequency response is parameterized by changing the width of the proof mass, increasing its inertia and viscous damping between the mass and substrate. The device is shown in the inset.

3.4.7 Sinusoidal Analysis

In sinusoidal analysis (or modal analysis), SugarCube computes the eigenmodes and frequencies of the system. This is obtained by solving the following eigenvalue problem

$$M\ddot{q} + Kq = 0 \tag{3.12}$$

where
   M and K are the multidisciplinary mass and stiffness matrices, respectively
   q is the state of the system

In SugarCube, the number of modes to compute is suggested or provided by the user. Once the analysis is performed, a pull-down menu appears in the plot window that can be used to select the desired mode to display. In Figure 3.19, we show an example of the second eigenmode of a folded flexure comb-drive resonator.

3.4.8 Transient Analysis

The fourth type of solver is transient (or continuous time) analysis. SugarCube computes the transient response of a system for a given effort, which may be nonlinear. Beginning by initially representing the system in the form of

$$M\ddot{q} + D\dot{q} + Kq = F(q) \tag{3.13}$$

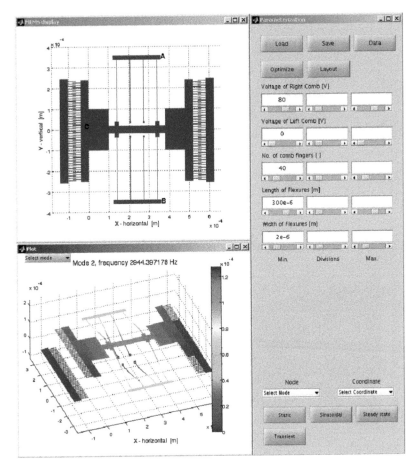

FIGURE 3.19
Sinusoidal (modal) analysis. A folded flexure comb resonator is shown oscillating in mode 2, which is out of plane. In this case, the modal frequency is 2.9 kHz. Designing around such frequencies might be important if they are considered to be failure modes.

where the quantities are defined earlier. We reduce the order of (3.5) by using our Krylov subspace method [29] to more efficiently compute the trajectory of motion of a user-specified node and coordinate.

For example, in Figure 3.20, we show a transient response of a sensor that is subject to temperature and electrostatic efforts. Although Sugar's lumped modeling is already a type of reduced-order modeling, our reduced-order transient analysis algorithm further reduces the system from 1152 to 50 degrees of freedom, which greatly reduces computation time without sacrificing a significant amount of relative accuracy.

3.4.9 Optimization

Sometimes it is not immediately obvious which, and by how much, geometric or material properties should be changed to achieve a particular performance metric. With SugarCube, we have implemented an algorithm that allows the user to input a particular performance to output a suggested parameter change. This allows the user to

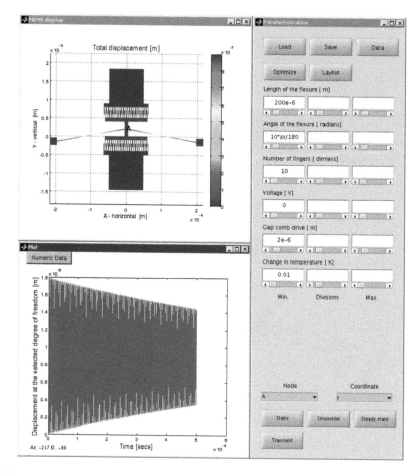

FIGURE 3.20
Transient response. Due to a step input of thermal stress and electrical static forces, its initial peak-to-peak amplitude of 1.8 μm quickly decays to a static deflection of 0.8 μm. Damped frequency is 1.7 kHz.

easily and quickly optimize MEMS with just a few button clicks. By switching to optimize mode, a list of available design parameters and analysis options is displayed. A user may select which parameters to optimize, choose the type of analysis, and specify the desired performance. Currently, our optimization feature applies only to static and sinusoidal analyses.

Given a desired performance and a range over which parameters are allowed to vary, SugarCube optimizes the parameter space to the desired performance by minimizing the following function:

$$f(P,C) = Desired\ perf. - Computed\ perf.\ (P,C), \tag{3.14}$$

where
  f is an objective function
  P is a vector of bounded parameters that are to be optimized
  C is the vector of remaining parameter values to be held constant

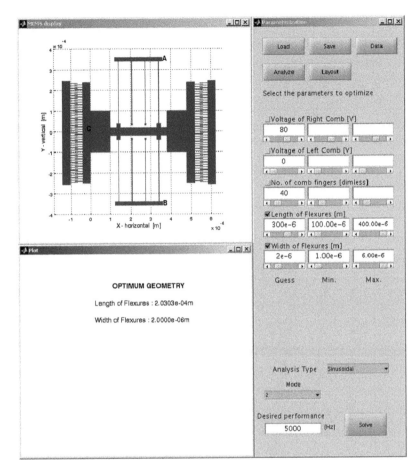

FIGURE 3.21
SugarCube optimization. SugarCube allows the user to input the desired performance and automatically obtain a modified design that meets the desired performance.

The objective function is automatically generated by SugarCube. For each iteration, the algorithm generates a new set of parameter values P, until (3.13) reaches a particular tolerance. Once optimized, the values of P are displayed.

For example, in Figure 3.21, we show a folded flexure comb-drive resonator that is optimized to resonate at a particular frequency. In the parameter list, the parameters that are chosen to be held constant have a single value, while parameters that are to be optimized are specified by minimum and maximum values and an initial guess. Each comb of the resonator has 40 fingers. The length and width of each finger are 60 and 2 μm, respectively. The design length and width of the folded flexure are 300 and 2 μm, respectively. We select the length and width of flexure as the design parameters to be optimized. Length varies from 100 to 400 μm and width varies from 1 to 6 μm. We specify the desired performance as the resonant frequency of mode 2 to be 5 kHz. SugarCube computes the optimum length and width to be 203 and 2 μm, respectively, see Figure 3.14. Note that the optimal parameters returned may not be unique. That is, there may be another set of parameters that yields the same performance. The search stops once a valid set is found.

3.4.10 Layout

The task of creating layout arrays of MEMS for fabrication is one of the most tedious tasks, often requiring days to weeks for skilled individuals. In SugarCube, the task of creating parameterized layout can be reduced to seconds with a single click of a button. SugarCube automates the process of checking for design rules, placing etch holes, supplying connecting layers for anchors and bonding pads topped with metal, supplying common ground tracers between structures, and optimizing chip real estate. Such features are expected to be novice-friendly and increase productivity.

For example, in Figure 3.22, we show the result of a GDS-II file created by SugarCube. The device used for this layout array is the temperature sensor shown in Figure 3.20. Similar to other features discussed earlier, to create a parameterized layout array in SugarCube, one simply selects which two parameters are to be swept along the row and column of

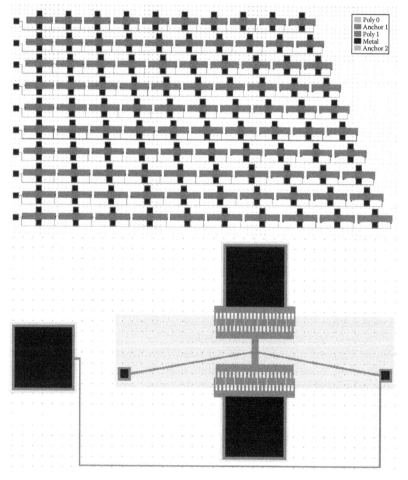

FIGURE 3.22
Layout in GDS-II format. A SugarCube generated array layout of the temperature sensor shown in Figure 3.20. The images shown are SugarCube's output file as loaded into a free GDS-II reader. In addition to plotting a performance manifold due to sweeping length and angle parameters, a simple push of the layout button yields a corresponding layout array that is ready for fabrication. Ground planes, tracers, bond pads, and anchor layer connects are automated (see enlarged image).

the array, then clicks the Layout button. In essence, each data point from a parameterized simulation can be validated by fabrication.

3.4.11 Summary

In this section, we discussed a CAD for MEMS tool called SugarCube that is novice-friendly and may improve the productivity of expert MEMS designers. This tool is easy to use and requires no programming, no training for most individuals, and no MEMS expertise. It is developed to allow novices (as well as experts) to easily explore what-if scenarios of ready-made MEMS. It is also useful for professionals who have little time or interest in learning how to master sophisticated simulation software for simulating variations of ready-made MEMS. SugarCube adds key MEMS-related automation features and an intuitive GUI to our MEMS simulation tool called Sugar. We have demonstrated SugarCube's features such as parameterization, simulation types, optimization, and layout generation of various MEMS.

SugarCube can be used online at nanoHUB.org through a standard Internet web browser. It has a wide collection of ready-made MEMS models available in its expandable library. The expandable library follows the Wiki concept. That is, users can upload their MEMS designs into SugarCube. A user can select MEMS from this library and analyze it using the parameterization feature. Features of its librarian include images with descriptions. Features of its analysis solvers (static, steady-state, modal, and transient) include default and easily modifiable options. Analysis data can be downloaded as 2D or 3D plots, or in numeric form. Other features such as geometry optimization (given desired performance) and the generation of layout arrays were briefly discussed.

SugarCube might be useful to (1) students for exploring MEMS performance for developing intuition and a working knowledge; (2) MEMS customers for modifying commercial MEMS designs for new products; (3) researchers for importing novel designs and models into SugarCube for quick and efficient parameterization, optimization, and layout generation; (4) customers of MEMS foundries as a one-stop-shop for online design, simulation, and layout submission for fabrication; or (5) researchers as a clearinghouse for quick and widespread dissemination and use of novel systems, as complement to conference or journal publications in which the system are not in readily useful forms. Such applications might allow SugarCube to improve and evolve over time.

3.5 Design and Simulation of N/MEMS

MEMSs have the potential to provide accurate mechanical characterization of nanostructures such as CNTs and nanowires. The performance of MEMS highly depends on its structural design, which determines the device's sensitivity and ranges of applied or sensed forces and displacements. The design and performance optimization of MEMS often require several weeks of specialized training in the use of CAD and engineering tools. In particular, an MEMS-based nanomaterial-testing device has been developed in Refs. [17,30,31], see Figure 3.23. The chosen stiffness of the actuator and the load sensor are a strong function of the properties of the specimen to be investigated. Such properties need to be tailored for the prescribed nanostructure specimen in order to obtain sufficient or optimal performance [31]. Hence, each nanoscale specimen may require the material-testing device to have a completely

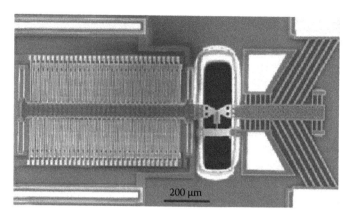

FIGURE 3.23
MEMS nanomaterial-testing system. (From Zhu, Y. and Espinosa, H.D., *Proc. Natl Acad. Sci. USA*, 102(41), 14503, 2005.)

different set of geometrical and material properties. Optimization using distributed analysis is computationally expensive, time consuming, and difficult to geometrically parameterize [31]. Analytical models of the complete device are not readily available, and only an FE analysis (FEA) simulation of the thermal actuator is provided in Ref. [31]. The simulation for comb-drive sensor is not provided. Here, we describe a parameterized computer model of this MEMS nanomaterial-testing device. What is interesting about our approach is that we integrate a CNT model with the microscale device for multiscale simulation. Such modeling integration allows the user to explore the performance of a proven MEMS tester applied to various nanoscale structures. We modeled both the MEMS and CNT using Sugar, and we export the design to SugarCube.

Our model allows the user to explore the electro-thermo-mechanical properties of the MEMS device and optimize its response by adjusting its geometry and material properties and by using variants of CNT samples. Our CNT model given earlier provides a structural mechanics–based lumped model of single-walled nanotubes that can be used to simulate a test specimen of desired chirality, diameter, and length. Our model is expected to benefit those who do not have design expertise with traditional CAE tools, ready access to such tools, or the time or desire to develop a new computer model from scratch. It will also help experts by providing a tool to quickly simulate variations of such MEMS devices and optimize them, in turn, facilitating development of tools for nanomaterial testing.

3.5.1 Sugar Model

We modeled the MEMS-based material-testing device in Sugar (see Figure 3.24). The complete device was modeled using basic Sugar building blocks, that is, anchors and 3D electro-thermo-mechanical flexures. The CNT specimens were modeled using the new CNT models available in Sugar. The design parameters were taken from Refs. [17,30,31]. We parameterized the geometry and material properties of both the device and the CNT test specimen to be used. Static, modal, and transient analyses can be performed on the device with applied force, voltage, or thermal loads. Once a device is configured in low-level Sugar, its netlist can be imported into the novice-friendly high-level SugarCube for easier exploration of design space performance, as well as easy creation of parameterized layout arrays for fabrication.

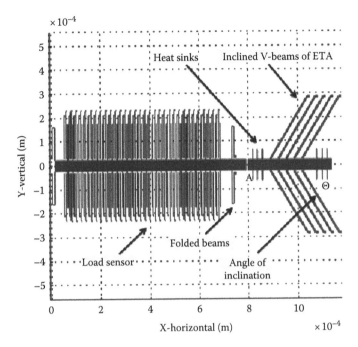

FIGURE 3.24
Nanomaterial-testing device modeled in Sugar.

The following geometrical properties are parameterized: number of fixed and moving fingers in the comb-drive sensor, length and width of the fingers, folded beams and inclined V-beams, angle of inclination of the V-beams, and number of heat sinks. The applied voltage or temperature conditions can also be manipulated. In Figure 3.24, "A" is the polysilicon or CNT test specimen under consideration.

3.5.2 SugarCube Model

SugarCube provides novices with simple high-level manipulation controls for ready-made M/NEMS using sliders that are governed by design rules. Our model was imported into SugarCube after we configured its basic geometry in Sugar using its electro-thermo-mechanical netlist language. A multitude of parameters may be chosen to be accessible through SugarCube. In the present example, we choose to demonstrate parameterization of the parameters for the thermal actuator (V-beams). As seen in Figure 3.25, sliders appear for V-beam length, width, number, angle, and temperature. We also parameterized the number of heat sinks and number of comb-drive sensor fingers. Static, modal, and transient analyses are available in SugarCube. Upon simulation, parameterized performance 1D values, 2D curves, or 3D manifolds are displayed in the lower-left window. Here, the plot shows displacement of the actuator–specimen junction with increase in the average temperature of the V-beams.

Another interesting aspect of SugarCube is that it can generate a parameterized layout array of the device in GDS-II format that can be directly send for manufacture. For example, Figure 3.26 shows an array of the nanomaterial testers, where the row is parameterized by the number of V-beams and the column by the angle of V-beams. The bonding pad layers and common ground tracers are automatically created for the layout. The layout image in Figure 3.26 is from using the free GDS-II viewer called CleWin [32]. SugarCube's easy layout

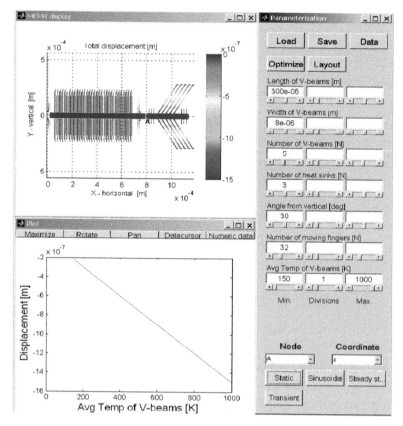

FIGURE 3.25
Nanomaterial-testing device in SugarCube.

generation can reduce the time conventionally spent on layout from days or hours to seconds and hence can provide significant time and cost savings. For instance, after choosing the two parameters to vary and pressing the Layout button, it takes SugarCube only 5 s to generate the GDS-II parameterized layout. MEMSCAP MUMP design rules are applied.

3.5.3 Carbon Nanotube Model in Sugar

We have developed a computer algorithm to model single-walled CNTs. Our lumped model is based on the elemental structural mechanics model developed by [11]. The model in Ref. [11] is based on the notion that the carbon bonds in nanotubes may be represented as geometrically framed structures, where the primary bonds between two nearest-neighboring atoms act like load-bearing members. That is, the carbon–carbon bonds are treated as solid rectangular flexure elements. We imported the parameters of this model into a constant stiffness electro-mechanical Sugar model. Our linear CNT model can simulate symmetrical CNTs (zigzag and armchair) and is parameterized by its chirality and length. We use matrix condensation to create a lumped model (see Figure 3.27). Matrix condensation is used to reduce the number of degrees of freedom of the model without affecting its accuracy. This enables quick modeling of large CNT specimen. In the present work, we integrate our CNT model to emulate more complete nanomechanical-testing conditions that help users choose more appropriate design parameters for the MEMS tester.

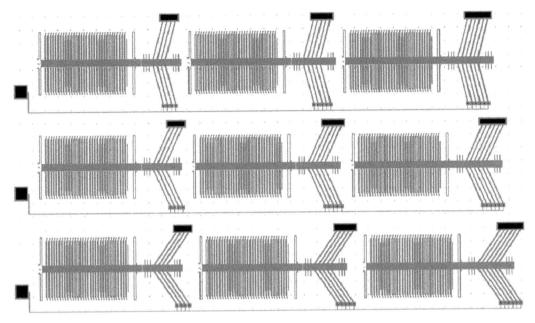

FIGURE 3.26
GDS-II layout of an array of nanomaterial-testing devices, viewed in Clewin 3.2.

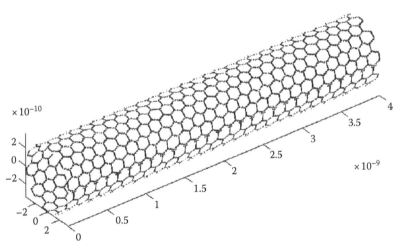

FIGURE 3.27
A (10, 10) armchair CNT modeled in Sugar.

3.5.4 First-Order Analysis of the Thermal Actuator

A thermal actuator is based on the principle that when a voltage is applied across the beam, thermal expansion is induced. The change in beam length can be calculated by

$$\Delta L = \alpha \int_0^L T(x)\,dx. \tag{3.15}$$

A Sugar routine for the electrothermal beam is developed based on the analytical model for the thermal flexure actuator, developed in Ref. [33]. For our model, we simplify by assuming that there is no heat conduction to the substrate. Therefore, under steady-state conditions, heat conduction out of the element is equal to the resistive heating of the element:

$$k\frac{d^2T}{dx^2} + J^2\rho = 0, \tag{3.16}$$

where
  J is the current density
  ρ is the resistivity of the beam
  k is the thermal conductivity of polysilicon

We assume that k is constant and is equal to the value when evaluated at room temperature (34 W/m K). Moreover, ρ varies linearly with temperature and is given by

$$\rho = \rho_0\left[1 + \lambda(T - T_S)\right] \tag{3.17}$$

where
  ρ_0 is the resistivity at ambient temperature (T_S) and is taken to be 3.4×10^{-5} Ωm
  λ is the linear temperature coefficient (1.25×10^{-3})

If we also assume $\lambda(T - T_S) \ll 1$, then Equation 3.16 can be rewritten as

$$k\frac{d^2T}{dx^2} + J^2\rho_0\left[1 + \lambda(T - T_S)\right] = 0. \tag{3.18}$$

The solution of the preceding differential equation is

$$T(x) = T_S + \frac{1}{\lambda} + be^{\tau x} + ce^{-\tau x}, \tag{3.19}$$

where $\tau = J\sqrt{\lambda\rho_0 / \kappa}$.

Also, b and c are constants that can be determined using the boundary conditions. For the thermal actuator, the pair of inclined beams can be modeled as three beams in a row, where the length of the middle beam is equal to the width of the shuttle. The boundary conditions require the continuity of both temperature and rate of heat conduction across the junction points.

The six boundary conditions are given as follows:

$$T_1(x)\big|_0 = T_S \tag{3.20}$$

$$T_3(x)\big|_{2L+L_c} = T_1(x)\big|_0 \tag{3.21}$$

$$T_1(x)\big|_L = T_2(x)\big|_L \tag{3.22}$$

$$T_2(x)\big|_{L+L_c} = T_3(x)\big|_{2L+L_c} \tag{3.23}$$

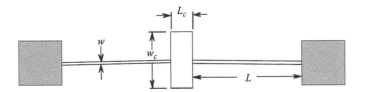

FIGURE 3.28
Geometric design parameters of a pair of V-beams. (From Lott, C.D., Electrothermomechanical modeling of surface-micromachined linear displacement microactuator, MS thesis, Brigham Young University, Provo, UT, 2001.)

$$w \frac{dT_1(x)}{dx}\bigg|_L = w_c \frac{dT_2(x)}{dx}\bigg|_L \tag{3.24}$$

$$w_c \frac{dT_2(x)}{dx}\bigg|_{L+L_c} = w \frac{dT_3(x)}{dx}\bigg|_{L+L_c} \tag{3.25}$$

where the parameters L, L_c, w, and w_c are defined in Figure 3.28. Using these boundary conditions in the above equations the average temperature of the inclined beam can be calculated as follows:

$$\Delta T = \left(T_S + \frac{1}{\lambda}\right) + \frac{b}{\tau L}\left(e^{\tau x} - 1\right) - \frac{c}{\tau L}\left(e^{-\tau x} - 1\right). \tag{3.26}$$

For our model, we apply the average temperature to the inclined beam to calculate the thermo-mechanical response of the device.

3.5.5 Thermo-Mechanical Response of the Device

In the following equations, the subscripts *TA*, *LS*, *SB*, *L*, and *S* refer to thermal actuator, load sensor, sink beams, folded beams, and specimen, respectively (see Figure 3.24). F is the applied load, K is the stiffness, and U is the elongation in different parts of the device. Also, l, b, and h are the dimensions, α is the coefficient of thermal expansion, θ is the beam angle defined from vertical, and E is the Young's modulus of the material used to model the device.

Figure 3.29 shows the schematic of the nanomaterial-testing device. It consists of an electrothermal actuator, a test specimen, and a load sensor in series. The governing equations

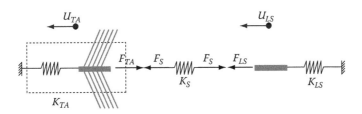

FIGURE 3.29
Free-body diagram of the thermal actuator showing internal forces and displacements. (From Zhu, Y. et al., *J. Micromech. Microeng.*, 16(2), 242, 2006.)

for the lumped model for mechanical analysis of the device, as given in Ref. [34], are given in the following. The force equivalence is given by Equations 3.27 through 3.29:

$$F_{TA} = F_S = F_{LS} \tag{3.27}$$

$$F_S = K_S \, \Delta U_S \tag{3.28}$$

$$F_{LS} = K_{LS} \, \Delta U_{LS} \tag{3.29}$$

The stiffness of the thermal actuator depends on the number of inclined beams (m) and the number of sink beams (n). It is calculated as follows:

$$K_{TA} = mK_{tb} + nK_{sb}, \tag{3.30}$$

where
  K_{tb} is the stiffness of an inclined beam
  K_{sb} is the stiffness of a sink beam

These are calculated using

$$K_{tb} = 2\frac{Ebh}{l}\sin^2\theta + 2\frac{Eb^3h}{l^3}\cos^2\theta \tag{3.31}$$

$$K_{sb} = 2\frac{Eb_{sb}^3h}{l_{sb}^3} \tag{3.32}$$

The stiffness of the specimen and the load sensor are given by K_S and K_{LS}, respectively, which are given by

$$K_S = \frac{E_S A}{l_s} \tag{3.33}$$

$$K_{LS} = \frac{2Eb_L^3h_L}{l_L^3} \tag{3.34}$$

The elongation of the specimen is

$$\Delta U_S = U_{TA} - U_{LS}. \tag{3.35}$$

If we assume α to be constant over the range of temperature change (2.5×10^{-6} K^{-1}), then the displacement of the thermal actuator is given by Equation 3.36, and the elongation of the specimen is given by (3.37):

$$U_{TA} = \frac{2m\sin\theta \times AE\alpha\,\Delta T - F_{TA}}{K_{TA}}, \tag{3.36}$$

$$\Delta U_S = \frac{2mAE\alpha\,\Delta T\sin\theta}{K_{TA} + K_S + (K_{TA}K_S / K_{LS})}. \tag{3.37}$$

3.5.6 Electro-Thermo-Actuator Model

We verified our Sugar model for the MEMS-based material-testing device with the results presented in Refs. [17,30,31]. The displacement of the electro-thermo actuator was compared with the FEA model given in Ref. [17]. An average temperature rise of 55°C was applied to the inclined beam elements. The maximum shuttle displacement was obtained to be 67.5 nm, which corresponded well with the displacement given by the FEA model (see Figure 3.30).

Next, we verified the results of the elongation test with the test data. In Ref. [17], they used a dog-shaped polysilicon test specimen of length = 4.7 μm, thickness = 1.6 μm, and width = 0.42 μm. We approximated the test specimen by modeling a rectangular polysilicon specimen of length = 4.7 μm, height = 1.6 μm, and width = 0.42 μm. The displacement of the actuator specimen junction was found to be 67.5 nm for a temperature increase of 55°C for the inclined beam elements. For a temperature increase of 180°C from ambient, this displacement of the actuator–specimen interface was found to be 270.1 nm.

For optimum performance of the device, specimen stiffness needs to be comparable to that of the load sensor [17]. Hence, for testing the polysilicon specimen, the width of the folded beams was increased to 35 μm. A temperature increase of 350°C was applied to the

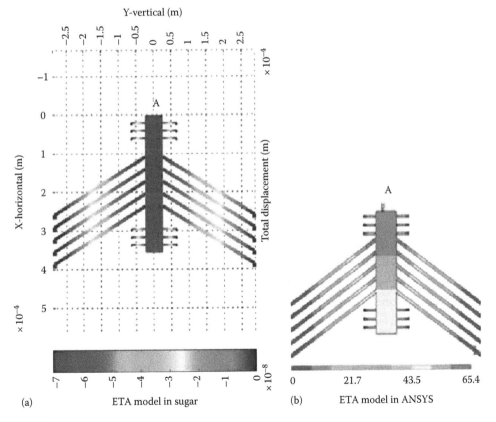

(a) ETA model in sugar   (b) ETA model in ANSYS

FIGURE 3.30
Displacement of a thermal actuator subject to a temperature increase of 55°C in the V-beams. Here, Y-direction displacement of point "A" is shown in nanometers for (a) our lumped model in Sugar and (b) Espinosa's distributed model. (From Espinosa, H.D. et al., *J. Microelectromech. Syst.*, 16(5), 1219, 2007.) *Note*: displacement map colors between Sugar (a) and ANSYS (b) are reversed.

inclined beams. Using the *cho_dq_view* routine in Sugar, the displacement of the actuator–specimen junction was found to be 387.02 nm, and elongation of the specimen was found to be 37.93 nm. From the analytical model given earlier, the specimen elongation was found to be 94.58 nm.

We also simulated our CNT model with the MEMS nanomaterial tester. The CNT test specimen of diameter = 0.78 nm and length = 5 nm was modeled using our CNT model described earlier. A specimen elongation of 172.8 nm was obtained from the Sugar CNT model for an average temperature rise of 150° of the inclined beams. For the same specimen, the elongation was found to be 149.6 nm from the analytical model.

The discrepancy in the results can be attributed to the following three reasons: Firstly, the geometry of the folded beams highly affects the stiffness of the load sensor and in turn the calculated specimen elongation. We have modeled the folded beams as rectangular beam elements, instead of arch-shaped structures. Secondly, we have approximated the polysilicon test specimen with a rectangular specimen, which would also introduce some discrepancy in results. Lastly, currently, a distributed temperature profile is not applied to the lumped elements in Sugar. Hence, the complete length of the V-beam is maintained at a high constant temperature, and the heat sinks and shuttle are maintained at room temperature. To be more accurate, proper effects of temperature gradients need to be addressed.

3.5.7 Summary

In this section, we discussed a method to simplify the designing of an MEMS-based nanoscale material-testing device. We also discussed the integration of CNTs with the mechanical and thermal properties of the microscale device. Our model allows the user to optimize the performance of the device by adjusting its geometry and material properties and variations of CNT test specimen. Our SugarCube-based model can generate the layout of an array of the device in GDS-II format for direct manufacture. This simulation tool will highly benefit both experts and novice users. Our effort significantly reduces design optimization time and overall manufacturing time for the device.

3.6 Distributed Element Parametric Design of MEMS

Comprehensive MEMS design and analysis often require a complicated mix of multiple modeling domains and numerical methods. Modeling domains might include electrical circuits, mechanical flexures, electromagnetic radiation, noise, packaging, temperature, pressure, noninertial forces, various parasitics, and coupling between the domains. An example of such coupling is electrical current passing through a flexure. As the structure heats and expands, its resistivity and resonance frequency will be affected. Although theoretically possible, it is not computationally efficient to represent every aspect of a system using large sets of partial differential equations. Depending on the level of analysis required, some solution methods provide good computational efficiency at the cost of losing high-order detail. It is this level of detail that an analyst typically considers when determining which methods to use when modeling various system components.

There are many stages in a design cycle. These might include modeling, simulation, optimization, layout generation, process design, system integration, fabrication, calibration, and performance testing. Due to the diverse methodologies involved in handling each stage, specialty CAD tools are often used. To name a few, there are tools that specialize in multiphysical distribution [7,19,20,41], in layout [35,36], in circuit analysis include [6,28,45], and in system level analysis [8,37,38,40]. At times, it may be necessary to create different versions of the same device if working between modeling domains. CAD for MEMS tools such as [41] and [5] have addressed this need by being able to plug into Cadence, MATLAB, Simulink, and others.

Without a hierarchical tool to facilitate seamless integration between systems of tools, a holistic approach to analysis can be difficult. This need is being addressed in iSugar with its ability to fully configure and control all aspects of lumped, distributed, and system-level integration within the iSugar tool itself. That is, the efficiency and versatility of our MEMS netlist language can be used to not only configure advanced structural designs, but also to specify Spice circuits, configure the geometry and boundary conditions for components that require distributed or FEA, control Simulink elements, and layout the resulting device for fabrication. Multiobjective optimization features are available in iSugar, including the ability to determine geometry given the desired performance. To facilitate user-modification, iSugar is an open source. Our tool should appeal to users that desire Sugar-style MEMS design with the addition of more sophisticated modeling capabilities.

3.6.1 Framework

Our objective with iSugar is to integrate Sugar's versatile design methodology with distributed analysis, control theory, digital signal processing, etc. Previously, Sugar's modeling capabilities were limited to parameterized lumped models, which meant that models for system components had to already exist. Although many MEMS can be decomposed into a small set of commonly used components, such as small deflection flexures, linear comb drives, and simple plates, more complex components such as those with unusually shaped structures, or those requiring fluid dynamics or electrodynamics, could not be fully accommodated in earlier versions of Sugar. The present version is seamlessly integrated through Sugar; that is, it is not necessary for the user to learn how to use the other tools that iSugar is integrated with to take advantage of their benefits. Although iSugar is readily available and is an open source, the tools that we have integrated it with (MATLAB, Simulink, and COMSOL) are available commercially.

COMSOL is a distributed analysis tool that is based on an FEA. It has a wide range of capabilities to model and simulate multiple energy domains, which is especially important in a field like MEMS. The accuracy of complicated models computed by COMSOL is usually better than those computed by Sugar; however, Sugar is usually more accurate for very simple models if they can accurately be expressed analytically, which can be directly implemented in Sugar. A useful feature in COMSOL that we exploit is COMSOL Script, which is based in MATLAB. That is, every operation in COMSOL can be performed from MATLAB's workspace. This allows users to effectively control all COMSOL capabilities from within iSugar. This also allows parameterized designs that are difficult or too time-consuming to configure within COMSOL to be easily configured in iSugar and then seamlessly imported into COMSOL.

Simulink is a system level simulation tool that is based in MATLAB. It uses graphical building blocks to configure systems. Simulink has a large library of building blocks that span a wide variety of modules including control theory, digital signal processing, COMSOL, Sugar, etc. For instance, Simulink can be used to impart feedback and control

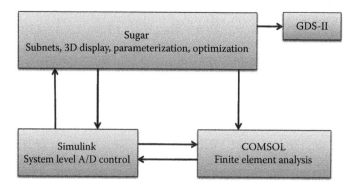

FIGURE 3.31

iSugar framework. Arrows indicate data flow directions. By hiding the complexities involved in FEA and layout packages, this framework simplifies and quickens the MEMS design and engineering path from idea to fabrication. The user just needs to create an MEMS design using Sugar's simple netlist description language. The system may be controlled from within the netlist or through Simulink®. Though seamless integration, COMSOL FEA models are automatically generated through iSugar, and the resulting COMSOL building block is available in Simulink. Finalized designs may be exported to layout in GDS-II format for fabrication.

signals, or environmental disturbances such as noninertial forces, temperature fluctuations, or noise, etc. Like COMSOL, Simulink operations can also be carried out in the MATLAB workspace, which we exploit with iSugar. The seamless integration of iSugar with SIMLINK allows for parametric optimization of the MEMS component as its performance is explored in a more complete system.

It is often the case that optimizing single components alone does not yield an optimized system with the components being assembled. However, iSugar allows the user to explore a more holistic approach to system analysis. We show iSugar's framework in Figure 3.31, which indicates data flow directions between its integrated packages.

3.6.2 Integration of Sugar with COMSOL

Configuring geometries in COMSOL and many other CAD tools is typically done by using geometric Boolean arithmetic. That is, there is a basic set of parameterizable shapes, such as spheres, boxes, etc., and by applying a combination of translations, rotations, unions, intersections, etc., a desired structure is obtained. This common method of construction can be difficult and time consuming for intricate structures like many MEMS. Parameterization is also difficult because shapes are usually positioned on a global rather than local reference frame; and sometimes the number of shapes may need to change. For instance, if the lengths of flexures change, then the elements that they are connected to may need to be repositioned automatically. Or if the number of comb fingers needs to change, shapes may need to be created or deleted automatically.

To overcome this limitation, we developed an algorithm called *cho2comsol* that converts geometry configured in Sugar into geometry that can be imported into COMSOL. We have previously reported on the efficiency and ease of configuring geometries using a parameterized Sugar netlist in Ref. [42], where an advanced MEMS with over 1000 shapes was configured using about 20 lines of netlist code. With our conversion algorithm, users can more efficiently define their intricate geometries in Sugar and import them into COMSOL for FEA. For example, in Figure 3.32, we show a microrobot design that was configured in

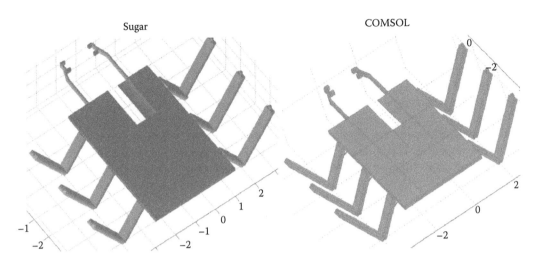

FIGURE 3.32
Sugar to COMSOL with 3D orientations. Creating intricate geometries in many FEA tools can be time consuming. And making them parameterizable can be difficult. However, doing so in Sugar is quick and easy. We show a microrobot from [44] that was easily configured in Sugar and then easily imported in COMSOL. (From Marepalli, P. and Clark, J. V., A system design framework based in MATLAB that integrates sugar, spice, SIMULINK, Fea comsol, and GDS-II layout, *International Conference on Modeling and Simulation of Microsystems*, Boston, MA.)

Sugar and then imported into COMSOL. It is important to note that modifying this design is very easy to do in Sugar, yet very difficult to do in COMSOL.

Using Sugar's set of parameterized geometries, we create corresponding geometries in COMSOL as follows. COMSOL has a scripting language based in MATLAB. Each geometry object in COMSOL may be defined by a geometry function that describes its shape. For example, rect2 for a 2D rectangle, circ2 for a 2D circle, etc. [43]. Parameters to these functions include dimensions, position, orientation, etc. Similarly, there are COMSOL Script commands for defining material properties such as Young's modulus, etc., and boundary conditions such as fixed, free, roller, electric potential, etc. Our Sugar-to-COMSOL algorithm can automatically generate the required COMSOL Script file for each geometric object that is configured in Sugar. However, since Sugar also does layout, some layout-specific geometries are optionally converted to COMSOL. For example, large wire bonding pads and tracers are often not converted over. This is because the dynamics of such objects is usually not required and its presence in COMSOL would be a large computational expense.

3.7 Integration of Sugar with Simulink®

3.7.1 Verification of Lumped Analysis

Although lumped analysis is much more computationally efficient than distributed analysis, this is usually done at the cost of refined information. For example, distributed analysis often provides temperature, charge, and stress distributions on structures, yet lumped analysis is often limited to the effective equivalent information lumped at the nodes. Moreover, lumped models are often created by reducing various types of physics involved in the problem to the bare minimum. Therefore, determining the accuracy and limits of lumped models is often necessary, and even more so, determining the accuracy and limits of a system of lumped models due to possible proximity effects is needed.

Such verification can be done more easily than before using iSugar. This is because we are able to not only import geometric and material properties from Sugar to COMSOL, but we also have automated the application of actuation efforts, meshing, and solver analyses. In this way, after configuring a design in Sugar, users may automatically verify their assembled models and simulations in iSugar. Although this process requires the user to have the COMSOL engine, iSugar's automation implies that the user is not required to have expertise in the use of COMSOL.

We show an example of iSugar automatic's verification in Figure 3.33. After a serpentine-flexure structure created in Sugar (with just nine lines of netlist text) is automatically exported into COMSOL, boundary conditions applied, meshed, and simulated, all with just a single command within the MATLAB workspace or within Simulink.

A goal of MEMS designers is to predict the performance of their systems under realistic operating conditions. Modeling such systems more completely than convention includes interface electronics, packaging, temperature variations, external vibrations, electromagnetic radiation, noninertial forces, etc. A system level simulation tool can be used to efficiently control such disturbances since such sources do not require as detailed modeling as the MEMS structure. In iSugar, we integrate Sugar with Simulink by implementing a Simulink Sugar block. These blocks can be used to perform different Sugar operations like simulating static, modal, and transient performance of MEMS, displaying the MEMS in their deflected states, etc.

In use, the user is able to interconnect one or more Sugar blocks of MEMS, one or more COMSOL blocks, and a host of other Simulink blocks to emulate a more complete system. In Figure 3.34, we show an example of a system level configuration in Simulink that connects control circuitry to an MEMS Sugar block. The output of the Sugar block is defined by the user. For instance, the output might be the mechanical deflection of node, resonance amplitude, capacitance of a comb drive, etc.

3.7.2 Integration of Sugar with Spice

In iSugar, we integrate Sugar with Spice by enabling the user to write pure Spice netlist syntax within a Sugar netlist. A preprocessor separates the Spice circuit part of the netlist from MEMS part. Once these partitions are identified, either one or both of the MEMS structure and or Spice circuitry can be imported and simulated in COMSOL. That is, COMSOL includes a Spice simulation engine.

3.7.3 Summary

In this section, we discussed our system design framework called iSugar that integrates lumped, distributed, and system level analyses. Sugar is the tool used for lumped analysis, COMSOL is used for distributed analysis, and Simulink for system level simulation. A common attribute in these tools is that their scripting is based in MATLAB, which we exploit in iSugar to seamlessly integrate these tools. With iSugar, users are also able

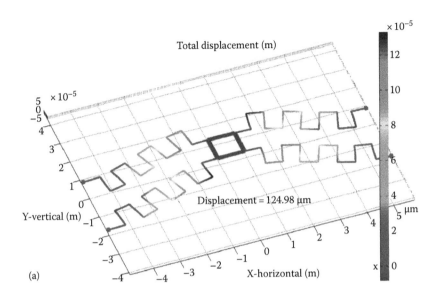

(a)

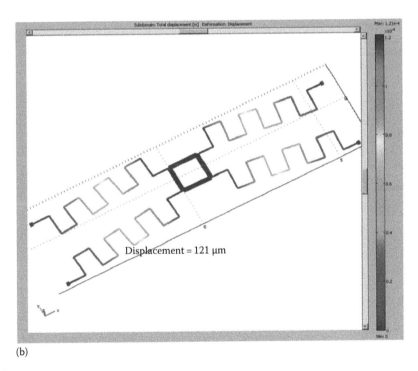

(b)

FIGURE 3.33
Verification of lumped analysis. An example of lumped analysis verification with distributed analysis is shown. A lumped model of a serpentine flexure is shown in (a) using just nine lines of netlist text. This lumped model can be automatically converted into COMSOL for distributed analysis, as shown in (b). This conversion process includes positioning and rotating structural elements, applying constraint and effort boundary conditions, meshing, selecting solver and solver parameters, and plotting generation the results. This conversion is done with a single command in MATLAB®; that is, no interaction with COSMOL is required by the user. In regard to the validation of static displacement for this model, the relative error of Sugar with respect to COMSOL is 3.3%.

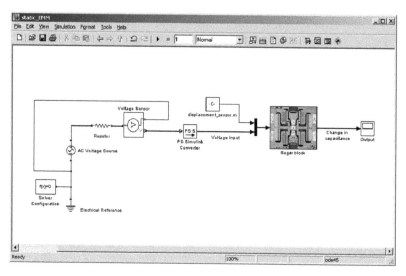

FIGURE 3.34
Integrating Sugar to Simulink® components. With the integration of Sugar to COMSOL and Simulink, iSugar shares the Sugar's ease of use, COMSOL's depth in simulating multienergy domain problems, and Simulink breadth in solving system level problems. In this example, we show how a Sugar block can be integrated to a system-level circuitry inside Simulink.

to integrate Spice analysis and layout in GDS-II format. The automation and control of these tools through iSugar are expected to enable greater efficiency and versatility in modeling MEMS.

3.8 Conclusion

We discussed an integration of distributed element, lumped element, and system level methods for the design, modeling, and simulation of N/MEMS. The benefits of lumped elements are computational efficiency and ease of parameterization; however, such lumped models must preexist. For models that do not preexist, or if the analysis of single component is necessary, then distributed element methods are often beneficial. However, distributed elements are computationally expensive, especially for devices with a multitude of multiphysical components. Often reduced-order methods are used to minimize the degrees of freedom for a more efficient model at the expense of trading off some accuracy. The integration of both distributed and lumped element methods can be useful for modeling micro- and nano-systems. Although not every N/MEMS can be represented using the methods discussed here, a large number of devices can be efficiently explored using these methods. The examples we provided included parameterized lumped CNTs; novice-friendly design, simulation, and layout of MEMS; an automated lumped to distributed verification method; and the design exploration of a nanomechanical material property tester.

References

1. R. Bansal and J. V. Clark, Simplifying the design, analysis, and layout of a N/MEMS material testing device, *Sensors and Transducers Journal*, 13(Special Issue): 87–97, December 2011.
2. P. Marapalli and J. V. Clark, An online tool for investigating the performance of ready-Made parameterized MEMS, A network-level layout tool for MEMS, *Nanotech 2011, International Nanotechnology Conference and Exhibition*, Boston, MA, June 13–16, 2011.
3. R. Bansal and J. V. Clark, Lumped model of a carbon nanotube, *Nanotech 2011, International Nanotechnology Conference and Exhibition*, Boston, MA, June 13–16, 2011.
4. P. Marepalli and J. V. Clark, Integration of sugar, comsol, spice, and simulink, *Nanotech 2011, International Nanotechnology Conference and Exhibition*, Boston, MA, June 13–16, 2011.
5. J. V. Clark and K. S. J. Pister, Modeling, simulation, and verification of an advanced micromirror using SUGAR, *Journal of Microelectromechanical Systems*, 16(6): 1524–1536, 2007.
6. L. W. Nagel, SPICE2: A computer program to simulate semiconductor circuits, ERL Memo. No. UCB/ERL, Vol. M75/520, 1975.
7. COMSOL, 1 New England Executive Park, Suite 350, Burlington, MA USA 01803, Inc., http://www.comsol.com
8. SIMULINK, 970 West 190th Street, Suite 530, Torrance, CA USA 90502, http://www.mathworks.com/products/simulink
9. MATLAB, 3 Apple Hill Drive, Natick, MA USA 01760-2098, http://www.mathworks.com
10. C. Li and T.-W. Chou, Vibrational behaviors of multiwalled-carbon-nanotube-based nanomechanical resonators, *Applied Physics Letters*, 84(1): 121–123, 2003.
11. C. Li and T.-W. Chou, A structural mechanics approach for the analysis of carbon nanotubes, *International Journal of Solids and Structures*, 40: 2487-2499, 2003.
12. H. Wan and F. Delale, A structural mechanics approach for predicting the mechanical properties of carbon nanotubes, *Meccanica*, 45: 43–51, 2010.
13. P.-K. Gudla, A. Kannan, Z. Tang, and N. Aluru, Carbon nanotube based NEMS with cantilever structure, doi: 10254/nanohub-r3933.2, 2009. http://nanohub.org/resources/3933
14. S. Melchor and J. A. Dobado, CoNTub: An algorithm for connecting two arbitrary carbon nanotubes, *Journal of Chemical Information and Computer Sciences*, 44: 1639–1646, 2004.
15. J. V. Clark, D. Bindel, N. Zhou, S. Bhave, Z. Bai, J. Demmel, and K. S. J. Pister, Sugar: Advancements in a 3D multi-domain simulation package for MEMS, *Proceedings of Microscale Systems: Mechanics and Measurements Symposium*, Portland, OR, 2001.
16. J.-S. Przemieniecki, *Theory of Matrix Structural Analysis*, Mc-Graw-Hill Book Company, New York, 1986.
17. H. D. Espinosa, Z. Yong, and N. Moldovan, Design and operation of a MEMS-based material testing system for nanomechanical characterization, *Journal of Microelectromechanical Systems*, 16(5): 1219–1231, 2007. Available at: http://clifton.mech.northwestern.edu/~espinosa/Papers/Espinosa_InsituDevice_JMEMS_2007.pdf
18. A. M. Fennimore, T. D. Yuzvinsky, W.-Q. Han, M.-S. Fuhrer, J. Cumings, and A. Zettl, Rotational actuators based on carbon nanotubes, *Nature*, 424: 408–410, 2003.
19. ANSYS, Inc., ANSYS, Inc. Southpointe, 275 Technology Drive, Canonsburg, PA USA 15317, http://www.ansys.com
20. Coventorware, 4000 Centregreen Way, suite 190, Cary, NC, USA 27513, http://www.coventor.com
21. IntelliSuite, IntelliSense Corp., 600 W. Cummings Park, Suite 2000, Woburn MA USA 01801, http://www.intellisense.com
22. Q. Jing and G. K. Fedder, NODAS 1.3: Nodal design of actuators and sensors, *Proceedings of IEEE/VIUF International Workshop on Behavioral Modeling and Simulation*, Orlando, FL, 1998.
23. S. D. Senturia, CAD challenges for microsensors, microactuators and microsystems, *Proceedings of the IEEE*, 86(8): 1611–1626, August 1998.

24. S. Senturia, N. Aluru, and J. White, Simulating the behavior of MEMS devices: Computational methods and needs, *IEEE Computational Science and Engineering*, 16(10): 30–43, January–March 1997.
25. MEMSolver, Kalpakam, SNNRA 15, TC 31/584-1, Pettah, Trivandrum, Kerala, India 695024, http://www.memsolver.com 32. CleWin - WieWeb Software, Delta mask B. V. Nijmansbos 56, 7543 GJ Enschede, The Netherlands, http://www.wieweb.com/nojava/layoutframe.html
26. Network for Computational Nanotechnology, http://nanoHUB.org
27. P. Marepalli, Advances in CAD for MEMS, MS thesis, Mechanical Engineering, Purdue University, West Lafayette, IN, May 2011.
28. L. W. Nagel and D. O. Pederson, SPICE (Simulation Program with Integrated Circuit Emphasis), Memorandum No. ERL-M382, University of California, Berkeley, CA, April 1973.
29. Z. Bai, D. Bindel, J. V. Clark, J. Demmel, K. S. J. Pister, and N. Zhou, New numerical techniques and tools in sugar for 3D MEMS simulation, *Technical Proceedings of the Fourth International Conference on Modeling and Simulation of Microsystems*, Hilton Head Island, SC, pp. 31–34, March 19–21, 2001.
30. Y. Zhu and H. D. Espinosa, An electromechanical material testing system for in situ electron microscopy and applications, *Proceedings of the National Academy of Sciences of the United States of America*, 102(41): 14503–14508, 2005. Available at: http://clifton.mech.northwestern.edu/~espinosa/Papers/PNAS-nMTS.pdf
31. Y. Zhu, A. Corigliano, and H. D. Espinosa, A thermal actuator for nanoscale in situ microscopy testing: Design and characterization, *Journal of Micromechanics and Microengineering*, 16(2): 242–253, 2006.
32. CleWin Layout Editor by WieWeb Software, Achterhoekse molenweg 76, 7556 GN Hengelo, The Netherlands, http://www.wieweb.com/nojava/layoutframe.html
33. Q. A. Huang and N. K. S. Lee, Analysis and design of polysilicon thermal flexure actuator, *Journal of Micromechanics and Microengineering*, 9: 64–70, 1999.
34. C. D. Lott, Electrothermomechanical modeling of surface-micromachined linear displacement microactuator, MS thesis, Brigham Young University, Provo, UT, 2001.
35. L-Edit-Tanner Research, 825 South Myrtle Avenue, Monrovia, CA USA 91016, http://www.tanner.com
36. Cadence Design Systems, Inc. Bagshot Road, Bracknell, Berkshire, RG12, OPH United Kingdom. www.cadence.com/eu/
37. Mathworks, 3 Apple Hill Drive, Natick, MA USA 01760-2098, http://www.mathworks.com
38. Ptolemy Project—Heterogeneous Modeling and Design. http://ptolemy.eecs.berkeley.edu
39. Coventorware's Architect and MEMS+ for Systems Design. 4000 Centregreen Way, suite 190, Cary, NC, USA 27513, http://www.coventor.com, http://www.coventor.com/memssystem/mems-system-design.html
40. MapleSim, 615 Kumpf Drive, Waterloo, ON, Canada, N2V 1K8, http://www.maplesoft.com/products/maplesim/index.aspx
41. Synple synple and EDA Linger, Intellisense Corp, 600W. Cummings Park, Suite 2000, Woburn MA 01801 USA, http://intelliSence.com
42. COMSOL Multiphysics Matlab Interface Guide, Version 3.5, COMSOL Inc.
43. J. V. Clark, D. Bindel, W. Kao, E. Zhu, A Kuo, N. Zhou, J. Nie, J. Demmel, Z. Bai, S. Govindjee, K. S. J. Pister, M. Gu, and A. Agogino, Addressing the needs of complex MEMS design, *Proceedings IEEE The Fifteenth Annual International Conference on Micro Electro Mechanical Systems*, Las Vegas, NV, January 20–24, 2002.
44. P. Marepalli and J. V. Clark, "Integration of Sugar, Comsol, Spice, and Simulink", Nanotech 2011, *International Nanotechnology Conference and Exhibition*, Boston MA, June 13–16, 2011.
45. Cadence Virtuoso Spectre Circuit Simulator, Cadence Design Systems, Inc. Bagshot Road, Bracknell, Berkshire, RG12, OPH United Kingdom. www.cadence.com/eu/

4

Nanorobotic Applications of the Finite Element Method

S. Sadeghzadeh
Iran University of Science and Technology, Tehran, Iran

M.H. Korayem
Iran University of Science and Technology, Tehran, Iran

V. Rahneshin
Iran University of Science and Technology, Tehran, Iran

A. Homayooni
Iran University of Science and Technology, Tehran, Iran

M. Moradi
Semnan University, Semnan, Iran

CONTENTS

4.1   Introduction 86
    4.1.1   Different Types of Nanorobots 87
    4.1.2   Applications of Various Nanorobots 89
4.2   Macroscale Nanorobots 90
    4.2.1   Schematic of System Configuration 90
    4.2.2   Different Methods of Nanomanipulation 91
    4.2.3   Dominant Forces in Nanoscale 92
4.3   Continuum Mechanics Role in the Nanorobotic Science 94
    4.3.1   Direct Solution of Nanorobotic Problem with FEM 94
4.4   Proposed Nanomanipulation Strategy 99
4.5   Multiscale Method 100
    4.5.1   Macro-Field Modeling, Finite Element Method 100
    4.5.2   Nano-Field Modeling, Molecular Dynamics Approach 103
        4.5.2.1   Need for Molecular Dynamics Simulations 103
        4.5.2.2   Molecular Dynamics 103
        4.5.2.3   Sutton–Chen Interatomic Potential 104
        4.5.2.4   Rafii-Tabar–Sutton Potential 104
        4.5.2.5   Equations of Motion 106
        4.5.2.6   Solution Procedure 106
        4.5.2.7   NVT Ensemble 107
        4.5.2.8   Coarse-Grained Molecular Dynamics 107
        4.5.2.9   Molecular Dynamics Simulation of Nanomanipulation Procedure 109
        4.5.2.10 Results Validation: A Comparison of CGMD and Macro-Model 109
        4.5.2.11 Some Challenges in the Way of Molecular Dynamics Simulations 111

4.5.2.12 Sutton–Chen Parametric Study: The Effects of Different
          Parameters on the Nanoparticle Deformation.................................. 114
4.5.2.13 Comprehensive Diagram for the Optimal Selection of Tip............. 115
4.5.3   Macro–Micro-Coupling Model... 116
4.5.4   Coupling of MF and NF.. 119
4.5.5   Algorithm for Establishment of Coupling, Problems of Statics..................... 122
        4.5.5.1   Problems of Damped Dynamics of MF 123
        4.5.5.2   Problem of Damped Dynamics of NF................................... 125
4.6   Application to the Macroscale Nanorobots.. 126
        4.6.1   Example 1: Roughness on Substrate.. 128
        4.6.2   Example 2: Material of Substrate 130
        4.6.3   Example 3: Notches on Substrate.. 130
4.7   Some Challenges in the Macroscale Nanorobotic Science 130
        4.7.1   Scanner Limitations... 131
        4.7.2   Hysteresis Behavior in the Micro-Actuators.............................. 132
        4.7.3   Modeling of Nonlinearities 133
                4.7.3.1   Results and Discussion 134
        4.7.4   Hysteresis Behavior in the Nanogenerators............................... 138
                4.7.4.1   Results and Discussion 140
4.8   Conclusion 144
References.. 145

4.1 Introduction

Nanorobotics is a field which calls for collaborative efforts between physicists, chemists, biologists, computer scientists, engineers, and other specialists to work toward this common objective. The ability to manipulate matter at the nanoscale is one core application for which nanorobots could be the technological solution. There are lots of work in the literature about the significance and motivation behind constructing a nanorobot. The applications range from medical or environmental sensing to space and military applications. Molecular construction of complex devices could be possible by nanorobots of the future.

This chapter focuses on the state of the art in the field of nanorobotics by considering various theories and experiments. Nanorobots are controllable machines at the nanometer or molecular scale that are composed of nanoscale components. With the modern scientific capabilities, it has become possible to attempt the creation of nanorobotic devices and interface them with the macro-world for control. There are countless such machines that exist in nature, and there is an opportunity to build more of them by mimicking nature. Even if the field of nanorobotics is fundamentally different from that of macrorobots due to the differences in scale and material, there are many similarities in design and control techniques that eventually could be projected and applied. Figure 4.1 introduces a general view on the nanorobotic science.

Reaching, identifying, and manipulating the nanoworld need some tools in macro-world. Teleoperated systems, especially the scanning probe microscopes (SPMs), are the usual systems that have been utilized for this purpose. Therefore, it is obvious that in a teleoperation system, both macro- and nanoscale processes can be extensively effective.

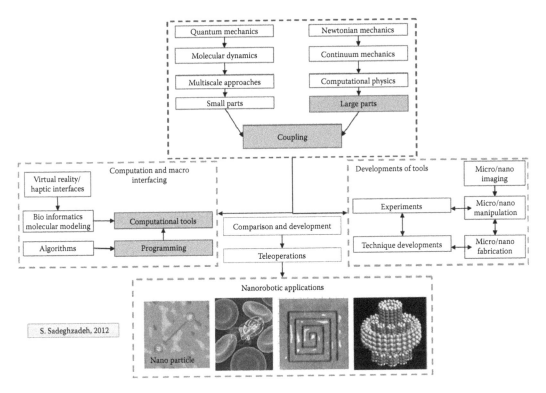

FIGURE 4.1
Flowchart of nanorobotic research.

This chapter reviews some descriptions and challenges in nanorobotic science. Then, a real view on the simulation of nanorobotic systems with macrodimensions has been discussed that is applicable onto the small-scale nanorobots too. After the introduction of different types of nanorobots, the applications of various nanorobots have been discussed. Focusing on the macroscale nanorobots, where the finite element method (FEM) can be applied, different methods of manipulation on the nanoscales using these nanorobots have been developed. Here, manipulation on the nanoscales using macroscale nanorobots is called nanomanipulation. With classification of effective parameters in the nanomanipulation, a relatively comprehensive model for these devices has been introduced. At the end, as an example of nano-sized nanorobots, the linear and nonlinear behavior of electrical nanogenerators have been studied.

4.1.1 Different Types of Nanorobots

Here, nanorobotic systems have been divided into two categories: nanoscale nanorobots and macroscale nanorobots.

Nanoscale nanorobots have been developed recently. Nonetheless, they are largely in the research-and-development phase [1]. It is very important that in the real world, some primitive molecular machines have been tested successfully. The nanoscale nanorobots can be divided into man-made and naturally occurring molecular machines. In some literatures, the molecular machines are divided into three broad categories: protein-based, DNA-based, and chemical molecular motors [2]. The ATP synthase, the kinesin, myosin, and flagellar molecular motors are some examples of main protein-based molecular machines [2]. DNA-based nanodevices that can move or change conformation have been developed rapidly since the first example was reported in Ref. [3]. So far, a number of different prototypes have been developed. These include devices that are driven by DNA hybridization and branch migration, walk and rotate, use Hoogsteen bonding to form multiplex (rather than double helix) DNA structures, and respond to environmental conditions. At least some of these concepts will need to be combined, and/or new ones developed before such devices would be of practical use.

These machines and many of other types have complicated structures that will allow them to use for various applications in the future. For example, some types of molecular motors that move unidirectionally along protein polymers (actin or microtubules) can drive the motions of muscles, as well as much smaller intracellular cargoes. Some others can rotate and actuate or sense a device in tangential direction. An example is a sensor having a switch approximately 1.5 nm across, capable of counting specific molecules in a chemical sample.

Since their discovery in 1991, carbon nanotubes have been investigated by many researchers all over the world. Their large length (up to several microns) and small diameter (a few nanometers) result in a large aspect ratio. They can be seen as the nearly one-dimensional form of fullerenes. Therefore, these materials are expected to possess additional interesting electronic, mechanic, and molecular properties. Especially in the beginning, all theoretical studies on carbon nanotubes focused on the influence of the nearly one-dimensional structure on molecular and electronic properties. Based on their special configuration, carbon nanotubes can be collated with other longitudinal structures such as DNA, nanorods, nanowires (NWs), etc. Nanoswitching, nanogripping, nanoactuating, and sensing can be performed utilizing the carbon nanotubes.

Despite the fact that these capabilities can be used in various industries, the first and most useful applications of nanoscale nanorobots might be in medical technology. For example, they could be used to identify and destroy cancer cells [4].

On the other hand, microscopes that have the capability of various operations at the nano- and sub-nanometric dimensions are called macroscale nanorobots. From this set, SPMs and especially atomic force microscope (AFM) are more famous. Macroscale nanorobots are robots that allow precision interactions with nanoscale objects or can manipulate with nanoscale resolution. Following the microscopy definition, even a large apparatus such as an AFM can be considered a nanorobotic instrument when configured to perform nanomanipulation. For this perspective, macroscale robots or microrobots that can move with nanoscale precision can also be considered nanorobots. In the early 1980s, SPMs dazzled the world with the first real-space images of the surface of silicon. Now, SPMs are used in a wide variety of disciplines, including fundamental surface science, routine surface roughness analysis, and spectacular three-dimensional imaging from atoms of silicon to micron-sized protrusions on the surface of a living cell. The SPM is an imaging tool with a vast dynamic range, spanning the realms of optical and electron microscopes. The AFM probes the surface of a sample with a sharp tip, a couple of microns long, and often less than 100 Å in diameter. The tip is located at the free end of a cantilever that is 100–200 μm long. Forces between the tip and the sample surface cause the cantilever to bend or deflect. A detector measures the cantilever deflection as the tip is scanned over the sample, or the sample is scanned under the tip. The measured cantilever deflections allow a computer to generate a map of surface topography. AFMs can be used to study insulators and semiconductors as well as electrical conductors [5].

4.1.2 Applications of Various Nanorobots

There are many applications for nanorobotic systems. First, let us start from the nanoscale nanorobots. As already mentioned, the first and most useful applications of nanomachines might be in medical technology. So far, however, there has been little discussion about possible applications of nanoscale nanorobots in medical science. Some examples can guide the researchers toward the new ideas. Retina implants are in development to restore vision by electrically stimulating functional neurons in the retina. An especial form of retina implants enables blind peoples to see shapes and objects [6]. In that work, some blind patients have had their sight partly restored after scientists in Germany developed an electronic eye implant. This can revolutionize the lives of 200,000 people worldwide who have retinitis pigmentosa (RP), a degenerative eye disease. So, this could clearly show the nanorobotics significance. Many examples have been mentioned in medical science. On the other hand, medical devices that contain nano- and microtechnologies will allow surgeons to perform familiar tasks with greater precision and safety, monitor physiological and biomechanical parameters more accurately, and perform new tasks that are not currently done. In addition, nano- and microtechnologies provide new solutions for increasing the speed and accuracy of identifying genes and genetic materials for drug discovery and development and for treatment-linked disease diagnostics products. Several new technologies are being developed to improve the ability to label and detect unknown target genes. The most significant work of nanorobots in medical field is killing the cancer cells. The device would circulate freely throughout the body and would periodically sample its environment by determining whether the binding sites were or were not occupied. Occupancy statistics would allow determination of concentration. Today's monoclonal antibodies are able to bind to only a single type of protein or other antigen and have not proven effective against most cancers. The cancer killing device suggested here could incorporate a dozen different binding sites and so could monitor the concentrations of a dozen different types of molecules. The computer could determine if the profile of concentrations fits a preprogrammed "cancerous" profile and would, when a cancerous profile was encountered, release the poison. Beyond being able to determine the concentrations of different compounds, the cancer killer could also determine local pressure. By using several macroscopic acoustic signal sources, the cancer killer could determine its location within the body much as a radio receiver on earth can use the transmissions from several satellites to determine its position (as in the widely used GPS system). The cancer killer could thus determine that it was located in (say) the big toe. If the objective was to kill a colon cancer, the cancer killer in the big toe would not release its poison. Very precise control over location of the cancer killer's activities could thus be achieved. The cancer killer could readily be reprogrammed to attack different targets (and could, in fact, be reprogrammed via acoustic signals transmitted while it was in the body). This general architecture could provide a flexible method of destroying unwanted structures (bacterial infestations, etc.).

Now, we want to determine the role of nanorobots in our knowledge about the nature of processes. The fundamental definition of nanoscale nanorobots may be as devices that interact the objects at the nanoscale of length. This means that they include the molecular machines. So, entrance to the interaction world of molecules may be possible with nanoscale nanorobots. They can be utilized to actuate two or more molecules to close together or repulse them. The more important capability is the sensor state. A nanoscale nanorobot may be used to identify and even measure accurately the interactions including forces such as the ionic and van der Waals forces. So, the interaction between two given molecules can be well understood by a set of laws governing them, which brings in a

TABLE 4.1

Various SPMs and Their Applications

Scanning tunneling microscope (STM)	1981	Inducing chemical reactions
Atomic force microscope (AFM)	1985	Manipulation, nanofabrication, etc.
Friction force microscopes (FFMs)	1989	Measures the adhesion and friction of surfaces
Lateral force microscopes (LFMs)	1989	Measure both normal and lateral forces
Scanning electrostatic force microscopy (SEFM)	1988	The electrostatic force is probed
Scanning force acoustic microscopy (SFAM)	1974, 1988	Flaw detection, internal stress investigation, elastic property characterization
Magnetic force microscopy (MFM)	1987	Applicable for electrically nonconductive samples
Scanning near-field optical microscopy (SNOM)	1928, 1989	Surface inspection with high spatial, spectral, and temporal resolving power
Scanning thermal microscopy (SThM)	1986	Thermal measurements at the nanoscale
Scanning electrochemical microscopy (SECM)	1990	Electrochemical reactions in solid–liquid interfaces
Scanning Kelvin probe microscopy (SKPM)	1991	Calculation of the work function of surfaces
Scanning chemical potential microscopy (SCPM)	1990	Measurement of thermoelectric potential variations
Scanning ion conductance microscopy (SICM)	1989	Soft nonconductive materials that are bathed in electrolyte solution
Scanning capacitance microscopy (SCM)	1990	Mapping and quantification of the dopant profile

definite level of predictability and controllability of the underlying mechanics. This is an example of possible role of nanorobots in our knowledge about the world and its realities.

Overall, the applications of SPMs can be divided into identification and operation tasks. Usually, these tasks are called imaging and manipulation, respectively. Table 4.1 lists a relatively comprehensive categorization of various SPMs and their applications. Table 4.1 is summarized from the text of Ref. [7].

As an advanced application of macroscale nanorobots, some new configurations have been developed for surgical purposes. Surgical robotics with nanoscale output is the result of these configurations. Robotic surgical systems are being developed to provide surgeons with unprecedented control over precision instruments. This is particularly useful for minimally invasive surgery. Instead of manipulating surgical instruments, surgeons use their thumbs and fingers to move joystick handles on a control console to maneuver two robot arms containing miniature instruments that are inserted into ports in the patient. The surgeon's movements transform large motions on the remote controls into micro-movements on the robot arms to greatly improve mechanical precision and safety.

4.2 Macroscale Nanorobots

4.2.1 Schematic of System Configuration

A common configuration for an AFM with nanomanipulation ability has been depicted in Figure 4.2. Upper piezotube is Z scanner, and underneath is XY one. The nanomanipulation

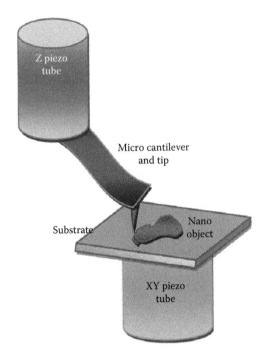

FIGURE 4.2
A common configuration for an AFM with nanomanipulation ability.

scheme has been done with fixing the Z scanner and moving XY scanner counter to the desired direction.

4.2.2 Different Methods of Nanomanipulation

Physical configuration of a nano-object manipulation using the AFM is shown in Figure 4.3. The nano-object is stationary at the beginning, but pushing force overcomes resistance forces at critical time. The particle moves to reach desired position in defined

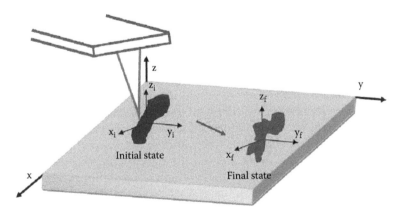

FIGURE 4.3
AFM tip moves nano-object on substrate.

TABLE 4.2

General Effective Parts and Parameters in the Manipulation Process

General Factors	1	2	3	4
Task	Push/pull	Pick and place	Cutting	Bending/buckling
Ambient	Air (simple, multi-object, online probability, …)	Vacuum (order of vacuum, sensitivity, …)	Liquid (simple liquids, blood, advanced chemical, …)	Bio and cell (soft, hard, genetic, …)
AFM specification	Operation modes (contact, noncontact, tapping)	Ambient conditions (wet, dry, low pH, high pH)	Cantilever shape (gripper, V-shape, rectangular)	Tip shape (triangular, rectangular, T-shape, …)
Object	Shape (sphere, rod, wire, cell, tube, …)	Dimensions (2D, 3D, small, big, …)	Mechanical behavior (rigid, elastic)	Physical field (metallic, biologic, magnetic, …)
Substrate	Roughness (low, high)	Mechanical behavior (rigid, elastic)	Physical field (metallic, biologic, magnetic, …)	Special conditions (wet, dry, low pH, high pH)
Dynamics of process	Dominant forces	2D/3D	Straight/curve path	Constant velocity/acceleration

path and trajectory. During the manipulation from the first to the final position, the nano-object may effectively undergo some physical deformations or even chemical reformations in the case of biological samples. So, the state, mentioned in the figure, includes all natures of the nano-object. Due to lack of real-time observation in the AFM, accurate modeling of manipulation and force estimation are very important issues. As mentioned in Section 4.1, researchers try to improve the model of process, and different new parameters are added to previous models. In the AFM, there are five effective parts in manipulation, and each of them has different parameters (Table 4.2). Several processes can be designed composing these parts and their parameters.

From Table 4.2, dynamics of the process has been explained here.

4.2.3 Dominant Forces in Nanoscale

There exist various nanoforces in the AFM-based nanomanipulation with a micro-probe. However, what are main forces and how they work remain not very clear [8]. Based on the recent researches [9] and considering effective factors such as humidity and electrostatic charge, the crucial nanoforces can be summarized as van der Waals' repulsive contact force and friction (three basic nanoforces). The capillary force could be considered that aroused by humidity or biological substrates, where the electrostatic force caused by the electrostatic charge. Based on their effect in nanomanipulation, these forces can be categorized into attractive, repulsive, and frictional forces [8]. The nanoforces among tip and particle can be described as shown in Figure 4.4, where superscripts o, t, and s correspond to probe, tip and substrate and subscripts f, rep, and atr correspond to the friction, repulsive, and attractive forces, respectively and are the nanofrictional forces, F_{atr}^{os} and F_{atr}^{ot} are attractive forces that consist of attractive van der Waals force, capillary force, and attractive electrostatic force, and F_{rep}^{os} and F_{rep}^{ot} are repulsive forces composed of repulsive contact force, repulsive van der Waals force, and repulsive electrostatic force, respectively [10].

Like gravitational force, van der Waals forces exist for every material in any ambient condition. These forces originate from electromagnetic forces between two dipoles and

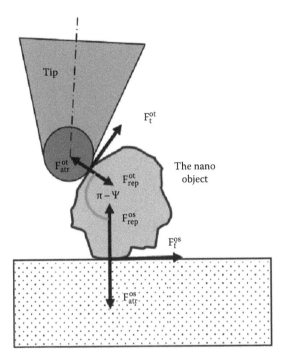

FIGURE 4.4
The nanoforces among tip, particle, and substrate.

depend on material types, separation distances, and object geometry [10]. As an example, for spherical tip-flat surface, the van der Waals force is as

$$f(h)_{wdv} = -\frac{A_H R_t}{6h^2} \tag{4.1}$$

where A_H, R_t, and h are Hamaker constant (about 10^{-19}), tip radius, and distance between the tip and substrate, respectively. The minus sign indicates attractive force and the plus sign indicates repulsive one [11]. When tip approaches to substrate, it reaches to a region of mechanical instability that the force gradient of the potential exceeds the spring constant of the cantilever [11]

$$\frac{df(h)_{wdv}}{dh} = K_Z \tag{4.2}$$

At this instability, the probe will jump into contact with the surface with a characteristic "snap-in" distance, d_s, as

$$d_s = \left(\frac{A_H R_t}{3K_Z}\right)^{1/3} \tag{4.3}$$

Snap in substrate phenomena using photodiode data can be detected [11]. The water layers on the surfaces of probe, object, or substrate result in the adhesion force. A liquid bridge occurs between tip surfaces at close contact [12]. The adhesion force between a nondeformable spherical particle of radius R and a flat surface in an atmosphere containing a condensable vapor is

$$F_S = 4\pi R_t \left(\gamma_{LV} \cos \varepsilon + \gamma_{SL} \right) \tag{4.4}$$

where ε is contact angle, the first term is due to the Laplace pressure of the meniscus (γ_{LV}: liquid–vapor surface energy), and the second one is due to the direct adhesion of the two contacting solids within the liquid (γ_{SL}: solid–liquid surface energy) [13].

As there will be some electrical charge accumulated on the surface of particles or the tip, the particle is prone to adhere on the tip, and manipulation may be failed. Since the particles are not picked up, the electrostatic force between the particle and the substrate is not important. However, after pushing, the charge on the particle is transferred to the tip which can cause an electrostatic force. Electrostatic force between the tip and substrate will be

$$F_e = \kappa R_t Z e^{\kappa h} \tag{4.5}$$

where
  κ is the Debye length
  Z is the characteristic parameter of the tip particle
  h is the distance [13]

In actual experiment condition, probe and substrate can be grounded to release the electrostatic charge for minimizing electrostatic force, and also the experiment condition can be kept dry to minimize the capillary force. Thus the crucial forces between the tip and the substrate are mainly van der Waals, friction, and repulsive contact force [13].

Using these forces and applying common continuum mechanics (CM)-based approaches, the equations of motion for present system may be derived. Then the FEM can be utilized to solve it. Many works may be cited that have been using this scheme. However, it needs more attention before applying. Let us consider it after the description of CM role in the nanorobotic science.

4.3 Continuum Mechanics Role in the Nanorobotic Science

Let us start from the simple shape of FEM formulation to describe the role of CM in the nanorobotic science. As already mentioned, nanoscale nanorobots have dimensions in the nano- or sub-nanoscales, where macroscale nanorobots have some components in the super- and some components in sub-nanometer scales. Then, can we apply the CM-based approaches for modeling all of nanorobots?

4.3.1 Direct Solution of Nanorobotic Problem with FEM

The conventional form of the equation of motion in FEM is

$$m\ddot{x} + c\dot{x} + kx = f \tag{4.6}$$

Since m, c, k, and f can all be influenced, they are adjustable parameters. Every term in this equation has the dimension of force (ML/T^2). From this, it follows that the dimensions of c and k are as follows:

$$[c] = \frac{M}{T}, \quad [k] = \frac{M}{T^2} \tag{4.7}$$

For the harmonic applied forces, such as $f = F_0 \sin(\omega t)$, the dimensionless nature of the angle (ωt) results in $[\omega] = 1/T$. The techniques of nondimensionalizing and scaling can be extremely powerful tools in analyzing the models. The basic idea is to apply a transformation to the variables and parameters such that simplified equations result. In practice, two methods are applied, dimensional analysis and scaling, each having its own merits. They are dealt with in the following sections, respectively. Dimensional analysis fully exploits the information contained in the physical dimensions of the variables and parameters. Scaling has a more restricted scope and aims at a reduction of the number of parameters. So, we use the dimensional analysis.

Consider a system with scalar variables $x_1, \dots, x_k$ and scalar parameters $p_1, \dots, p_m$. So, the total number of quantities involved is $N = k + m$. We now form the products

$$x_1^{r_1}, \dots, x_k^{r_k}, \quad p_1^{r_{k+1}}, \dots, p_m^{r_{k+m}} \tag{4.8}$$

and ask for which choices of the r_i these products are dimensionless. The answer follows from replacing each x_i and p_i with its fundamental dimensions. This procedure is famous as Buckingham method [14]. Since Buckingham denoted the dimensionless quantities by π_i, this theorem is often referred to as the π-theorem of Buckingham. So, we use $x^{r_1} t^{r_2} m^{r_3} c^{r_4} k^{r_5} F_0^{r_6} \omega$ for our problem. Substituting the dimensions, we arrive at the products

$$L^{r_1} T^{r_2} M^{r_3} \left(\frac{M}{T}\right)^{r_4} \left(\frac{M}{T^2}\right)^{r_5} \left(\frac{ML}{T^2}\right)^{r_6} \left(\frac{1}{T}\right)^{r_7} \tag{4.9}$$

Collecting powers of M, L, and T, we obtain the following three linear equations for the r_i:

$$r_1 + r_6 = 0$$
$$r_2 - r_4 - 2r_5 - 2r_6 - r_7 = 0 \tag{4.10}$$
$$r_3 + r_4 + r_5 + r_6 = 0$$

Here, we meet with three equations for seven unknowns, so four unknowns can be treated as free parameters. For example, the choices $(r_1, r_2, r_3, r_4) = (1,0,0,0),(0,1,0,0),(0,0,1,0),(0,0,0,1)$ yield the following dimensionless quantities:

$$x^* = xkF_0^{-1}, \quad t^* = \omega t, \quad m^* = m\omega^2 k^{-1}, \quad c^* = c\omega k^{-1} \tag{4.11}$$

Then, the dimensionless equation can be obtained as

$$m^* \ddot{x}^* + c^* \dot{x}^* + x^* = \sin t^* \tag{4.12}$$

More attention to the first relation in Equation 4.11 shows that the scaling coefficient for the lengths is kF_0^{-1}. Usually, the force range at the nanoscale is about the nanonewton, and the norm of stiffness matrices is of order 10, and it increases with the scale growth. Thus, the value of kF_0^{-1} remains about 10^{10} from nano- to macroscales. This guarantees the possibility of conversion of a nanoscale problem to a macroscale one.

This procedure may be useful for many cases, especially for the micro-actuators or sensors that are widely utilized in the nanorobots. But is there any limitation for this scheme? The CM, as it is obvious from its name, can be utilized for the continuous domains. So, definition of continuous and discrete domains is very important. The macroscale is a continuous domain, and when we move from macro- to nanoscale, we are in the continuous domain until the mean free path (MFP) of the substrate does not pass. Then, the characteristic length of the considered system and the value of MFP of utilized material play the principle role in the possibility of utilizing the CM.

Although it is impossible to define an MFP for the whole metals, some ranges can be determined. MFP in high-mobility semiconductor, commercial semiconductor devices, and polycrystalline metallic films is 10–100 μm [15], 100–1000 nm [16], and 10–100 nm [17], respectively. With a little caution, these values may be considered as 1–10 μm, 10–100 nm, and 1–10 nm, respectively. Thus, for a lot of nanorobotic systems, where the semiconductor materials play a special role, MFP is very small and domain is discontinuous. So, the CM could not be applied directly.

Up to last decades, CM-based approaches such as the mass–spring–damper model were the commonly used approaches for mechanical systems. These models were used even for nanorobots. Mentioned discontinuity, nonlinear, and multifield or scale behaviors strongly questioned utilization of these approaches and other simple theories that were used. In account of this, for a comprehensive model of nanorobots, a multiscale method is needed.

The great progress achieved on mentioned nanorobotic sciences has opened up a new frontier, whose aim is the more accurate prediction of behavior of systems at nanoscale. Since some components of nanorobots are smaller than their MFPs, the study of the aforementioned two nanorobotic systems, macroscale and nanoscale nanorobots, could not be accomplished using CM-based approaches. However, for macroscale ones, the continuum-based approaches can be used directly or coupled with some molecular mechanics (MM)-based approaches. For an example, FEM can be coupled with the molecular dynamics (MD) to predict more accurate behaviors. In atomic scales, the CM analysis methods, including the FEM or other numerical methods, do not provide the correct physics. But nevertheless, for many systems, and for small (or even very large) time ranges, the noted methods are considered to be good approaches. For the kinds of problems that can be solved through the methods based on CM, it is very important to define and present an appropriate criterion. The ratio of a system's characteristic length (which is often the largest length used in a particular setting) to its MFP is generally considered as a criterion. If a system's characteristic length is longer than its MFP, the use of methods based on CM is allowed; otherwise, caution should be exercised, and instead of using these methods, MM-based approaches including the MD, Monte Carlo, and other methods should be applied. Now, what should be done if a system has various mixed sections, and parts of it have dimensions larger than the MFP and parts of it smaller? Often, multiscale methods are presented for solving such systems. In multiscale approaches, the use of methods that are based on atomic models guarantees to include the correct nonlinear behavior of the system without the need for additional parameters; while for larger dimensions that do not have tangible nonlinear behavior, applying the methods based on CM is effective, and there is no need to spend a lot of time to solve the atomic model for these parts. Therefore, the CM models and the atomic models interact with each other in such a way as to model the whole system and achieve the goal which has been defined and set.

During the last decade, to investigate and analyze materials with large sizes, some methods have been developed that are different from those for nanoscale. As was pointed out in the previous section, the common conventional models have many limitations. Despite their reasonable computational costs, the CM methods are totally incapable of describing phenomena in the nanoscale. On the other hand, MD models are very limited with respect to the time dimension; in other words, the modeling of objects in the micrometer and microsecond dimensions is only possible by means of supercomputers. Moreover, the two aforementioned models are not capable of studying the material's electron structure. Meanwhile, the simulation of real systems (whether large-scale or small-scale) by the use of methods based on MM is basically impossible even when utilizing supercomputers. So, the researchers are trying to use the advantages of the aforementioned methods and to cover the existing flaws by combining them together. The outcome of these efforts has been the development of multiscale methods. Multiscale methods are divided into the "hierarchical" and "concurrent" groups.

In the hierarchical models, the properties are calculated at one scale and then passed on to another scale. In other words, the information obtained from one model is enriched by another model. These approaches include two groups, in which, the information from microscale systems is transferred to the CM model. The first group are the models based on the Cauchy–Born hypothesis [18,19], in which the information acquired from the atomic structure of the object clearly reveals themselves in the calculation of the elastic stress and tensor of the material's properties. The second group of these models is based on the virtual atom cluster [20], in which the structure of the material is enriched on the basis of the information obtained from MM. This method has been used for the study of carbon nanotubes.

In the concurrent models, several models exist simultaneously in the multiscale simulation, and information is exchanged among them concurrently. These methods are the result of efforts that have tried to combine the MD models with the CM models. This group of multiscale models has been in use for about a decade, and three major efforts could be cited in the development of these approaches. The quasi-continuum (QC) method [21,22], which was presented by Ortiz et al., is now the most applied method, and many studies have been conducted on this method. The only flaw in this approach is that the existing mesh should be decreased to the size of the atomic structure. The other model is the bridging domain method [23], which has been developed by Belytschko et al. In this method, in part of the simulation zone, the continuum domain and the atoms exist together; therefore, the validity of this model in the bridging domain needs to be investigated extensively. The last approach, known as the bridging scale method [24,25], has been presented by Liu and his students. In this method, it is assumed that the continuum solution is not exact and that the resulting error can be removed through MD. Due to the complexity of the governing relations in this model, even the founders of this method did not want to develop it further.

The most important issue in the development of these hybrid methods has been the formulation of a comprehensive computational coupling along the interface. This fact has been revealed in a brief review of the developed and presented methods. In the coupling models, the continuity of the material's characteristics should be preserved during the transition from the atomic forces to the stress–strain field of CM. Coupling models have been developed for many problems, including the crack problem, and they have often been named finite element–atomistic (feat) coupling procedure, which is the combination of a MD system and a finite element system. Likewise, a general formulation of the ordinary finite element, which allows the macro-field (MF) nodes to be examined as coarse and fine nano-field (NF) atoms, has resulted in another computational scheme for the coupling of the continuum and the atomic environments, called the coarse-grained molecular dynamic (CGMD).

The QC method that has been studied by Miller and Tadmor [26] is explicitly based on the complete description of a material's environment. Although, in this approach, to get a higher computational performance, the regions in which the atoms have been discretized can be classified into groups in order to form a local continuum.

The coupled atomistic/dislocation dynamics (CADD) method of Shilkrot et al. [27] has been presented for the simulation, detection, and justification of the separations between the atomic and the continuum regions. This model had first been offered for the simulation of materials at 0°K, but recently, it has been developed to deal with the effects of finite temperature as well.

The general characteristic of these approaches, for the atomic and continuum coupling, has been the fine graining and manipulation of MF mesh configuration for conformity with atomic length scales and also the kinematic coupling of finite element nodes to discrete atoms along an interface. Henceforth, the approaches that make a one-to-one coupling between the atoms and finite element are called direct coupling (DC). When DC procedures are followed, the major problem that arises is the inherent difference between the atomic and the continuum computational models. The physical state of the atomic region is described by means of the nonlocal inner-molecular forces between discrete atoms with specific position and moment, while the physical state of the continuum region is described by using the stress–strain fields which are statistical averages of the atomic attractions at larger scales of length and time. Generally, the ordinary coupling between discrete and continuum values can only be obtained by taking a statistical average of the scales in which the discreteness of the atomic structure can be approximated in the QC form. Although, much better ways could be offered for the development of methods of coupling of the continuum domain with discrete domain, nevertheless, the application and development of these methods for the static and dynamic problems related to mechanical engineering is highly important. So far, no concentrated work has been devoted to this subject, while the nature of the problem and the challenges associated with these types of problems (due to particular initial conditions and the mixed issues in the presence of force) are totally different from those pursued in other problems. Here, it will be attempted to discuss problems such as the inevitable existence of drawbacks in the discrete domain compared to the continuum domain and to present an approach for escaping all these difficulties.

Up to now, the most significant (and it could be said, the only) approach that has been used for the modeling of macrodimensions in the coupling models has been the FEM, while much better and more accurate methods, and even more exact numerical methods, exist for this purpose. Thus, in describing the problem, instead of the FEM, the more general form of finite element, i.e., the MF solution method, is used. Based on this notion, we try to present a model that can be attached to the FEM without any restriction and can be used with other methods as well. In the end, to evaluate the accuracy of the presented model, the FEM (which we have developed ourselves) is used.

Then, this chapter has been focused on the modeling of macroscale nanorobots using the coupling of FEM with MD.

4.4 Proposed Nanomanipulation Strategy

There exist various strategies for nanomanipulation process in literature [28,29]. The manipulation process cannot be observed in real time. During the pushing of objects, imaging is impossible because imaging and manipulation tools are the same. As a solution, the surface and targeted clusters could be imaged before and after the manipulation. Using the obtained images, the positions of clusters relative to the basic reference point can be determined [30]. Due to the lack of real-time images, using the force feedback data during the process is crucial for proper manipulation. The manipulation strategy for the pushing of a nanocluster is shown in Figure 4.5. Using a suitable model, the force feedback data during the manipulation strategy can be calculated accurately.

In this problem, both the substrate and the nanocluster are stationary at the beginning. Then, the probe moves down to approach the substrate. The van der Waals force increases until the snap occurs at the point of instability. At this point, the tip jumps to the substrate.

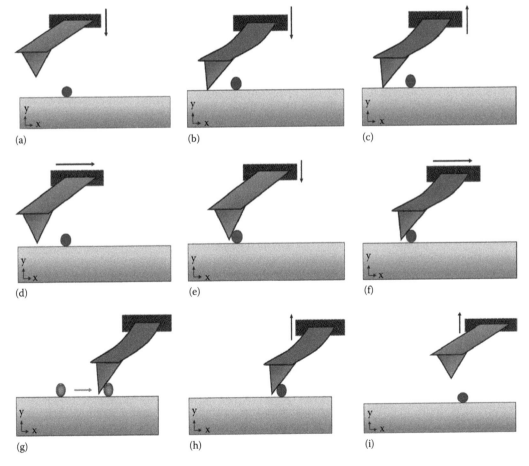

FIGURE 4.5
Nanomanipulation strategy using the AFM: (a) auto parking, (b) snap in substrate, (c) pull away from substrate, (d) approach to nanocluster, (e) snap in nanocluster, (f) offset in Z direction, (g) pushing, (h) pull away from nanocluster, and (i) going to reference point.

FIGURE 4.6
Three expected results: (a) rigid nanocluster, (b) flexible nanocluster, and (c) soft nanocluster.

This phenomenon can be detected via photodiode data. Then, the tip starts moving upward. Deflection in the cantilever increases until the pulling force overcomes the attraction force. In view of the adhesion force between the tip and substrate, the retraction force is larger than the attraction force. Next, the tip moves reaching the desired cluster, horizontally. Furthermore, the van der Waals force between the tip and cluster increases until the snap to the cluster takes place. Then, the substrate movement follows, and the pushing force on the cluster increases.

The tip may cross the cluster and the process may fail. To ensure the desired contact, a small normal preload force, F_{z0}, is exerted by providing normal deflection offset, Z_{P0}, on the AFM probe. Then, the substrate moves with constant velocity, and the cluster sticks to it and moves with the substrate. The lateral motion of the cluster helps increase the pushing force, FT. Finally, the pushing force reaches the magnitude of the critical force required to overcome the adhesion forces between the cluster and substrate. The cluster's movement with the substrate stops, when the cluster has reached the desired position. At this time, depending on the dynamic mode diagram of the cluster, suggested behavior will be expected of the cluster. The probe moves upward and goes to the initial reference position when the process is completed (Figure 4.5).

The pushing force imposes a deflection along the path of movement during the manipulation. Based on the cluster-substrate properties and the pushing force, three different deflection results can be expected (Figure 4.6). Although rigid clusters can be moved without deformation, flexible clusters may undergo considerable deformation during the moving process, and the soft clusters may be damaged when the pushing force exceeds the yield strength of the cluster.

4.5 Multiscale Method

4.5.1 Macro-Field Modeling, Finite Element Method

Extensive work has been done on the development of FEM for various systems. Since in mechanical systems, usually the macro-section has a moving part and a sensing part, and these parts often operate by means of the piezoelectric property, we try to deal with the macro-section from this perspective. Rajeev Kumar et al. [31] investigated a finite element model for the active control of induced thermal vibration in layered composite shells with piezoelectric sensors and actuators (piezothermoelastic). Then, they presented a finite element formulation for the modeling of static and dynamic responses of multilayered composite shells with integrated piezoelectric sensors and actuators and subjected to mechanical, electrical, and thermal loadings [32].

In 2008, Ziya and Thomas [33] have presented a finite element formulation for the vibrations of layered piezoceramic plates, which accounts for the effects of hysteretic behavior. The hysteretic behavior has been simulated in the dielectric domain by using the FEM and applying the Ishlinskii's model. In 2008, Balamurugan and Narayanan [34] have used a nine-noded piezo-laminated degenerated shell element in order to model and analyze multilayered composite shell structures together with sensors and piezoelectric actuators.

The FEM, usually by using the principle of minimum potential energy, extracts the governing equations pertaining to a sample element from the internal components of a system, and then by applying a set of principles, and through the superposition of the hardness, damping, and mass matrices of the internal components presents the system's mechanical characteristics as a whole. Depending on the type of problem and type of loading, the obtained problem can be solved by various numerical approaches. In extracting the equations of motion, the energy that is considered to be minimized is often important. Different coordinate systems could also be important for transferring the problem to different coordinates. The three coordinate systems that are normally used include the global coordinate system (x, y, z), natural coordinate system (ξ, η, ζ), and the current local coordinate system (x', y', z'). These coordinate systems have been illustrated in Figure 4.7.

The global coordinate system is the basic coordinate system. All the analytical information which is defined by the user is entered into this system. The natural coordinate system simplifies the defining of the shape functions and the two-dimensional numerical integration. The current local coordinate system is very important, because it is used for describing the element's geometry, and if necessary, the nonisotropic characteristics (e.g., in composite layers) are described relative to this coordinate system. The coordinates of any arbitrary parameter, at any arbitrary point, can be expressed by the use of nodal coordinates and isoparametric shape functions in the following way:

$$P = \sum_{i=1}^{Q} \aleph_i(\xi, \eta, \zeta) P(X_i) \tag{4.13}$$

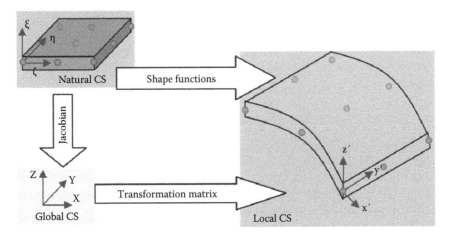

FIGURE 4.7
Coordinate systems for the nine-noded Lagrange shell element for electromechanical systems in the FEM.

where

P is the noted parameter

א is the isoparametric functions

Q is the number of nodes of the considered element

X is the position of the nodes

After writing the required geometrical relations for the considered element, it is necessary to describe the structural configuration of the material used in the system. As a more general case, here, we assume that the system is excited by piezoactuators and then sensed by them. Based on the piezoelectric association with the displacement and electric fields, the basic nonlinear piezoelectric equation is obtained by the direct and the inverse piezoelectric effects. The equations of the nonlinear direct and inverse piezoelectric effect are as follows:

$$\sigma = \mathbb{Q}\epsilon - \mathbb{N}E \tag{4.14}$$

$$\mathbb{D} = \mathbb{N}\epsilon - \mathbb{Z}E + P \tag{4.15}$$

In the earlier relations, σ, ϵ, E, $\mathbb{D}$, and P are the stress field, mechanical strain field, electric potential field, the electric strain (displacement) field, and remanent polarization, respectively. Also, the $\mathbb{Q}$, $\mathbb{N}$, and $\mathbb{Z}$ matrices are the matrices of elasticity, piezoelectric stress constants, and the permittivity constants, respectively. By using the presented structural relationships, the strain energy, electric energy, and the kinetic energy of an element are expressed as follows:

$$UE = \frac{1}{2}\iiint_{\forall} \epsilon^T \mathbb{Q}\epsilon\, d\forall - \iiint_{\forall} \epsilon^T \mathbb{N}E\, d\forall$$

$$EE = \iiint_{\forall} E^T \mathbb{N}\epsilon\, d\forall + \frac{1}{2}\iiint_{\forall} E^T \mathbb{Z}E\, d\forall + \iiint_{\forall} E^T P\, d\forall \tag{4.16}$$

$$TE = \frac{1}{2}\int_{\forall} \rho\dot{u}^T\dot{u}\, d\forall$$

where $\forall$ denotes the volume of the element. By using these energies and writing the minimum energy principle, the equations of motion for the finite element system can be presented as follows:

$$[M_{uu}]_e\, \ddot{q}_e + [C_{uu}]_e\, \dot{q}_e + \left[K_{uu} - K_{u\phi}K_{\phi\phi}^{-1}K_{\phi u}\right]_e q_e = F_{qe} - \left[K_{u\phi}K_{\phi\phi}^{-1}\right]_e \left(F_{\phi e} + N_H\right) \tag{4.17}$$

where $[M_{uu}]_e$, $[K_{uu}]_e$, $[K_{u\phi}]_e$, F_{qe}, $[K_{\phi\phi}]_e$, $F_{\phi e}$, $[C_{uu}]_e$, N_H, and q_e are, respectively, the element's mass matrix, stiffness matrix, electromechanical coupling hardness matrix, mechanical load, dielectric hardness matrix, electric force vector, structural damping matrix, the hysteresis related force, and the vector of change of degrees of freedom in the considered system. The hysteresis model has been considered only for the modeling of nonlinearities of MF, and for the multiscale model, it was assumed that the nonlinear behavior has been compensated completely.

4.5.2 Nano-Field Modeling, Molecular Dynamics Approach

4.5.2.1 Need for Molecular Dynamics Simulations

Mathematical models could be used for the interpretation of physical phenomena. The goal of these models is to provide simple ways of showing the causes of various behaviors. In the field of numerical problems, there exist two basic models: the classical Newtonian mechanics and the Quantum mechanics. These models might be used for problems with macro-, nano-, or sub-atomic scales. In the models based on classical Newtonian mechanics, Newton's equations are used to analyze the dynamics of all parts of the system. Having high precision in the interpretation of physical phenomena and being applicable in various scales make these methods very interesting among researchers. Considering the desired scale, for which one can use the Newtonian mechanics, there are different methods of applying the Newton's equations, such as CM, dislocation dynamics, and molecular methods [35].

In the molecular models, all the atoms of a particular material react through inter-atomic potentials. These models are capable of simulating the ultrafine structures of materials in nanoscale. These models are categorized into two distinct branches of sto-chastic and deterministic methods. As an example of the first category, one can men-tion the Monte Carlo method. Among deterministic methods, in which the governing equation of motion is solved explicitly, MD and statics are of more interest [36]. Since in nanoscale problems, the temperature effects and velocity play important roles, MD is a more applicable method.

Typically, classical models have lots of limitations when applied. Despite their low computational costs, continuum-based methods are incapable of describing different phenomena in nanoscale fields. On the other hand, MD models are limited with respect to the time scale. For this reason, modeling on the order of micrometer and micro-second is only possible by supercomputers. Meanwhile, the two foregoing models are not capable of studying the electrical structure of materials. Whether in large or small scale, the simulation of real systems, with MM-based models, is impossible even with the use of supercomputers. Therefore, researchers are in search of some comprehensive synthetic methods that have the advantages, but not the drawbacks, of the original methods. The outcome of these research works has been the introduction of multiscale methods.

4.5.2.2 Molecular Dynamics

In order to establish the relations of MD for a system of particles, some parameters are required. Particle mass, initial position of atoms, initial velocity of atoms, potential energy between the atoms and external force fields, and the equations of motion are five effective parameters that must be determined for the system in order to have a unique solution [37].

In a state of equilibrium, atoms can have different arrangements based on the material phase. It is common for the atoms of gases and liquids not to have a specific structure. Contrary to gases and liquids, solids possess a higher potential energy level relative to the kinetic energy. This fact forces the atoms of solid, especially of metals, to form spe-cific network configurations. For metals, there are two types of solid networks that are more common, namely, face-centered cubic (FCC) and base-centered cubic (BCC) lattices. In the current work, the FCC lattice is applied for the metal atoms to represent the initial positions.

In order to carry out atomic simulations, a way of expressing atomic rules is inevitable. In computer simulations, these rules are known as "interatomic potential energy." The dynamics of the atoms is obtained from prescribed two-body or many-body interatomic potentials, $H_I(r_{ij})$, from which the Newtonian forces experienced by these atoms are derived [38]:

$$F_i = \sum_{j>i} \nabla_n H_I(r_{ij}) \tag{4.18}$$

where r_{ij} is the distance between the atoms i and j.

4.5.2.3 Sutton–Chen Interatomic Potential

For the simulation of a metal system, the simple two-body potentials, like the Lenard-Jones potential, are not intelligent choices, as they lack the desired capability to encompass all the physical properties of metals. Therefore, the Sutton–Chen (SC) multibody long-range potential, first introduced by Sutton and Chen [38], is used in the current study. The general form of the SC potential is [38]:

$$H_I^{SC} = \varepsilon \left[\frac{1}{2} \sum_i \sum_{j \neq i} V(r_{ij}) - c \sum_i \sqrt{\rho_i} \right], \quad V(r_{ij}) = \left[\frac{a}{r_{ij}} \right]^n, \quad \rho_i = \sum_{j \neq i} \left[\frac{a}{r_{ij}} \right]^m \tag{4.19}$$

where ε is a parameter with the dimensions of energy, a is a parameter with the dimensions of length and is normally taken to be the equilibrium lattice constant, m and n are positive constants with $n > m$. Table 4.3 lists the SC parameters for some metals.

4.5.2.4 Rafii-Tabar–Sutton Potential

Rafii-Tabar and Sutton have further generalized the SC potential to model the interactions of unlike atomic clusters in FCC random binary metallic alloys [39]. Rafii-Tabar–Sutton

TABLE 4.3

Parameters of the SC Potential

Element	m	n	ε (eV)	c
Ni	6	9	1.5707e–02	39.432
Pt	8	10	1.9833e–02	34.408
Au	8	10	1.2793e–02	34.408
Ag	6	12	2.5415e–03	144.41
Cu	6	9	1.2382e–02	39.432
Ir	6	14	2.4489e–03	334.94
Pb	7	10	5.5765e–03	45.778
Pd	7	12	4.1790e–03	108.27
Rh	6	12	4.9371e–03	144.41
Al	6	7	3.3147e–02	16.399

Source: Sutton, A.P., J. Chen. *Phil. Mag.*, 61, 139, 1990.

(RTS) multibody long-range potential, which is an extended form of the SC potential, and capable of modeling the interactions of dissimilar materials, is used in the current study. The general form of the RTS potential for the binary A-B dissimilar materials is [39]:

$$H_I^{RTS} = \frac{1}{2} \sum_i \sum_{j \neq i} V(r_{ij}) - d^{AA} \sum_i \hat{p}_i \sqrt{\rho_i^A} - d^{BB} \sum_i (1 - \hat{p}_i) \sqrt{\rho_i^B} \qquad (4.20)$$

with

$$V(r_{ij}) = \hat{p}_i \hat{p}_j V^{AA}(r_{ij}) + (1 - \hat{p}_i)(1 - \hat{p}_j) V^{BB}(r_{ij}) + [\hat{p}_i(1 - \hat{p}_j) + \hat{p}_j(1 - \hat{p}_i)] V^{AB}(r_{ij}) \qquad (4.21)$$

$$\rho_i^A = \sum_{j \neq i} \Phi^A(r_{ij}) = \sum_{j \neq i} [\hat{p}_j \Phi^{AA}(r_{ij}) + (1 - \hat{p}_j) \Phi^{AB}(r_{ij})] \qquad (4.22)$$

$$\rho_i^B = \sum_{j \neq i} \Phi^B(r_{ij}) = \sum_{j \neq i} [(1 - \hat{p}_j) \Phi^{BB}(r_{ij}) + \hat{p}_j \Phi^{AB}(r_{ij})] \qquad (4.23)$$

where $\hat{p}_i$ is the site occupancy operator, defined as follows:

$$\hat{p}_i = \begin{cases} 1 & \text{if site i is occupied by an A atom} \\ 0 & \text{if site i is occupied by an B atom} \end{cases} \qquad (4.24)$$

The functions $V^{xy}(r)$ and $\Phi^{xy}(r)$ are defined as follows:

$$V^{xy}(r) = \varepsilon^{xy} \left[\frac{a^{xy}}{r} \right]^{n^{xy}} \qquad (4.25)$$

$$\Phi^{xy}(r) = \left[\frac{a^{xy}}{r} \right]^{m^{xy}} \qquad (4.26)$$

And the constants are expressed as follows:

$$d^{AA} = \varepsilon^{AA} C^{AA} \quad d^{BB} = \varepsilon^{BB} C^{BB}$$

$$m^{AB} = \frac{1}{2}(m^{AA} + m^{BB}) \quad n^{AB} = \frac{1}{2}(n^{AA} + n^{BB}) \qquad (4.27)$$

$$a^{AB} = \sqrt{a^{AA} a^{BB}} \quad \varepsilon^{AB} = \sqrt{\varepsilon^{AA} \varepsilon^{BB}}$$

In the previous relations, ε is a parameter with the dimensions of energy, "a" is a parameter with the dimensions of length, and it is normally taken to be the equilibrium lattice constant, and "m" and "n" are positive constants with $n > m$. The RTS potential has the advantage that all its parameters can be easily obtained from the SC elemental parameters of metals.

4.5.2.5 Equations of Motion

In the present study, the Hamilton's principle is used to represent the equations of motion of each single atom:

$$\int \delta(PE + KE)dt = 0 \tag{4.28}$$

Through simplification, the equation of motion for the ith particle is finally in the form of

$$m^i \ddot{X}_i + c^i \dot{X} + \frac{\partial H_{RTS}}{\partial X_i} = F^i_{ext} \tag{4.29}$$

The first and second terms in the preceding equation denote the kinetic and dissipative energies, respectively. The third term is an interpretation of potential energy, which is related to the interatomic potential. Using the velocity Verlet computational algorithm, the equations of motion can be solved numerically for the system of particles, and the positions and velocities of each single atom can be calculated in desired time steps.

4.5.2.6 Solution Procedure

The common method for solving the differential equations of motion in MD is the well-known finite difference method (FDM). Knowing the positions, velocities, and other essential data of the system at the time "t," the goal of the FDM is to calculate these parameters for the next time step, $t + \Delta t$. Therefore, all the differential equations of motion are solved in a step-by-step manner. Finite-difference-based methods utilize the Taylor series for expanding the position function of particles [36,37].

Various finite difference algorithms exist for solving the differential equations of atomic motion. They differ in the way of calculating r.. Since the real nanoscale systems consist of numerous numbers of atoms, the simulation procedure must encompass large degrees of freedom to show the real behavior of the system under consideration. Therefore, an algorithm with both low computational costs and good precision is in great demand. The Verlet algorithm is one of the most popular methods among nanomechanics researchers. This algorithm combines the forward and backward Taylor expansions. With some simplifications, the relations for velocities and positions of atoms at time 't' are expressed as follows:

$$v(t) = \frac{r(t + \Delta t) - r(t - \Delta t)}{2\Delta t} \tag{4.30}$$

$$r(t + \Delta t) = r(t) + \Delta t \times v(t) + \frac{1}{2}(\Delta t)^2 \times a(t) + O(\Delta t^4) \tag{4.31}$$

In a more advanced form of the Verlet algorithm, known as the velocity Verlet, the velocity is calculated using the half-step acceleration:

$$v(t + \Delta t) = v(t) + \Delta t \times a\left(t + \frac{1}{2}\Delta t\right) + O(\Delta t^2) \tag{4.32}$$

This leads to the final form of the velocity Verlet algorithm as follows:

$$r_i(t+\Delta t) = r_i(t) + \Delta t \times v_i(t) + \frac{1}{2}(\Delta t)^2 \times a_i(t) \tag{4.33}$$

$$v_i\left(t+\frac{1}{2}\Delta t\right) = v_i(t) + \Delta t \times \frac{a_i(t)}{2} \tag{4.34}$$

$$a_i(t+\Delta t) = \frac{F_i(t+\Delta t)}{m_i} \tag{4.35}$$

$$v_i(t+\Delta t) = v_i\left(t+\frac{1}{2}\Delta t\right) + \Delta t \times \frac{a_i(t+\Delta t)}{2} \tag{4.36}$$

The velocity Verlet algorithm is the most popular computational method in MD simulations. By using the second derivative formula in this method, the computational precision gets to about fourth order. This results in a high computational speed as well as high precision.

4.5.2.7 NVT Ensemble

This ensemble is usually used for the simulation of closed systems that just exchange energy with the environment. In this ensemble, the system under consideration must be invariable in temperature, volume, and the number of particles [37].

To realize this ensemble, there must be some ways of making the aforementioned qualities constant. A fixed number of atoms can be realized through periodic boundary conditions. Using the total pressure of the system, its volume can be fixed as well. Temperature stabilization in computer simulations is not an easy task. However, it can be realized in three different ways, namely, the stochastic Langevin-type method, the constraint method, and the extended system method [36].

The simple constraint method can be carried out by using a velocity scaling procedure in each time step. This tends to limit the total kinetic energy of the system in order to stabilize the system's initial temperature. This ensemble can be produced as follows:

$$\vec{v}_i \rightarrow s\vec{v}_i \tag{4.37}$$

where

$$s = \sqrt{\frac{T_0}{T_i}} \tag{4.38}$$

where
T_0 is the desired system temperature
T_i is the temperature in ith time step

4.5.2.8 Coarse-Grained Molecular Dynamics

Electromechanical processes normally occur on the order of nano-, micro-, milli-, and even a few seconds. In addition, they have higher-than-nano dimensions. Therefore, their real dimensions and time ranges cannot be determined through the use of MD method.

However, by the use of "coarse graining," larger dimensions in longer time ranges could be modeled. Now, since a remarkable method known as the CGMD has been presented for this purpose, while applying it, a general description of this approach is also provided. The CGMD method is based on the notion that, if instead of one atom, a larger number of atoms can be taken as a unit, then a larger volume of material as well as more simulation time can be considered [40]. Even by utilizing the world's largest and most advanced super-computers, the MD simulations cannot be performed for more than several microseconds. Various procedures have been proposed for the CGMD methods [41–43]. The only crucial issue in these models is the manner of predicting and estimating the system's potential. Achieving a good potential for the system can be guaranteed through a process of trial and error and by comparing the radial distribution function (RDF) of the system with that which is observed in the MD process, although other ways also exist for this achievement. If, on the average, the nominal mass and distance of atoms are on the order of m_c and L_c, and the nominal mass and distance of the CGMD samples are on the order of m and L, respectively, the following relations could be considered for the time steps that are used:

$$\Delta t_{max}\big|_{MD} \sim 1\sqrt{\frac{m}{kT}}, \quad \Delta t_{max}\big|_{CGMD} \sim l_c\sqrt{\frac{m_c}{kT}} \tag{4.39}$$

Here, for using the CGMD approach, the SC potential has been rewritten, and the RDF diagrams of two cases of MD and CGMD have been compared with each other, and the obtained CGMD model has been validated. Figure 4.8 shows the comparison between the RDFs of the two noted cases.

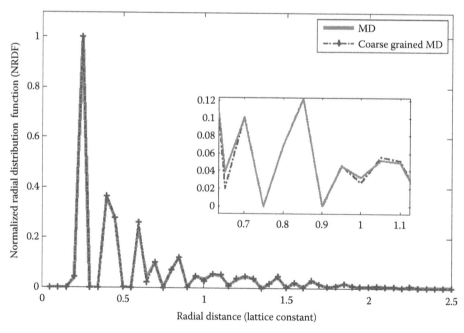

FIGURE 4.8
Comparison of the RDFs of the MD and CGMD models.

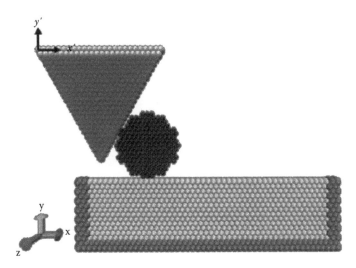

FIGURE 4.9
Arrangement of the manipulation system in the CGMD and the definition of the local and global coordinate systems.

4.5.2.9 Molecular Dynamics Simulation of Nanomanipulation Procedure

Figure 4.9 shows the initial system configuration for the nanomanipulation procedure. Since the manipulation procedure is planar, the atom's movement is limited to the x–y plane. After the relaxation phase, the nanoparticle is subjected to the proposed nanomanipulation strategy. The manipulator, which is part of an AFM tip, starts moving with constant velocity, dictated by the constant velocity of the uppermost atoms, i.e., the tip base region in Figure 4.9. The manipulator is flexible, so its atoms can move freely in both the x and y directions and make different vibration modes along the tip length. The substrate is flexible as well; however, the gray atoms (as depicted in Figure 4.9) are fixed to avoid movement during manipulation. The manipulator, nanoparticle, and the substrate can be made of a variety of materials (Table 4.3).

To validate the obtained results, it is necessary to compare them with the already verified results. For this purpose, the results of the present work are compared with the modification to those obtained from the model presented in Ref. [44].

4.5.2.10 Results Validation: A Comparison of CGMD and Macro-Model

Using CGMD, a comparison has been made between the results and those of the macro-model in Ref. [44]. The system geometry has been depicted in Figure 4.10. The tip, nanoparticle, and substrate are made of Rh, Ni, and Au, respectively. The nanoparticle has the same size in both models, but the substrate and the tip are considered as depicted in Figure 4.10.

In the CGMD model, to ensure the same conditions as the macro-model, the nanoparticle is considered to have a low degree of flexibility, and the surface interactions are modified so that they follow the same behavior of the JKR contact model. However, the manipulator tip is considered to be flexible to ensure the real conditions.

In the macro-model, the manipulation procedure has been simulated by giving the base of the AFM tip (i.e., the stage in Figure 4.10) a constant speed along the "x" direction. The manipulator is the triangular tip of the AFM, which is considered to be rigid. The

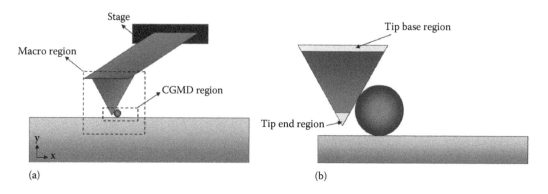

FIGURE 4.10
(a) Macro and CGMD regions and (b) geometry of tip and nanoparticle in CGMD model.

deflections of the AFM tip, when manipulating the nanoparticle, are due to the flexibility of AFM cantilever. The nanoparticle is considered to have a spherical shape and no flexibility (it is rigid). The nanoscale interactions between the substrate and nanoparticle are modeled with the JKR contact model. At the start of the manipulation procedure, it takes a few seconds for the nanoparticle (and AFM tip) to start moving, because the pushing force is not large enough to overcome the resistant frictional force. Once the movement starts, the AFM tip and nanoparticle are assumed to move together and therefore trace the same line in the traveled distance–time diagram.

Considering the fact that in the macro-model, the tip and particle are assumed to be rigid and the force is supposed to be applied from the moment the probe contacts the particle, in Figure 4.11, the diagram marked with square shows the displacement of the particle and probe tip, and the diagram indicated by the black circle shows the displacement of the moving substrate. The comparison of these diagrams with the results obtained from the CGMD model has yielded very interesting results. As it can be observed in this figure, there is relatively good agreement between the diagrams of the CGMD and the results of the macro-model. There are also differences that we claim arise from the not-so-correct assumptions made in the macro-model and as a result of the higher accuracy of the CGMD model.

There are several important points regarding the previous diagrams. The macro-model has focused on the estimation of the time of nanoparticle movement subsequent to the application of force. In the macro-model, after the application of force, the friction forces resist against the exerted force until they get to be equal to the applied force. From this moment forward, the particle starts moving. Therefore, the frictional characteristics that are considered between the particle and substrate could be very important in determining this manipulation time. In the CGMD model, we do not necessarily impose any specific frictional characteristics on the model. It is the forces arising from the potentials between dissimilar atoms, and the effects of damping considered in the CGMD model that play the most significant roles in determining the time of movement of nanoparticle. In view of the presented diagrams, the CGMD model has underestimated a little, the time of movement of nanoparticle, as compared to the macro-model. The other noteworthy point is that the amount of traveled distance of nanoparticle is less than that of the tip. This has two major causes: first, the existing delay in the start of nanoparticle movement, and second, the initial distance between the probe tip and nanoparticle. This short distance is not necessarily due to the initial position of the tip but may be the result of initial deformations of the tip end.

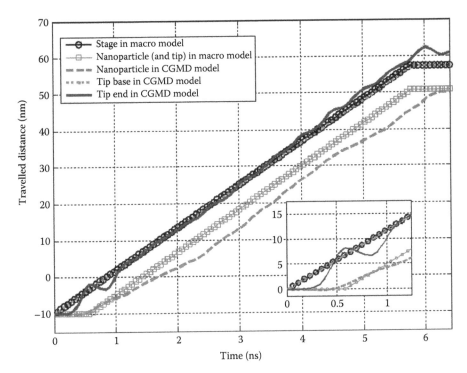

FIGURE 4.11
Comparison between the results of macro-model and CGMD model for the manipulation of a nickel nanoparticle.

In the enlarged portion of the diagram of Figure 4.11, it is clear that, at the onset of particle movement, the tip end vibrates in the CGMD, which is due to the local attraction and repulsion between the tip end and particle and also due to the immediate application of velocity at the tip base; and of course, as a result of damping effect and also the local equilibrium of potentials, these vibrations are gradually eliminated. With regards to the comparisons and offered explanations, the correctness of the results of the proposed model is verified, and it can be used for other studies including the study of the effects of different parameters.

4.5.2.11 Some Challenges in the Way of Molecular Dynamics Simulations

Like what happens in real systems, preparing for a successful manipulation in computer simulations is not an easy task. In some cases, the tip damage tends to the manipulation failure. In some others, the nanoparticle penetration into the substrate or distortion of nanoparticle is the cause of manipulation malfunction. In any case, adjusting the parameters so that it tends to a successful manipulation process demands the adequate experience as well as enormous effort. These problems might well be accentuated when the manipulator tip must perform many repeated manipulation procedures. For instance, in an "automatic nanomanipulation" procedure [45], since the tip must frequently manipulate several different particles, the destructive effects of manipulation procedure endured by the tip are more possible. Accordingly, the tip damage, which might be in the form of either initiation of the crack or the permanent deformation of the tip end, or even in some more severe cases, the tip fracture, is one of the biggest obstacles to a successful

FIGURE 4.12
Some examples of tip damage.

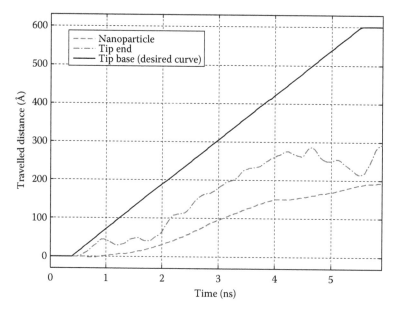

FIGURE 4.13
Traveled distance of tip end, nanoparticle, and tip base in the case of tip damage.

positioning task. Figure 4.12 depicts some examples of the tip damage. The tip end and the nanoparticle traveled distances for one of the Figure 4.12 examples are depicted in Figure 4.13. The deviation of the nanocluster manipulation curve from the desired curve can be clearly seen.

Since the aim of a manipulation process is the positioning of nanoscale objects, the nano-object crushing during the procedure of manipulation, which was mentioned earlier, is of high importance as well. However, the objective is not to study the particle deformation now; this phenomenon can be studied in the same way as well [46]. For instance, look at Figures 4.14 and 4.15. Figure 4.14 depicts some examples of nanoparticle deformation during the nanomanipulation procedure. The traveled distance of the nanoparticle and the tip end and also the desired curve are depicted in Figure 4.15. In this figure, the tip end

FIGURE 4.14
Some examples of nanoparticle crushing.

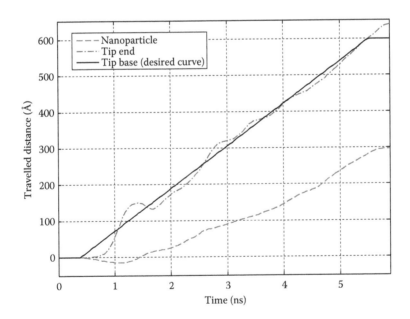

FIGURE 4.15
Traveled distance of nanoparticle, tip end, and tip base in the case of nanoparticle crushing.

almost pursues the movement of the tip base (or desired curve). However, the deviation of nanoparticle curve from the desired one can be clearly seen.

Considering the foregoing diagrams, it seems that the tip damage, among the aforementioned obstacles, has the most negative effects on the nanomanipulation success, as it possesses the maximum deviation between the nanoparticle position and the desired position.

4.5.2.12 *Sutton–Chen Parametric Study: The Effects of Different Parameters on the Nanoparticle Deformation*

In this part, the effects of two parameters on the success of manipulation procedure have been studied. These parameters are ε_p and ε_t. The criterion considered for the success of the manipulation process, called the success parameter (SP), is defined as the change in the ratio of two equal orthogonal radii of the nanoparticle:

$$SP = \frac{R_a}{R_b} \tag{4.40}$$

In the preceding parameter, R_a and R_b are the large and small diameters of the nanoparticle, respectively (Figure 4.16).

When the nanoparticle is spherical, the SP ratio will be equal to 1.0, and as the nanoparticle deforms, the value of SP will deviate from 1.0. Although this parameter indicates particle deformation, however, also in some rare cases (while the nanoparticle has deformed and its shape is no longer spherical), this parameter might be equal to 1.0. In any case, if the change of nanoparticle box is concerned, with regard to Figure 4.16b, the change of nanoparticle shape can be generally defined as relation (4.40).

For the general case, R_a is the distance between the farthest atoms. The x' axis is defined along these atoms. With a coordinate transformation from the xy axes to x'y', R_b is defined as the distance between the highest and lowest points in the y' direction. Using these definitions, all the atoms are swept with a rectangle area $R_a R_b$.

Since particle deformation is one of the crucial parameters in the success of a nanomanipulation process, to analyze and evaluate this phenomenon, a new parameter called SP has been introduced. This parameter can be very useful in the displacement of biological particles, particularly in the manipulation and making of special biological objects.

Table 4.4 summarizes the results of different simulations with various amounts of ε_p and ε_t. The base materials of the tip and nanoparticle are cupper (Cu) and silver (Ag), respectively. The substrate is made of Au. By fitting an exponential curve in the form of

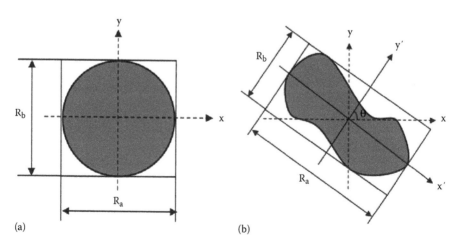

(a)   (b)

FIGURE 4.16
(a) Initial form of the nanoparticle with equal diameters and (b) with unequal diameters.

TABLE 4.4

Results of Various Simulations with the Change of ε_p and ε_t

Values of SP with the Change of ε_p (Ag Tip: $\varepsilon_t = 2.5415e{-}03$)		Values of SP with the Change of ε_t (Cu Nanoparticle: $\varepsilon_p = 1.2382e{-}02$)	
ε_p	SP	ε_t	SP
0.000707	2.0981	0.0000684	1.1000
0.002707	1.9872	0.0002333	1.2104
0.005707	1.6512	0.000563	1.1710
0.007707	1.4046	0.0012225	1.2389
0.010707	1.3083	0.0022118	1.2937
0.012707	1.1969	0.0028712	1.3640
0.015707	1.1670	0.0033659	1.3681
0.017707	1.1448	0.0038605	1.3790
0.020707	1.0195		
0.023707	1.0245		
0.026707	1.0232		

$y = ae^{bx} + ce^{dx}$ to the case that changes the values of 1.697, 88.73, 0.545, and 17.29 are obtained for the coefficients a, b, c, and d, respectively. The R-square value of this curve is equal to 0.986. The data and the fitted curve have been shown in Figure 4.17.

Like the previous case, by fitting an exponential curve in the form of $y = ax^b + c$ to the case that changes the values of 6.41, 0.5563, and 1.095 are obtained for the coefficients a, b, and c, respectively. The R-square value of this curve is equal to 0.9318. The data and the fitted curve have been shown in Figure 4.17.

Figure 4.17 shows that, with the increase of the ε_p coefficient, the value of the SP parameter, and consequently the deformation of the particle, decreases. The increase of the SP increases the cohesion of the nanoparticle's atom, and therefore, ε_p is proportionate to the hardness of the particle, and the obtained result is very reasonable. Also, it can be observed in Figure 4.18 that, in general, with the increase of ε_t, the SP parameter, and consequently the particle deformation, increases, although the local data changes do not necessarily have an ascending trend. At any rate, the increase of the SP is not favorable to the nanomanipulation process. It seems that the tip flexibility (softness of the tip) can be beneficial for not deforming the nanoparticle, as long as it does not hinder the other objectives of the manipulation. This conclusion also seems reasonable because part of the energy which is supposed to deform the nanoparticle will be absorbed through the deformation of the tip, and the nanoparticle will remain more rigid.

4.5.2.13 Comprehensive Diagram for the Optimal Selection of Tip

Some simulations for the manipulation of several different particles have been performed by using various manipulator tips. The following diagram (Figure 4.19) covers all the metals of the SC table. By using this diagram, one can easily select the appropriate manipulator tip for the considered nanoparticle so that the smallest SP number and thus the least amount of physical change could be resulted. As an example, if the considered nanoparticle is made of gold, it is suggested that a platinum tip be selected. The use of a silver tip is the worst choice and will result in relatively large deformation of nanoparticle.

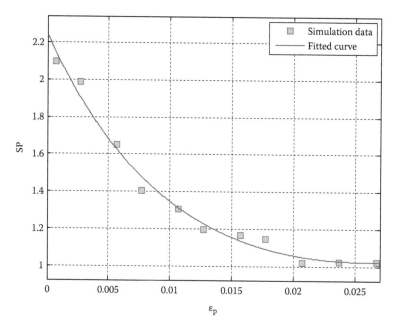

FIGURE 4.17
Data and the fitted SP curve with the change of ε_p.

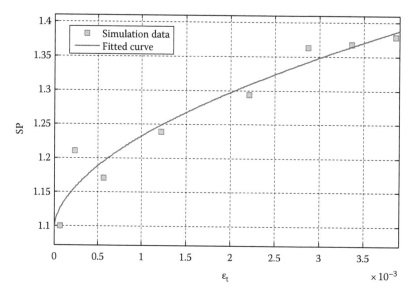

FIGURE 4.18
Data and the fitted SP curve with the change of ε_t.

4.5.3 Macro–Micro-Coupling Model

In the research works that have been conducted so far, various methods of NF–MF coupling on different basis have been presented. The mentioned nanomechanics field provides a new point of focus in the study of mechanics of materials, especially the simulation of basic atomic mechanisms including the propagation and expansion of flaws. These

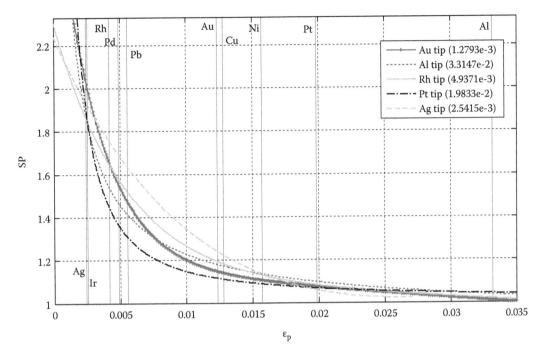

FIGURE 4.19
Change of SP parameter versus the changes of ε_t and ε_p.

simulations are generally based on MM methods, including the tight bonding, ab initio, and the density function theory for the classical NF or on molecular statics methods. These predictions of material behavior in nanometer scales make multiscale analyses possible, which could help in the understanding of failure mechanisms. However, it should be pointed out that with the increase of the size of the system, the modeling calculations of atomic processes increase significantly.

Although, the multifield behavior of nanorobots has not yet been discussed in full, in several research works, it has been attempted to open some windows into the dynamics modeling problems in the area of nanomechanics. In this regard, Park and Liu [47] have examined several issues in the area of multiscale problems related to solids.

In the present method, a direct link between the single NF atoms and MF nodes by means of the statistical averaging of local atomic volume displacements associated with every MF node in their common zone has been replaced with the previous methods. Moreover, considering the mechanics of the problem and using a system of equations in the matrix form, a dynamic algorithm has been presented for dynamically solving the problem. The fourth-order Runge–Kutta method is used for solving the problem dynamically. The MF and NF computational systems are independent of each other and only relate through an iterative update of their boundary conditions. This method presents an improved coupling approach which is inherently applicable for three-dimensional domains. In addition, it prevents the resolving of the continuum model into atomic resolution and allows finite temperature cases to be applied. One of the prominent features of the present work is the presentation of reliable solutions for problems that have natural, forced, body, and interfacial degrees of freedom.

To generalize the problem, the macro-nano-related problems are divided into two groups of closed and open systems. The group of problems, where all the side boundaries of the nano domain overlap the interfacial degrees of freedom, is called "closed systems," and the group of problems, where the side boundaries of the nano region, in addition to the interfacial degrees of freedom, possess limited (and in some cases, unlimited) degrees of freedom, is called "open systems." Figures 4.20 and 4.21 illustrate the general cases of the

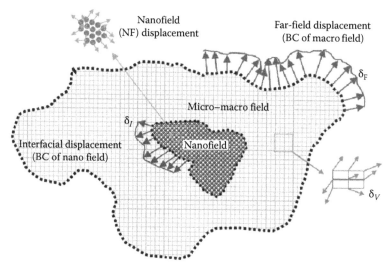

FIGURE 4.20
General case definition of closed mechanical systems.

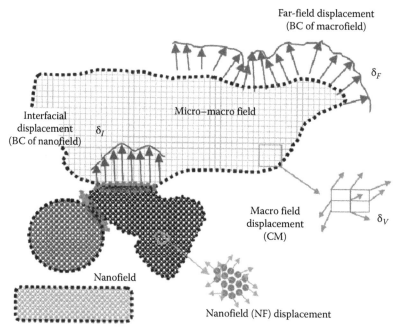

FIGURE 4.21
General case definition of open mechanical systems.

closed and open systems, respectively. In the closed system, usually one NF and one MF exist. For example, in the crack propagation problem, the fine region of crack propagation is designated as the NF, and the coarse region in which the crack is growing is designated as the MF. In the open system, several MFs could be interacting with several NFs. If the numbers of MF and NF are equal to M and N, respectively, and the area of each field is indicated by Ω, then for the closed system, we can write:

$$\begin{cases} \Omega_i \cap \Omega_j = \varnothing, & i, j \in NF \\ \Omega_i \cap \Omega_j = \Pi \text{ or } \varnothing, & i, j \in MF \\ \Omega_i \cap \Omega_j \neq \varnothing, & i \in NF, j \in MF \end{cases} \tag{4.41}$$

And for the open system,

$$\begin{cases} \Omega_i \cap \Omega_j = \Pi \text{ or } \varnothing, & i, j \in NF \\ \Omega_i \cap \Omega_j = \Pi \text{ or } \varnothing, & i, j \in MF \\ \Omega_i \cap \Omega_j = \Pi \text{ or } \varnothing, & i \in NF, j \in MF \end{cases} \tag{4.42}$$

In the previous relations, $\varnothing$ and Π denote the empty and nonempty spaces, respectively. With this notation, it can be easily proved (considering the presented definitions) that a closed system is a special case of an open system. Therefore, in an open system, there may be more than one NF, and each of the NFs may be in contact with one another in different ways. These contacts (e.g., in the nanomanipulation process using nanorobots) may not occur during a certain time range, and after that duration, these contacts may be established. Here, exclusively mechanical systems are taken into account, and therefore, the mentioned contacts are of the second order only, and volumetric sharing is not considered.

The concept of multiscale coupling methods can be very useful in cases where we want to model a relatively large region of the material in order to study the whole deformation field, but the atomic and subatomic scales are needed only in specific and limited regions of the material. A practical example of a closed system can be demonstrated in the modeling of crack nucleation and propagation. As was mentioned earlier, for such problems, various works have been presented. The present model has a special application in open systems; systems where practically no interface may even exist between the macro- and nano-environments in some cases and in a certain range of work, while after a certain time duration (which could be known or unknown), a relationship may form between these two environments. Through the use of coupling models for closed environments, the size limitation of atomic modeling could be minimized, such that an inner region (with complex dynamic processes and large deformation gradients) could exist inside an outer region (with small deformation gradients). It is not like this in open systems, where the effect of size will be considerable. To demonstrate the effectiveness of the model, the special case of a conic region for NF has been investigated. Also, in the MF model, an elastic beam with piezoelectric properties has been considered.

4.5.4 Coupling of MF and NF

For the coupling of MF and NF in closed systems, four regions are considered throughout the system shown in Figure 4.22. These four regions, in the order of proceeding

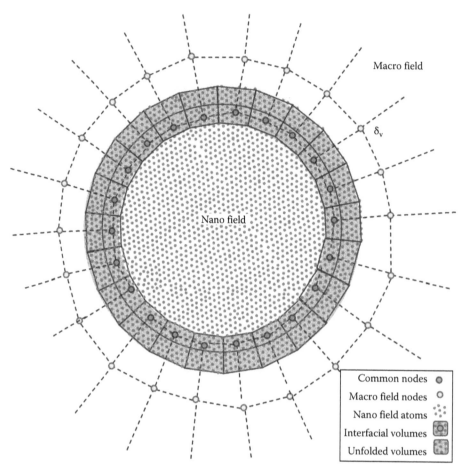

Macro field

δ_v

Nano field

Common nodes	◉
Macro field nodes	○
Nano field atoms	⦂⦂⦂
Interfacial volumes	▣
Unfolded volumes	▨

FIGURE 4.22
Common region between NF and MF in the closed system.

from micro- to nano-environments, consist of MF, unfolded volume (UV), interfacial volume (IV), and NF. The IV region is in fact a region where the terminal atoms of an NF model have surrounded an MF node in the model. The IU region is the region between the end nodes of MF and the end of the NF model. The two regions of MF and NF need no further explanation. In view of the presented cases, IVs estimate the mean displacements of NF in the center of mass of these displacements. These averages are later used as the initial conditions of displacements in the relevant interfacial nodes. It should be mentioned that the IV need not match the macro-element that surrounds it, with respect to the size and shape. Normally, a macro-element, in the interfacial section, consists of hundreds to thousands of atoms. By taking an effective average for the atomic points, the discreteness of the atomic structure can be sufficiently homogenized so that the MF region responds to the excitations of the atomic region as an expanded volume of itself.

For the analysis of open systems, in addition to the four regions of MF, UV, IV, and NF, two regions of free boundaries of NF (denoted by FB) and common boundaries of NF (denoted by CB) are also defined. Regardless of the type of initial state these two regions may have, each one has the potential of undergoing different changes during the analysis

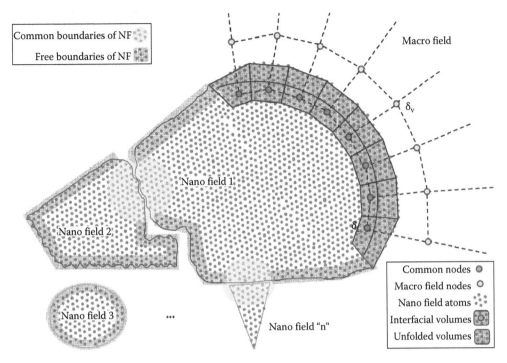

FIGURE 4.23
Common region between NF and MF in the open system.

time range. The CB region is usually circumscribed around a circular zone because in small dimensions, for considering the forces which in this zone are accounted among different sections of the NF, the concept of "cut-off radius" is used. Moreover, the MF region is also divided into four sections of "free far fields," "internal volume," "boundary field," and "interfacial field" (Figure 4.23).

It seems necessary here to describe the method of analysis of the FB and CB regions. In order for the FB region to behave freely (at surface), changes should be made to the model. This is the philosophy behind the establishment of the FB region. The existence of free surface creates unwanted effects in the NF system. In comparison with the cases in which the boundary is affected by an external load, this occurrence in FB is not so critical. In addition to unwanted effects, since atoms at the free surface or close to it do not have a complete set of neighboring atoms, the coordination between the atoms falls apart. To remedy this lack of coordination, and to make the atoms stable in the interfacial NF region, two approaches can be adopted. The first approach is to offer an additional volume of atoms away from the center, which forms the surface NF region. The second approach is to consider a number of the same NF system atoms as an unfolded volume. In case of using the first approach, although the surface NF region eliminates the effects of the free surface, it applies an unwanted virtual stiffness to the system, which elastically constrains the deformation of the inner NF region. To counteract this effect, the unwanted virtual hardness should be compensated. Since the effects of surface in solids are controllable, to a large extent, by the inner volume, it is suggested to use the second approach. Of course, in places where the limitation of size exists (like the tip of a cone-shaped region), the use of the first approach is inevitable.

During the simulation, the average of k numbers of IV, for obtaining the displacement of the center of mass, is defined as $\vec{\delta}_{CM,k}^{MD}$, which, to get the statistical displacement vector $\vec{\delta}_{I,k}^{MD}$, is averaged along M time ranges of NF:

$$\vec{\delta}_{I,k}^{MD} = \left\langle \vec{\delta}_{CM,k}^{MD} \right\rangle_{Time} = \frac{1}{M} \sum_{j=1}^{M} \left(\vec{r}_{CM,k}(t_j) - \vec{r}_{CM,k}(0) \right) \qquad (4.43)$$

In the earlier relation, $\vec{r}_{CM,k}(t_j) = (1/N_k) \sum_{i=1}^{N_k} \vec{r}_i(t_j)$ is the center of mass of the kth IV, which has N_k atoms in the position $\vec{r}_i$ at time t_j of the jth NF time range.

In open systems, in order for the NF region to behave freely (at surface) or to be subjected to specific external force, some alterations should be made in the model. This is the philosophy behind the establishment of the UV region. In the best case, when the free movement of the surface is intended, the existence of the free surface produces unwanted effects in the NF system. This event, in cases where an external force is considered instead of the free movement, will be much worse. In addition to unwanted effects, since atoms at the free surface or close to it do not have a complete set of neighboring atoms around them, the coordination between the atoms falls apart. To reduce this lack of coordination, and to make the atoms stable in the interfacial NF region, an additional volume of atoms far from the center, which forms the surface NF region, is offered. On the other hand, although the surface NF region eliminates the effects of the free surface, it applies an unwanted virtual stiffness to the system, which elastically constrains the deformation of the inner NF region. Due to the particular complexity of the problem, in this report, a simple and, at the same time, effective procedure is presented for the calculation of virtual hardness.

4.5.5 Algorithm for Establishment of Coupling, Problems of Statics

In general, the coupling of MF and NF is accomplished through schemes based on the establishment of iterative equilibrium between these two regions. In these schemes, the iterations begin with the displacements of the MF and NF interface. These displacements are obtained as statistical average from the atomic positions of every IV and by averaging in the time duration of NF. These average displacements are then applied in the MF region, as displacement boundary conditions ($\vec{\delta}_I$). Then, the obtained MF boundary value problem is solved to yield the new interfacial reaction forces, i.e., $\vec{R}_I$. Then, these forces are applied to the atoms located in IVs, and so, the fixed-reaction boundary conditions are defined in the NF system. During the iterations in which MF is solved, the reaction boundary conditions are fixed, and they are applied to the NF region to guarantee the correct application of the elastic field from the MF domain. In solving the problems of statics, the iteration cycle of NF and MF continues until the system reaches a lasting equilibrium of displacements and forces between the continuum and atomic fields. While for the problems of dynamics, after the establishment of static equilibrium (using the aforementioned method), the system should be solved dynamically (generally via numerical methods). This issue constitutes one of the substantial complexities of the present work.

Several works have been presented on the statics of coupling models. Since the statics of this problem is a special case of its dynamics, the statics of the problem is not discussed.

It should be pointed out that in the mechanics of open systems, normally there are several systems in the NF and that in a certain time range, these systems may have no common region on any order, and in another time range, they may have common region on every order. Therefore, the problem will go through changes in time dimension, which are often significant. As a result, the statics of the system (despite the fact that it should be established prior to the analysis) is not so important, and only the dynamics of the system is important and should be investigated.

4.5.5.1 Problems of Damped Dynamics of MF

Here, through a unique algorithm, the manner of analysis of mechanics' dynamic problems (including the dynamics of nanomanipulation of AFM) will be presented.

By considering the dynamics in the MF model, dynamic continuum equations in the nth MF step at time t_n are given as follows:

$$\in P(t_n), \dot{P}(t_n), \ddot{P}(t_n), \ldots) = R_f(t_n) \tag{4.44}$$

where $\in$ is the term related to the equations extracted from the system's internal energy, which, in different methods, are functions of $P(t_n)s$ (the studied variables of the system) and of different orders of their derivatives. Also, at different times, function $R(t_n)$ is the function resulting from external loads applied to the system and proportionate to the orders of the system.

In problems that possess natural and forced boundary conditions and, at the same time, are supposed to be used in multiscale coupling models, the degrees of freedom should be divided into several groups. The first group includes the degrees of freedom that are governed by the natural boundary conditions. This group will be designated by F. The free boundary condition is the most usual condition of this group. The second group includes the degrees of freedom that are governed by the forced boundary conditions. This group will be designated by B. In many problems related to dynamics of solids, the clamped boundary condition can be regarded as a forced boundary. Also, in the area of fluid dynamics, the no-slip conditions at the surface can be mentioned. The third group is the degrees of freedom that are included in the inner points of the domain. This group will be designated by V. The fourth group is the degrees of freedom that are supposed to be coupled with the common degrees of freedom in MD. This group will be designated by I. The matrices related to the degrees of freedom of general equation (4.44) are reduced, based on these definitions. Therefore, $\in$ is broken down as $[\in_{\alpha\beta}]$, in which $\alpha, \beta = V, F, I, B$, V indicates the internal MF region, F is the far-field variables, I is the variables of the interface, and B is the variables of the MF boundary conditions. Using these definitions, the dynamic continuum equations in the nth step of MF at time t_n are expressed as follows:

$$\in_f(P(t_n), \dot{P}(t_n), \ddot{P}(t_n), \ldots) = R_f(t_n), \quad f = V, F, I, B \tag{4.45}$$

It should be noted here that $P(t_n)$ includes all the degrees of freedom in every set of equations. The general state of a multiscale problem includes the cases of external forces and the problem of initial value. First, the initial value problem, and then the external force, and the overall combined state are discussed.

4.5.5.1.1 Initial Value Problem

The general equations of motion of the macro-system's displacement, when no external forces exist, are as follows:

$$\mathcal{E}(P(t_n), \dot{P}(t_n), \ddot{P}(t_n), \ldots) = 0 \tag{4.46}$$

In any case, whether the problem pertains to the subject of solids or fluid dynamics, the general dynamic displacement vector could be expressed as follows:

$$P(t) = u(t) + g(t) \tag{4.47}$$

where $g(t)$ and $u(t)$ are the vector of initial degrees of freedom and the elastic vector of the whole system, respectively. By substituting the general dynamic displacement vector in the equation of motion and using the principle of superposition, we have

$$\mathcal{E}(u(t_n), \dot{u}(t_n), \ddot{u}(t_n), \ldots) = -\mathcal{E}(gt_n), \dot{g}(t_n), \ddot{g}(t_n), \ldots) \tag{4.48}$$

And thus, the initial value problem is converted into the external force problem. If the effects of orders higher than the second derivative of the degrees of freedom are disregarded, and the effects of the considered orders are assumed as linear (like many common methods, including the finite element), then the equations can be rewritten as follows:

$$\mathcal{E}_K(u(t_n)) + \mathcal{E}_D(\dot{u}(t_n)) + \mathcal{E}_A(\ddot{u}(t_n)) = -\mathcal{E}_K(g(t_n)) - \mathcal{E}_D(\dot{g}(t_n)) - \mathcal{E}_A(\ddot{g}(t_n)) \tag{4.49}$$

By arranging the total displacement vector, the state form of the equations can be expressed as

$$\frac{d}{dt}\vec{Q} = \begin{bmatrix} 0 & I \\ -\mathcal{E}_A^{-1}\mathcal{E}_K & -\mathcal{E}_A^{-1}\mathcal{E}_D \end{bmatrix}\vec{Q} - \begin{bmatrix} 0 \\ \mathcal{E}_A^{-1} \end{bmatrix}(\mathcal{E}_K(g(t_n)) + \mathcal{E}_D(\dot{g}(t_n)) + \mathcal{E}_A(\ddot{g}(t_n))) \tag{4.50}$$

where
$\vec{Q} = \{u_V, u_F, u_B, u_I, \dot{u}_V, \dot{u}_F, \dot{u}_B, \dot{u}_I\}^T$ is the state vector
$\mathcal{E}_K$, $\mathcal{E}_D$, and $\mathcal{E}_A$ matrices are the resolved zero to second-order terms of the general equation of motion

The solution of unknown elastic displacements within the MF region, i.e., $\{u_V, u_F, u_B, u_I\}$, can be obtained by solving the preceding state equation and by applying the initial condition IC = 0 in the first step and applying the condition IC = IC$_i$ in the ith step of the macro-solution. Then, the interfacial forces are obtained through the following relations:

$$R_I(t_n) = \mathcal{E}_I(P(t_n), \dot{P}(t_n), \ddot{P}(t_n), \ldots) \tag{4.51}$$

4.5.5.2 Problem of Damped Dynamics of NF

The dynamics of the atom i with mass $m^{(i)}$, at the position $r^{(i)}$, and in the NF regions are described through the Newton's equations of motion. Thus, for different regions belonging to the general NF, we have

$$m_i \ddot{\vec{r}}_i = \vec{f}_i + f_i^D \quad \text{In the inner NF region} \tag{4.52a}$$

$$m_i \ddot{\vec{r}}_i = \vec{f}_i + \frac{\vec{R}_I^k}{N_I^k} + f_i^D, \quad (\vec{r}_i, \dot{\vec{r}}_i)\Big|_{\text{first step}} = (\vec{P}_{\Gamma(i)}, \dot{\vec{P}}_{\Gamma(i)})\Big|_{\text{it's last step}} \quad \text{Interfacial NF (for the kth IV)} \tag{4.52b}$$

$$m_i \ddot{\vec{r}}_i = \vec{f}_i + \frac{\vec{f}_{cuv}^k}{N_{UV}^k} + \frac{\vec{f}_{Auv}^k}{N_{UV}^k} + f_i^D \quad \text{In the surface NF region} \left(\text{for the kth UV}\right) \tag{4.52c}$$

$$m_i \ddot{\vec{r}}_i = f_i + \frac{\vec{f}_{cf}^k}{N_{FB}^k} + f_i^D \quad \text{In the free NF region (for the kth FB)} \tag{4.52d}$$

$$m_i \ddot{\vec{r}}_i = f_i + \frac{\sum_{i=1}^{Atom_{RC}} \left(\vec{f}_c^k\right)_i}{N_{CB}^k} + f_i^D \quad \text{In the common NF region (for the kth CB)} \tag{4.52e}$$

The atoms in the inner NF region only experience the atomic force $\vec{f}_i = \sum_j \vec{f}_{ij}$ and the frictional forces of f_i^D, which result from their neighboring atoms. The atoms existing in the interfacial NF region (which belong to the kth IV) also experience an additional force $(\vec{R}_I^k)$ which is distributed among N_I^k atoms. In addition, the continuity of the fields of the zero- and first-order degrees of freedom should be guaranteed in it. The atoms existing in the surface NF region (which belong to the kth UV) also experience an opposing force $(\vec{f}_{cuv}^k)$ which is distributed among N_S^k atoms. Moreover, they tolerate the force of $\vec{f}_{Auv}^k$ due to the crushing of the system, which itself should be divided by N_S^k atoms. In the common boundaries of the NF region, some magnitude of force, which arises from the forces of atoms inside the cut-off radius of the contact surface of two NFs, should be considered. It should be mentioned that, except in the areas of direct contact between surfaces (where the effect of friction is modeled with the inclusion of some impacts), in the earlier equations, the viscous friction force f_i^D is uniformly applied to the atoms inside the IVs and UVs. Throughout the integration of the earlier equations for a period of $\Delta t_M = M\Delta t$ (where M is the number of time steps and Δt is the time step duration), the new average displacements are determined by Equation 4.43. The new atomic displacements for the next MF step at time $t_{n+1} = t_n + \Delta t_M$ are again applied in Equations 4.51 and 4.52a for the calculation of forces in the next iteration step. The complete algorithm for the simulation of coupling has been given in Figure 4.24.

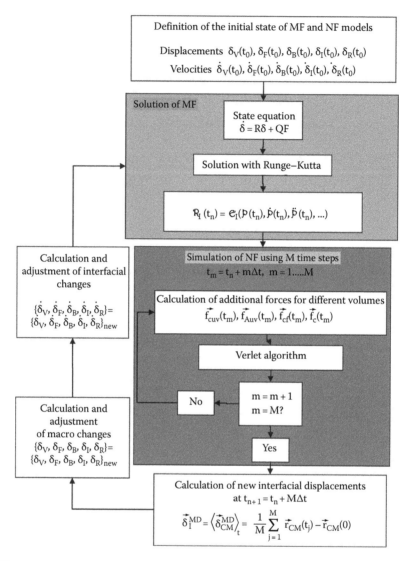

FIGURE 4.24
Dynamic analysis algorithm of MF and NF coupling.

In the following sections, present approach has been applied to the nanomanipulation scheme using the AFM.

4.6 Application to the Macroscale Nanorobots

In SPMs, especially AFM, a piezotube has been used to move the stage in space directly and a micro-cantilever to sense or actuate partially the stage. Figure 4.25 schematically shows the single nanoparticle manipulation scheme. Although the accurate dynamic model of macroscale nanorobots has not yet been discussed in full, several works may be cited.

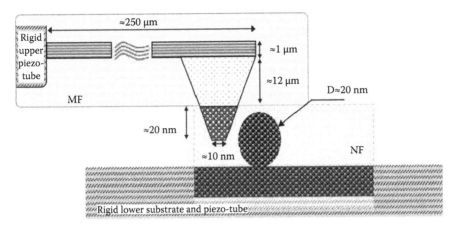

FIGURE 4.25
Various parts of MF and NF.

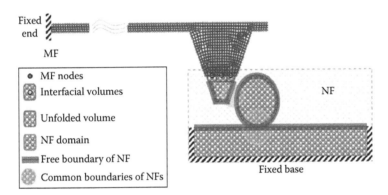

FIGURE 4.26
Common region between NF and MF in the nanomanipulation process as an open system.

In this regard, Park and Liu [47] have examined several issues in the area of multiscale problems related to solids. Here, problem is completely different. Complicated dynamic behavior of a multibody system in the multifield is considered. As depicted in Figure 4.26, the micro-cantilever, topside of the tip, and a large part of substrate belong to the MF, and particle, end of the tip, and a limited area of substrate near the particle belong to the NF. Since the dynamics of piezotubes is relatively rigid with regard to the dynamics of micro- and nano-fields, the piezotubes have been assumed rigid. Nonetheless, their behavior may be studied in a valuable work published by authors [48].

Firstly, presented method should be validated. Before this, a macro-base mechanical model was introduced by the first author and his colleagues in Refs [48,49], where presented model has been validated with experimental works. Here, a comparison with that work and a CGMD approach has been made with the purpose of validating present method. CGMD approach has been validated last by authors [49]. Figure 4.27 shows the comparison. For macro-part, 1000 quadratic finite element has been used. In the CGMD approach, a constant velocity same as the velocity of cantilever base is applied to the tip base. Therefore, some discrepancies are completely prospected. However, we claim that present multiscale method includes more dynamics of the system, and so it is more and more accurate than the CGMD approach.

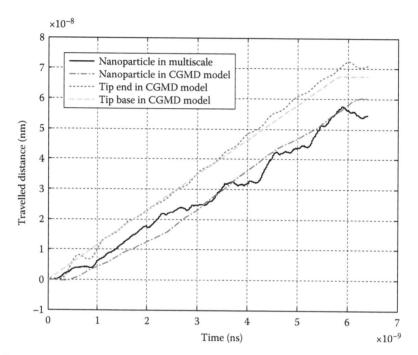

FIGURE 4.27
Comparison with a CGMD approach. (From Moradi, M. et al., *Micro Nano Lett.*, 5(5), 324, 2010.)

As it can be observed in Figure 4.27, there is relatively good agreement between the diagrams of the presented model and the results of the CGMD approach. In this figure, the considered multiscale model scheme and various states of system during the nanomanipulation have been sketched clearly.

Using the present procedure, some nanomanipulation behaviors could be studied, identified, and formulated based on a real view. Usually, the variation of physical properties (force, displacement, etc.) can be used for identification in small scales. In the following, three examples have been introduced for more clarification.

4.6.1 Example 1: Roughness on Substrate

To study the high sensitivity of the results to the unexpected material roughness, a standard nanomanipulation system with six configurations has been considered. In the first configuration (the simple case), the substrate surface is smooth, and in the other five cases (types 1–5), some small size particles are attached onto the substrate.

Various configurations of considered surface are depicted in Table 4.5. For all considered roughs, each particle is approximately 1 nm in diameter. Figure 4.28 shows the horizontally traveled distance by a nickel nano particle on a gold substrate. As can be observed in this figure, after the tip has contacted the particle (at nearly 0.4 ns), the diagram of the first case (smooth substrate) has shifted lower than the other diagrams. This is due to the fact that the considered roughness pulls the particle, and since the substrate material is relatively stiff, the roughness is digested by the substrate. Hence, after 2 ns, all the cases show relatively the same behavior.

TABLE 4.5

Configuration of Holes Used to Study the
Roughness Effects on the Nano Manipulation

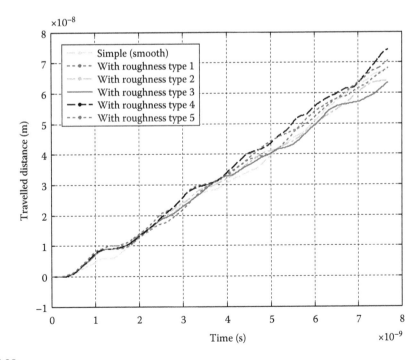

Type	Roughness Configuration	Hole Configuration
1		
2		
3		
4		
5		

FIGURE 4.28

Horizontally traveled distance by a nickel nanoparticle on a gold substrate with some predefined roughness (in the six noted configurations).

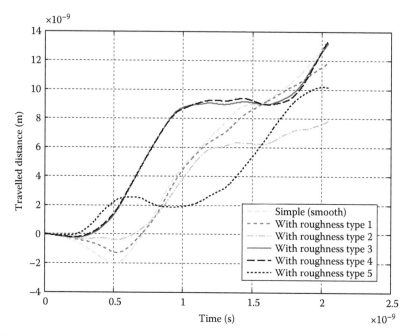

FIGURE 4.29
Horizontally traveled distance by a nickel nanoparticle on a Rhodium substrate with some predefined roughness (in the six noted configurations).

4.6.2 Example 2: Material of Substrate

Now, let us look at a harder material as the substrate. Figure 4.29 shows the horizontally traveled distance by a nickel nano particle on a Rhodium substrate, in the six noted configurations. In the time range of 0.4–0.6 ns, the nickel nanoparticle slides on the hard substrate toward the tip. Then, an unpredictable behavior is observed. A simple comparison between Figures 4.28 and 4.29 indicates that the substrate material is a more influential factor than was previously thought.

4.6.3 Example 3: Notches on Substrate

Due to chemical processes before the manipulation task, some notches may be created. Let us consider a system with the same configuration as in the previous roughness study, but with a hole instead of the roughness. In type 1, a hole is considered by taking out an atom from the substrate lattice (all removed particles are approximately 1 nm in diameter). Various configurations of considered surface have been depicted in Table 4.5. Figure 4.30 shows the horizontally traveled distance by a nickel nanoparticle on a gold substrate, in the six noted configurations. To illustrate the effect of substrate material, a diagram similar to Figure 4.29 may be depicted.

4.7 Some Challenges in the Macroscale Nanorobotic Science

As already mentioned, macroscale nanorobots are some advanced devices for manipulating objects in the nanoscale. Many of macroscale nanorobot (especially in SPMs) applications in nanoworld have been hampered in the past by the large nonlinearities encountered in

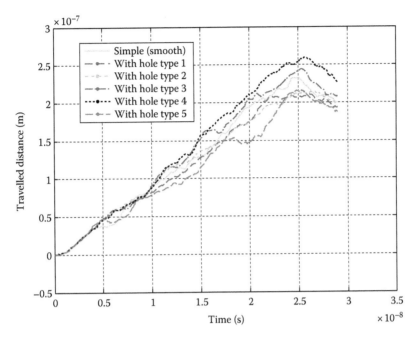

FIGURE 4.30
Horizontally traveled distance by a nickel nanoparticle on a gold substrate with some predefined indentations (in the six noted configurations).

tip steering. Although AFM is famous for its specific characteristics in nanofabrication and nanomanipulation, and other various applications, it has limitations that still exist in experimental results, too. These limitations are classified as (1) scanner limitations and (2) other limitations (nonlinearities from electric device and thermal field [drift]). Thermal drift can be compensated using some advanced compensators [50]. This classification clarifies the significance of scanner limitations, especially creep and hysteresis of scanners. Thus, these effects have been discussed here.

4.7.1 Scanner Limitations

Piezoelectric actuators can provide sub-nanometer displacement and achieve a high bandwidth. There have been extensive works in literature on dynamic behavior of piezoelectric actuators. Piezoelectric actuators come in different forms and shapes including tube actuators [5,51]. The tube actuator that is involved offers a compact design, three degrees of freedom motion, and low cost of construction. These desirable features have made tube actuators widely used in precision instruments such as SPMs.

Earliest study of AFM creep was performed by S. Vieira [52], and first model for AFM creep was introduced by Richter et al. [53]. They had modeled and confirmed creep as function of impressed voltage on input to the mechanical subsystem, reinforcing Vieira's earlier experimental results. After these studies, many researchers modeled the creep effect by superposition multiple set of linear spring–mass model, and further it was involved in hysteresis effect. A lot of studies about hysteresis refer to basic conception of hysteresis in general cases. One of the earliest and the most detailed analysis was introduced with Prandtle [54]. Then Frenc Preisach researches on ferromagnetism led to an alternative phenomenological description which is more general [55]. Some works used the Preisach model for hysteresis controlling and compensation [56–58]. Subsequently many works such as Tan in Ref. [59] proposed the use of new methods to adapt hysteresis model in real time. All of these works that are not inferred here have no well usability. Also, models were not in accordance with experimental evidence and were not analytic. Therefore, it is obvious that they are computationally intensive for reasonable accuracy, and their behavior is not unique to each device. Recently, authors developed a creep and hysteresis model that was generic, computationally efficient, and mathematically traceable such that it was applicable to random input functions such as undesired voltages, hysteresis loop with nonhomogeneous forms, asymmetric shapes, at existence of a thermal field, and other unusual conditions [60]. They also introduced a compensation algorithm that was very applicable and independent of the considered device, velocity of input, and variation of device characteristics. However, in that work and many others, the scanner has been modeled with some mass–spring–damper blocks, whereas the continuous behavior of system (i.e., usually as plate, hollow cylinder, etc.) needs to model with a CM-based model such as FEM. Here, a new compact formulation for hysteresis of a micro- or nano-actuator or generator has been developed. This formulation has been utilized in an FEM code, and some significant results have been introduced.

As already mentioned, the equation of motion for a finite element could be expressed as relation (4.17) that has repeated as follows:

$$[M_{uu}]_e \ddot{q}_e + [C_{uu}]_e \dot{q}_e + \left[K_{uu} - K_{u\phi}K_{\phi\phi}^{-1}K_{\phi u}\right]_e q_e = F_{qe} - \left[K_{u\phi}K_{\phi\phi}^{-1}\right]_e \left(F_{\phi e} + N_H\right) \quad (4.53)$$

For actuator case, the mechanical, electrical, and hysteresis forces have calculated. The mechanical force depends on the applied mechanical displacement or direct force, whereas two others are related to the applied voltage. Then, the problem can be solved using the general solution approaches for FEM-related problems. Based on the assumption of compensation of nonlinear behavior of micro-actuators in the nanomanipulation scheme, the nonlinear behavior has ignored in the coupling of MF and NF. Here, we return to the MF model. For compensation of nonlinearities, a comprehensive model is needed.

4.7.2 Hysteresis Behavior in the Micro-Actuators

Up to now, lumped models and FEMs have been used for nonlinear problems. In Ref. [61], the mass–spring approach has been used with feedback-linearized inverse feed forward for creep and hysteresis compensation. In Refs [62–64], using an estimator, transfer function of a macro-cantilever has been derived, and its nonlinearities have been modeled using two well-known approaches. In Ref. [65], using a developed model of Bouc–Wen approach, the piezotube hysteresis has been simulated based on a mass–spring model. There exist some valuable works that considered this problem experimentally [65,66]. Using the FEM for linear and nonlinear problem of hysteresis, some works have studied the hysteresis behavior using FEM [33,67,68].

Here, a self-contained model for hysteretic nonlinear dynamic of macro-, micro-, and nano-sized electromechanical systems has been introduced. Focusing on the macro- and nano-electromechanical systems (MEMS/NEMS), where necessarily, some applications do not relate to nano-actuating or sensing; some useful results have been achieved. The most significant application is manipulating the nanoworld, of course.

Here, the equations of motion and related boundary conditions subjected to the mechanical or electrical fields have been obtained by Hamilton principle, and then they are reshaped in the form of FEM equations. Exact solution of linear problem in some especial cases was compared with the presented approach, and we have seen them in good agreement. The present approach has the advantage of providing solutions for piezoactuators with some different types of boundary conditions. The considered problem is the same as the actuator of AFM that has been used in the nanomanipulation scheme.

4.7.3 Modeling of Nonlinearities

Many literatures presented and developed the hysteresis models. Each of them had appropriate application into one or more fields with some advantages and disadvantage. For example, the Bouc–Wen model [69] establishes an analytic relation between the applied voltage and polarization. Preisach [70] and Prandtl–Ishlinskii (PI) [71] models use a summation of numerical operators. Here, for more simplicity, the numerical PI method was used. PI operator is a subclass of the Preisach hysteresis model and employs combination of several rate-independent backlash or linear play operators as shown in Figure 4.31.

When input increases, $z_r(t) = \max\{z_r(t_i), u - r\}$, and when input decreases, $z_r(t) = \max\{z_r(t_i), u + r\}$. Thus, we can write

$$\begin{cases} z_r(t) = \max\left\{u(t) - r, \min\left\{z_r(t_i), u(t) + r\right\}\right\} \\ t_i < t \le t_{i+1} \end{cases} \tag{4.54}$$

Using a summation of these operators, the hysteretic behavior can be calculated as

$$y_H(k) = \sum_{i=1}^{N_H} w_i z_{r_i}(k) \tag{4.55}$$

where
N_H is order of hysteresis
$w_i > 0$ are weight coefficients for each operator

For calculation of electric forces, electric charge is

$$Q_3 = \frac{\varepsilon_{33}}{h} \alpha \varphi_A \tag{4.56}$$

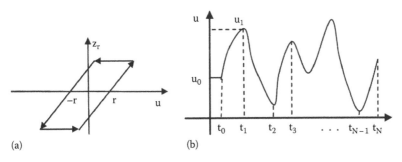

(a) (b)

FIGURE 4.31
PI operator (a) and input diagram in the ascendant and descendent intervals (b).

where φ_A and α are the applied voltage and the linear aspect ratio. It shows the direct effect level of applied voltage on the orientation of dipoles. It is equal to 1.0 for the linear system. In Equation 4.16, we can write the nonlinear term as

$$p = \frac{\varepsilon_{33}}{h} \varphi_h \tag{4.57}$$

where φ_h is the voltage equivalent to hysteresis that is determined with the PI model.

4.7.3.1 Results and Discussion

For more brevity, we present just some results here. In addition to the presented results, convergence study of responses, static and free-vibration results, and dynamic responses for various boundary conditions are significant. All of these are calculated and used for linear model validation. However, only dynamic response of a plate with simply supported boundary condition in all edges has been compared with the exact solution (using Green's function). These comparisons have been depicted in Figure 4.32.

Obviously, the presented method and exact responses are coincided together. This comparison builds confidence in the approach presented here for linear part. Nevertheless, comparison of nonlinear part remained yet. Figure 4.33 shows the comparison of an experimental work [62] with present approach for a $15 \times 2 \times 0.3\,\mathrm{mm}$ cantilever that composed of two layers, PI150 and cupper. It should be noted that piezo layer has excited with $40 \times \sin(2\pi t)$. This figure shows the accuracy of estimated hysteresis loop clearly.

Before anything, for clarification, dynamic response of a stepped micro-cantilever has been introduced. Figures 4.34 through 4.36 show the dynamic responses of a stepped micro-plate that its patch covered the half of plate in length direction perfectly. Mechanical

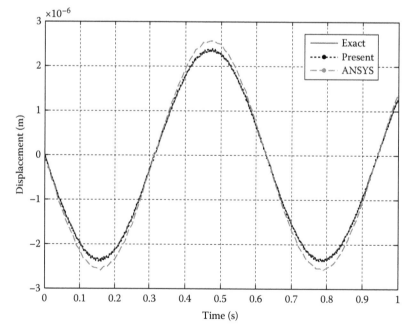

FIGURE 4.32
Dynamic response for plate; comparison between the exact and present models.

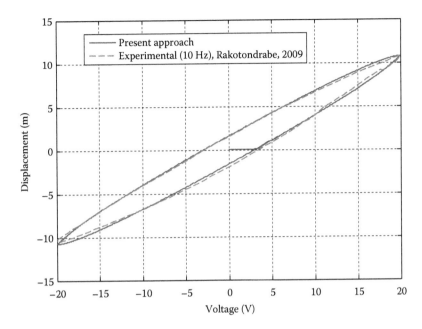

FIGURE 4.33
Comparison of an experimental work with present approach. (From Rakotondrabe, M. et al., Nonlinear modeling and estimation of force in a piezoelectric cantilever, *IEEE/ASME International Conference on Advanced Intelligent Mechatronics*, Zurich, Switzerland, September 4–7, 2007.)

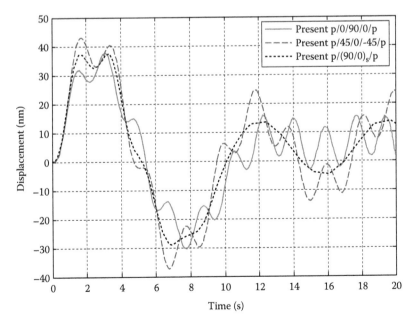

FIGURE 4.34
Dynamic response of stepped micro-plate due to the mechanical excitation.

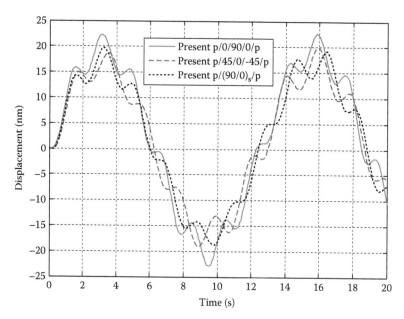

FIGURE 4.35
Dynamic response of stepped micro-plate due to the electrical excitation.

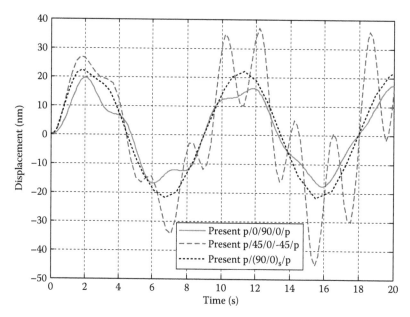

FIGURE 4.36
Dynamic response of stepped micro-plate due to the electromechanical excitation.

load is a sinusoidal line load as $0.1 \sin(7 \times 10^5 \ t)$ on the end edge of micro-plate. Electrical load is a voltage as $0.1 \sin(5 \times 10^5 \ t)$ applied at the upper surface of the patch.

Usually, in the nanorobotic systems, such as AFM devices, the micro-plate is excited close to its first natural frequency. So frequency of applied loads has been selected moderately high. Figure 4.37 shows the dynamic response of that system in linear and nonlinear manners.

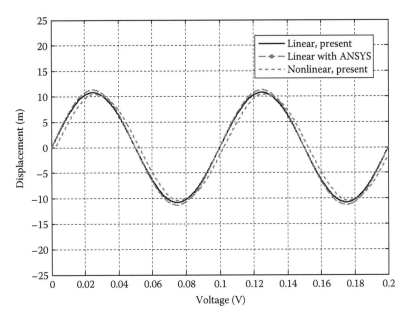

FIGURE 4.37
Comparison of dynamic response of linear and nonlinear problems due to a sinusoidal excitation.

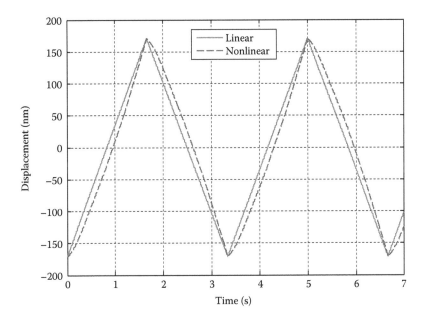

FIGURE 4.38
Comparison of dynamic response of linear and nonlinear problem due to a triangular excitation.

With these validations, the linear and nonlinear results have been guaranteed and can be used for MEMS and NEMS. Here, we focused on the MEMS. Let a standard micro-cantilever be used as an actuator in AFM. A triangular voltage is applied to a $3 \times 10 \times 250\,\mu m$ bimorph cantilever. Figure 4.38 shows the comparison of linear and nonlinear responses.

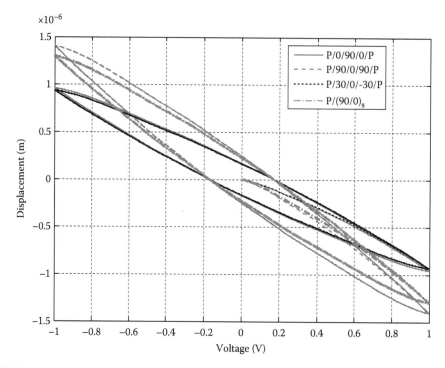

FIGURE 4.39
Comparison of hysteresis loops for various lamination of composite configuration.

An accurate look at Figure 4.38 represents the most significant property of this result. It can be seen that for some times, difference between linear and nonlinear responses exceeds 20nm. This means that if we want to manipulate a nanoparticle with 20nm in diameter, then giving a specific path, the particle may be lost, and the nanomanipulation will be failed. We expect that the lamination has no considerable effect on the shape of hysteresis loop. Figure 4.39 shows the comparison of hysteresis loops for various lamination of composite configuration.

Piezo layers are demonstrated with "p" and "s" means symmetry. It is truly viewed that stiffest (p/90°/0°/90°/p) and hardest (p/0°/90°/0°/p) configurations have biggest and smallest amplitudes of response, respectively. In this figure, no changes can be seen in the shape of loops. They are same in shapes, and lamination changes rotate the loop. This rotation is in clockwise direction with stiffness decreasing.

4.7.4 Hysteresis Behavior in the Nanogenerators

Energy production and then harvesting have been considered since ancient times. Advanced materials are a novel energy source that may be the dominant used energy of future. Energy generation utilizing piezoelectric materials is more considerable example of energy sources in advanced material field. It has been well studied over the past two decades [72,73]. In particular, because of their piezoelectric response, large aspect ratio, superior mechanical properties relative to bulk zinc oxide, ZnO NWs [72] have an especial potential application for nanodevices [74] and low production costs. The future of nanotechnology research is in building integrated nanosystems from individual components.

Piezotronic components based on ZnO NWs and nanobelts have several important advantages that will help make such integrated nanosystems possible.

For example, ZnO nanostructures can tolerate large amounts of deformation without damage, allowing their use in flexible electronics such as folding power sources. The large amount of deformation permits a large volume density of power output. ZnO materials are biocompatible, allowing their use in the body without toxic effects. The flexible polymer substrate used in nanogenerators would allow implanted devices to conform to internal structures in the body. Nanogenerators based on the structures could directly produce power for use in implantable systems. Thus, in comparison to conventional electronic components, the nano-piezotronic devices operate much differently and exhibit unique characteristics [72]. There are some reports on the generation of electric potential by bending ZnO NWs [73,75,76]; however, using piezoelectric nanomaterials for energy harvesting applications is still a matter of debate [73,77]. Multiscale modeling can facilitate the development of nanodevices that incorporate ZnO NWs by predicting the overall piezoelectric response as a function of structural geometry. Piezoelectricity of nanomaterials has been a subject of current research in the scientific community. There are reports on enhancement of piezoelectricity in nanomaterials compared to bulk materials [73]. Using FEM simulations followed by an experimental setup, it has been shown that electrical voltage can be generated due to mechanical bending of ZnO NWs [73]. Presence of Schottky diode between the nanomaterial and AFM tip used for deflecting the nanomaterial plays the main role in voltage generation [76].

With these descriptions, what will be the role of nonlinearities in the nano-sized piezoelectric devices? Really, it has not been studied yet. Here, a simple model for this purpose is developed based on the presented model. The objective of this study is to develop a multiphysics analytical model that predicts the nonlinearities of ZnO nano-cantilevers contacted with a relatively rigid body, as used in some experimental cases. One of the most applicable examples is nanomanipulation of nano-objects that have piezoelectricity behavior.

Modeling the nonlinearity behavior of electric potential generated by the nano-electrical generators would be a major step toward the design of self-powered MEMS/NEMS devices. As mentioned in Section 4.7, many of electromechanical system applications in nanoworld have been hampered in the past by the large nonlinearities. This included the NEMS, too.

Here, the hysteresis effects on the nano-sized piezo-driven systems are presented. The equations of motion and related boundary conditions subjected to the mechanical or electrical fields are obtained by Hamilton principle, and then they are solved using the FEM. Here, for considering the efficiency of nanobars and nanobeams, the generated voltage with respect to the applied load has been depicted in some figures. At the end, a hysteresis loop has been depicted that had been constructed from the generated voltage due to the applied lateral force.

Here, a nano-sized plate was considered as the general nanogenerator system. Figure 4.40 shows a double-layered plate reinforced with a 13×3 nanoplate array. Dimension of exciting and fixed plates may exceed even the micrometer sizes. It depends on the nature of the problem discussed. However, embedded nanoplates have nanoscale dimensions. We consider micro-excited and fixed plate ($10 \times 10 \times 1\,\mu m$) reinforced with a 13×3 nanoplate array, which have 200, 50, and 10 nm in length, width, and thickness, respectively.

It is worth mentioning that for considered systems and especially for nonlinear problem, singularity of matrices seems inevitable. Here, size of nanodevice is another problem. Thus, nondimensionalization is needed. However, some computational experience, effort, and more patients result in significant success.

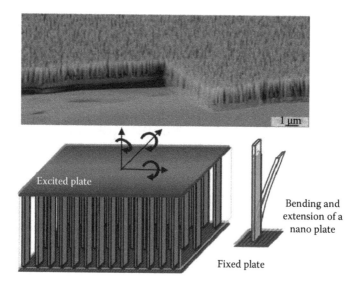

FIGURE 4.40
A double-layered plate reinforced with a 13×3 nanoplate array and a tested case. (From Zhiyong, F. and Jia, G.M., *J. Nanosci. Nanotechnol.*, 5(10), 1561, 2005.)

4.7.4.1 Results and Discussion

Results of this part are categorized in two set. First, consider the linear system. For generality, a beam model including the extension capability is used here. Firstly, consider the bar state, where the extension is dominant. It is assumed that the longitudinal direction has the stronger piezoelectricity. For considering the efficiency of nanobars and nanobeams, the generated voltage with respect to the applied load has been depicted in some figures. Figure 4.41 shows the generated voltage with respect to the applied longitudinal force in the end of nanobar with 200 nm in length, and 50 nm in width. It obviously shows that the generated voltage increases with applied force increasing. In addition, thickness increasing leads to the less voltage generation. It may be introduced from the less strain in the bar and thus less voltage generation.

Natural frequency of nanodevices is very high. In account of this, for dynamic response studies, we excite the bar with a load with 300 nm in amplitude and with 10 THz in frequency. Figure 4.42 shows the dynamic response of generated voltage in nanobar with respect to the time. Figure 4.43 shows the phase diagram of this response. Based on the Lyapunov approach of stability, presented diagram demonstrates the stability of generated voltage.

Figure 4.44 shows the generated voltage in a nanobeam with respect to the applied lateral force. Same as the bar case, it is obvious that the generated voltage increases with applied force increasing. In addition, thickness increasing leads to the less voltage generation. It may be introduced from the more mechanical stiffness in the beam and thus less voltage generation.

A simple comparison between Figures 4.41 and 4.44 shows that a bar with the dominant piezoelectricity in the longitudinal direction has more and more efficiency rather than the beam with same dimensions. Thus, in the following, the bar will be selected for nonlinearity prediction. Figure 4.45 compares the linear and nonlinear generated

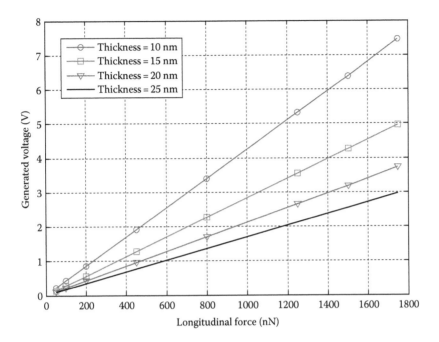

FIGURE 4.41
The generated voltage with respect to the applied longitudinal force in the end of nanobar.

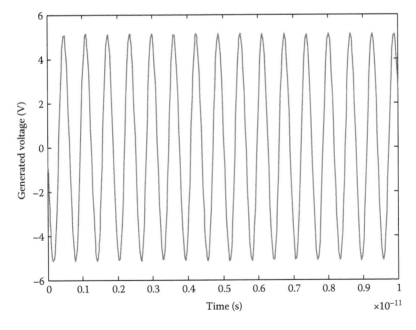

FIGURE 4.42
Dynamic response of generated voltage in nanobar with respect to the time.

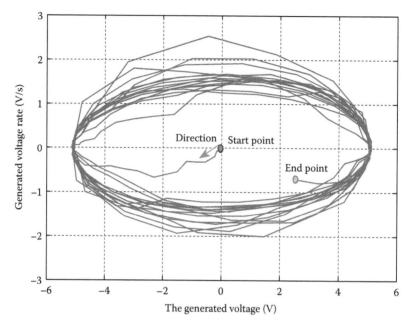

FIGURE 4.43
Phase diagram of generated voltage in nanobar.

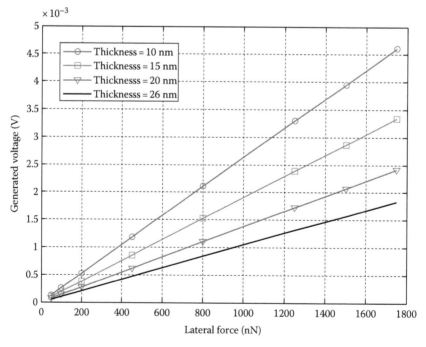

FIGURE 4.44
Generated voltage in a nanobeam with respect to the applied lateral force.

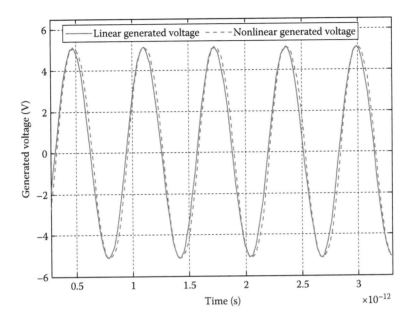

FIGURE 4.45
Comparison of the linear and nonlinear generated voltage in a nano-sized bar.

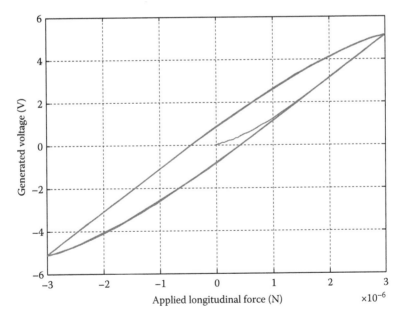

FIGURE 4.46
The hysteresis loop constructed from the generated voltage due to the applied longitudinal direction of a bar.

voltage in a dynamic manner. Discrepancies show the importance of nonlinear model for prediction of some noises in devices that use nano-sized piezo-electromechanical devices. High-frequency transistors, gates, and in future, nanochips are some examples of them. Figure 4.46 shows the hysteresis loop constructed from the generated voltage due to the applied longitudinal direction of a bar.

4.8 Conclusion

After some definitions, classifications, and descriptions for the nanorobotic science and related systems, the correct modeling approaches based on the CM have been found out. Overall, the conclusion of this chapter can be summarized in four parts as follows:

1. *Coupling of CM and MM*: As was previously pointed out, the CM-based methods that analyze different systems do not provide the correct physics in atomic domains, and on the other hand, to achieve the required precision in small dimensions, the methods based on MM and discrete solutions have to be applied. Here, a method has been presented for the coupling of the continuum and discrete environments together. In the present model, the undesirable effects of free surfaces, common surfaces, and surfaces close to the interface with the MF have been removed. To generalize the issue, the macro-/nano-related problems were divided into two groups of closed and open systems. Some applied examples have been modeled with the presented approach. With respect to the obtained results, it was demonstrated that the presented method can be applied for the simulation of systems with considerable dimensions and for relatively large time ranges. The use of CGMD in this method has made this capability possible. Also the use of FEM, and the presentation of an element which is capable of being applied for the electromechanical systems, has greatly expanded the range of application of this method.

2. *Nanomanipulation of nanoparticle with the CGMD*: Before using the CM for a nanoscale with dynamic manner, the macroscale nanorobot has been modeled with the CGMD to compare with the multiscale method.

   To show the failures of the manipulation process, some examples of these failures were presented, and it has been shown that nanocluster crushing was one of the most crucial failures which must be taken deeply into consideration in real manipulation systems. Therefore, studies were conducted on the effect of different parameters on the deformation of nanocluster to study the success of manipulation procedure. It was observed from the obtained results that by increasing the ε_p and ε_t (the parameters of SC potential for nanocluster and tip) coefficients, the cluster deformation decreases and increases, respectively. The obtained result is reasonable. Moreover, for the benefit of designers and manufacturers of nanomanipulator systems, a diagram was developed, in which the simultaneous effects of ε_p and ε_t parameters on the success of manipulation of nanocluster have been shown.

3. *Nonlinear behavior of MEMS*: By studying the linear and nonlinear behavior of micro-sized nano-piezoactuators, applicability of the obtained hysteresis loop in a nanomanipulation scheme has been described exactly. It has been shown that an expensive nanomanipulation scheme may be failed in the presence of hysteresis loops. In addition, the effects of the lamination on the hysteresis loop are investigated. We found that stiffest and hardest configurations have biggest and smallest amplitudes of response, respectively. In addition, no changes can be seen in the shape of loops. They are same in shapes, and lamination changes rotate the loop. This rotation is in clockwise direction with stiffness decreasing.

4. *Linear and nonlinear behavior of NMES*: In addition, by studying the linear and nonlinear behavior of nano-sized piezogenerators, considering the efficiency of nanobars and nanobeams, the generated voltage with respect to the applied load

have been depicted in some figures. It has been shown that the generated voltage increases with applied force increasing. Thickness increasing leads to the less voltage generation. It may be introduced from the less strain in the bar and thus less voltage generation. In addition, the dynamic response of generated voltage in nanobar presented with respect to the time. Stability of generated voltage is illustrated with the phase diagram of response. For the beam case, the same results have been seen. A simple comparison between bar and beam cases has shown that a bar with the dominant piezoelectricity in the longitudinal direction has more and more efficiency rather than the beam with same dimensions. The linear and nonlinear generated voltage is compared in a dynamic manner. Discrepancies show the significance of nonlinear model for prediction of some noises in devices that use nano-sized piezo-electromechanical devices. High-frequency transistors, gates, and in future, nanochips are some examples of them. At the end, a hysteresis loop has been depicted that had been constructed from the generated voltage due to the applied lateral force.

References

1. Wang J. 2009. Can man-made nanomachines compete with nature biomotors? *ACS Nano* 3(1): 4–9.
2. Hamdi M. and Ferreira A. 2011. *Design, Modeling and Characterization of Bio-Nanorobotic Systems*, Springer, New York.
3. Yang X., Vologodskii A. V., Liu B., Kemper B., and Seeman N. C. 1998. Torsional control of double-stranded DNA branch migration. *Biopolymers* 45: 69–83.
4. Balasubramanian S. et al. 2011. Micromachine-enabled capture and isolation of cancer cells in complex media. *Angew. Chem. Int. Ed.* 50: 4161–4164.
5. Wiesendanger R. 1994. *Scanning Probe Microscopy and Spectroscopy*, Cambridge University Press, Cambridge, U.K.
6. Zrenner E. et al. 2010. Subretinal electronic chips allow blind patients to read letters and combine them to words. *Proc. R. Soc. B* 278: 1489–1497.
7. Bharat B. (Ed.). 2010. *Springer Handbook of Nanotechnology*, 3rd edn., Springer Verlag, Heidelberg, Germany.
8. Tian X. et al. 2007. A study on theoretical nano forces in AFM based nanomanipulation. *Proceeding of the 2nd IEEE International Conference on Nano/Micro Engineered and Molecular Systems*, Bangkok, Thailand, pp. 4738–4744.
9. Israelachvili J. N. 1991. *Intermolecular and Surface Forces*, Academic Press, London, U.K.
10. Sitti M. and Hashimoto H. 2000. Controlled pushing of nanoparticles: Modeling and experiments. *Proc. IEEE/ASME Trans. Mechatron.* 5(2): 199–211.
11. Serafin J. M. and Gewirth A. A. 1997. Measurement of adhesion force to determine surface composition in an electrochemical environment. *J. Phys. Chem. B* 101: 10833–10838.
12. Tafazzoli A. and Sitti M. 2004. Dynamic modes of nanoparticle motion during nanoprobe-based manipulation. *Proceedings of the 4th IEEE Conference on Nanotechnology*, Munich, Germany, pp. 35–57.
13. Bhushan B. 2005. *Nanotribology and Nanomechanics: An Introduction*, Springer-Verlag, Heidelberg, Germany.
14. Kline S. J. 1986. *Similitude and Approximation Theory*, Springer-Verlag, New York.
15. Durkop T., Kim B. M., and Fuhrer M. S. 2004. Properties and applications of high-mobility semiconducting nanotubes. *J. Phys. Condens. Matter.* 16: R553–R580.

16. Jablonski A., Mrozek P., Gergely G., Menhyárd M., and Sulyok A. 1984. The inelastic mean free path of electrons in some semiconductor compounds and metals. *Surf. Interface Anal.* 6: 291–294.
17. Vancea J., Reiss G., and Hoffmann H. 1987. Mean-free-path concept in polycrystalline metals. *Phys. Rev. B* 35(12–15): 6435–6438.
18. Ericksen J. L. and Gurtin M. 1984. The cauchy-born hypothesis for crystals, in *Phase Transformations and Material Instabilities in Solids*, M. Gurtin (Ed.), Academic Press, New York, pp. 50–66.
19. Sauer R. A. and Li S. 2007. An atomistically enriched continuum model for nanoscale contact mechanics and its application to contact scaling. *J. Nanosci. Nanotechnol.* 8: 1–17.
20. Qian D. and Gondhalekar R. H. 2004. A virtual atom cluster approach to the mechanics of nanostructures. *Int. J. Multiscale Comput. Eng.* 2: 277–289.
21. Tadmor E. B., Ortiz M., and Phillips R. 1996. Quasicontinuum analysis of defects in solids. *Phil. Mag. A* 73: 1529–1563.
22. Shenoy V. B., Miller R., Tadmor E. B., Phillips R., and Ortiz M. 1998. Quasicontinuum models of interfacial structure and deformation. *Phys. Rev. Lett.* 80: 742–745.
23. Xiao S. P. and Belytschko T. 2004. A bridging domain method for coupling continua with molecular dynamics. *Comput. Meth. Appl. Mech. Eng.* 193: 1645–1669.
24. Liu W. K., Karpov E. G., Zhang S., and Park H. S. 2004. An introduction to computational nanomechanics and materials. *Comput. Meth. Appl. Mech. Eng.* 193: 1529–1578.
25. Liu W. K. 2006. Bridging scale methods for nanomechanics and materials. *Comput. Meth. Appl. Mech. Eng.* 195: 1407–1421.
26. Miller R. and Tadmor E. B. 2002. The quasicontinuum method: Overview, applications and current directions. *J. Comput. Aided Mater. Des.* 9: 203.
27. Shilkrot L. E., Miller R., and Curtin W. 2002. Coupled atomistic and discrete dislocation plasticity. *Phys. Rev. Lett.* 89: 025501.
28. Mahboobi S. H., Meghdari A., Jalili N., and Amiri F. 2009. Qualitative study of nanocluster positioning process: Planar molecular dynamics simulations. *Curr. Appl. Phys.* 9: 997–1004.
29. Mahboobi S. H., Meghdari A., Jalili N., and Amiri F. 2010. Qualitative study of nanoassembly process: 2-D molecular dynamics simulations. *Sci. Iranica* 17(Transaction F-1): 1–11.
30. Requicha A. G. 1999. *Nanorobotics: Handbook of Industrial Robotics*, 2nd edn., John Wiley & Sons, New York, pp. 199–210.
31. Kumar R., Mishra B. K., and Jain S. C. 2008. Thermally induced vibration control of cylindrical shell using piezoelectric sensor and actuator. *Int. J. Adv. Manuf. Technol.* 38: 551–562.
32. Kumar R., Mishra B. K., and Jain S. C. 2008. Static and dynamic analysis of smart cylindrical shell. *Finite Elem. Anal. Des.* 45: 13–24.
33. Ziya K. K. and Thomas J. R. 2008. Nonlinear modeling of composite plates with piezoceramic layers using finite element analysis. *J. Sound. Vib.* 315: 911–926.
34. Balamurugan V. and Narayanan S. 2008. A piezolaminated composite degenerated shell finite element for active control of structures with distributed piezosensors and actuators. *Smart Mater. Struct.* 17(035031): 18pp.
35. Hirth J. P. and Lothe J. 1982. *Theory of Dislocations*, John Wiley & Sons, New York.
36. Rapaport D. C. 2004. *The Art of Molecular Dynamics Simulation*, Cambridge University Press, Cambridge, U.K.
37. Allen M. P. and Tildesley D. J. 1987. *Computer Simulation of Liquids*, Clarendon Press, Oxford, U.K.
38. Sutton A. P. 1990. Long-range Finnis–Sinclair potentials. *Phil. Mag.* 61: 139–146.
39. Rafii-Tabar H. 2000. Modelling the nano-scale phenomena in condensed matter physics via computer-based numerical simulations. *Phys. Rep.* 325: 239–310.
40. Rudd R. E. 2001. Concurrent multiscale modeling of embedded nanomechanics. *Mater. Res. Soc. Symp. Proc.* 677(AA1.6): 1–12.
41. Marrink S. J., de Vries A. H., and Mark A. E. 2004. Coarse grained model for semiquantitative lipid simulations. *J. Phys. Chem. B* 108: 750–760.

42. Rudd R. E. and Broughton J. Q. 2005. Coarse-grained molecular dynamics: Nonlinear finite elements and finite temperature. *Phys. Rev. B* 72: 104–144.
43. Xiantao Li. 2010. A coarse-grained molecular dynamics model for crystalline solids. *Int. J. Numer. Methods Eng.* 83: 986–997.
44. Moradi M., Fereidon A. H., and Sadeghzadeh S. 2011. Dynamic modeling for nanomanipulation by AFM: Polystyrene nanorod case study. *Sci. Iranica Trans. F Nanotechnol.* 18(3): 808–815.
45. Mokaberi B. and Requicha A. A. G. 2004. Towards automatic nanomanipulation: Drift compensation in scanning probe microscopes. *Proceedings of IEEE International Conference on Robotics and Automation*, New Orleans, LA, pp. 416–421.
46. Korayem M. H., Rahneshin V., and Sadeghzadeh S. 2012. Coarse-grained molecular dynamics simulation of automatic nanomanipulation process: The effect of tip damage on the positioning errors. *Comp. Mat. Sci.* 60: 201–211.
47. Park H. S. and Liu W. K. 2004. An introduction and tutorial on multiple-scale analysis in solids. *Comput. Methods Appl. Mech. Eng.* 193: 1733–1774.
48. Korayem M. H., Sadeghzadeh S., and Homayooni A. 2011. Semi-analytical motion analysis of nano-steering devices, segmented piezotube scanners. *Int. J. Mech. Sci.* 53(7): 536–548.
49. Moradi M., Fereidon A. H., and Sadeghzadeh S. 2010. Aspect ratio and dimension effects on nanorod manipulation by atomic force microscope (AFM). *Micro Nano Lett.* 5(5): 324–327.
50. Korayem M. H. and Sadeghzadeh S. A. 2010. Comprehensive nanomechanical drift identification and prediction for nanoimaging and nanomanipulation by AFM. *Sci. Iranica Trans. F Nanotechnol.* 17(2): 133–147.
51. Mokaberi B. and Requicha A. A. G. 2008. Compensation of scanner creep and hysteresis for AFM nanomanipulation. *IEEE Trans. Autom. Sci. Eng.* 5(2): 197–206.
52. Vieria S. 1986. The behavior and characterization of some piezoelectric ceramics used in the STM. *IBM J. Res. Dev.* 30: 553.
53. Richter H., Misawa E. A., Lucca D. A., and Lu H. 2001. Modeling nonlinear behavior in a piezoelectric actuator. *J. Int. Soc. Precis. Eng. Nanotechnol.* 25(1): 128–137.
54. Prandtle L. 1924. Ein Gedankenmodell zur kinetischen plastischen korpern. *Proceedings of First International Congress on Applied Mechanics*, Delft, the Netherlands, pp. 43–54.
55. Preisach F. 1935. Under die magnetische nachwirkung. *Z. Phys.* 94: 277–302.
56. Liu H., Lu B., Ding Y., Tang Y., and Li D. 2003. A motor-piezo actuator for nano-scale positioning based dual servo loop and nonlinearity compensation. *J. Micromech. Microeng.* 13: 295–299.
57. Choi S. B., Han S. S., and Lee Y. S. 2005. Fine motion control of a moving stage using a piezoactuator associated with a displacement amplifier. *Smart Mater. Struct.* 14(1): 222–230.
58. Ge P. and Jouaneh M. 1996. Tracking control of a piezoceramic actuator. *IEEE Trans. Control Syst. Technol.* 4(3): 209–216.
59. Li C.-T. and Tan Y.-H. 2005. Adaptive output feedback control of systems preceded by the preisach-type hysteresis. *IEEE Trans. Syst. Man. Cybern. B: Cybern.* 35(1): 130.
60. Korayem M. H. and Sadeghzadeh S. A new modeling and compensation approach for creep and hysteretic loops in nanosteering by SPM's piezotubes. *Int. J. Adv. Manuf. Technol.* 44(11–12): 1133–1143.
61. Leang K. K. and Devasia S. 2007. Feedback-linearized inverse feedforward for creep, hysteresis, and vibration compensation in AFM piezoactuators. *IEEE Trans. Control Syst. Technol.* 15(5): 927–935.
62. Rakotondrabe M., Yassine H., and Philippe L. 2007. Nonlinear modeling and estimation of force in a piezoelectric cantilever. *IEEE/ASME International Conference on Advanced Intelligent Mechatronics*, Zurich, Switzerland, September 4–7, pp. 1–6.
63. Rakotondrabe M. and Clevy C. L. 2008. Hysteresis and vibration compensation in a nonlinear unimorph piezocantilever. *IEEE/RSJ International Conference on Intelligent Robots and Systems*, Acropolis Convention Center, Nice, France, September 22–26, 2008.
64. Rakotondrabe M. and Clevy C. L. 2010. Complete open loop control of hysteretic, creeped, and oscillating piezoelectric cantilevers. *IEEE Trans. Autom. Sci. Eng.* 7(3): 440–450.

65. Mokaberi B. and Requicha A. G. G. 2008. Compensation of scanner creep and hysteresis for AFM nano-manipulation. *IEEE Trans. Autom. Sci. Eng.* 5(2): 197–206.
66. Leang K. K. and Devasia S. 2006. Design of hysteresis-compensating iterative learning control for piezo-positioners: Application to atomic force microscopes. *Mechatronics* 16: 141–158.
67. Gaul L. and Becker J. 2009. Model-based piezoelectric hysteresis and creep compensation for highly-dynamic feedforward rest-to-rest motion control of piezoelectrically actuated flexible structures. *Int. J. Eng. Sci.* 47(11–12): 1193–1207.
68. Chung S. H. and Fung E. H. K. 2010. A nonlinear finite element model of a piezoelectric tube actuator with hysteresis and creep. *Smart Mater. Struct.* 19: 19.
69. Ikhouane F., Mañosa V., and Rodellar J. 2007. Dynamic properties of the hysteretic Bouc-Wen model. *Syst. Control Lett.* 56: 197–205.
70. Ge P. and Jouaneht M. 1997. Generalized preisach model for hysteresis nonlinearity of piezoceramic actuators. *Precis. Eng.* 20: 99–111.
71. Krejci P. and Kuhnen. 2001. Inverse control of systems with hysteresis and creep. *IEEE Proc. Control Theory Appl.* 148: 185–192.
72. Zhou J., Fei P., Gao Y., Gu Y., Liu J., Bao G., and Wang Z. L. 2008. Mechanical-electrical triggers and sensors using piezoelectric micowires/nanowires. *Nano Lett.* 8: 2725.
73. Wang Z. L. and Song J. 2006. Piezoelectric nanogenerators based on zinc oxide nanowire arrays. *Science* 312(5771): 242–246.
74. Wen B., Sader J. E., and Boland J. J. 2008. Mechanical properties of ZnO nanowire. *Phys. Rev. Lett.* 101(17): 175502.
75. Wang X. D., Song J. H., Liu J., and Wang Z. L. 2007. Direct-current nanogenerator driven by ultrasonic waves. *Science* 316: 102–105.
76. Wang Z. L. 2008. Towards self-powered nanosystems: From nanogenerators to nanopiezotronics. *Adv. Funct. Mater.* 18: 3553–3567.
77. Alexe M., Senz S., Schubert M. A., Hesse D., and Gösele U. 2008. Energy harvesting using nanowires? *Adv. Mater.* 20: 4021–4026.
78. Zhiyong F. and Jia G. M. 2005. Zinc oxide nanostructures: Synthesis and properties. *J. Nanosci. Nanotechnol.* 5(10): 1561–1573.

5

Simulations of Dislocations and Coherent Nanostructures

Anandh Subramaniam

Indian Institute of Technology, Kanpur, India

Arun Kumar

Indian Institute of Technology, Kanpur, India

CONTENTS

5.1 Introduction 150
5.2 Microstructure, Finite Element Analysis, and Simulations at the Nanoscale 151
    5.2.1 Microstructure 151
    5.2.2 Finite Element Analysis of Structures, Processes, and Phenomena 151
    5.2.3 Finite Element Techniques for Nanoscale Problems 152
5.3 Computational and Theoretical Background for the Case Studies 152
    5.3.1 Basic Considerations 152
        5.3.1.1 Mesh 152
        5.3.1.2 Boundary Conditions 152
        5.3.1.3 Choice of Domain 153
    5.3.2 Eigenstrains 153
    5.3.3 Edge Dislocations in Finite Domains 154
        5.3.3.1 Image Forces 155
    5.3.4 Heteroepitaxy 158
    5.3.5 Coherent Precipitates 160
5.4 Case Studies 162
    5.4.1 Edge Dislocations in Nanocrystals and Hybrids 163
        5.4.1.1 Image Forces 165
    5.4.2 Neutral Equilibrium and Stability of Edge Dislocations 166
        5.4.2.1 Dislocation-Free Nanocrystals 169
    5.4.3 Heteroepitaxial Structures 170
        5.4.3.1 Growth of Thin Films and Critical Thickness 170
        5.4.3.2 Thin Substrates, Stripes, and Islands 173
    5.4.4 Precipitation 176
        5.4.4.1 Coherent to Semicoherent Transition of Precipitates 177
        5.4.4.2 Precipitation in Nanocrystals 178
    5.4.5 Improvements 180
5.5 Conclusions (and Scope for the Future) 181
Appendix 5A: Material Properties Used in the Simulations 181
References 182

5.1 Introduction

Finite element method (FEM) has proved to be an invaluable tool not only in engineering analysis but also an indispensable technique for discovery of new mechanisms and phenomena across various disciplines of engineering and sciences. FEM can model phenomena across length scales starting from the nanoscale to the scale of the component, thus bridging the length scales in a multiscale study [1–3]. FEM can also couple with other techniques like molecular dynamics (MD), dislocation dynamics (DD), kinetic Monte Carlo (KMC), etc. The results of these techniques (MD, DD, etc.) can feed into an FEM model in a hierarchical manner [4–7].

Over the years, standard finite element software (Abaqus, Ansys, COMSOL multiphysics, etc. [8]) have attained a high degree of "maturity" and have become capable of handling a variety of complex problems. These software are also amenable to some degree of customization via subroutines. Given this, FEM can now be used by a wider range of researchers to uncover fundamental aspects of the science of materials. FEM can be used not only as an approximate solver for analytically intractable/complex problems but also to simulate structures and processes, with a view to uncover new phenomena.

The usual advantages of FEM are retained in the studies at the nanoscale as well. These include the ability to handle (1) complex geometry; (2) complex distributions of materials; and (3) varying distribution of loads, boundary conditions, and body forces.

The instructional level of the chapter is pitched so that a researcher/student entering the field can easily comprehend the power of FEM in solving problems at the nanoscale. Stress will be on physical aspects of the problems, and the reader can refer to standard texts on FEM (the authors have found Burnett's book a good one for beginners [9]) to delve deeper into the numerical aspects and their implementation using software. The reader may additionally refer to the books by Cook et al. [10], B. Szabó and I. Babuška [11], Logan [12], and Madenci and Guven [13].

This chapter aims at introducing a researcher/student to the utility of FEM in solving selected problems related to mechanics of materials problems at the nanoscale. Focus will be on the mechanics of crystalline structures, interfaces, and crystal defects. Two important objectives of the chapter will be (1) to illustrate the power of FEM in simplifying analytically complex problems and (2) to discover new criteria and phenomenon. From a student's perspective, it will be seen that a variety of problems can be solved using standard software, and hence, the gestation period involving coding can be reduced. After an overview of the literature in the area of simulation of structures and material processes at the nanoscale, case studies will be used to highlight the utility of finite element methodologies in simplifying theoretically complicated problems of materials science. Case studies will include (1) simulation of processes and (2) computation of useful parameters from the simulations. The simulations will highlight many effects uniquely occurring at the nanoscale. Processes which will be considered include (1) displacement of an edge dislocation in thin plates, (2) simulation of the growth of heteroepitaxial films, and (3) the growth of coherent precipitates. Parameters computed from the simulations will encompass (1) region of stability of a dislocation in a thin plate; (2) range of neutral equilibrium; (3) image force experienced by the dislocation; (4) stress state, energy, and bent configuration of the domain; (5) size of the nanocrystal at which it can become spontaneously edge dislocation free; (6) critical thickness for the energetic feasibility of an interfacial misfit edge dislocation; (7) critical size for an epitaxial islands; and (8) critical size of the precipitate for coherent to semicoherent transition.

5.2 Microstructure, Finite Element Analysis, and Simulations at the Nanoscale

FEM is a versatile tool, which can be used for the study of structures and processes in the macro-, micro-, and nanoscales. However, when FEM is applied for solving nanoscale problems, the limitations of the technique have to be clearly understood and delineated, along with the advantages of using the same. Whenever the accuracy of an FEM model is limited or its applicability is questionable, it is expected that the shortcomings are listed and necessity for use of the model clearly explained.

Given the vast amount of literature available on most of topics covered in this chapter, effort has been made to restrict the references to important books and review articles, wherever possible.

5.2.1 Microstructure

The term microstructure in the usual sense implies structure seen under magnification (revealing micron-sized features). Hence, structures seen at other length scales have to be referred as macrostructure, nanostructure, etc. The most important use of the term microstructure is its correlation with properties of the material (i.e., structure–property correlation). From this point of view, a functional definition of microstructure is [14]

Microstructure = (phases + defects + residual stress) and their distributions

These entities of the microstructure can exist across length scales (e.g., the scale of the vacancy is the atomic scale, the scale of the GP zones may be in the nanometer scale, while the scale of the grains may be in microns). Further, the role of microstructure in determining the behavior and properties of a material has to be comprehended [15]. The first step toward this grand goal is to study individual entities of the microstructure and their interactions with each other.

5.2.2 Finite Element Analysis of Structures, Processes, and Phenomena

FEM has proved to be an invaluable tool for the study of various structures, processes, and phenomena. Traditionally, most of the structures studied using FEM were typically large-scale structures like vehicles, bridges, buildings, etc. The deformation and failure of these structures under various types of loading (e.g., wind loading on tall building, vehicular traffic on bridges, deformations due to a car crash, etc.) were the primary concern in these studies.

Many manufacturing processes like rolling, extrusion, machining, welding, etc., have been simulated using FEM [16,17]. Various phenomena like elastic and plastic deformation, twinning, fracture, fatigue, and creep have been modeled successfully using FEM [18]. FEM is an ideal tool for studying not only monolithic materials but also complex hybrids [19].

The term structure has to be differentiated with respect to the term material. The use of the term material implies that no specific geometry is being considered. On the other hand, when the term structure is used, a specific geometry is kept in focus. Copper is a material, while a plate of Cu with a given length, breadth, and thickness is a structure. Young's modulus is a material property, while stiffness is property of a structure (e.g., a spring).

5.2.3 Finite Element Techniques for Nanoscale Problems

In the context of nanoscale materials, structures and defects studied include voids, cracks, dislocations, twins, interfaces (surface, grain boundary, twin boundary, interphase boundary), precipitates, and epitaxial systems (films, islands, groves, etc.).

A variety of processes have also been investigated using FEM. These include compression of nanopillars [20], deformation of nanotubes [21–23], and nanoindentation [24]. Continuum models of nanostructured materials have also been developed [25]. Some examples of phase transformations studied are precipitation [26,27], martensitic transformation [28], and magnetic transformations [29]. FEM has also been used in conjunction with other techniques like molecular statics and dynamics [30], phase field method [31], DDs [32], and KMC methods [33]. Atomic-scale FEMs have also been developed [34].

Few important questions have to be addressed in finite element simulations at the nanoscale: (1) At what length scale is the approximation of a continuum valid? (2) What properties of the material should be used? (i.e., will it be bulk properties or will confinement and surface effects on properties have to be included?) (3) How much error is introduced by assuming bulk properties? These questions have been answered to some extent in literature, but much more work is required for specific cases.

5.3 Computational and Theoretical Background for the Case Studies

5.3.1 Basic Considerations

As in any finite element study, some of the basic tenets of the method should be paid attention to, and none of the cardinal principles should be violated, if the simulations have to yield correct results. We shall briefly list these here, without going into the details.

5.3.1.1 Mesh

Mesh size (h) and order of the mesh (p) refinement have to be carried out to make sure that the results are meshing invariant. A fine uniform mesh costs in terms of the computation time, but in the nanoscale, often one can get away with this kind of meshing, as the number of nodes may still be small in number. If there is a constraint of the size of the mesh arising from the kind of simulation being carried out (e.g., in some of the case studies to be considered, the mesh size has to be of the magnitude of the Burgers vector, at least in one dimension), then further "h" refinement may not be possible, and the validity of the results may be cross-checked by other means (theory, simulations using other techniques or experiments).

5.3.1.2 Boundary Conditions

There are three origins of boundary conditions: (1) those physically imposed in the problem, (2) those arising from symmetry, and (3) those that are required for a unique numerical solution (e.g., one node locked in "x," "y," and "z" degrees of freedom). It should be noted that the symmetry has to be complete (i.e., with respect to domain shape, material properties, loading, etc.), if symmetry boundary conditions have to be imposed. The category (3) boundary conditions have to be chosen carefully so as to (i) not violate the physics of the problem (i.e., not impose a constraint that is physically incorrect) and (ii) make the solutions numerically sound (i.e., should be close to the region where the gradients in the relevant physical parameters are large, as "errors will tend to propagate" from these invariant points). In the case studies we consider in this chapter, essential boundary conditions have been used (i.e., those specifying displacements). As it will be seen, these are often the bare minimum required to determine a unique solution.

5.3.1.3 Choice of Domain

Two important factors that have to be considered in the choice of the domain should be such that it is (1) able to capture the essential physics of the problem and (2) easy for comparisons of the results obtained with theory, other computations, and experimental results. For example, if one is interested in simulating the stress field of a straight edge dislocation in an infinite domain (and make comparisons with theoretical formulae), then it is clear that an infinite domain cannot be constructed. Keeping in view the earlier-given two factors, the next best choice would be to (1) use a circular domain under plane strain conditions (this makes it easy for comparison with theoretical results), (2) make the radius of the domain much larger than the Burgers vector, (3) position the dislocation at the center of the domain (i.e., symmetrically), and (4) compare results close to the dislocation line (ignoring regions close to the surface). Additionally, the validity of the simulation maybe checked using the formulae for stress fields in a finite cylindrical domain. On the other hand, if one is interested in the calculation of the image force (refer Section 5.3.3) and its comparison with theoretical formulae, then a square or rectangular domain is preferred as this would clearly demarcate the surface toward which the force is being calculated.

Another point worthy of mention here is regarding the use of 2D simulations. As 3D simulations are computationally costly, they should be resorted to only when the problem requires a true 3D construct. Wherever possible, qualitative features of the problem should be gathered from 2D simulations before a 3D simulation is attempted.

5.3.2 Eigenstrains

The importance of eigenstrains in the mechanics of solids can be comprehended from the fact that Professor Toshio Mura [35] starts his book on "micromechanics of defects in solids," with the theory of eigenstrains. The term has been used to describe nonelastic strains associated with thermal expansion, phase transformations, inclusions, etc. An equivalent term to eigenstrains is "stress-free strains" (the term residual strain has also been used in some contexts). It is important to note that there can be strains without stresses (e.g., heating a rod, which is free to expand) and there can be stresses without net strains (e.g., heating a thick rod constrained between two rigid walls). The corresponding term (to eigenstrains) referring to stresses is "eigenstress." Eigenstress is the self-equilibrated internal stresses (residual stress) resulting from eigenstrains in a free-standing body (i.e., under no external forces or constraints). Needless to say, eigenstrains imposed on the entire body will not result in any stresses. Figure 5.1 shows the eigenstrains arising from heating ($\varepsilon_{ij}^* = \delta_{ij} \alpha \Delta T$, where δ_{ij} is the Kronecker delta, α is the coefficient of thermal expansion, and ΔT is the change in temperature). In Figure 5.1a, a free-standing body is heated, which results in no stresses. In Figure 5.1b, material A has a higher "α" than material B, and hence, the constraint of the surrounding medium results in eigenstresses. The strain in this case (Figure 5.1b) is the resultant of thermal and elastic strains.

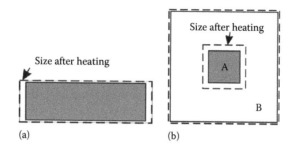

FIGURE 5.1
Schematics (exaggerated) showing heating of (a) a free-standing body and (b) a body with two materials (A with a higher coefficient of thermal expansion as compared to B).

Under isotropic plane stress conditions, element stresses due to a change in temperature ΔT are given by [10]

$$\left\{\begin{array}{c} \sigma_x \\ \sigma_y \\ \tau_{xy} \end{array}\right\} = \frac{E}{(1-v^2)} \begin{bmatrix} 1 & v & 0 \\ v & 1 & 0 \\ 0 & 0 & (1-v)/2 \end{bmatrix} \left([\mathbf{B}]\{d\} - \left\{\begin{array}{c} \alpha\,\Delta T \\ \alpha\,\Delta T \\ 0 \end{array}\right\} \right)$$

where
   Matrix $[\mathbf{B}]$ is the strain-displacement matrix ($\{\varepsilon\} = [\mathbf{B}]\{d\}$)
   $\{d\}$ is the nodal displacement degrees of freedom of an element

The coefficient of thermal expansion is assumed to be constant over the temperature range. Matrix $[\mathbf{B}]$ is evaluated at the coordinates of the element, where stresses are to be calculated. The reader may consult standard finite element texts for further details regarding the implementation of eigenstrains in a finite element code.

5.3.3 Edge Dislocations in Finite Domains

A vast amount of literature is available regarding theoretical and computational aspects of dislocations. For all aspects related to dislocations, the reader may refer to the book by Hirth and Lothe [36]. The book *Computer Simulations of Dislocations* by Bulatov and Cai [37] covers wide-ranging aspects of simulations of dislocations, including atomistic methods, KMC methods, and phase field methods.

   The residual stress state in a free-standing body satisfies the following condition: $\int_V \sigma_{ij}dV = 0$ (wherein, V is the volume of the domain). This implies that if there are regions with tensile residual stresses, then there must be regions of compressive residual stresses. Edge dislocations are associated with both dilatational and shear strains, while screw dislocations have pure shear fields. This implies that screw dislocations will not interact with other defects in the material having pure dilatational fields (e.g., some substitutional atoms in a crystal). The state of stress (σ_{xx} and τ_{xy}) in the presence of an edge dislocation at the center of a *finite cylindrical domain* is given by [36]

$$\sigma_x = -\frac{Eb}{4\pi(1-v^2)}\frac{y}{(x^2+y^2)^2}\left\{ 3\left(1-\frac{r^2}{r_2^2}\right)x^2 + \left(1-\frac{3r^2}{r_2^2}\right)y^2 \right\} \tag{5.1}$$

$$\tau_{xy} = -\frac{Eb}{4\pi(1-v^2)} \frac{x}{(x^2+y^2)^2} \left\{ -\left(1-\frac{r^2}{r_2^2}\right)x^2 + \left(1+\frac{r^2}{r_2^2}\right)y^2 \right\} \tag{5.2}$$

where
  E is the modulus of elasticity
  r_2 is the radius of circular domain
  v is the Poisson's ratio
  b is the modulus of the Burgers vector

The results of edge dislocation in infinite homogeneous media are obtained by letting r_2 approaches infinity.

The plot of Equations 5.1 and 5.2 is given in Figure 5.2. The following points should be noted: (1) the plots show left-right mirror symmetry and top-bottom "inversion" mirror symmetry (i.e., tensile stresses are reflected as compressive stresses); (2) in an infinite domain, the entire top half-space would be tensile, and the bottom half-space (with the "extra" half-plane) would be compressive, while in the finite domain, this is not the case; (3) the dislocation can interact with externally applied forces and with other internal stresses in the material; (4) the formulae are derived from the linear theory of elasticity and break down near the dislocation line (called the core region of the dislocation).

The energy of the dislocation per unit length (E_{dl}) is approximately given by

$$E_{dl} \approx \frac{Gb^2}{4\pi(1-v)}\left(2 + \ln\frac{R}{r_o}\right) \tag{5.3}$$

where
  R is the size of the domain
  r_o is the radius of the core of dislocation
  G is the shear modulus of the material

The first term in the brackets arises from the core energy term. It is difficult to compute the core energy, and hence the formula is an approximate one (the core energy is typically about 10% of the total energy of the dislocation). R in the equation is the "effective radius" of the dislocation (the energy would become infinite in an infinite domain).

5.3.3.1 Image Forces

A dislocation in a semi-infinite domain (and off-center in finite domains) feels a force toward a free surface, called the image force. Image force is a configurational force, which is computed using an image dislocation of opposite sign on the other side of the free surface, as shown in Figure 5.3a, for a semi-infinite domain. The magnitude of the force decays asymptotically as given by [38]

$$F_{image} = \frac{-Gb^2}{4\pi(1-v)d} \tag{5.4}$$

where
  F_{image} is the image force experienced by the dislocation (N/m)
  d is the distance of the dislocation from a free surface

It is clear that image forces become significant only when the dislocation is positioned closed to a surface.

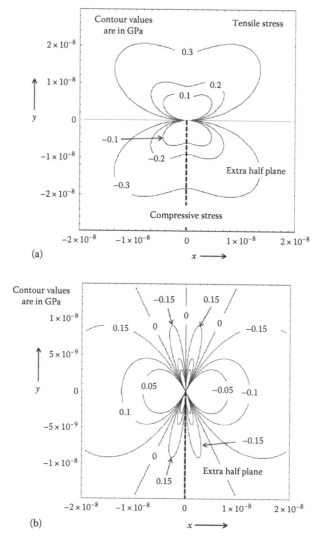

FIGURE 5.2

Contour plot of (a) σ_{xx} and (b) τ_{xy} stress fields, as computed from Equations 5.1 and 5.2. The "extra half-plane" is schematically overlaid on the plots.

In the presence of multiple proximal free surfaces, the net force is the resultant of forces arising from the superposition of many images (Figure 5.3b shows two images along the "x" direction). The resultant image force can be computed as a superposition of the individual image forces as [39]

$$F_{image} = \frac{-Gb^2}{4\pi(1-\nu)}\left[\frac{1}{d} - \frac{1}{L-d}\right] = \frac{-Gb^2}{\pi(1-\nu)}\left[\frac{2x}{L^2 - 4x^2}\right] \qquad (5.5)$$

where
  d is the distance of the dislocation from a nearest free surface
  x is the distance of the dislocation from the midplane of the domain (as shown in Figure 5.3b).

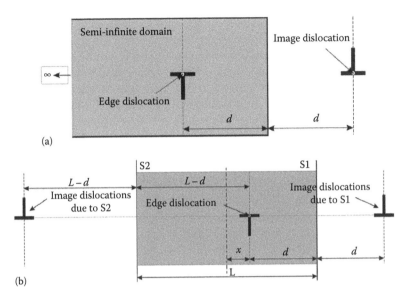

(a)

(b)

FIGURE 5.3
Use of "image dislocation(s)" for the computation of the attractive force toward a free surface(s) in (a) a semi-infinite domain and (b) finite body.

It should be noted that for an edge dislocation, the image dislocation does not annul the shear stresses (τ_{xy}) on the surface. That is, unlike an image screw dislocation, an image edge dislocation does not lead to a traction free surface. However, the image force calculated by using the image dislocation is valid for both screw and edge dislocations.

For an edge dislocation, the component of the image force parallel to the slip plane is the glide force, and the component perpendicular to the slip plane is the climb force (for a screw dislocation, the Burgers and line vectors of the dislocation do not define a unique slip plane, and hence, screw dislocations "cannot climb").

At the heart of plastic deformation (by slip) is the motion of dislocations (finally leaving the crystal/grain). Dislocations can move if shear stresses exist parallel to the slip plane. The origin of this shear stress could be (1) externally applied forces, which translate into shear stresses at the slip plane level; (2) internal residual stresses (arising from sources like other dislocations, coherent precipitates, twins, etc.); and (3) a free surface (or a less stiffer material) in the vicinity of the dislocation.

If the image force exceeds the Peierls stress of the material, then the dislocation can spontaneously move (in the absence of externally applied stresses) and would leave the crystal. Surface regions of materials can become dislocation free by this mechanism, and if the crystallite is very small, it can become completely dislocation free. Formulae have been proposed for the computation of the Peierls stress, when the core is planar. If the core is nonplanar, then atomistic simulations can only yield good results for the value of Peierls stress. For the case of the planar core, the Peierls stress can be given by an exponential formula [40]:

$$\tau_P = \frac{G}{\alpha}\exp\left(-\frac{2\pi d}{b\alpha}\right) \tag{5.6}$$

where
d is the interplanar spacing (of the slip plane)
$\alpha = (1 - \nu)$ for an edge dislocation and $\alpha = 1$ for a screw dislocation

Alternate equations for the computation of Peierls stress also exist.

Though the concept of an image force in its original sense implies the attractive force experienced by a dislocation toward a free surface (which can be computed using the construction of an image dislocation), it is necessary to generalize the concept to accommodate other situations. If the free surface (with air/vacuum adjacent to the material under consideration) is replaced by an interface with an elastically harder or softer material, the dislocation will feel a repulsive or attractive configurational force, respectively. In special cases, the force may be repulsive for certain position of the dislocation in the domain and may become attractive for certain other positions. Under these circumstances, the "usual" concept of an image dislocation cannot be used for the computation of the configurational force experienced by the dislocation. Additionally, it will be shown later that when a dislocation is positioned near a free surface, the image construction does not yield correct results, and hence, we may have to go beyond the concept of "image construction."

5.3.4 Heteroepitaxy

Due to potential applications in a variety of devices like field-effect transistors, heterostructure bipolar transistors, mid- and far-infrared photodetectors, and resonant tunneling diodes, there has been a second wave of intense investigations of epitaxial systems. The books by Freund and Suresh [41], Grovner [42], and Ayers [43] not only give a broad overview of the subject but also cover specific aspects of technological importance. The book by Matthews [44] and the review articles by Jain et al. [45,46] are also well-written expositions on the subject.

In epitaxial growth, a crystalline layer (overlayer) grows on a crystalline substrate in a manner such that there is matching of at least one set of atomic planes between the overlayer and the substrate. Usually when one material (say GeSi) is deposited on another material (say Si single crystal) epitaxially, the word heteroepitaxy is used, and in these cases, the matching of atomic planes is usually not perfect leading to coherency (or epitaxial) strains and stresses. Examples of epitaxial systems include InGaAs (overlayer) on GaAs (substrate), Au/Ag, Co/Ni, etc.

Three modes of growth of an overlayer can be identified:

1. Van der Merve growth (2D): Here, there is layer by layer growth with "full wetting" of the substrate by the film. The cohesive force between the adsorbed atoms (adatoms) is weaker than the adhesive force with the surface $\left(\gamma_{Substrate}^{Surface} > \left(\gamma_{Overlayer}^{Surface} + \gamma_{Substrate-Overlayer}^{Interface} \right) \right)$

2. Stranski–Krastanov growth mode (2D–3D): Initially there is a layer by layer growth, followed by island formation.

3. Volmer–Weber growth (3D): Islands directly grow on the substrate ("partial wetting" of the substrate by the overlayer). Here, the cohesive force between the adatoms is stronger than the adhesive force with the surface $\left(\gamma_{Substrate}^{Surface} < \left(\gamma_{Overlayer}^{Surface} + \gamma_{Substrate-Overlayer}^{Interface} \right) \right)$

The film in the Van der Merve growth mode is biaxially strained, and the energy of the epitaxial film of thickness h is given by

$$E_h = 2G \left[\frac{1+v}{1-v} \right] f_m^2 h \tag{5.7}$$

where "f_m" is misfit strain. This implies that as the film grows thick, the energy per unit interfacial area increases. In such a system (Van der Merve growth mode), the film is completely under one state of stress (compression if the relevant lattice parameter of the film is larger and tension if the situation is reversed). In island growth (Volmer–Weber mode), the situation is more complicated. In the preceding equation, the substrate is assumed to be rigid, and the energy of the substrate is ignored. The system is treated as 2D infinite one, and edge effects are also ignored.

As the thickness of the epitaxial layer increases (on continued deposition), the energy due to epitaxial stresses becomes large, and the stresses relax partially by the formation of interfacial misfit dislocations (i.e., the interface becomes semicoherent). Based on the symmetry of the system, two or three arrays of parallel dislocations can form at the interface. There are many mechanisms by which the interfacial misfit edge dislocations can form. Two of the important ones are (1) formation of a misfit segment in threading dislocations and (2) extension of a dislocation loop to the interface (thus leading to a misfit segment). The thickness (h_c) at which the formation of a misfit segment becomes energetically feasible is given by

$$h_c = \frac{b}{8\pi f_m (1+\nu)} \left[\ln \frac{2h_c}{r_0} + 1 \right] \tag{5.8}$$

where "r_0" is the inner cut-off radius and is given by [47]

$$r_0 = \frac{(\pi b)}{2\sqrt{2(1-\nu)}}$$

Matthews and Blakeslee [48] had used a force balance approach (i.e., the epitaxial stresses tending to extend the misfit segment balance the line tension force of the dislocation) to derive the following equation for critical thickness:

$$h_c = \frac{b}{2\pi f} \frac{\left(1 - \nu \cos^2 \alpha\right)}{(1+\nu)\cos \lambda} \left(\ln \frac{h_c}{b} + 1 \right) \tag{5.9}$$

where
  h_c is the critical thickness
  α is the angle between the dislocation line and its Burgers vector
  λ is the angle between the slip direction and that direction in the film plane, which is perpendicular to the line of intersection of the slip plane and the interface

A misfit edge dislocation with its Burgers vector parallel to the interface gives maximum stress (and hence energy) relief. However, in some cases (e.g., GeSi film on Si substrate), the interfacial dislocation is of mixed character (60° misfit dislocation in GeSi/Si system), and only the edge component gives stress relief. The screw component, in spite of not giving any stress relief, costs energy to the system. Similarly, in the Nb (film)/sapphire (substrate) epitaxial system, the edge dislocation has a Burgers vector inclined to the interface, and the component parallel to the interface gives stress relief (while the perpendicular component only costs energy to introduce).

In many systems (e.g., $Ge_x Si_{1-x}$ film on Si substrate), the composition of the film can be varied to vary the misfit (this is done to engineer the bandgap, which depends on the strain) and hence alter the critical thickness at which the misfit dislocation forms. If the misfit between the film and the substrate is too large, domain matching epitaxy may occur [49], or the interface may be incoherent.

5.3.5 Coherent Precipitates

To get a general idea about phase transformations, including precipitation, the reader may consult the book by Porter and Esterling [50]. The chapter by Matthews in the series on *Dislocations in Solids* (edited by F.R.N. Nabarro) [51] comprehensively covers all important aspects of coherent structures and the formation of misfit dislocations.

The precipitation being considered in this section is that of a solid phase from a solid matrix. At constant temperature and pressure, the Gibbs free energy (G) determines the stability of a system. When a system is heated or pressurized, phase transformation may take place to lower the Gibbs free energy. In a reconstructive first-order transformation, the process starts by the nucleation of a second phase from the parent phase, followed by its growth. The energies involved in the process are (1) reduction in volume free energy due to transformation, (2) increase in interfacial energy, and (3) strain energy due to volume mismatch between the parent and the product phases. It should be noted that the reference state for the interface energy is the infinite crystalline solid.

Due to change in temperature or pressure, a solid solution may become supersaturated and may lead to precipitation of a solid phase in a solid matrix. When a given volume of material transforms from a parent phase to a product phase, the following aspects are expected to change in the volume being considered: (1) the composition, (2) the crystal structure, and (3) the shape and size. These aspects will determine the lattice parameters of the precipitate and the strain associated with the transformation. The strain associated with the transformation can be schematically understood using the Eshelby construction (Figure 5.4): (1) a volume of material is cut out and allowed to transform, (2) the shape and/or size of the transformed material will be different from the parent material (in Figure 5.4c, only a shape change is considered for simplicity), (3) the transformed material is reintroduced into the hole it originally occupied, and (4) the system equilibrates (continuity of the material is not broken). The shape/size change during phase transformation can be modeled as eigenstrains (stress-free strains). The stress state of a growing precipitate can be simulated by successively considering larger and larger volumes (V) which undergo transformation.

In order that the interfacial energy between the parent and product phase (the precipitate) is small, there is a tendency to keep the interface coherent. This usually happens when the size of the precipitate is small. That is, interfacial energy is kept low at the expense of strain energy arising from coherency. The strain energy in the presence of a coherent precipitate is given by [15]

$$E_{ppt}^{strain} = \frac{8\pi r_p^3 f_m^2 G_{ppt}}{3} \frac{(1+\nu_{ppt})}{(1-\nu_{ppt})} \tag{5.10}$$

where
r_p is the radius of coherent precipitate
f_m is the misfit present at the interface

Interfacial diffusion may play a prominent role in the growth of the precipitate, which may be stress assisted. FEM computations can give us the precise stress state at the interfacial region (both for coherent and semicoherent interfaces), which can form an input into the diffusion calculations.

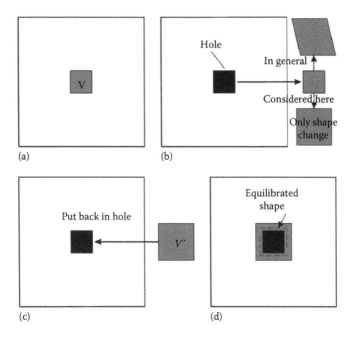

(a)   (b)

(c)   (d)

FIGURE 5.4
Schematic of the Eshelby construction to understand the origin of the stresses due to phase transformation of a volume (*V*): (a) region *V* before transformation, (b) the region *V* is cut out of the matrix and allowed to transform, (c) the transformed volume (*V'*) is inserted into the hole, and (d) the system is allowed to equilibrate. The continuity of the system is maintained during the transformation.

On growth (beyond a critical size), the strain energy cost in maintaining the precipitate coherent becomes large, and misfit dislocations form at the interface, to partially relieve the coherency stresses (Figure 5.5 shows a schematic). This transition of the interface usually occurs when the size of the precipitate is in the nanoscale. In the initial stages of the growth, the precipitate may be spherical and may become polyhedral on further growth (like in the precipitation of γFe from Cu–Fe matrix). The change in shape may coincide with the coherent to semicoherent transformation of the interface. For precipitates in the shape of disks, plates, etc., one interface may become incoherent, while others may remain coherent. It is to be noted that incoherent/semicoherent interfaces are more glissile as compared to coherent interfaces.

Misfit dislocations are structural dislocations, and structural dislocations in general can accommodate (1) linear misfit, (2) tilt, or (3) twist. The term "structural dislocations" is

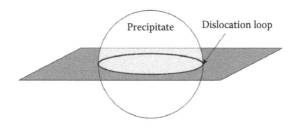

FIGURE 5.5
Schematic showing a spherical precipitate with an interfacial misfit loop. The precipitate is embedded in a homogenous matrix.

often related to "geometrically necessary dislocations," but we shall avoid the second term in the current context. The process of precipitation can be visualized as the 3D analog of epitaxial film growth (with respect to the origin of stresses and formation of misfit dislocations).

The conditions leading to the formation of an interfacial misfit dislocation loop are a little more involved. Necessary (global energy-based criterion) and sufficient (local stress-based criterion) conditions need to be satisfied for the formation of the dislocation loop.

The radius of the precipitate (assuming it to be spherical) at which a misfit loop becomes energetically favorable is given by [51]

$$r^* = \frac{b}{8\pi f_m \left(1 - v_{ppt}\right)} \left(1 + \frac{4G_{ppt}}{3K_{ppt}}\right) \left[\ln\left(\frac{8r^*}{b}\right) + 1\right] \tag{5.11}$$

where K is the bulk modulus of the precipitate. Other equations for r^*, similar to the one given earlier, can also be found in literature [52]. Isotropic conditions have been assumed here. Note that the material properties of the matrix are not present in the equation.

A precipitate of a lower radius than this (i.e., $r_c < r^*$) can also support an interfacial loop due to local force balance (i.e., the line tension force tending to shrink the loop equals the coherency stresses trying to expand the loop), even though it is not energetically favorable to do so. That is, if, by some means, the interfacial loop forms in a precipitate of radius r_c, then it would be stable. This radius (r_c) is given by [51]

$$r_c = \frac{b}{16\pi f_m \left(1 - v_{ppt}\right)} \left(1 + \frac{4G_{ppt}}{3K_{ppt}}\right) \left[\ln\left(\frac{8r_c}{b}\right) + 2\right] \tag{5.12}$$

The sufficient condition for the formation of an interfacial loop is based on the specific mechanism by which the loop forms. Preexisting dislocations may move (by climb or glide) to the interface, or new loops may be nucleated (or nucleated loops may move to the interface). The preexisting dislocation may be (1) inside the precipitate, (2) outside the precipitate, or (3) at the interface (but not in the equilibrium position). Processes which aid the formation of the loop include radiation (giving rise to point defects) and plastic deformation. Some of the mechanisms giving rise to the interfacial loop are [51] (1) expansion of a loop inside the precipitate, (2) climb of an interfacial loop, (3) punching-in of shear loops, (4) trap of external loops, (5) climb of Orowan loops, (6) cross-slip of screw dislocations from the matrix, (7) Gleiter's mechanism, etc.

The situation with respect to the formation of misfit dislocation loops around precipitates may become inverted, and dislocations may act like preferential sites for heterogeneous nucleation of precipitates. In both cases, the precipitate and dislocation are associated with each other to give a low-energy configuration.

5.4 Case Studies

A few case studies are taken up here to demonstrate the utility of finite element simulations in the nanoscale. These are expected to illustrate the utility of numerical simulations in solving a variety of problems related to materials science. These examples cover structures and processes, wherein strain energy is the dominant term and can be simulated using eigenstrains in the finite element model. An important point to be noted from the simulations is the ease with which some difficult problems of mechanics can be solved using FEM. The examples cited are by no means comprehensive, in either methodology developed or in the kind of problems, which can be solved using FEM. Additional information regarding the case studies can be found elsewhere [53–58]. In the case studies considered, the reader should note that the domains are in the nanoscale. Material properties used in the simulations are listed in Appendix 5.A.

One important question which needs to addressed, before we proceed to the case studies is "if theoretical models and equations are available (as described in Section 5.3), then why do we need to perform finite element computations/simulations?" The following points address this question. The usual theoretical equations for a given problem in hand (e.g., the equation for the stress fields of an edge dislocation derived using the linear elasticity theory) are derived for some ideal conditions (or assumptions). For example, the domain may have been assumed to be infinite or cylindrical. The material may have been assumed to be uniform and homogeneous. Additionally, a "simple" set of boundary conditions may have been imposed. If there are deviations to these "ideal" conditions, the theoretical analysis may become cumbersome or intractable. In finite element analysis, once a simulation methodology has been validated for simple cases (wherein theoretical and experimental results are available), it can be extended to complex configurations with ease. This aspect will become clear via the case studies considered.

Some important points to be noted in the finite element simulations are as follows: (1) material properties have been taken to be bulk values, and the validity of the simulations can be improved by feeding-in values corresponding to those in a nanocrystal; (2) surface tension effects have been ignored and can be introduced by considering more "elaborate" simulations; (3) interfaces are considered to be sharp (which may deviate from reality); (4) some simulations are performed using isotropic material properties (i.e., with two elastic constants) for the sake of easy comparison with theoretical expressions, and the simulations can readily be extended to anisotropic properties; (5) the structure and energy of the core of the dislocation have been ignored in the numerical models considered (extended finite element models can be used to account for this to some extent [59]); (6) often ideal domain shapes have been used to illustrate the methodology (and for easy conceptual visualization), and the models can be extended to complex domain shapes, with multiple material components; (7) eigenstrains are imposed as thermal strains using a standard commercial software [60]; (8) the simulations are "static" in nature (i.e., kinetic energy, transients, wave propagation, etc., are ignored); (9) a series of equilibrated states can be played out in pseudo-time to simulate the evolution of a system, which may actually take place in real time (keeping in view the fact that there is no kinetics in the simulations); and (10) for energy plots in the presence of a dislocation, Peierls energy oscillations [38] are ignored (these oscillations are small in comparison with the strain energy of the system).

5.4.1 Edge Dislocations in Nanocrystals and Hybrids

The stress fields arising from an edge dislocation in a nanocrystal can be simulated by imposing eigenstrains corresponding to the insertion of an "extra" half-plane of atoms (as shown schematically in Figure 5.6a). Material properties correspond to that for aluminum ($a_0 = 4.04$ Å, slip system: <110>{111}, $b = \sqrt{2}a_0/2 = 2.86$ Å, $G = 26.18$ GPa, $\nu = 0.348$ [61]). Isotropic elastic moduli (E, G, ν) are computed by averaging single crystal data (C_{11}, C_{12}, and C_{44}) [62].

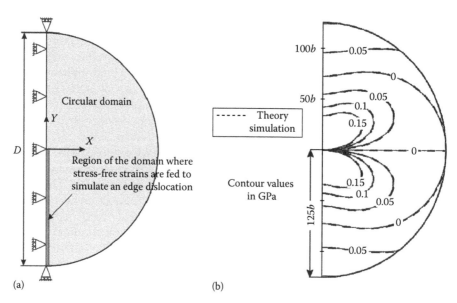

FIGURE 5.6

(a) Schematic of the numerical model used for the simulation of the stress state an edge dislocation in a nanocrystal, (b) simulated σ_{xx} contours (with $D = 250b$) and its comparison with the theoretical equation (5.1). The boundary conditions imposed along the y-axis are due to symmetry. (From Khanikar, P., Kumar, A., and Subramaniam, A, *Philos. Mag.*, 91, 730, 2011. With permission.)

The mesh size is $b \times b$ (in "x" and "y" dimensions). The elements used for meshing are four-noded quadrilateral elements. Plane strain conditions are assumed. As the mesh size is already of the order of the lattice parameter, no further mesh refinement is possible. It is obvious that at the level of a few nodes, the assumption of a continuum is not valid; however, at larger length scales, the model is expected to yield good results.

Isotropic moduli have been used for comparison with simpler theoretical equations. In a material like Al, the anisotropy factor $(A = (2C_{44})/(C_{11} - C_{12}) = 1.23)$ is close to one, and hence, the error introduced by using assumptions of isotropy is small. In Al, the stacking fault energy is high, and hence, the splitting of a perfect dislocation into partials need not be considered.

It is seen that there is a good match of the simulated stress contours with the theoretical plot (of Equation 5.1). The energy of the system computed using the model (as in Figure 5.6 with $D = 200b$) is 6.6×10^{-10} J/m. This matches reasonably well with the energy computed using the theoretical equation (5.3) of 1.2×10^{-9} J/m.

This method of simulation does not take into account the structure and energy of the core of the dislocation. This is expected to introduce error in the calculated stress values close to the dislocation line ($\sim$few b). The energies computed from the simulations will also be an underestimate by about 10% [36,63]. However, in many of the parameters determined using the simulations (e.g., image forces, regions of stability, etc.), the variation in energy is important and not the correct value of the total energy (i.e., a constant additive term of core energy does not change the conclusions drawn). The core energy of the dislocation is expected to be constant, except when the dislocation is positioned a few "b" from a free surface or an interface.

If the domain shape is rectangular (or a complex shape), theoretical description of the stress state becomes cumbersome (or even intractable). However, as the current methodology involves the equivalent of the insertion of an "extra half-plane," it is expected to work for complex geometries, boundary conditions, and material combinations. If a dislocation is positioned in the "central region" of a bulk crystal, then this is essentially equivalent to a dislocation in an infinite medium, and two neighboring positions of the dislocation ("b" apart) are equivalent. In nanocrystals on the other hand, (1) free-surface effects play an important role, (2) neighboring positions are not equivalent, and (3) domain deformations in the presence of the dislocation can become significant.

5.4.1.1 Image Forces

The model considered for the computation of image force (glide component) in a nano-crystal is as in Figure 5.7. The energy of the system for various positions of the dislocation in the domain (along "x") is determined, and the image force is computed as the slope of the energy versus position of the dislocation plot. The climb component of the image force can similarly be determined, by varying the position of the dislocation along y-direction.

The methodology of calculation of image force has the following features: (1) no fictitious image needs to be constructed (often a sequence of infinite images need to be constructed in the presence of proximal parallel interfaces); (2) deformation of the free surface is taken into account, and shape of the deformed surface can be determined; and (3) due to deformation of the surface, the standard theoretical equations (5.4 and 5.5) are no longer valid, and the current methodology needs to be used. The model should not be used when the dislocation is within a few "b" from the free surface (due to interaction of the core with the free surface and due to the fact that surface tension has not been taken into account in the model).

The method is easy to extend to forces experienced toward interfaces with harder and softer materials (schematic of one such case is shown in Figure 5.8). It should be noted that interfaces with elastically harder materials are repulsive, and softer materials are attractive. As for the case of the free surface, when the dislocation is within a few "b" from the interface, the method described here is not expected to yield correct results.

The image force computed from the model in Figure 5.7 along with a comparison with the theoretical equation (5.5) is shown in Figure 5.9. Figure 5.10 shows the deformation of the free surface when the dislocation is at a distance of 20b from the free surface. Figure 5.11 shows

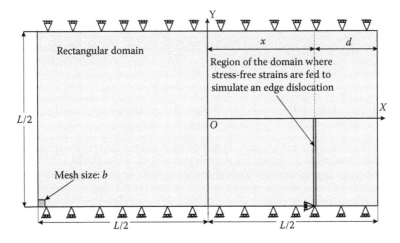

FIGURE 5.7
Schematic of domain and boundary conditions used for the computation of image force experienced by an edge dislocation. $L = 500b$. (From Khanikar, P., Kumar, A., and Subramaniam, A, *Philos. Mag.*, 91, 730, 2011. With permission.)

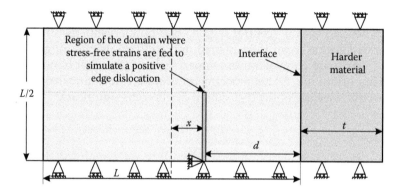

FIGURE 5.8

Schematic of a model to compute the "image force" experienced by an edge dislocation in the presence of a harder material (nickel). $L = 500b$. (From Khanikar, P., Kumar, A., and Subramaniam, A, *Philos. Mag.*, 91, 730, 2011. With permission.)

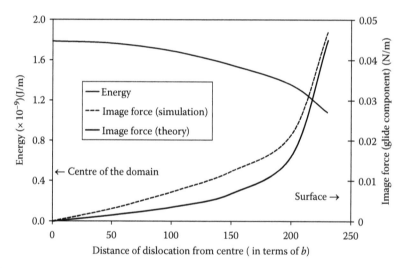

FIGURE 5.9

Energy (from finite element simulation) of an edge dislocation (in J/m depth) and the image force (along x-direction) experienced by it, as a function of its position along the x-axis. (From Khanikar, P., Kumar, A., and Subramaniam, A, *Philos. Mag.*, 91, 730, 2011. With permission.)

the configurational force experienced by the dislocation toward the interface (generalized image force), computed from the model in Figure 5.8. It can be seen that for a thickness of $5b$, the force is attractive toward the interface. When the thickness of the harder material (t) is $50b$, the force becomes repulsive as the dislocation is positioned closer to the interface. For even thicker harder material ($t = 100b$), for displacements of up to about $175b$ from the center, the force remains constant and nearly zero (before turning repulsive for $x > 175b$).

5.4.2 Neutral Equilibrium and Stability of Edge Dislocations

In the previous subsection, we had mentioned about the effect of surface deformation on the image force experienced by the edge dislocation. Here, we consider an extreme form of domain deformation due to an edge dislocation. If an edge dislocation is present in a thin free-standing plate (as in Figure 5.12a), the plate will bend due to the stresses introduced

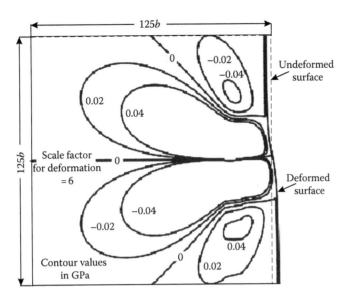

FIGURE 5.10

Plot of σ_x contours and deformations to a free surface (exaggerated by a scale factor of 6), due to the presence of an FEM simulated edge dislocation, at a distance of $20b$ from the surface. (From Khanikar, P., Kumar, A., and Subramaniam, A, *Philos. Mag.*, 91, 730, 2011. With permission.)

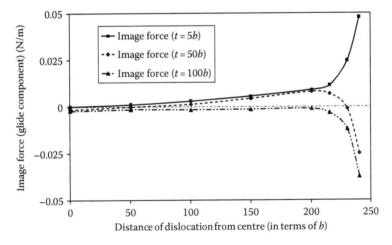

FIGURE 5.11

Glide component of image force (FEM results) in the presence of an elastically harder material. Computations are for the model as in Figure 5.8, for three thicknesses ($t = 5b$, 50, and $100b$). (From Khanikar, P., Kumar, A., and Subramaniam, A, *Philos. Mag.*, 91, 730, 2011. With permission.)

by the dislocation. The plate considered is a 2D nanocrystal. If we plot the energy of the system as a function of the variable "x" (as in Figure 5.12a), the plot shown in Figure 5.12b is obtained. The plot shows a very interesting feature—the energy of the system does not change for considerable positions of the dislocation in the domain (with $y = 0$), within a numerical accuracy of about 0.1%. This is as if the free surface has vanished to infinity!

The stress contours (along with domain configuration) for a few values of "x" are shown in Figure 5.13. Over all the configurations in the figure, the system exists in a state of

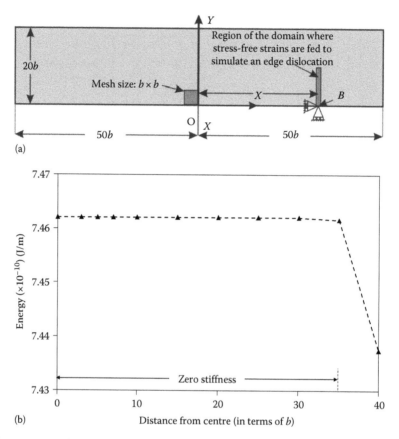

FIGURE 5.12
(a) Schematic of the model used for the simulation of an edge dislocation in a thin free-standing Al plate (nano-crystal) and (b) energy of the system as a function of "x." (From Kumar, A. and Subramaniam, A., *Philos. Mag. Lett.*, 91, 272, 2011. With permission.)

neutral equilibrium. A rigid ball on a plane is in a state of neutral equilibrium. Similarly, an Anglepoise lamp (which is a structure) is also in a state of neutral equilibrium with respect to various positions of the lamp head. The earlier-given simulations show that for specific geometries, "material structures" can also exist in a state of neutral equilibrium. The term "material structures" has been used in this context, as the domain has a specific geometry (which changes with location of the dislocation), which qualifies it as a structure, and the defect is crystallographic in origin, which qualifies it as a material. That is, material structures have both "material" and "structural" characteristics.

Plastic deformation is permanent deformation in the absence of external constraints on the system. As the dislocation is positioned at various "x" coordinates in the plate, the deformations caused are permanent and hence can be termed as plastic deformation. In the "zero-stiffness" regime, the dislocation can move without a change in energy; that is, the process is reversible. The stresses and strains in the system are a result of the elastic response of the material in the presence of an edge dislocation. Hence, this phenomenon can be "curiously" phased as "reversible plastic deformation due to elasticity."

Dislocations are thermodynamically unstable defects (unlike vacancies) and would leave the crystal when kinetics permits [64]. "Can dislocations be mechanically stable defects in finite crystals?" Eshelby had answered this question for the case of a screw dislocation

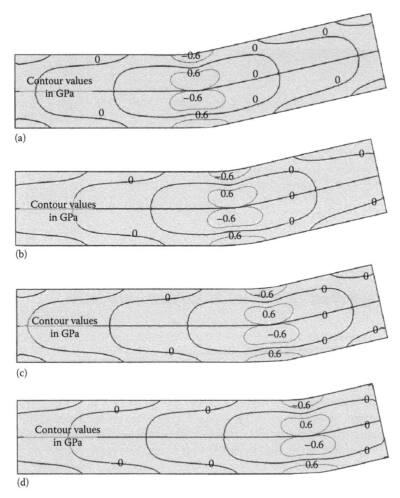

FIGURE 5.13
Deformed configurations of the domain for (a) $x = 0b$, (b) $x = 10b$, (c) $x = 20b$, and (d) $x = 30b$. σ_x stress contours are overlaid in on the deformed contours (deformations exaggerated by a scale factor of 3). (From Kumar, A. and Subramaniam, A., *Philos. Mag. Lett.*, 91, 272, 2011. With permission.)

in a thin cylinder, which is free to twist [65]. It turns out that the edge dislocation can be stable in domain shown in Figure 5.14, for a wide range of positions within the crystal (region enclosed by the curve P'Q'RQP). The region of stability has been determined by plotting the energy of the system for various positions of the edge dislocation. For plates up to a thickness of about $85b$ (i.e., L/d of ~1.18, keeping the length constant at $100b$), the dislocation can be stable in a finite crystal. The partial relaxation of strain energy due to bending gives rise to this stability. Eshelby had stated that solving the problem of an edge dislocation in a thin plate would be a difficult problem analytically due to the bending of the plate. It is seen that using FEM, this can be solved easily.

5.4.2.1 Dislocation-Free Nanocrystals

As pointed out earlier, if the resultant image force on a dislocation exceeds the Peierls force (Peierls force = Peierls stress × b), then the dislocation will move (and can leave the crystal).

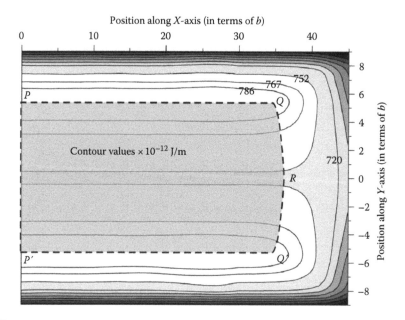

FIGURE 5.14

Plot of the energy of the system for different positions of the edge dislocation in a domain of thickness $20b$. The region of stability is enclosed by the curve PQRQ'P' and has been shaded. Point R is at $(35b, 0)$.

For a free-standing (finite) single crystal to become completely free of dislocations, the minimum image force experienced by the dislocation has to exceed the Peierls force. In symmetrical domains (e.g., with square, rectangular, circular, etc., cross sections), the dislocation image forces arising from various free surfaces will cancel each other when the dislocation is at the center of the domain. It is to be noted that irregular domains have no such special position. The dislocation will experience a minimum image force, when just "off-center" in the domain (at a distance of a Burgers vector from the center of the domain). If this minimum value of image force exceeds the Peierls force, then even the dislocation "just off-center" can glide out of the crystal, thus making the crystal dislocation free. This is expected to occur only in small crystals, as in large crystals, the surfaces would be "far away" from the center of the domain. Thus by reducing the crystal size and comparing the minimum image force with Peierls force, we can determine the size of the crystal at which it will become spontaneously edge dislocation free (as shown in Figure 5.15). The value of critical size for Al is determined to be ~36 nm. Few points have to be noted in this context: (1) just off-center the strength of the two images is nearly equal, and hence, the net force is very small; (2) as the domain size is reduced, the energy of the dislocation decreases, and hence, the image force experienced by the dislocation at a given distance "d" from the surface decreases; and (3) typically, we have to reduce the crystallite size to nanoscale before the reduced image force can exceed the Peierls force. FEM is an invaluable tool to determine this size, as we have already seen that the theoretical formulae do not work well due to domain deformations.

5.4.3 Heteroepitaxial Structures

5.4.3.1 Growth of Thin Films and Critical Thickness

First, we consider one-dimensional growth of film with complete coverage over the substrate (Van der Merve growth mode). We simplify the problem by considering 2D plane

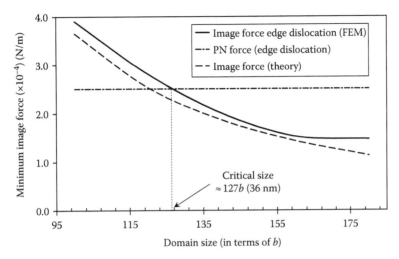

FIGURE 5.15
Comparison of the minimum image force with the Peierls force for domains of decreasing size. For sizes of crystals below the critical size, the crystal would become spontaneously edge dislocation free. Peierls force is calculated from theory. (From Wang, J.N., *Mater. Sci. Eng.* A, 206, 259, 1996.)

strain condition. In reality (as we have noted in Section 5.3.4), misfit strain is biaxial, and hence, the simulation has to be carried out in 3D. The growth of Nb on a sapphire substrate is considered here as an illustrative example.

An epitaxial overlayer can be simulated by imposing stress-free strains to a region in the domain corresponding to the film (schematic of the model in Figure 5.16). A growing film can be simulated by playing out a sequence of simulations with increasing height (h) of region A on which stress-free strains are imposed. The stress state of the system after the

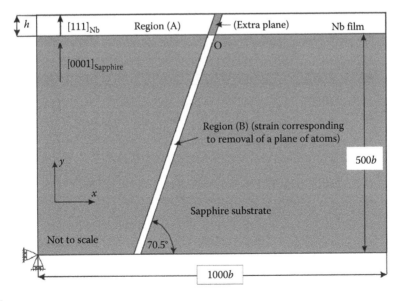

FIGURE 5.16
Numerical model used for the simulation of the stress state of an epitaxial film. A growing film can be simulated by increasing the height (h) of the region corresponding to the film.

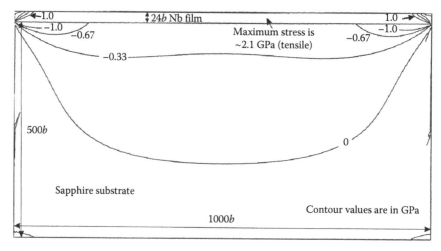

FIGURE 5.17
Stress state of the epitaxial system after the growth of about 12 layers.

growth of about 12 layers is shown in Figure 5.17. The following points have to be noted: (1) the film is uniformly strained, expect for regions close to the free lateral surface; (2) in a large system, which can be considered as 2D infinite, the edge effects can be ignored, and the film can be considered to be uniformly stressed; and (3) most of the substrate is stress free, except close to the free lateral surface.

The stress state of the system in the presence of a misfit edge dislocation (with its Burgers vector inclined at an angle of 19.5°) is shown in Figure 5.18. The critical thickness at which a misfit dislocation becomes energetically feasible is determined by comparing the energy of the system with and without a misfit dislocation, as "h" is increased (shown in Figure 5.19). The value of critical thickness, determined from the plot, is ~39 Å.

The following advantages of the finite element simulation are to be noted over the standard theoretical formulations: (1) the substrate is not assumed to be rigid, (2) separate material properties have been used for the film and the substrate, (3) the energy of the whole

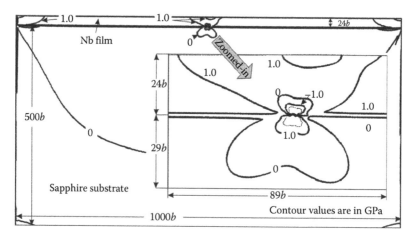

FIGURE 5.18
Stress state of the system (plot of σ_{xx} contours) in the presence of a misfit edge dislocation.

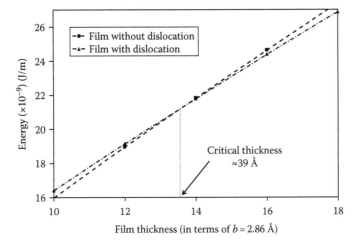

FIGURE 5.19
Determination of the critical thickness for the feasibility of a misfit edge dislocation, by comparison of the energy of the system with and without the dislocation.

system is considered in determination of the critical thickness, (4) the complete Burgers vector inclined to the interface is considered (and not just the component parallel to the interface, and (5) effects of the free lateral surface are captured. Additionally, the energy partitioning between the substrate and film can be easily calculated from the model, and the energy density (per unit area of the interface) and its variation with position along the interface can be computed with ease.

Using an inclined Burgers vector not only gives us the correct energy of the dislocation (the component parallel to the interface gives strain relief, while the perpendicular component costs energy to the system without proving strain relief) but also gives the true stress state of the system resulting from the interaction of the dislocation stress fields with the epitaxial stress fields.

It is to be noted that in standard theoretical formulations, (1) the substrate is assumed to be rigid, and the energy stored therein is ignored; (2) material properties of the film alone are considered; and (3) if the substrate is assumed to be rigid, then we are forced to put "half" a dislocation (as either the tensile or the compressive part of the dislocation has to lie in the substrate).

5.4.3.2 Thin Substrates, Stripes, and Islands

A uniform epitaxial film on a thick substrate represents the simplest geometry in some sense. The geometry of the system can play a significant role in determining the stress state of the system (and hence the critical thickness/size). Here, we consider three illustrative geometries, which bring in further considerations: (1) thin substrate (which can bend due to epitaxial stresses), (2) finite stripes, and (3) islands.

The models used for the simulations are shown in Figure 5.20. The methodology of simulating the growth of an epitaxial stripe or island is similar to that for epitaxial films. Plane strain conditions are assumed for the thin substrate and strip cases (Figure 5.20a and b), while axisymmetry is assumed for the island case (Figure 5.20c). The stress states for the three cases are as in Figure 5.21.

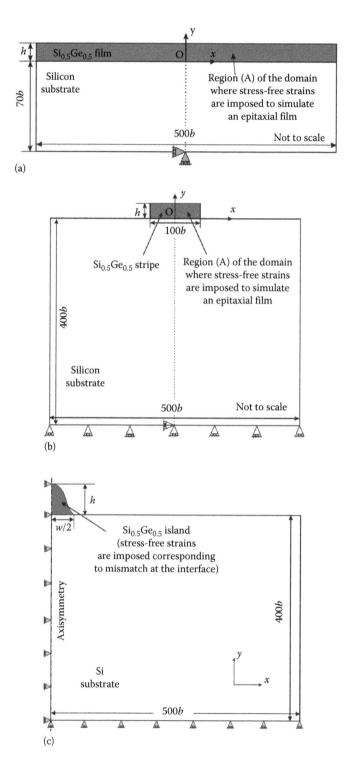

FIGURE 5.20
Models used for the simulation of three kinds of epitaxial systems: (a) film on a thin substrate, (b) finite epitaxial stripes, and (c) island. The dimensioning is to be noted on the figures.

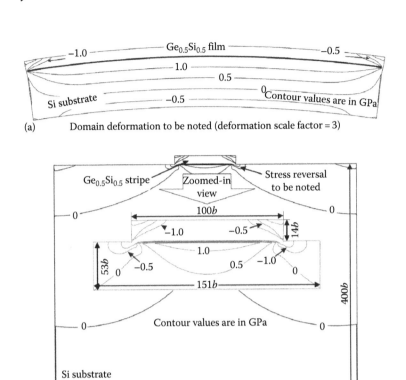

(a) Domain deformation to be noted (deformation scale factor = 3)

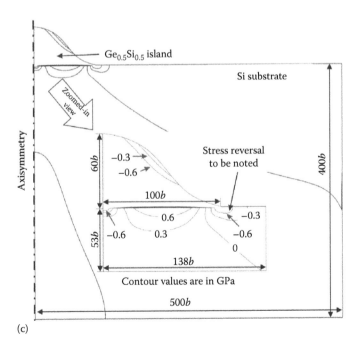

FIGURE 5.21
Stress states (plot of σ_x contours) of the three systems shown in Figure 5.20: (a) thin substrate (film thickness, $h = 14b$), (b) stripe, and (c) island.

Finite stripes and islands are fundamentally different with respect to films in the following aspects. (1) In films, the entire film is under one state of stress (compressive) and the substrate under another (tensile). This is not the case in stripes and islands (region of the substrate close the edges is in an altered state of stress). (2) The film is assumed to be infinite, and hence edge effects are ignored. Stripes and islands are inherently finite, and hence, the thickness at which it is energetically feasible to support a misfit edge dislocation is position dependent (along the interface).

In the case of islands, multiple islands grow simultaneously on the substrate, and hence, there would be interaction in the strain fields arising from individual islands. The model developed can easily be extended to simulate the growth of multiple islands and to compute the strain interaction effects.

5.4.4 Precipitation

To illustrate the methodology of simulation of the stress state of a system with a coherent precipitate, we consider the example of Co precipitation from a Cu-4wt.%Co solid solution. Both the matrix and the precipitate have cubic closely packed crystal structure (with cube on cube orientation), and hence, the precipitation involves only dilatational strains. In many cases (e.g., cubic to monoclinic transformation), the precipitation may involve shear strains as well. To simulate the stress state of a coherent precipitate and to compute the value of r^*, the model in Figure 5.22 is used. Axisymmetric conditions are assumed, with axis of symmetry being the Y-axis.

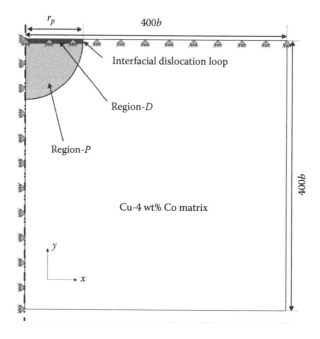

FIGURE 5.22
Schematic of the model used for the simulation of the stress state of a coherent precipitate and to compute the critical size r^*. Eigenstrains corresponding to lattice mismatch is imposed in region P. In region D, eigenstrains equivalent to the insertion of a disk of atoms is used as input. Region D is a subset of region P.

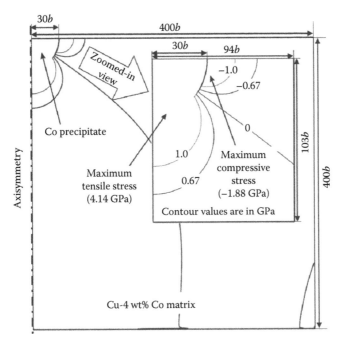

FIGURE 5.23
State of stress (σ_{yy} plot) on the growth of a precipitate $30b$ in radius.

Stress-free strains are imposed in region P corresponding to the misfit strain between the matrix and the precipitate, which is computed as follows:

$$f_m = \left(\frac{a_{ppt} - a_{matrix}}{a_{ppt}} \right) \tag{5.13}$$

The stress state of the system in the presence of a coherent precipitate $30b$ in radius is shown in Figure 5.23.

The stress state of the system due to a growing precipitate can be simulated by increasing the radius of the region A. If one intends to simulate a precipitate in an infinite matrix, then for the largest size of the precipitate, the domain size should be successively increased until the change in energy is small (say less than 1%). An interfacial misfit edge dislocation loop is simulated by feeding in strains corresponding to that of introduction of a disk of atoms. The Burgers vector for the example considered is [111], and the strain imposed in region P is 0.019 ($= f_m$).

Separate material properties have been used for the precipitate and matrix.

5.4.4.1 Coherent to Semicoherent Transition of Precipitates

The stress state of the system (plot of σ_{yy}) in the presence of an interfacial loop is shown in Figure 5.24. To determine r^*, the energy of the precipitate is tracked before and after the introduction of the misfit edge dislocation loop. When the precipitate size is small, the energy of the system increases on the introduction of an interfacial misfit edge dislocation loop. After r^* is exceeded, the dislocated state is the preferred state. Hence, similar to the case of an epitaxial film, the critical size (r^*) can be determined to be the crossover point

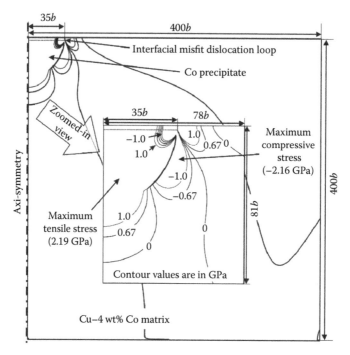

FIGURE 5.24
State of stress (σ_{yy} plot) on the growth of a precipitate of radius 35b in the presence of an interfacial misfit edge dislocation loop.

in the energy versus radius of precipitate plot. The critical size (r^*) determined from the simulations is ~32b (~80 Å).

5.4.4.2 Precipitation in Nanocrystals

The following aspects come into play when one considers precipitation in small domains (nanocrystals): (1) the strain energy associated with the coherent precipitate is altered with respect to precipitation in a bulk material, (2) the solute required for the growth of the precipitate may be limited by the restricted amount material available in the nanocrystal, (3) surface (and domain) deformations may lead to partial relaxation of the strain energy, and (4) heterogeneous nucleation of the precipitate may occur preferentially at the surface.

Figure 5.25 shows two models used for the simulation of precipitation in a small domain. The "small" dimension is chosen to be the diameter of the domain (while the height is assumed to be long compared to the diameter). Sufficient solute (Co in the Cu-4 wt% Co system) is assumed to be present in the domain (i.e., the precipitation process is not hampered by lack of solute). The stress state of the system (plot of σ_{yy}) on growth of precipitate to a radius of 30b is shown in Figure 5.26a. Figure 5.26b shows the deformation of the domain in the presence of the precipitate. A comparison of the energy of the "large" domain with the "small" domain is shown in Figure 5.27. It is seen that the small domain has a lower energy as it is able to relax by domain deformations. This becomes evident by comparison of the energy plots for the two domains (as in Figures 5.25 and 5.22), which is illustrated in Figure 5.27.

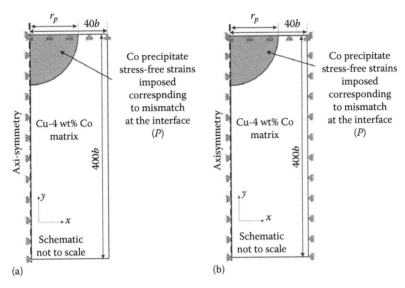

FIGURE 5.25
FEM used for the simulation of a precipitate in a small domain: (a) with free surfaces and (b) with constrained surfaces.

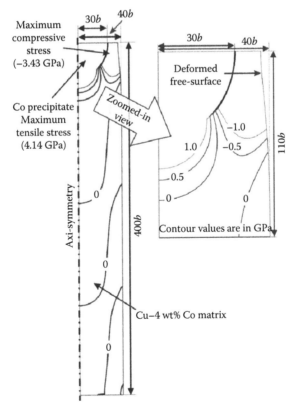

FIGURE 5.26
Stress state (plot of σ_{xx}) in the presence of a coherent precipitate of $30b$ radius in the small domain (model as in Figure 5.25a). Domain deformations are exaggerated by a scale factor of 10.

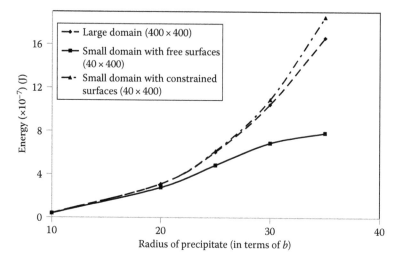

FIGURE 5.27

Comparison of the energy of the precipitate in domains of two sizes ($400b \times 400b$ and $40b \times 400b$). The energy of the smaller domain is compared for the two boundary conditions as shown in Figure 5.25a and b.

5.4.5 Improvements

We have considered a few cases, which outline the utility of finite element simulations at the nanoscale. These simulations are based on some assumptions, and the quality of results can be improved by using (1) less constraining assumptions and (2) FEM in conjunction with other techniques in a multiscale simulation. Here, we list a few of these possible improvements (some of these have been considered earlier):

1. Isotropic material parameters have been used for easy comparison with simple theoretical formulae. Anisotropic material properties should be used, which can be done with ease in the computational models.

2. Core of the dislocation has been replaced with a linear elastic material of size $\sim b$. To improve on this model, an elastic core can be assumed [66], or an atomistic model of the core can be used to feed results into the FEM model [67]. Details of the core structure of interfacial dislocations should also be taken into account.

3. For disordered alloys, a linear interpolation of material properties has been used. This can be replaced by actual experimental measurements of the properties.

4. Bulk material properties have been used in the simulations. Material properties may be different at the nanoscale and in regions close to the surface. Hence, position-dependent material properties (of the nanocrystal) can be used as input to improve the results of the simulations.

5. Interfaces have been assumed to be abrupt and flat. Real interfaces may be diffuse and may have their own set of properties. Typically, surfaces and internal interfaces are not flat and have some roughness associated with them. This could be taken into account by introducing appropriate changes in the simulations.

6. Emphasis has been on 2D simulations (plane strain and axi-symmetric), to illustrate the development of new methodologies. In specific cases, depending on

the parameters to be computed and level of accuracy required, 3D simulations may have to be used to capture reality.

7. Theory of linear elasticity has been assumed in all the simulations. Under certain circumstances, the theory of nonlinear elasticity may have to be used [68].

5.5 Conclusions (and Scope for the Future)

FEM has proved to be a powerful tool to perform simulations at the nanoscale. A variety of structures like dislocations, precipitates, epitaxial films, etc., can be simulated using FEM. Processes can also be simulated using a sequence of static simulations, which include growth of epitaxial films, displacement of edge dislocations, and growth of coherent precipitates. These simulations can be used for the computation of important parameters like image force, bending angle of thin plates in the presence of an edge dislocation, critical size for coherent to semicoherent transition (in epitaxial interfaces and coherent precipitates), etc. Furthermore, new effects and phenomena can be discovered using the finite element studies, which include "neutral equilibrium in material structures" and "reversible plastic deformation due to elasticity."

The assumptions involved and the limitations of any finite element numerical model should be clearly stated, and needless to say, the models should not be used in cases wherein the validity of the simulations is questionable.

The important goal of finite element simulations at the nanoscale is to simulate entire microstructures and components composed of these microstructures. The microstructural entities that need be considered in a single simulate include clusters of vacancies, dislocations, grain boundaries, twins, precipitates, voids/cracks, multiple phases, etc. These simulations are expected to capture effects across various length scales. Further, the evolution of the microstructure under external stimuli (load, temperature) also needs to be determined.

Appendix 5A: Material Properties Used in the Simulations

Material (Structure)	Lattice Parameter(s) (Å)	Burgers Vector (Å)	E (GPa)	G (GPa)	Poisson's Ratio (ν)
Al (FCC)	4.04	2.86	70.57	26.18	0.348
Ni (FCC)	3.52	2.49	236.63	92	0.286
Nb (BCC)	3.30	2.86	100	36.23	0.380
Sapphire (hexagonal)	$a = 4.76\ c = 12.99$	—	390	153.54	0.270
Si (DC)	5.43	3.84	165.86	68.12	0.218
$Ge_{0.5}Si_{0.5}$ (DC)	5.545	3.92	150.62	62.25	0.210
Cu (FCC)	3.61	2.552	197.16	75.3	0.309
Co (FCC)	3.54	2.503	329.52	129.2	0.275

Source: Brandes, E.A., ed., *Smithells Metals Reference Book*, Butterworths, London, U.K., 1983.

References

1. E. Weinan, *Principles of Multiscale Modeling*, Cambridge University Press, Cambridge, New York, 2011.
2. Y. Efendiev and T.Y. Hou, *Multiscale Finite Element Methods: Theory and Applications*, Springer, New York, 2010.
3. V. Nassehi and M. Parvazinia, *Finite Element Modelling of Multiscale Transport Phenomena*, Imperial College Press, London, U.K., 2011.
4. J. Fish, *Multiscale Methods: Bridging the Scales in Science and Engineering*, Oxford University Press, New York, 2009.
5. G.C. Sih, ed., *Multiscaling in Molecular and Continuum Mechanics: Interaction of Time and Size from Macro to Nano*, Springer, the Netherlands, 2007.
6. A. Maiti, *Microelectron. J.* 39 (2008) 208.
7. T.E. Karakasidis and C.A. Charitidis, *Mater. Sci. Eng. C* 27 (2007) 1082.
8. K.H. Huebner, D.L. Dewhirst, D.E. Smith, T.G. Byrom. *The Finite Element Method for Engineers*, John Wiley and Sons, New York, 2001.
9. D.S. Burnett, *Finite Element Analysis: From Concepts to Applications*, Addison-Wesley Publishing Company, Reading, MA, 1987.
10. R.D. Cook, D.S. Malkus, M.F. Plesha, and R.J. Witt, *Concepts and Application of Finite Element Analysis*, John Wiley & Sons Pvt. Ltd., Singapore, 2003.
11. B. Szabó and I. Babuška, *Introduction to Finite Element Analysis: Formulation, Verification and Validation*, John Wiley & Sons Ltd., Chichester, U.K., 2011.
12. D.L. Logan, *A First Course in the Finite Element Method*, Nelson, Toronto, ON, Canada, 2007.
13. E. Madenci and I. Guven, *The Finite Element Method and Applications in Engineering Using ANSYS*, Springer, New York, 2006.
14. A. Subramaniam, *Materials Science and Engineering: An Introductory E-book*, http://home.iitk.ac.in/~anandh/E-book/Materials_Science_&_Engineering_Introductory_E-book.ppt, 2011.
15. R. Phillips, *Crystals, Defects and Microstructures*, Cambridge University Press, Cambridge, New York, 2001.
16. K. Mori, ed., *Simulation of Materials Processing: Theory, Methods and Applications*, Swets and Zeitlinger, Lisse, the Netherlands, 2001.
17. J. Mackerle, *Eng. Comput. Int. J. Comput.-Aided Eng. Softw.*, 23 (2006) 250.
18. F. Roters, P. Eisenlohr, T.R. Bieler, and D. Raabe, *Crystal Plasticity Finite Element Methods: In Materials Science and Engineering*, Wiley-VCH, Weinheim, Germany, 2010.
19. F.L. Matthews, G.A.O. Davies, D. Hitchings, and C. Soutis, *Finite Element Modelling of Composite Materials and Structures*, CRC Press, Boca Raton, FL, 2000.
20. A. Jérusalem, A. Fernández, A. Kunz, and J.R. Greer, *Scripta Materialia* 66 (2012) 93.
21. M.N. Nahas and M.A. Rabou, *Int. J. Mech. Mechatronics IJMME-IJENS* 10 (2010) 19.
22. L. Nasdala and G. Ernst, *Comput. Mater. Sci.* 33 (2005) 443.
23. X.Y. Wang and X. Wang, *Composites B* 35 (2004) 79.
24. A.C. Fischer-Cripps, *Nanoindentation*, Springer, New York, 2011.
25. G.M. Odegarda, T.S. Gates, L.M. Nicholson, and K.E. Wise, *Compos. Sci. Technol.* 62 (2002) 1869.
26. F. Tancret, P. Guillemet, F.F.D. Chabert, R.L. Gall, and J.F. Castagné, *Solid State Phenomena* 172–174 (2011) 881.
27. B. Li and M. Luskin, *SIAM J. Numer. Anal.* 35 (1998) 376.
28. B. Peultier, T. Benzineb, and E. Patoor, *J. Phys. IV Proc.* 115 (2004) 351.
29. J. Bastos, B. Sadowski, N. Sadowski, and M. Dekker, *Electromagnetic Modeling by Finite Element Methods*, Marcel Dekker Inc., New York, 2003.
30. R.E. Miller and E.B. Tadmor, *MRS Bull.* 32 (2007) 920.
31. K. Ammar, B. Appolaire, G. Cailletaud, F. Feyel, and S. Forest, *Comput. Mater. Sci.* 45 (2009) 800.
32. S. Groh and H.M. Zbib, *J. Eng. Mater. Technol.* 131 (2009) 041209-1.

33. P.D. Spanos and A. Kontsos, *Probab. Eng. Mech.* 23 (2008) 456.
34. B. Liu, Y. Huang, H. Jiang, S. Qu, and K.C. Hwang, *Comput. Methods Appl. Mech. Eng.* 193 (2004) 1849.
35. T. Mura, *Micromechanics of Defects in Solids*, Martinus Nijhoff Publishers, The Hague, the Netherlands, 1982.
36. J.P. Hirth and J. Lothe, *Theory of Dislocations*, McGraw-Hill, New York, 1968.
37. V.V. Bulatov and W. Cai, *Computer Simulations of Dislocations*, Oxford Series on Materials Modeling, Oxford University Press, Oxford, U.K., 2006.
38. D. Hull and D.J. Bacon, *Introduction to Dislocations*, Pergamon Press, London, U.K., 1984.
39. P. Khanikar, A. Kumar, and A. Subramaniam, *Adv. Mater. Res.* 67 (2009) 33–38.
40. J.N. Wang, *Mater. Sci. Eng. A* 206 (1996) 259–269.
41. L.B. Freund and S. Suresh, *Thin Film Materials: Stress, Defect Formation and Surface Evolution*, Cambridge University Press, Cambridge, U.K., 2003.
42. C.R.M. Grovenor, *Microelectronic Materials*, Adam Hilger IOP Publishing Ltd., Bristol, U.K., 1989.
43. J.E. Ayers, *Heteroepitaxy of Semiconductors: Theory, Growth and Characterization*, Taylor & Francis Group, Boca Raton, FL, 2007.
44. J.W. Matthews, ed., *Epitaxial Growth*, Academic Press, New York, 1975.
45. S.C. Jain, A.H. Harker, and R.A. Cowley, *Philos. Mag. A* 75 (1997) 1461.
46. S.C. Jain, J.R. Willis, and R. Bullough, *Adv. Phys.* 39 (1990) 127.
47. E. Kasper, H.J. Herzog, and G. Abstreiter, Layered structure and epitaxy, in *Materials Research Society Symposium Proceedings*, J.M. Gibson, G.C. Osbourn, and R.M. Tromp, eds., Materials Research Society, Pittsburgh, PA, Vol. 56 (1986) p. 347.
48. J.W. Matthews and A.E. Blakeslee, *J. Cryst. Growth* 27 (1974) 118.
49. J. Narayan and B.C. Larson, *J. Appl. Phys.* 93 (2003) 278.
50. D.A. Porter and K.E. Easterling, *Phase Transformations in Metals and Alloys*, Chapman & Hall, London, U.K., 1992.
51. J.W. Matthews, Misfit dislocations, in *Dislocations in Solids*, F.R.N. Nabarro, ed., North-Holland Publishing Company, New York, 1979, p. 463.
52. L.M. Brown, G.R. Woolhouse, and U. Valdrè, *Philos. Mag.* 17 (1968) 781.
53. P. Khanikar, A. Kumar, and A. Subramaniam, *Philos. Mag.* 91 (2011) 730.
54. A. Kumar and A. Subramaniam, *Philos. Mag. Lett.* 91 (2011) 272.
55. A. Kumar and A. Subramaniam, *Philos. Mag.*, in press.
56. P. Khanikar and A. Subramaniam, *J. Nano Res.* 10 (2010) 93.
57. A. Kumar and A. Subramaniam, *Int. J. Nanosci.* 10 (2011) 351.
58. A. Kumar, G. Kaur, and A. Subramaniam, to be Communicated to *Philos. Mag.*
59. T. Belytschko and R. Gracie, *Int. J. Plast.*, 23 (2007) 1721.
60. Abaqus Standard, Version 6.81, Dassault Systemes, 2008.
61. E.A. Brandes, ed., *Smithells Metals Reference Book*, Butterworths, London, U.K., 1983.
62. R.E. Newnham, *Properties of Materials*, Oxford University Press, New York, 2005.
63. W. Bollmann, *Crystal Defects and Crystalline Interfaces*, Springer-Verlag, Berlin, Germany, 1970.
64. P. Haasen, *Physical Metallurgy*, Cambridge University Press, Cambridge, U.K., 1996.
65. J.D. Eshelby, *J. Appl. Phys.* 24 (1953) 176.
66. V.A. Lubarda and X. Markenscoff, *Arch. Appl. Mech.* 77 (2007) 147.
67. R. Gracie and T. Belytschko, *Int. J. Numer. Meth. Eng.* 78 (2009) 354.
68. A. Ibrahimbegović, *Nonlinear Solid Mechanics: Theoretical Formulations and Finite Element Solution Methods*, Springer, Dordrecht, the Netherlands, 2009.

6

Continuum and Atomic-Scale Finite Element Modeling of Multilayer Self-Positioning Nanostructures

Y. Nishidate

University of Aizu, Aizu-Wakamatsu, Fukushima, Japan

G.P. Nikishkov

University of Aizu, Aizu-Wakamatsu, Fukushima, Japan

CONTENTS

6.1   Introduction 186
      6.1.1   Self-Positioning Nanostructures 186
      6.1.2   Previous Works 188
6.2.   Continuum Mechanics Solutions 189
      6.2.1   Transformation of Constitutive Matrix 189
      6.2.2   Generalized Plane Strain Solution 192
      6.2.3   Generalized and Ordinary Plane Strain Solutions 196
6.3   Finite Element Modeling 196
      6.3.1   Finite Element Equation System 196
      6.3.2   Stiffness Matrix 197
      6.3.3   Load Vector 199
      6.3.4   Stress Update 200
      6.3.5   Finite Element 201
      6.3.6   Time Integration Scheme 202
      6.3.7   Finite Element Modeling of Anisotropic Structures 203
      6.3.8   Anisotropic Constitutive Law 205
6.4   Atomic-Scale Finite Element Modeling 206
      6.4.1   Atomic-Scale Finite Element Method 206
      6.4.2   AFEM for Geometrically Nonlinear Problems 208
      6.4.3   Modeling of Crystalline Structures 209
      6.4.4   Atomic Interactions Potential 210
      6.4.5   Validating Tersoff–Nordlund Potential for GaAs and InAs Structures 213
      6.4.6   PCG Algorithm for Solution of AFEM Equation System 214
6.5   Strain Estimation Methods in the AFEM 217
      6.5.1   Moving Least Squares Approximation 217
      6.5.2   Radial Basis Function Interpolations 219
      6.5.3   Moving Kriging Interpolation 220
      6.5.4   Neighbor Atom Search 222

6.6   Comparison of Continuum Mechanics Solution with FEM 223
6.7   Comparison of Continuum Mechanics Solutions with AFEM Results 226
       6.7.1   Atomic-Scale Modeling of Self-Positioning Nanostructures.......................... 226
       6.7.2   Effects of Thickness on the Curvature Radius 228
       6.7.3   Effects of Material Anisotropy on the Curvature Radius.............................. 229
       6.7.4   Results for Strains 230
                6.7.4.1   Moving Least Squares Approximation.............................. 231
                6.7.4.2   Kriging Interpolation 233
6.8   Conclusion 235
References.. 236

6.1 Introduction

Self-positioning is a phenomenon that occurs in structures which are subjected to a strain/stress imbalance. Multilayer thin films consisting of different materials are rolled up and form nanohinges and nanotubes. Complicated three-dimensional (3D) nanostructures can be fabricated by utilizing the self-positioning phenomenon. In this research, modeling of the self-positioning nanostructures is performed by the continuum mechanics theory, the finite element method (FEM), and the atomic-scale FEM taking into account cubic crystal anisotropy.

The continuum mechanics solution has been derived for multilayer thin film structures subjected to initial strains under generalized plane strain conditions. The finite element modeling has been applied for estimation of the curvature radius of self-positioning hinges. An atomic-scale finite element procedure has been developed for modeling of self-positioning nanostructures. The results are compared with each other through modeling of bilayer self-positioning nanostructures.

6.1.1 Self-Positioning Nanostructures

Fabrication of nanoscale structures has attracted substantial attention for several years. However, fabrication and manipulation of nanoscale structures are usually difficult to control. Simple and robust procedures for formation of 3D nanostructures are desired for designing smart integrated micro-/nano electromechanical systems (MEMS/NEMS). One promising approach is the method utilizing multilayer structures with crystal lattice mismatching [1–7]. Figure 6.1 illustrates a common example of nanohinge fabrication.

From the bottom, the initial structure consists of a substrate, a sacrificial layer, and lattice mismatched layers (layers 1 and 2). Multilayer structures can be grown on substrates using techniques such as the molecular beam epitaxy or the chemical vapor deposition. After selective etching of the sacrificial layer, lattice mismatched layers are released from the substrate and a rolled-up nanostructure is fabricated. For epitaxially grown samples, typical source of the deformation is crystal lattice mismatching strains in different material layers (Figure 6.2).

In the rolled-up structures, the main structural parameter of interest is the radius of curvature because the final shape and strains in the equilibrium state can be expressed through this parameter according to the continuum mechanics theory. Researchers have devoted extensive efforts for investigation of characteristics of the self-positioning

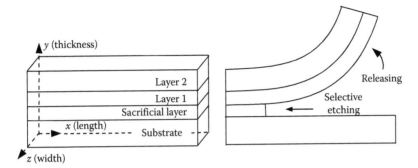

FIGURE 6.1
Fabrication procedure of rolled-up nanohinges. In this example, the lattice period of layer 1 material is larger than that of layer 2.

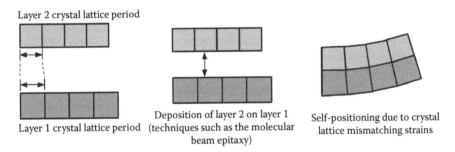

FIGURE 6.2
An explanation of the self-positioning due to crystal lattice mismatching.

structures. For example, several experimental studies have been performed which investigate growth temperature and surface oxidation effects on crystalline structures and curvature radius [8], crystalline structures by x-ray microbeam diffraction [9], effect of surface oxidation on deformations [10], and probing of residual strains by μ-Raman spectroscopy [11,12]. The curvature radius of self-positioning structures can be controlled by selecting material properties, crystal orientations, layer thicknesses, and atomic arrangements. There are many potential applications of the rolled-up nanostructures [13,14]. Self-positioning hinged structures can be used as self-folding membranes for fabrication of 3D nanoscale structures like *origami* (Japanese paper craft work) [15–17]. Applications of the self-positioning nanostructures are still under progress, but several applications have already been developed. For example, nanotubes to convey liquid of reservoirs for on-chip fluid dynamics simulations [18], optical ring resonators [19,20], electromechanical sensors [21], electrochemical capacitors [15], and on-chip microtube refractometers [22] have been developed. Self-positioning hinges can be used for fabrication of 3D cubes (can be utilized as microcontainers) [23,24] and 3D sensors [25]. An experimental investigation has discovered that the Si/SiGe microtubes are extremely flexible and free from plastic deformation even when it has been bent 180° or more [26]. Advances in the fabrication technology of rolled-up nanostructures [19,27–30] will allow development of other potential applications.

Analytical continuum mechanics and computational modeling are essential to explore characteristics of the self-positioning nanostructures to support development of practical nanodevices. Especially, computational modeling is vital for understanding characteristics of rolled-up nanostructures. However, only a few examples of computational modeling can be found. In this chapter, we present continuum mechanics solutions, a finite element algorithm, and an atomic-scale finite element procedure for modeling of self-positioning nanostructures taking into account their anisotropic properties. The continuum mechanics solution is derived under the conditions of generalized plane strain. The conventional FEM has been performed. The atomic-scale finite element procedure has been developed based on the Tersoff–Nordlund interatomic potential model for atomic-scale analysis of self-positioning nanostructures. The atomic-scale finite element procedure is applied to investigation of size and material crystal orientation effects on the self-positioning.

6.1.2 Previous Works

Since rolled-up nanostructures were introduced [1–6], researchers investigated their properties and some ways for development of practical applications. Also, analytical and computational modeling has been performed for estimating deformations of self-positioning nanostructures. Several continuum mechanics solutions are derived based on elasticity theories for estimation of structure deformations. Computational FEM has been performed for more flexible and detailed analyses.

A closed-form solution has been applied for multilayer structures under plane stress conditions [31]. The solution corresponds to structures that have small width, and hence the z stress component is assumed to be zero (definition of axes is shown in Figure 6.1). However, this solution is not appropriate for estimation of curvature radius for structures that have considerable width. Another solution has been obtained under plane strain conditions [32]. The solution is derived for structures that have large widths and bending constraint in one direction assuming that z component of total strain is zero. This is a reasonable idealization in some cases since the width of self-positioning structures is typically large compared to the structure thickness. There is another closed-form solution based on strain energy minimization [33]. Recently, this approach has been extended for ultra-thin films [34,35] by modifying classical Stoney and Timoshenko formula for bending of bilayer films [36,37]. The solution includes surface stress effects in the classical theory. However, it is still possible to improve the assumptions since structures can deform in the width direction as long as displacement is not restricted by other objects. In order to obtain a new solution under more realistic assumptions, we considered a solution under generalized plane strain conditions [38]. The generalized plane strain solution allows presence of the strain in z direction.

Computational FEM has also been performed to estimate curvature radius of structures with more realistic details [39]. The self-positioning problem is formulated as a geometrically nonlinear finite element problem with assumptions that rotational and translational displacements are large but strains are sufficiently small. Analysis shows that the curvature radius predicted by the computational FEM yields plane stress and plane strain solutions for extreme cases (small and large widths, respectively) and intermediate curvature radius values for intermediate widths. Typically, semiconductor materials such as GaAs and InAs are used for fabrication of self-positioning structures. These materials imply material anisotropy depending on crystal orientations. We developed a finite element procedure for modeling effects of the material anisotropy, and the results were compared with experimental data [40].

Advances in computers and atomic-scale computational algorithms allow us to deal with hundreds of thousands of atoms [41]. Several finite element algorithms have been developed for multiscale computational analysis [42–46]. The molecular dynamics (MD) simulation has been applied for investigation of the ultra-thin rolled-up nanotubes [47]. However, small time steps should be selected for the MD to maintain simulation stability, and thus it is a formidable task to simulate atomic systems consisting of large amount of atoms. Furthermore, a damping force or averaging of atomic positions during some time period is necessary to determine the final equilibrium configuration of atoms if the MD is used. Therefore, we consider that the AFEM [42,43] is more suitable for modeling quasi-static nanostructures. We developed an atomic-scale computational procedure for analysis of self-positioning nanostructures [48–50]. The AFEM is appropriate to find static equilibrium configuration of atoms using the total energy minimization.

6.2. Continuum Mechanics Solutions

Continuum mechanics solutions are derived under the ordinary and the generalized plane strain conditions taking into account material anisotropy. The material anisotropy is characterized by the orientation angle of cubic crystal structures. The solutions are obtained by transforming the constitutive matrix with respect to the specified crystal orientation.

6.2.1 Transformation of Constitutive Matrix

For linear elastic materials, the constitutive relations are described by the following generalized Hooke's law:

$$\sigma_{ij} = C_{ijkl}\varepsilon_{kl}^e \tag{6.1}$$

where
  σ_{ij} is the stress tensor
  C_{ijkl} is the fourth order elasticity tensor
  ε_{kl}^e is the elastic fraction of strain tensor

Taking into account symmetry of the tensors, relation (6.1) can be simplified in the following matrix–vector form:

$$\{\sigma\} = [C]\{\varepsilon^e\}. \tag{6.2}$$

Here, the stress vector contains independent components of the stress tensor

$$\{\sigma\}^T = \{\sigma_x,\ \sigma_y,\ \sigma_z,\ \sigma_{xy},\ \sigma_{yz},\ \sigma_{zx}\} \tag{6.3}$$

and the strain vector comprises elastic strains expressed through total strain components and initial strains:

$$\{\varepsilon^e\}^T = \{\varepsilon_{xx} - \varepsilon^0,\ \varepsilon_{yy} - \varepsilon^0,\ \varepsilon_{zz} - \varepsilon^0,\ \varepsilon_{xy},\ \varepsilon_{yz},\ \varepsilon_{zx}\} \tag{6.4}$$

Initial strain ε^0 is induced by the crystal lattice mismatching or thermal effects. For cubic crystals, the constitutive matrix [C] is expressed through three independent material constants C_{11}, C_{12}, and C_{44} as follows:

$$[C] = \begin{bmatrix} C_{11} & C_{12} & C_{12} & & & \\ C_{12} & C_{11} & C_{12} & & & \\ C_{12} & C_{12} & C_{11} & & & \\ & & & C_{44} & & \\ & & & & C_{44} & \\ & & & & & C_{44} \end{bmatrix}. \tag{6.5}$$

The original constitutive matrix [C] is related to the coordinate system x_i, axes of which coincide with the crystallographic orientations [100], [010], and [001]. The constitutive matrix can be transformed into another coordinate system x'_i using the following matrix multiplications:

$$[C'] = [K][C][K]^T \tag{6.6}$$

where the transformation matrix [K] is composed of four 3×3 matrices

$$[K] = \begin{bmatrix} K^{(1)} & 2K^{(2)} \\ K^{(3)} & K^{(4)} \end{bmatrix} \tag{6.7}$$

$$K^{(1)}_{ij} = \alpha^2_{ij} \tag{6.8}$$

$$K^{(2)}_{ij} = \alpha^2_{ij}\alpha^2_{i\,\mathrm{mod}(j+1,3)} \tag{6.9}$$

$$K^{(3)}_{ij} = \alpha^2_{ij}\alpha^2_{\mathrm{mod}(i+1,3)j} \tag{6.10}$$

$$K^{(4)}_{ij} = \alpha_{ij}\,\alpha_{\mathrm{mod}(i+1,3)\mathrm{mod}(j+1,3)} + \alpha_{i\,\mathrm{mod}(j+1,3)}\alpha_{\mathrm{mod}(i+1,3)j} \tag{6.11}$$

Here, $1 \leq i, j \leq 3$; $\mathrm{mod}(i, 3)$ is the modulo function which returns the remainder of i divided by 3; and α_{ij} is a direction cosine between the original and the transformed material coordinate systems as follows

$$\alpha_{ij} = x'_i \cdot x_j \tag{6.12}$$

where dot indicates the vector inner product.

Suppose that the crystal orientation of multilayer structures is characterized by the rotation angle θ as shown in Figure 6.3. Axes x' and z' correspond to the length and the width directions of the structure, and axis y is normal to layer planes. The structure is bent around z' axis due to displacement constraints. Directions of the coordinate axes x' and z' can be different from the crystallographic axes x ([100]) and z ([001]).

Transformation from the original coordinate system (xyz) to the material coordinate system ($x'y'z'$) can be performed by rotation around the y-axis. Relations (6.7) through (6.11)

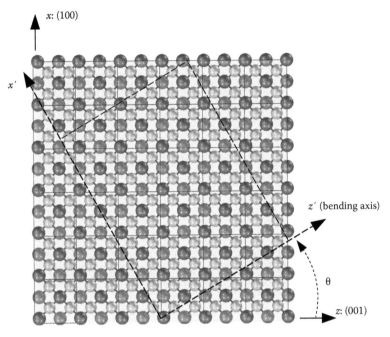

FIGURE 6.3
Original coordinate axes x, y, and z are aligned to the crystallographic axes [100], [010], and [001]. The coordinate axis z' coincides with the bending axis of the self-positioning.

and direction cosines (6.12) lead to the transformation matrix $[K]$ for rotation around the y-axis as follows:

$$[K] = \begin{bmatrix} \cos^2\theta & & \sin^2\theta & & & -2\cos\theta\sin\theta \\ & 1 & & & & \\ \sin^2\theta & & \cos^2\theta & & & 2\cos\theta\sin\theta \\ & & & \cos\theta & -\sin\theta & \\ & & & \sin\theta & \cos\theta & \\ \cos\theta\sin\theta & & -\cos\theta\sin\theta & & & \cos^2\theta-\sin^2\theta \end{bmatrix}. \quad (6.13)$$

Substituting the original elasticity matrix (6.5) and the transformation matrix (6.13) into Equation 6.6, the elasticity matrix $[C']$ for the crystal orientation with angle θ can be obtained

$$[C'] = \begin{bmatrix} C'_{11} & C_{12} & C'_{13} & & & C'_{16} \\ C_{12} & C_{11} & C_{12} & & & \\ C'_{13} & C_{12} & C'_{11} & & & -C'_{16} \\ & & & C_{44} & & \\ & & & & C_{44} & \\ C'_{16} & & -C'_{16} & & & C'_{44} \end{bmatrix}. \quad (6.14)$$

where

$$C'_{11} = \frac{1}{4}\{3C_{11} + C_{12} + 2C_{44} + (C_{11} - C_{12} - 2C_{44})\cos 4\theta\}, \tag{6.15}$$

$$C'_{13} = \frac{1}{4}\{C_{11} + 3C_{12} - 2C_{44} - (C_{11} - C_{12} - 2C_{44})\cos 4\theta\}, \tag{6.16}$$

$$C'_{16} = \frac{1}{4}(C_{11} - C_{12} - 2C_{44})\sin 4\theta, \tag{6.17}$$

$$C'_{44} = \frac{1}{4}\{C_{11} - C_{12} + 2C_{44} - (C_{11} - C_{12} - 2C_{44})\cos 4\theta\}. \tag{6.18}$$

It can be seen that angular dependence of material properties is characterized by coefficient $(C_{11} - C_{12} - 2C_{44})$ multiplied by $\cos 4\theta$ or $\sin 4\theta$. For isotropic materials, constitutive matrix components are equal to

$$C_{11} = \lambda + 2\mu$$

$$C_{12} = \lambda \tag{6.19}$$

$$C_{44} = \mu$$

where λ and μ are Lame's constants. Substitution of (6.19) produces $(C_{11} - C_{12} - 2C_{44}) = 0$ and, naturally, leads to angular independence of elastic properties for isotropic materials.

6.2.2 Generalized Plane Strain Solution

The generalized plane strain solution has been derived for multilayer structures, which is characterized by zero total force in the bending axis direction [38]. The solution can be obtained for anisotropic multilayer structures using the transformed constitutive matrix characterized by direction of cubic crystals [51]. Let us consider an elastic multilayer structure shown in Figure 6.4 under the generalized plane strain condition.

The structure consists of n layers of cubic crystal material with thickness t_i, where $i = 1$, 2, …, n from the bottom. Under the plane section assumption, the structure deforms with zero shear strains, and thus coefficient C'_{16} is not important in the transformed elasticity matrix (6.14).

Taking into account that the stress σ_y is zero, Hooke's law is rewritten as follows:

$$\sigma_x = E'\varepsilon_x + G'\varepsilon_z - (E' + G')\varepsilon_0, \tag{6.20}$$

$$\sigma_z = G'\varepsilon_x + E'\varepsilon_z - (E' + G')\varepsilon_0 \tag{6.21}$$

where

$$E' = \frac{C_{11}C'_{11} - C_{12}^2}{C_{11}} \tag{6.22}$$

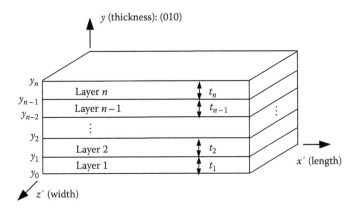

FIGURE 6.4
Multilayer structure and definition of its directions and axes. Axis z' coincides with the bending axis of the self-positioning.

and

$$G' = \frac{C_{11}C'_{13} - C^2_{12}}{C_{11}} \tag{6.23}$$

For isotropic materials, these constants are expressed through the Young's modulus E and the Poisson's ratio ν

$$E' = \frac{E}{1-\nu^2} \tag{6.24}$$

and

$$G' = \frac{\nu E}{1-\nu^2} = \nu E' \tag{6.25}$$

Under the plane section assumption, the strain ε_x can be expressed as a linear function of the local y coordinate

$$\varepsilon_x = c + \frac{y - y_b}{R} \tag{6.26}$$

where
   c is the uniform strain component
   y_b is the y level where the bending component of strain is zero
   R is the curvature radius which is the reciprocal of curvature K

In the generalized plane strain conditions, the total strain ε_z is equal to an unknown constant d:

$$\varepsilon_z = d \tag{6.27}$$

The total strain ε_y can be expressed through other total strain components ε_x and ε_z:

$$\varepsilon_y = -\frac{C_{12}}{C_{11}}(\varepsilon_x + \varepsilon_z - 2\varepsilon^0) + \varepsilon^0 \tag{6.28}$$

It is convenient to rewrite the stresses σ_x (6.20) and σ_z (6.21) using the expressions for strains (6.26) and (6.27):

$$\sigma_x = E'\left(c + \frac{y - y_b}{R}\right) + G'd - (E' + G')\varepsilon^0, \tag{6.29}$$

$$\sigma_z = G'\left(c + \frac{y - y_b}{R}\right) + E'd - (E' + G')\varepsilon^0, \tag{6.30}$$

Relations (6.29) and (6.30) are used to construct the following equilibrium equations under the generalized plane strain conditions.

Force due to bending fraction of stress σ_x:

$$\sum_{i=1}^{n} \int_{y_{i-1}}^{y_i} \frac{E_i'(y - y_b)}{R} dy = 0; \tag{6.31}$$

Force due to uniform fraction of stress σ_x:

$$\sum_{i=1}^{n} t_i \left[E_i'c + G_i'd - (E_i' + G_i')\varepsilon_i^0 \right] = 0; \tag{6.32}$$

Bending moment created by the normal stress σ_x with respect to bending axis z':

$$\sum_{i=1}^{n} \int_{y_{i-1}}^{y_i} \left[E_i'\left(c + \frac{y - y_b}{R}\right) + G_i'd - (E_i' + G_i')\varepsilon_i^0 \right](y - y_b)dy = 0; \tag{6.33}$$

Total force in z-direction:

$$\sum_{i=1}^{n} \int_{y_{i-1}}^{y_i} \left[G_i'\left(c + \frac{y - y_b}{R}\right) + E_i'd - (E_i' + G_i')\varepsilon_i^0 \right]dy = 0; \tag{6.34}$$

Parameter y_b is directly obtained from the equilibrium equation (6.31):

$$y_b = \frac{\sum_{i=1}^{n} E_i't_i(y_i - y_{i-1})}{2\sum_{i=1}^{n} E_i't_i} \tag{6.35}$$

The other parameters c, d, and $K = 1/R$ are obtained from the following equation system:

$$\begin{bmatrix} a_{11} & a_{12} & 0 \\ a_{21} & a_{22} & a_{23} \\ a_{31} & a_{32} & a_{33} \end{bmatrix} \begin{Bmatrix} c \\ d \\ K \end{Bmatrix} = \begin{Bmatrix} b_1 \\ b_2 \\ b_3 \end{Bmatrix}, \tag{6.36}$$

$$a_{11} = a_{32} = \sum E_i' t_i,$$

$$a_{12} = a_{31} = \sum G_i' t_i,$$

$$a_{21} = \sum E_i' t_i \left(y_i^m - y_b \right),$$

$$a_{22} = a_{33} = \sum G_i' t_i \left(y_i^m - y_b \right),$$

$$a_{23} = \frac{1}{3} \sum E_i' t_i \left[4 \left(y_i^m \right)^2 - y_i y_{i-1} - 3 y_b \left(2 y_i^m - y_b \right) \right],$$

$$b_1 = b_3 = \sum \left(E_i' + G_i' \right) t_i \varepsilon_i^0, \quad \text{and}$$

$$b_2 = \sum \left(E_i' + G_i' \right) t_i \varepsilon_i^0 \left(y_i^m - y_b \right),$$

where $y_i^m = (y_i + y_{i+1})/2$.

Equation system (6.36) leads to the generalized plane strain solution for anisotropic multilayer structures composed of materials with the cubic crystal symmetry

$$c = \frac{\sum_{i=1}^{n} t_i \left[\left(E_i' + G_i' \right) \varepsilon_i^0 - G_i' d \right]}{\sum E_i' t_i}, \tag{6.38}$$

$$d = \frac{(a_{12} - a_{11}) a_{23} b_1 - (a_{21} b_1 - a_{11} b_2) a_{22}}{(a_{11} a_{22} - a_{12} a_{21}) a_{22} - \left(a_{12}^2 - a_{11}^2 \right) a_{23}}, \tag{6.39}$$

$$K = \frac{1}{R} = -\frac{3 \sum_{i=1}^{n} t_i (y_i + y_{i-1} - 2 y_b) \left[E_i' c - \left(E_i' + G_i' \right) \varepsilon_i^0 - G_i' d \right]}{2 \sum_{i=1}^{n} E_i' t_i \left[y_i^2 + y_i y_{i-1} + y_{i-1}^2 - 3 y_b (y_i + y_{i-1} - y_b) \right]}, \tag{6.40}$$

where y_b is defined by Equation 6.35.

For bilayer structures ($n = 2$), the expression for curvature K is reduced to

$$K = -\frac{6 t_1 t_2 (t_1 + t_2) \left[E_2' \left(E_1' + G_1' \right) \varepsilon_1^0 - E_1' \left(E_2' + G_2' \right) \varepsilon_2^0 + \left(E_1' G_2' - E_2' G_1' \right) d \right]}{E_1'^2 t_1^4 + E_2'^2 t_2^4 + 2 E_1' E_2' t_1 t_2 \left(2 t_1^2 + 2 t_2^2 + 3 t_1 t_2 \right)}, \tag{6.41}$$

The positive curvature or curvature radius corresponds to the multilayer structure bending downward in the positive y-direction.

6.2.3 Generalized and Ordinary Plane Strain Solutions

A curvature estimate for multilayer systems under the ordinary plane strain conditions is obtained as follows if the value $d = \varepsilon_z$ is zero in Equations 6.38 and 6.40:

$$c_{PS} = \frac{\sum_{i=1}^{n} t_i \left(E_i' + G_i' \right) \varepsilon_i^0}{\sum E_i' t_i}, \tag{6.42}$$

$$K_{PS} = -\frac{3 \sum_{i=1}^{n} t_i (y_i + y_{i-1} - 2y_b) \left[E_i' c_{PS} - \left(E_i' + G_i' \right) \varepsilon_i^0 \right]}{2 \sum_{i=1}^{n} E_i' t_i \left[y_i^2 + y_i y_{i-1} + y_{i-1}^2 - 3y_b (y_i + y_{i-1} - y_b) \right]}, \tag{6.43}$$

Therefore, a curvature estimate is derived for bilayer systems under the ordinary plane strain condition:

$$K_{PS} = -\frac{6 t_1 t_2 (t_1 + t_2) \left[E_2' \left(E_1' + G_1' \right) \varepsilon_1^0 - E_1' \left(E_2' + G_2' \right) \varepsilon_2^0 \right]}{E_1'^2 t_1^4 + E_2'^2 t_2^4 + 2 E_1' E_2' t_1 t_2 \left(2t_1^2 + 2t_2^2 + 3t_1 t_2 \right)}, \tag{6.44}$$

In bilayer systems, the relative difference of the curvature given by the generalized plane strain solution K_{GPS} (6.41) and the ordinary one K_{PS} (6.44) is

$$\frac{K_{GPS} - K_{PS}}{K_{PS}} = -\frac{E_1' G_2' - E_2' G_1'}{E_2' \left(E_1' + G_1' \right) \varepsilon_1^0 - E_1' \left(E_2' + G_2' \right) \varepsilon_2^0} d, \tag{6.45}$$

Therefore, the relative difference between the generalized plane strain solution and the ordinary plane strain solution is determined by the value of the transverse strain d, material properties, and lattice mismatching strains.

6.3 Finite Element Modeling

The FEM is successfully applied in various fields of science and engineering, especially in the field of structural mechanics. A geometrically nonlinear finite element procedure is developed for modeling the self-positioning nanostructures with effects of material anisotropy. It is assumed that rotational and translational displacements are large, but strains are sufficiently small.

6.3.1 Finite Element Equation System

There are many finite element formulations for geometrically, materially nonlinear and coupled nonlinear problems [52–61]. In finite element procedures, displacements are obtained as a solution of a linear equation system. In self-positioning structures, large rotational displacements are involved during deformation. Problems with large deformations are known as geometrically nonlinear problems. In order to simulate nonlinear displacements, finite element equation systems are constructed with respect to displacement increment at each current configuration. In our modeling, the geometrically nonlinear problem is formulated by assuming (1) rotational and translational displacements are large, but (2) strains are small. In order to obtain reasonable approximation of the nonlinear displacements, incremental decomposition of loading is necessary. Our approach to modeling of geometrically nonlinear structures is based on the updated Lagrangian formulation. In the updated Lagrangian formulation, the global finite element equation system for displacement increments has the following appearance:

$$^t K \Delta u = {}^{t+\Delta t} f - {}^t r + {}^{t+\Delta t} h \tag{6.46}$$

where

$^t K$ is the global stiffness matrix at current time t
Δu is the nodal displacement increment vector
$^{t+\Delta t} f$ is the external load vector for nodes
$^t r$ is the internal force vector at nodes
$^{t+\Delta t} h$ is the fictitious force vector at nodes which is used for modeling of initial strain influence

The global finite element equation system is constructed as assemblage of equation systems for all finite elements. The finite element equation at the element level has similar appearance

$$^t K_e \Delta u_e = {}^{t+\Delta t} f_e - {}^t r_e + {}^{t+\Delta t} h_e \tag{6.47}$$

where

$^t K_e$ is the element tangent stiffness matrix at time t
Δu_e is the nodal displacement increment vector
$^{t+\Delta t} f_e$ is the load vector at time $t + \Delta t$
$^t r_e$ is the vector of nodal internal forces that corresponds to stress state at time t
$^{t+\Delta t} h_e$ is the fictitious force vector at time $t + \Delta t$ due to initial strains

In the finite element formulation, displacement boundary conditions are not satisfied during derivation of the equations. Because of this, displacement boundary conditions should be implemented after assembly of finite element equations into the global equation system.

6.3.2 Stiffness Matrix

In updated Lagrangian formulation for geometrically nonlinear problems, the element tangent stiffness matrix $^t K_e$ consists of the linear component $^t K_l$ and the nonlinear addition $^t K_{nl}$ due to stresses [56]:

$$^t K_e = {}^t K_l + {}^t K_{nl} \tag{6.48}$$

The virtual displacement and the potential energy minimization lead to an equation for calculation of the linear component of element tangent stiffness matrix as follows:

$$^t K_l = \int_{^t V} B_m^T C B_n dV \tag{6.49}$$

where
  B is the displacement differentiation matrix
  C is a constitutive matrix which is sometimes referred to as the elasticity matrix
  $^t V$ is the current volume

The combination of unknown variables with interpolation functions at nodes can be used to create continuous fields:

$$x_i(\xi, \eta, \zeta) = \sum_{j=1}^{m} N_j(\xi, \eta, \zeta) x_i^j \tag{6.50}$$

where
  x_i is the variable along ith axis at local coordinates ξ, η, ζ
  x_i^j is the nodal value of a quantity at the jth element node
  N_j is an interpolation function called the shape function in the FEM
  m is the number of nodes in an element

Derivatives of the variable field are obtained by differentiated interpolation functions and unknown variables at nodes. In 3D finite element analyses, the displacement differentiation matrix for local node number i can be written as follows:

$$B_i = \begin{bmatrix} \partial N_i/\partial x & & \\ & \partial N_i/\partial y & \\ & & \partial N_i/\partial z \\ \partial N_i/\partial y & \partial N_i/\partial x & \\ & \partial N_i/\partial z & \partial N_i/\partial y \\ \partial N_i/\partial z & & \partial N_i/\partial x \end{bmatrix}. \tag{6.51}$$

Mechanical properties of materials are characterized by elastic material properties. The elasticity matrix denotes relation between stresses and strains, and in the 3D case, it has the following appearance for isotropic linear elastic materials:

$$C = \begin{bmatrix} \lambda+2\mu & \lambda & \lambda & & & \\ \lambda & \lambda+2\mu & \lambda & & & \\ \lambda & \lambda & \lambda+2\mu & & & \\ & & & \mu & & \\ & & & & \mu & \\ & & & & & \mu \end{bmatrix}. \tag{6.52}$$

where λ and μ are Lame's constants. In 3D analyses, the constants can be expressed through the elasticity modulus E and the Poisson's ratio ν:

$$\lambda = \frac{\nu E}{(1+\nu)(1-2\nu)}$$

$$\mu = \frac{E}{2(1+\nu)}.$$

(6.53)

Since matrices B and C involve zeros, coefficients of the element stiffness matrix ${}^tK_l$ are written as follows after multiplication of three matrices in Equation 6.49:

$$\left({}^tK_l\right)_{ii}^{mn} = \int_{tv}\left[(\lambda+2\mu)\frac{\partial N_m}{\partial^t x_i}\frac{\partial N_n}{\partial^t x_i}+\mu\left(\frac{\partial N_m}{\partial^t x_{i+1}}\frac{\partial N_n}{\partial^t x_{i+1}}+\frac{\partial N_m}{\partial^t x_{i+2}}\frac{\partial N_n}{\partial^t x_{i+2}}\right)\right]dV$$

$$\left({}^tK_l\right)_{ij}^{mn} = \int_{tv}\left[\lambda\frac{\partial N_m}{\partial^t x_i}\frac{\partial N_n}{\partial^t x_j}+\mu\frac{\partial N_m}{\partial^t x_j}\frac{\partial N_n}{\partial^t x_i}\right]dV.$$

(6.54)

where

m, n are local node numbers
i, j are indices related to coordinate axes $(x_1, x_2, x_3)=(x, y, z)$
N_m are nodal shape functions depending on local natural coordinates ξ, η, and ζ

Cyclic rule is used in the previous equation if coordinate indices become greater than 3. The element stiffness matrix consists of linear and nonlinear components as in Equation 6.48. Coefficients of the nonlinear element stiffness matrix K_{nl} are given as follows:

$$\left({}^tK_{nl}\right)_{ij}^{mn} = \int_{tv}{}^t\sigma_{kl}\frac{\partial N_m}{\partial^t x_k}\frac{\partial N_n}{\partial^t x_l}\delta_{ij}\,dV$$

(6.55)

where

${}^t\sigma_{kl}$ are Cauchy stress components in the global coordinate system
δ_{ij} is the Kronecker delta symbol

The repeated indices k and l in the right-hand side imply summation over x_1 to x_3. The global stiffness matrix is obtained as assemblage of the element stiffness matrices. Shape functions are defined in the local coordinate system ξ, η, and ζ. Derivatives of shape functions with respect to global coordinates are obtained through transformation of derivatives of shape function N with respect to local coordinates by Jacobian matrix.

6.3.3 Load Vector

The global load vectors are assembled of element load vectors. Components of element external load vector ${}^{t+\Delta t}f$ are calculated as follows:

$$\left({}^{t+\Delta t}f\right)_i^m = \int_{tv}N_m\,{}^{t+\Delta t}p_i^V\,dV + \int_S N_m\,{}^{t+\Delta t}p_i^S\,dS$$

(6.56)

where

N_m is the shape function at a node with local node number m
p_i^V and p_i^S are ith components of nodal equivalents of volume and surface loads, respectively

After deformation of structures, the internal force appears due to stresses. The nodal internal force ${}^t r$ is calculated by integration of the Cauchy stress components σ_{ik}:

$$\left({}^t r\right)_i^m = \int_{tv} {}^t\sigma_{ik} \frac{\partial N_m}{\partial^t x_k} dV. \tag{6.57}$$

Equilibrium state is achieved if the internal force vector and the external force vector are balanced over the whole structure. Another loading vector ${}^{t+\Delta t}h$ is calculated for modeling of fictitious forces due to initial strains as follows:

$$\left({}^{t+\Delta t}\mathbf{h}\right)^m = \int_{tv} \mathbf{B}_m^T \mathbf{C}\boldsymbol{\varepsilon}^t dV \tag{6.58}$$

where $\boldsymbol{\varepsilon}^t$ is the vector of initial strains or thermal loading. For example, assuming that the thermal expansion coefficient is α and temperature is T, the components of $\boldsymbol{\varepsilon}^t$ are

$$\boldsymbol{\varepsilon}^t = \{\alpha T,\ \alpha T,\ \alpha T,\ 0,\ 0,\ 0\}. \tag{6.59}$$

The fictitious force due to the initial strains can be modeled by substituting the initial strains value instead of thermal loading vector. Suppose that the initial strains are $\boldsymbol{\varepsilon}_x^0$, $\boldsymbol{\varepsilon}_y^0$, and $\boldsymbol{\varepsilon}_z^0$ in each global coordinate direction. Then, the vector $\boldsymbol{\varepsilon}^t$ is

$$\boldsymbol{\varepsilon}^t = \left\{\varepsilon_x^0,\ \varepsilon_y^0,\ \varepsilon_z^0,\ 0,\ 0,\ 0\right\}. \tag{6.60}$$

6.3.4 Stress Update

The Cauchy or true stress is used in the updated Lagrangian formulation of the FEM for the solution of geometrically nonlinear problems with small strains. Incremental decomposition of stresses and strains provides the following relation for the Cauchy stress update [57]:

$$^{t+\Delta t}\sigma_{ij} = {}^t\sigma_{ij} + C_{ijkl}\ {}^{t+\Delta t}\Delta e_{kl}. \tag{6.61}$$

where
$C_{ijkl} = \lambda\delta_{ij}\delta_{kl} + \mu(\delta_{ik}\delta_{jl} + \delta_{il}\delta_{jk})$ is the constitutive tensor
$^{t+\Delta t}\Delta e_{kl}$ is the Almansi strain increment which relates the previous length-increment vector to the current length-increment vector

Incremental Almansi strain components are defined as follows:

$$^{t+\Delta t}\Delta e_{kl} = \frac{1}{2}\left(\frac{\partial \Delta u_i}{\partial x_j} + \frac{\partial \Delta u_j}{\partial x_i}\right). \tag{6.62}$$

Here, Δu is the displacement increment vector that determines the next position from the current position. Derivatives of the displacement increment with respect to the global coordinates can be performed as follows:

$$\left\{\begin{array}{c} \left[\dfrac{\partial \Delta u_i}{\partial x}\right] \\[2mm] \dfrac{\partial \Delta u_i}{\partial y} \\[2mm] \dfrac{\partial \Delta u_i}{\partial z} \end{array}\right\} = J^{-1} \left\{\begin{array}{c} \left[\sum \dfrac{\partial N_m}{\partial \xi} \Delta u_i^m\right] \\[2mm] \sum \dfrac{\partial N_m}{\partial \eta} \Delta u_i^m \\[2mm] \left[\sum \dfrac{\partial N_m}{\partial \zeta} \Delta u_i^m\right] \end{array}\right\}$$

(6.63)

where J is the Jacobian matrix.

6.3.5 Finite Element

Finite element discretization is performed by dividing a structure into some reasonable set of finite elements. We used the 20-node hexahedral serendipity element for discretization of 3D self-positioning nanostructures. The 20-node hexahedral element is shown in Figure 6.5.

In the 20-node hexahedral element, interpolation in a local region is performed by tri-quadratic shape functions. Each shape function has properties that $N_i(\xi, \eta, \zeta) = 1$ at ξ_i, η_i, and ζ_i, but zero at other nodes. Suppose that the local node numbers are assigned as shown in Figure 6.5a. Then the shape functions N_i can be written as follows:

$$N_i = \begin{cases} \dfrac{1}{4}(1-\xi^2)(1+\eta_0)(1+\zeta_0) & i = 2,6,14,18 \\[3mm] \dfrac{1}{4}(1-\eta^2)(1+\xi_0)(1+\zeta_0) & i = 4,8,16,20 \\[3mm] \dfrac{1}{4}(1-\zeta^2)(1+\xi_0)(1+\eta_0) & i = 9,10,11,12 \\[3mm] \dfrac{1}{8}(1-\xi_0)(1+\eta_0)(1+\zeta_0)(\xi_0+\eta_0+\zeta_0-2) & \text{corner nodes} \end{cases}$$

$$\xi_0 = \xi\xi_i, \quad \eta_0 = \eta\eta_i, \quad \zeta_0 = \zeta\zeta_i$$

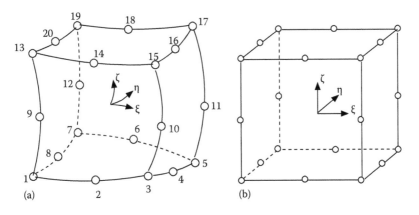

FIGURE 6.5
(a) The 20-node hexahedral element in the physical space with node numbering. There may be different shapes of finite elements in physical space to express the geometry of object. (b) Appearance of the element in the local natural coordinate space. The element nodes are aligned within the range from −1.0 to 1.0 in all three ξ, η, and ζ directions.

The stiffness matrices (6.54 and 6.55) and the load vectors (6.56 through 6.58) should be integrated to construct the finite element equation system. Matrices and vectors are typically evaluated numerically using Gauss integration rule over hexahedral regions in finite element calculations.

The Gauss quadrature with $3 \times 3 \times 3$ is usually used for stiffness matrix integration for 3D quadratic elements. For more efficient integration, the special 14-point Gauss integration rule exists, which provides sufficient precision of integration of the 3D quadratic element. For stress calculations, it should be taken into account that displacement gradients have quite different precision at different points inside finite elements. The highest precision of displacement gradients are obtained at $2 \times 2 \times 2$ reduced integration points for the quadratic hexahedral element [62]. Hence, it is efficient to calculate displacement gradients, such as strains and stresses, at these reduced integration points.

6.3.6 Time Integration Scheme

There are possibilities to improve initial solutions obtained at each configuration, because the solution is calculated by the initial tangent stiffness matrix and load vector which approximate geometrically nonlinear behavior of structures as piecewise linear displacements. The updated Lagrangian formulation accumulates incremental displacements at each configuration, and thus an error at previous finite element equation systems may have certain influence on the result. In order to improve the solution, the Newton–Raphson iteration algorithm is employed to reach more accurate equilibrium state at each configuration. The Newton–Raphson iteration procedure can be written as the following sequence of steps:

$$^{t+\Delta t}K^{(0)} = {}^{t}K$$

$$^{t+\Delta t}u^{(0)} = {}^{t}u$$

$$\psi^{(0)} = {}^{t+\Delta t}f - {}^{t}r$$

do

$$^{t+\Delta t}K^{(i-1)}\, \Delta u^{(i)} = \psi^{(i-1)}$$

$$^{t+\Delta t}u^{(i)} = {}^{t+\Delta t}u^{(i-1)} + \Delta u^{(i)}$$

$$^{t+\Delta t}K^{(i)} = K({}^{t+\Delta t}u^{(i)})$$

$$^{t+\Delta t}r^{(i)} = r({}^{t+\Delta t}u^{(i)})$$

$$\psi^{(i)} = {}^{t+\Delta t}f - {}^{t+\Delta t}r^{(i)}$$

$$\alpha = |\psi^{(i)}| / |\psi^{(0)}|$$

$$\beta = |\Delta u^{(i)}| / |{}^{t+\Delta t}u^{(i)}|$$

while $\alpha > \varepsilon$, and/or $\beta > \varepsilon$.

$$(6.64)$$

The iterations are terminated when the norm of residual vector ratio α or displacement increment ratio β become less than a specified tolerance ε. A common practice is to set the ε from 0.1% to 1% [61]. The schematic of a convergence process in the Newton–Raphson iteration procedure is shown in Figure 6.6.

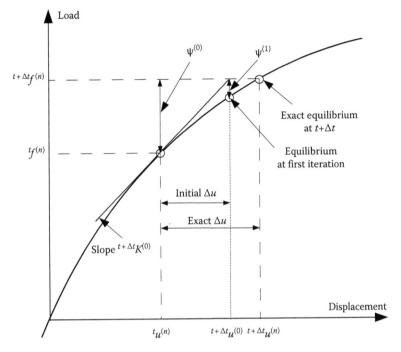

FIGURE 6.6

A procedure of solution improvement by the Newton–Raphson algorithm. An initial displacement increment Δu is obtained by using current tangent stiffness matrix $^t K$ and residual vector $\psi^{(0)}$. By using updated tangent stiffness matrix and residual vector $\psi^{(1)}$, the numerical solution gets closer to the exact solution $^{t+\Delta t} u$.

There are two kinds of Newton–Raphson iteration algorithm. One is the full Newton–Raphson method and the other is the modified Newton–Raphson method [56]. In the modified Newton–Raphson method, the tangent stiffness matrix is recalculated once at the second iteration and later used during further iterations. This allows to avoid updating stiffness matrix at each iteration, but it may be less precise and may take more iterations to reach convergence. It was reported that the modified Newton–Raphson iteration method is more efficient than other time integration schemes such as the Newmark method [63]. We employed the full Newton–Raphson method to obtain the best precision possible for our numerical experiments.

6.3.7 Finite Element Modeling of Anisotropic Structures

Here we present the finite element procedure for 3D modeling of anisotropic elastic geometrically nonlinear structures under influence of initial strains. In order to derive the finite element equation system, three coordinate systems are employed, as shown in Figure 6.7.

1. x_1, x_2, x_3 is the global Cartesian coordinate system (fixed in space, used for the whole structure)
2. ξ_1, ξ_2, ξ_3 is the local element coordinate system (nonorthogonal, movable, one for each finite element)
3. $\bar{x}_1$, $\bar{x}_2$, $\bar{x}_3$ is the material coordinate system (orthogonal, movable, defined at any point inside a finite element)

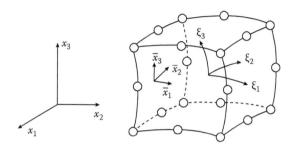

FIGURE 6.7
Global Cartesian coordinate system x_1, x_2, x_3, local element coordinate system ξ_1, ξ_2, ξ_3, and material coordinate system $\bar{x}_1, \bar{x}_2, \bar{x}_3$.

The global coordinate system is employed in the global finite element equation system. The local coordinate systems are used for interpolation within finite elements. Material coordinate axes are involved in anisotropic constitutive relations. The material coordinate system can be introduced at any point of the structure. It rotates with the material with respect to the fixed global coordinate system. Since the local coordinate system ξ_i is not orthogonal, its axes cannot be used directly as axes of the material coordinate system. However, it is convenient to use local coordinate axes ξ_i for building those of the material coordinate system $\bar{x}_i$ at any point inside the finite element. Unit vectors e_{ξ_i} tangent to the local coordinates ξ_i have the following components in the global coordinate system:

$$v_{\xi_i} = \left\{ \frac{\partial x_1}{\partial \xi_i}, \frac{\partial x_2}{\partial \xi_i}, \frac{\partial x_3}{\partial \xi_i} \right\},$$

$$e_{\xi_i} = \frac{v_{\xi_i}}{|v_{\xi_i}|}. \tag{6.65}$$

Approximation of unknown field can be performed by Equation 6.50, so derivatives of the global coordinates with respect to local coordinates can be expressed as

$$\frac{\partial x_j}{\partial \xi_i} = \sum_{k=1}^{m} \frac{\partial N_k}{\partial \xi_i} x_j^k. \tag{6.66}$$

Let us adopt that the unit vector $e_{\bar{x}_1}$ of the material coordinate $\bar{x}_1$ coincides with the direction of the unit vector e_{ξ_1} (tangent to the local coordinate ξ_1). Then, two other unit vectors of the material coordinate system can be determined as vector products:

$$e_{\bar{x}_1} = e_{\xi_1},$$

$$e_{\bar{x}_3} = e_{\xi_1} \times e_{\xi_2}, \tag{6.67}$$

$$e_{\bar{x}_2} = e_{\bar{x}_3} \times e_{\bar{x}_1}.$$

Direction cosines α_{ij} for transformations from the global coordinate system x_i to the material coordinate system $\bar{x}_i$ are expressed through the unit vectors $e_{\bar{x}_i}$:

$$\alpha_{ij} = \cos(\bar{x}_i x_j) = e_{\bar{x}_{ij}}. \tag{6.68}$$

Transformations of vectors from the global coordinate system to the material coordinate system and back are performed in the following ways:

$$\bar{x}_i = \alpha_{ij} \, x_j,$$

$$x_i = \alpha_{ji} \, \bar{x}_j.$$

(6.69)

6.3.8 Anisotropic Constitutive Law

Referring to a fixed orthogonal coordinate system, the stress tensor σ_{ij} and the strain tensor ε_{ij} for an anisotropic elastic material are related through Hooke's law (6.1). The elasticity tensor C_{ijkl} contains 81 coefficients. Because of the symmetry of stress and strain tensors, the elasticity tensor has the symmetry properties $C_{ijkl} = C_{jikl} = C_{ijlk}$, which allows representation of Hooke's law in so-called contracted form using matrix–vector notations

$$\sigma = C\varepsilon,$$

$$\sigma = \{\sigma_{11} \ \sigma_{22} \ \sigma_{33} \ \sigma_{12} \ \sigma_{23} \ \sigma_{31}\}^T,$$

$$\varepsilon = \{\varepsilon_{11} \ \varepsilon_{22} \ \varepsilon_{33} \ 2\varepsilon_{12} \ 2\varepsilon_{23} \ 2\varepsilon_{31}\}^T,$$

(6.70)

where
  C is the contracted 6×6 elasticity matrix
  σ, ε are the contracted 1×6 stress and strain vectors

The contracted form of Hooke's law is convenient for using in a finite element computation since it reduces the number of array dimensions. For triclinic crystal symmetry, the elasticity matrix C is fully populated and symmetric, thus having 21 independent components. We perform numerical modeling of anisotropic structures with cubic crystal symmetry. The elasticity matrix for materials with cubic crystal symmetry has the following appearance:

$$[C] = \begin{bmatrix} C_{11} & C_{12} & C_{12} & & & \\ C_{12} & C_{11} & C_{12} & & & \\ C_{12} & C_{12} & C_{11} & & & \\ & & & C_{44} & & \\ & & & & C_{44} & \\ & & & & & C_{44} \end{bmatrix}.$$

(6.71)

where C_{11}, C_{12}, and C_{44} are material-dependent constants. Transformations from the global coordinate system to the material coordinate system for the stress and elasticity tensors are performed in the full tensor form:

$$\bar{\sigma}_{ij} = \alpha_{ip} \, \alpha_{jq} \, \sigma_{pq},$$

$$\bar{C}_{ijkl} = \alpha_{ip} \, \alpha_{jq} \, \alpha_{kr} \, \alpha_{ls} \, C_{pqrs}.$$

(6.72)

In order to transform elasticity matrix according to a material orientation angle using Equation 6.72, the elasticity tensor C_{ijkl} should be contracted to elasticity matrix C_{ij}, and vice versa. During calculations of element matrices and vectors, it is more efficient to use matrix–vector notation. To perform these operations, it is useful to introduce two index vectors:

$$m = \{1\ \ 2\ \ 3\ \ 1\ \ 2\ \ 3\},$$
$$n = \{1\ \ 2\ \ 3\ \ 2\ \ 3\ \ 1\}. \tag{6.73}$$

Contraction from the stress tensor to the stress vector and expansion from the stress vector to the stress tensor are done as follows:

$$\sigma_p = \sigma_{m_p n_p}, \quad p = 1\ldots6,$$
$$\sigma_{m_p n_p} = \sigma_p, \quad p = 1\ldots6. \tag{6.74}$$

The expansion operation should be accompanied by filling symmetrical terms of the stress tensor $\sigma_{ij} = \sigma_{ji}$. For the elasticity tensor, contraction and expansion operations are carried out in the following ways:

$$C_{pq} = C_{m_p n_p m_q n_q}, \quad p = 1\ldots6,\ q = 1\ldots6,$$
$$C_{m_p n_p m_q n_q}, = C_{pq}, \quad p = 1\ldots6,\ q = 1\ldots6. \tag{6.75}$$

To finish expansion of the elasticity tensor, it is necessary to fill symmetric terms $C_{ijkl} = C_{jikl} = C_{ijlk} = C_{jilk}$. After transformation of constitutive tensor into 6×6 elasticity matrix, the matrix is used as elasticity matrix for calculation of stiffness matrix (6.49) and fictitious force vector (6.58).

6.4 Atomic-Scale Finite Element Modeling

The continuum mechanics solutions and the finite element modeling are based on continuous medium relations. They do not take into account atomic-scale effects. It is interesting to investigate effects of the structure size on the curvature radius. The AFEM was first developed for multiscale analysis of carbon nanotubes (CNTs) [42,43]. The AFEM can model atomistic nature of materials consisting of large number of atoms. The method resembles the conventional FEM. Therefore, several computational algorithms for the FEM can be applied to the AFEM. In the AFEM models, the nodes correspond to the atoms. Therefore, it reproduces real atomic structures, and atomic-scale effects are naturally involved.

6.4.1 Atomic-Scale Finite Element Method

The AFEM equation system is derived from the total energy minimization. Since the underlying principle is general for any kind of atomic structure, the resulting equilibrium configuration is reliable for any type of atomic systems. The AFEM equation system is derived from the approximation of total energy E around current configuration x_i

$$E(x) \approx E(x^{(i)}) + \frac{\partial E}{\partial x}\bigg|_{x=x^{(i)}} \cdot (x - x^{(i)}) + \frac{1}{2}(x - x^{(i)})^T \frac{\partial^2 E}{\partial x^2}\bigg|_{x=x^{(i)}} \cdot (x - x^{(i)}), \qquad (6.76)$$

and its subsequent minimization

$$\frac{\partial E}{\partial x} = 0. \qquad (6.77)$$

Substituting Equation 6.76 into Equation 6.77, the global AFEM equation system can be expressed in a form similar to the conventional finite element equation system

$$Ku = f, \qquad (6.78)$$

where
  K is a global stiffness matrix
  u is a displacement vector
  f is a load (force) vector

In the AFEM, the global stiffness matrix K and the load vector f are composed of second and first derivatives of the system energy with respect to atomic positions

$$K = \begin{bmatrix} \left(\dfrac{\partial^2 E}{\partial x_1 \partial x_1}\right)_{3\times3} & \cdots & \left(\dfrac{\partial^2 E}{\partial x_1 \partial x_n}\right)_{3\times3} \\ \vdots & \ddots & \vdots \\ \left(\dfrac{\partial^2 E}{\partial x_n \partial x_1}\right)_{3\times3} & \cdots & \left(\dfrac{\partial^2 E}{\partial x_n \partial x_n}\right)_{3\times3} \end{bmatrix}, \qquad (6.79)$$

$$f = \begin{bmatrix} \left(-\dfrac{\partial E}{\partial x_1}\right)_{1\times3} \\ \vdots \\ \left(-\dfrac{\partial E}{\partial x_1}\right)_{1\times3} \end{bmatrix}, \qquad (6.80)$$

where
  E is the total energy of the atomic system
  n is the number of atoms (AFEM nodes)
  x_i is the coordinate vector of ith atom

The total energy E coincides with a potential energy of the problem of finding static equilibrium configuration of atomic structures. Without any external loads, the total energy is replaced by the potential energy. Some expressions of the energy and its derivatives should be obtained to formulate the AFEM global equation system. In order to calculate the AFEM equation system (6.78), energy E should be at least twice differentiable with respect to the position.

6.4.2 AFEM for Geometrically Nonlinear Problems

Equation 6.76 implies that the solution has less precision as current configuration gets farther from the equilibrium atomic configuration. The self-positioning nanostructures deform with large translational and rotational displacements. Hence, it is desirable to divide loading into multiple steps and apply them gradually. The following iteration procedure based on the Newton–Raphson algorithm is applied to the geometrically nonlinear problem:

$$\text{do}$$

$$K = K(x^{(i)})$$

$$f = f(x^{(i)})$$

$$\alpha = g(K, f)$$

$$f = \alpha f(x^{(i)}) \tag{6.81}$$

$$\Delta u = K^{-1} f$$

$$x^{(i+1)} = x^{(i)} + \Delta u$$

$$u^{(i+1)} = u^{(i)} + |\Delta u|$$

$$\text{while } |\Delta u| / u^{(i+1)} > \varepsilon$$

At each updated configuration, the tangent stiffness matrix and the load vector should be calculated using Equations 6.79 and 6.80. In the iteration algorithm (6.81), ε is an error tolerance and g is a function for estimating the load relaxation factor α. In general, initial configuration of atoms leads to large forces due to steep energy descent, and thus α may be small at the first step and gets closer to 1 as current position gets closer to the equilibrium atomic configuration where energy descent is small (Figure 6.8). Constant α factor

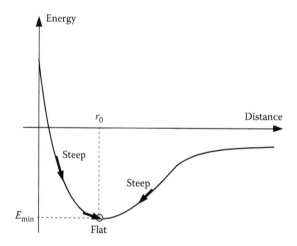

FIGURE 6.8

An atomic interaction potential energy function depending on the distance between atoms. The energy has the minimum value E_{min} at the equilibrium distance r_0. A stronger load relaxation may be applied to suppress large displacements at those steps of solution where slope (derivative) of the energy is steep.

throughout overall solution can cause unnecessary iterations or can lead to divergence, because derivatives (slope) of the interaction potential energy usually become smaller as current configuration gets closer to the equilibrium where loading becomes zero.

A semiautomatic method can be considered to determine the factor, because it is difficult to select an appropriate load relaxation factor during iterations. The user specifies an admissible displacement length to suppress displacements at the solution step within the length where the energy approximation around current configuration may be satisfied (i.e., displacement length $\ll 1$). The relaxation factor α is estimated at each solution step using the tangent stiffness matrix and current load vector. We suggest the following estimate of α

$$u_{mean} = \frac{1}{n} \sum_{i=0}^{n} \left| \left(\frac{f_1^i}{K_{11}^{ii}}, \frac{f_2^i}{K_{22}^{ii}}, \frac{f_3^i}{K_{33}^{ii}} \right) \right|,$$

$$u = \frac{u_{mean}}{a}, \tag{6.82}$$

$$\alpha = \begin{cases} 1, & u \le \delta, \\ \delta/u, & u > \delta, \end{cases}$$

where

n is the number of atoms

f_j^i is the jth component of full load vector acting on the ith atom

K_{jj}^{ii} correspond to diagonal entries of current tangent stiffness matrix

a is a characteristic length of an atomic system

δ is a displacement suppression factor representing admissible mean displacement length

The AFEM equation system is sparse due to using empirical potential energy models with cutoff radius of interaction. Therefore, the displacement approximation using diagonal entries of the stiffness matrix can be used as rough estimate of displacement before solving the equation systems. In the displacement approximation (6.82), we selected characteristic length a as an initial lattice period and the constant δ with the value 2×10^{-4}. Stronger load relaxations are applied if smaller δ is selected, or calculated mean value of the solution guess u_{mean} increases. A larger δ can be selected for problems in which displacements are well approximated by linear functions, but smaller values are necessary for strong geometrically nonlinear problems.

6.4.3 Modeling of Crystalline Structures

Input data for the AFEM should be created in accordance with the crystalline structure of a target material. In this study, a bilayer self-positioning nanostructure consisting of GaAs top and InAs bottom layers is modeled by the AFEM. Our studies focus on lattice mismatching layers, and interactions with sacrificial layer and substrate are not considered here. Alignment of the coordinate axes directions is same as defined in the continuum mechanics solutions. The crystalline structure of GaAs and InAs is the zincblende crystal type. The arrangement of atoms in the zincblende crystals is shown in Figure 6.9. The arsenide atoms occupy the crystal corners and the face centers, and four gallium/indium atoms are located inside with their positions at (0.25, 0.25, 0.25), (0.75, 0.25, 0.75), (0.25, 0.75, 0.75), and (0.75, 0.75, 0.25) in the unit crystal [64].

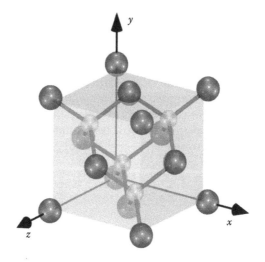

FIGURE 6.9
The atomic configuration and their bonding in the zincblende crystalline structures. There are eight corner and six face center atoms (bigger) and four inner atoms (smaller).

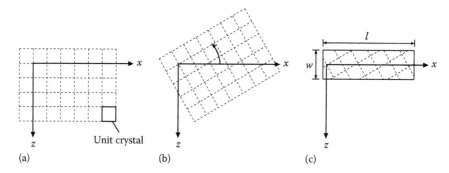

FIGURE 6.10
Modeling atomic structures with different material axes orientations. (a) An initial structure with unit crystals aligned to global axes. (b) The structure rotation around the y axis by specified orientation angle. (c) Removal of atoms outside a rectangular region.

GaAs and InAs possess material anisotropy depending on their crystal orientations. The material crystal orientation is modeled by rotating crystals around the y-axis (thickness direction). Figure 6.10 shows the procedure of the AFEM mesh creation for modeling GaAs and InAs material anisotropy. First, an original structure with zero material orientation angle is prepared. Then the structure is rotated around the y-axis in accordance with the material orientation angle. Finally, atoms outside the rectangular solution domain are removed.

We model structures with material orientation angles 0°, 15°, 30°, 45°, 60°, 75°, and 90°. Appearances of crystalline structures are shown in Figure 6.11 for these orientation angles.

6.4.4 Atomic Interactions Potential

The AFEM employs empirical interatomic potential function which describes atomic interactions. Several empirical interatomic potential models have been developed to

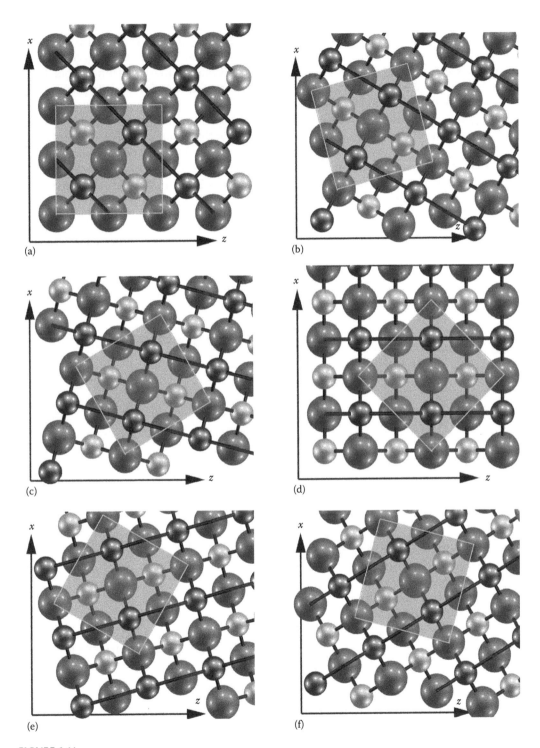

FIGURE 6.11
Top view of zincblende crystal atom configurations for different orientation angles. Orientation angles 0°, 45°, and 90° have periodic boundary in the z (width) direction. (a) 0°, (b) 15°, (c) 30°, (d) 45°, (e) 60°, and (f) 75°.

study the behavior of atomic systems [65–73]. The multibody potential models such as Stillinger–Weber [68], Tersoff [69–71], and Brenner [72,73] are more preferred than other pair potential models like Lennard–Jones [65,67] and Morse potential [66] because two body models are inapplicable to strongly covalent systems like semiconductors [71]. For example, Liu [42,43] and Leung [74] used the potential function and its parameters developed by Brenner [72,73] for AFEM analyses of CNTs.

The target material to be modeled should be determined prior to selection of an interatomic potential model since parameters are sometimes not available for modeling of the material. In this study, we model In–Ga–As systems. The Brenner potential model is widely used and successfully applied for modeling several types of atomic structures, especially carbon systems. However, its parameters for indium, gallium, and arsenide systems are not available. Similar multibody-type potential has been proposed by Tersoff [69–71]. In the Tersoff model, the total potential energy E is given by the following function:

$$E = \sum_i E_i = \frac{1}{2} \sum_i \sum_{j \neq i} V_{ij}, \quad V_{ij} = f_C(r_{ij})[f_R(r_{ij}) + b_{ij} f_A(r_{ij})],$$

$$f_R(r_{ij}) = A_{ij} \exp(-\lambda_{ij} r_{ij}), \quad f_A(r_{ij}) = -B_{ij} \exp(-\mu_{ij} r_{ij}),$$

$$f_C(r_{ij}) = \begin{cases} 1, & r_{ij} \leq R_{ij}, \\ 0.5 + 0.5 \cos\left[\dfrac{r_{ij} - R_{ij}}{S_{ij} - R_{ij}} \pi\right], & R_{ij} \leq r_{ij} \leq S_{ij}, \\ 0, & S_{ij} \leq r_{ij}, \end{cases} \tag{6.83}$$

$$b_{ij} = \left(1 + \beta_{ij}^{n_{ij}} \zeta_{ij}^{n_{ij}}\right)^{-1/(2n_{ij})}, \quad \zeta_{ij} = \sum_{k \neq i, j} f_C(r_{ik}) g(\theta_{ijk}),$$

$$g(\theta_{ijk}) = 1 + \frac{c_{ik}^2}{d_{ik}^2} - \frac{c_{ik}^2}{d_{ik}^2 + (h_{ik} - \cos\theta_{ijk})^2},$$

where
- E_i is the potential energy of atom i
- V_{ij} is the potential energy of a bond i–j
- r_{ij} is the distance from atom i to atom j
- f_C is the cutoff function to disregard effects from distant atoms
- f_R is a reactive component
- f_A is an attractive component
- b_{ij} is a bonding term to represent multiatom interaction effects characterized by bonding angles

The appearance of potential function (6.83) is slightly different from the original Tersoff potential function due to the subsequent parameterization. There are two parameter sets available for indium, gallium, and arsenide systems. One is developed by Ashu [75] and another by Nordlund [76]. Their parameters are almost same, but the parameter set developed by Nordlund is more suitable for our purposes. The parameter values for indium, gallium, and arsenide systems are listed in Table 6.1.

TABLE 6.1

Tersoff Potential Energy Function Parameters for Indium, Gallium, and Arsenide Systems Fit by Nordlund

	InGa	**InIn**	**InAs**	**AsAs**	**GaAs**	**GaGa**
n	3.43739	3.40223	0.7561694	0.60879133	6.31741	3.4729041
c	0.0801587	0.084215	5.172421	5.273131	1.226302	0.07629773
d	19.5277	19.2626	1.665967	0.75102662	0.790396	19.796474
h	7.26805	7.39228	−0.5413316	0.15292354	−0.518489	7.1459174
B	0.705241	2.10871	0.3186402	0.00748809	0.357192	0.23586237
Λ (Å^{-1})	2.5616	2.6159	2.597556	2.384132239	2.82809263	2.50842747
μ (Å^{-1})	1.58314	1.68117	1.422429	1.7287263	1.72301158	1.490824
A (eV)	1719.7	2975.54	1968.295443	1571.86084	2543.2972	993.888094
B (eV)	221.557	360.61	266.571631	546.4316579	314.45966	136.123032
R (Å)	3.4	3.5	3.5	3.4	3.4	3.4
S (Å)	3.6	3.7	3.7	3.6	3.6	3.6

Source: Nordlund, K. et al., *Comput. Mater. Sci.*, 18, 283, 2000.

6.4.5 Validating Tersoff–Nordlund Potential for GaAs and InAs Structures

The parameters obtained by Nordlund correspond to basic elastic and melting crystal properties. These parameters are developed for investigations of damage at Si/Ge, AlAs/GaAs, and InAs/GaAs interfaces. It is noted that the parameters should be used with care for other purposes, and thus we performed several tests to confirm parameters' suitability for our modeling of self-positioning nanostructures. The first test measured elastic properties, and the second test calculated the crystal lattice periods of GaAs and InAs. They are compared to values from a literature. Using the AFEM with the Tersoff potential and the Nordlund parameters, the elastic properties of GaAs and InAs are estimated by applying distributed external load f at the end of specimen shaped into a thin rod along its longitudinal direction (Figure 6.12).

Strains and stresses are calculated at a position sufficiently far from the free end where external load is applied. Taking into account that the specimen is thin in its transverse directions, the Young's modulus E and the Poisson's ratio ν are determined by

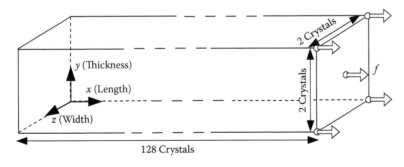

FIGURE 6.12

Conditions of specimen for testing the Tersoff potential function with Nordlund parameters for GaAs and InAs structures.

TABLE 6.2

GaAs and InAs Properties: Comparison of Estimation by the AFEM
with Values from a Literature

	GaAs			InAs		
	Experiment	AFEM	δ (%)	Experiment	AFEM	δ (%)
E (GPa)	85.3	81.0	−5.04	51.8	51.4	−0.77
ν	0.312	0.313	0.32	0.352	0.357	1.42
LP (nm)	0.56533	0.56389	−0.25	0.60584	0.60592	0.01

Source: Bhattacharya, P., *Properties of Lattice-Matched and Strained Indium Gallium Arsenide*, Institution of Electrical Engineers, London, U.K., 1993.

$$E = \frac{\sigma_x}{\varepsilon_x}, \quad \nu = -\frac{\varepsilon_y}{\varepsilon_x}. \tag{6.84}$$

For cubic crystals with axes aligned to the cube edges, the Young's modulus and the Poisson's ratio are estimated from the constitutive tensor components C_{11} and C_{12} as follows:

$$E = \frac{(C_{11} - C_{12})(C_{11} + 2C_{12})}{C_{11} + C_{12}}, \quad \nu = \frac{C_{12}}{C_{11} + C_{12}}. \tag{6.85}$$

The lattice period is estimated at the center of a cube structure consisting of several crystals in all directions. Elastic properties and lattice period estimated by the AFEM modeling for GaAs and InAs are compared with experimental values in Table 6.2.

The maximum difference of 5% is observed for Young's modulus of GaAs; however, in general the correspondence of estimated elastic properties to their experimental values is acceptable. Estimated lattice periods are in very good agreement with experimental values. Therefore, we concluded that the AFEM with Tersoff potential and parameters developed by Nordlund is suitable for simulation of the atomic-scale behavior of nanostructures composed of GaAs and InAs.

6.4.6 PCG Algorithm for Solution of AFEM Equation System

In the AFEM procedure, solution of the equation system (6.78) is the main computational bottleneck as in the ordinary FEM. The order of complexity is $O(N^3)$ in direct algorithms, and hence a fast algorithm with lower complexity can considerably reduce the computational time. The preconditioned conjugate gradient (PCG) method is employed in our study, since the PCG is a fast iterative algorithm for solution of large sparse equation systems. In the AFEM, the sparseness of equation system depends on the number of neighboring atoms participating in atomic interactions. It is determined by the atomic configuration of crystals and the cutoff distances defined in interatomic potential models. Our AFEM equation system is constructed using the Tersoff–Nordlund potential function. The number of neighbors is up to 4 for GaAs and InAs, and hence the equation system is extremely sparse. The sparse row format effectively reduces the matrix storage space for sparse linear systems. In this format, only nonzero entries of the stiffness matrix are stored in one-dimensional array, and two additional arrays are used to indicate starting point of row entries and column index of nonzero elements in each row.

For a given equation system $Ax = b$, the PCG algorithm is written as the following procedure:

$$n = \text{maximum iterations}$$

$$m = \text{CalculatePreconditionMatrix}\ (A)$$

$$r_0 = d_0 = b - Ax_0$$

$$b = |b|$$

$$r = |r_0| / b$$

$$\text{do } i = 0, \ i < n \text{ or } r > \varepsilon$$

$$\alpha = \frac{|r_i|}{d_i^T A d_i}$$

$$x_{i+1} = x_i + \alpha d_i \qquad (6.86)$$

$$r_{i+1} = r_i - \alpha A d_i$$

$$\beta = \frac{|r_{i+1}|}{|r_i|}$$

$$d_{i+1} = r_{i+1} - \beta d_i$$

$$r = \frac{|r_{i+1}|}{b}$$

$$i = i + 1$$

$$\text{end do}$$

$$x = x_i / m_i$$

where m is the vector containing diagonal entries of a preconditioning matrix. The order of complexity is $O(N^2)$ for one PCG iteration due to matrix–vector multiplications, but it can be almost linear complexity for strongly sparse linear systems. The matrix–vector product $v = Au$ for sparse row matrices may be calculated in the following procedure:

$$\text{do } j = 1, \ N$$

$$v[j] = 0$$

$$\text{do } i = prow[j], \ prow[j+1] - prow[j]$$

$$v[j] = v[j] + A[i]v[pcol[i]] \qquad (6.87)$$

$$\text{end do}$$

$$\text{end do}$$

where
  pcol is the vector which contains the positions of all nonzero entries
  prow is the vector which indicates the index of first nonzero entry

The preconditioning matrix M is selected to obtain solution faster, and it is used for multiplying both sides of the equation system:

$$MAx = Mb \qquad (6.88)$$

The matrix M should be obtained through a computationally cheap procedure. The simplest preconditioning matrix is the inverse diagonal preconditioner which is constructed from the reciprocal of diagonal entries of the matrix A as $M_{ii} = 1/A_{ii}$. In order to obtain a better preconditioner, we adopted the norm scaling preconditioning algorithm [78]. This algorithm produces a diagonal preconditioner which is found to scale average row norm of the preconditioned matrix $B = MA$. The row norm in ith row of the preconditioned matrix is defined as follows:

$$|B_i| = \sqrt{\sum_j B_{ij}^2}$$

$$= \sqrt{\sum_j \left(\frac{A_{ij}}{D_{ii}D_{jj}}\right)^2} \qquad (6.89)$$

where $D_{ii} = 1/\sqrt{M_{ii}}$. In the norm scaling algorithm, the norm of ith row $|B_i|$ is scaled to have $|B_i| = 1$. Then, the diagonal matrix D can be obtained as follows:

$$D_{ii} = \sqrt{\sum_j \left(\frac{A_{ij}}{D_{jj}}\right)^2} \qquad (6.90)$$

In order to solve Equation 6.90, the following iterative procedure is introduced:

$$d[i] = \sqrt{A_{ii}}$$

$$\text{do}$$
$$\quad \text{do } i = 0, \ N$$
$$\quad\quad r = 0$$
$$\quad\quad \text{do } j = prow[i], \ prow[i+1]$$
$$\quad\quad\quad r = r + \left(A[j]/d[pcol[j]]\right)^2$$
$$\quad\quad \text{end do}$$
$$\quad\quad d[i] = \sqrt{r}$$
$$\quad \text{end do}$$
$$\quad l = 0$$
$$\quad \text{do } i = 0, \ N$$
$$\quad\quad r = 0$$
$$\quad\quad \text{do } j = prow[i], \ prow[i+1]$$
$$\quad\quad\quad r = r + \left\{A[j]/\left(d[pcol[j]]d[i]\right)\right\}^2$$
$$\quad\quad \text{end do}$$
$$\quad\quad l = l + \sqrt{r}$$
$$\quad \text{end do}$$
$$\text{while } |l/N - 1| > \varepsilon$$

(6.91)

where ε is the error tolerance, selected here as $\varepsilon = 10^{-7}$. In our problems, the norm scaling preconditioning reduces computational time by about 10% comparing to conventional reciprocal diagonal preconditioner. The efficiency is almost same as reported in the publication [78].

6.5 Strain Estimation Methods in the AFEM

In the conventional FEM, finite elements are used as interpolation functions between nodes which can be used for estimation of derivatives. Therefore, the number of nodes typically corresponds to element geometry in the FEM. In the AFEM, connectivities between atoms are used to calculate the stiffness matrix, and there is no solid element as used in the FEM. Direct differentiation is impossible since nodal displacements are discrete. One of the possibilities is to construct triangle or tetrahedral elements for two-dimensional (2D) and 3D cases, but division of the domain into tetrahedral element is complicated in the 3D space. Another possibility is to employ scattered data interpolation methods used in mesh free FEMs [79–87]. Mesh free FEMs are developed to reduce difficulties for construction of finite elements from nodes which are distributed arbitrarily. In the mesh free FEMs, interpolation functions are constructed by special interpolation techniques. Usually, local interpolation functions are constructed by selecting neighboring nodes based on distances.

The moving least squares (MLS) approximation may be the most popular method which is incorporated in several mesh free FEMs [79,82]. The MLS is an extension of the least squares approximation, and it is intuitive and relatively easy to understand. In general, there are fundamental difficulties to specify essential boundary conditions since the MLS shape functions do not go through exact nodal values. Because of this, the other interpolation methods have been considered and applied to mesh free FEMs. The other popular interpolation method is the kernel approximation in the reproducing kernel particle method (RKPM) [80,81]. Equivalence between the kernel approximation and MLS has been discussed [84,88], and it has been shown that these methods produce mathematically same shape functions. Radial basis function (RBF) is an interpolation method which satisfies Kronecker delta properties. The RBF shape functions are produced to satisfy $f(x_i) = f_i$ where $f(x_i)$ is the approximation of the value f_i at the coordinate x_i. Various RBFs have been developed for scattered data interpolation: multiquadratic (MQ) [89–92], polynomials [93–95], and hybrid interpolations [85,96]. Among these functions, Hardy's MQ [89] is the best RBF interpolation for a set of test functions [97]. However, the quality of MQ interpolation strongly depends on selection of a shape parameter [90–92]. Recently, the Kriging interpolation method has been applied to mesh free FEMs [98–100]. The Kriging interpolation is a geostatistical technique for spatial data interpolations in geology and mining. The procedure for calculation of the Kriging interpolation is similar to the MLS, but the Kriging interpolation possesses Kronecker delta properties. We implemented MLS, RBF, and Kriging interpolations for calculation of strains in the AFEM procedure.

6.5.1 Moving Least Squares Approximation

In the least square approximations, unknown coefficients of a given set of polynomials are found by minimizing the sum of differences between nodal values with the set of polynomials. Typically, the least square approximation is a linear function fitting for a given data set, while the MLS approximate it in each partial region. Strains are obtained through differentiations of interpolation functions of displacements. In the moving least squares approximation, the displacement u^h approximation at a position x is calculated as linear combination of fitting polynomial terms and their coefficients as follows:

$$u^h(x) = p^T(x)a(x) \tag{6.92}$$

where p is the term vector consisting of nonlinear functions p_i of x. For example, if the basis is a quadratic function, the term vector may be

$$p^T = \{1,\, x,\, y,\, z,\, x^2,\, y^2,\, z^2,\, xy,\, yz,\, zx\} \tag{6.93}$$

The coefficient vector a is obtained from solution of the following equation system:

$$M(x)a(x) = N(x)u \tag{6.94}$$

where u is a vector consisting of a component of displacement components at nodes. Matrices M and N are obtained as follows:

$$M(x) = P^T W(x)P, \tag{6.95}$$

$$N(x) = P^T W(x), \tag{6.96}$$

$$P = \begin{bmatrix} p^T(s_1) \\ \vdots \\ p^T(s_n) \end{bmatrix}, \tag{6.97}$$

$$W(x) = diag\{w_1(x),\, \cdots\, w_n(x)\} \tag{6.98}$$

where W is the diagonal matrix consisting of weight function values at nodes. There are several possibilities for selection of the weight function. Popular weight functions are the Gaussian and cubic/quartic spline functions. We use a quartic spline weight function w_i of distance r_{ij}

$$w_i(x_j) = 1 - 6\left(\frac{r_{ij}}{d}\right)^2 + 8\left(\frac{r_{ij}}{d}\right)^3 - 3\left(\frac{r_{ij}}{d}\right)^4 \tag{6.99}$$

$$r_{ij} = \sqrt{(x_j - x_i)^2 + (y_j - y_i)^2 + (z_j - z_i)^2} \tag{6.100}$$

where d is the largest distance among pairs of nodes in the partial region. Strains are calculated by partial derivatives of the MLS approximation (6.92) with respect to atom positions:

$$\frac{\partial u^h}{\partial x} = \frac{\partial p^T}{\partial x}a \tag{6.101}$$

6.5.2 Radial Basis Function Interpolations

The radial basis function interpolation of a displacement u^h is given by

$$u^h(x) = R^T(x)a \tag{6.102}$$

$$R^T(x) = \{R_1(x), \cdots, R_n(x)\}$$
$$= \{R(|x - s_1|), \cdots, R(|x - s_n|)\}. \tag{6.103}$$

where
  R is a radial basis function (RBF)
  a is the vector consisting of coefficients which are calculated to satisfy Kronecker delta
    properties of the RBF interpolation
  s_i is the coordinate of ith node
  n is the number of nodes

The Kronecker delta properties of interpolation stand for $u^h(x_i) = u_i$ where x_i is the coordinate vector and u_i is the displacement at ith node. The coefficient vector a is obtained from the following equation system:

$$R_0 a = u \tag{6.104}$$

$$R_0(x) = \begin{bmatrix} R(s_1, s_1) & \cdots & R(s_1, s_n) \\ \vdots & \ddots & \vdots \\ R(s_n, s_1) & \cdots & R(s_n, s_n) \end{bmatrix} \tag{6.105}$$

$$u^T = \{u_1, \cdots, u_n\} \tag{6.106}$$

Therefore, strains are obtained by differentiating Equation 6.114:

$$\frac{\partial u^h}{\partial x} = \frac{\partial R^T(x)}{\partial x} a \tag{6.107}$$

Various RBFs have been proposed for scattered data interpolations. Existing RBFs are MQ, inverse MQ, thin-plate spline, Gaussian, polynomials, and their combinations. Among known RBFs, the Hardy's MQ [89] is the most popular since Franke reported that the MQ exhibits the best performance on majority of his test functions [97]. Zhang [83] used several RBFs: the polynomial forms developed by Wendland [93] and Wu [94], combinations of logarithm and polynomial by Buhmann [95], the MQ, the reciprocal MQ (RMQ), the Gaussian, and the thin plate spline. In their experiments, the MQ exhibits the best performance. The original form of MQ has the following appearance:

$$R_i(r_j) = \left(r_i^2 + c^2\right)^q \tag{6.108}$$

where
  c is the shape parameter
  q is the order ($q = 0.5$ in the Hardy's MQ)

The shape of MQ gets closer to cone as the shape parameter c gets closer to 0, and it gets closer to flat surface with large c values [90,91]. It is clear that parameter c controls the quality of the RBF interpolations. Several suggestions have been made for selection of the optimal shape parameter c. For example, Franke [97] proposed the following equation for selection of c:

$$c = 1.25 \frac{d_i}{\sqrt{n}} \tag{6.109}$$

where
    d_i is the distance between the *i*th node with its nearest neighbor
    n is the number of neighbors

Wang [85] used the MQ augmented with polynomials for 2D mesh free FEMs. They evaluated the MQ by varying the exponent *q* instead of the shape parameter *c*. With fixed $c = 1.42$, the best result was obtained for $q = 1.03$. Several researchers used their variations of the MQ [87,96,101,102]. In the RBF interpolation, results of interpolation are sensitive to the shape of RBF. Rippa [92] developed an algorithm for automatic selection of *c*, but it is an empirical approach and computationally expensive. Currently, a common algorithm for automatic determination of the optimal *c* is not available. Appropriate shape parameter may be selected according to intuition and experiments. Researchers adopt existing or their own functions which may be suitable for their problem.

6.5.3 Moving Kriging Interpolation

In the moving Kriging interpolation, a displacement u^h is expressed by the relation

$$u^h(x) = p^T(x)a + r^T(x)b, \tag{6.110}$$

where
    p is the vector consisting of the terms of a fitting polynomial
    r is the vector consisting of the correlation functions
    a and *b* are vectors containing coefficients for the vectors *p* and *r*

Without the term $r^T(x)b$, the interpolation is similar to the MLS. Actually, the term $r^T(x)b$ corresponds to a correction which forces going through nodal values. In order to construct the Kriging interpolations, the coefficient vectors *a* and *b* are calculated from the following equations:

$$A = (P^T R^{-1} P)^{-1} P^T R^{-1}, \tag{6.111}$$

$$a = Au, \tag{6.112}$$

$$b = R^{-1}(I - PA)u. \tag{6.113}$$

Here, *u* is the vector consisting of the displacement components at all nodes in a partial region, matrix *P* contains the term vector *p* at given nodes in the region, and matrix *R* is composed of the correlation function values for a pair of given nodal coordinates

$$P = \begin{bmatrix} p^T(s_1) \\ \vdots \\ p^T(s_n) \end{bmatrix},$$

(6.114)

$$R = \begin{bmatrix} 1 & R(s_1, s_2) & \cdots & R(s_1, s_n) \\ R(s_2, s_1) & 1 & & \vdots \\ \vdots & & \ddots & R(s_{n-1}, s_n) \\ R(s_n, s_1) & \cdots & R(s_n, s_{n-1}) & 1 \end{bmatrix}$$

(6.115)

where
  s_i is the coordinate of the ith node
  n is the number of nodes
  R is the correlation function

The polynomial term vector p and the correlation vector r are given by

$$p^T(x) = \{t_1(x), \cdots, t_m(x)\}$$

(6.116)

$$r^T(x) = \{R(s_1, x), \cdots, R(s_n, x)\}$$

(6.117)

where t_i is the ith term of a given polynomial and m is the number of terms. There are numerous possibilities for selection of the polynomial and the correlation function. The polynomial controls the order of interpolation while the correlation function determines a relation between nodes. In this study, the following quadratic polynomial basis is selected as follows:

$$t^T(x) = \{1,\ x,\ y,\ z,\ x^2,\ y^2,\ z^2,\ xy,\ yz,\ zx\}$$

(6.118)

with the following quartic spline correlation function

$$R(x_i,\ x_j) = 1 - 6\left(\theta \frac{r_{ij}}{d}\right)^2 + 8\left(\theta \frac{r_{ij}}{d}\right)^3 - 3\left(\theta \frac{r_{ij}}{d}\right)^4,$$

(6.119)

$$r_{ij} = \sqrt{(x_j - x_i)^2 + (y_j - y_i)^2 + (z_j - z_i)^2},$$

(6.120)

and its parameters developed in publication [103]

$$\theta = \begin{cases} 0.1329n - 0.3290 & 3 \leq n < 10 \\ 1.0 & 10 \leq n \end{cases}$$

(6.121)

where
  d is the largest distance of a pair of nodes in a partial region
  θ is the positive correlation parameter

Strain components are calculated by differentiation of the Kriging interpolation (6.110) with respect to coordinates x:

$$\frac{\partial u^h}{\partial x} = \frac{\partial p^T}{\partial x} a + \frac{\partial r^T}{\partial x} b \qquad (6.122)$$

Vectors a and b do not depend on the coordinate vector x. Thus, strains are calculated from derivatives of given polynomial terms in the vector p and correlation functions in r with respect to the direction of differentiation x.

6.5.4 Neighbor Atom Search

Atomic interactions are calculated based on real atomic structures in the AFEM. Therefore, neighboring atoms should be found in accordance with the connectivities between atoms in the materials of our interest. In the GaAs and InAs zincblende crystals with periodic boundary in z direction, the number of neighboring atoms are from 2 to 4. It is difficult to obtain smooth fitting and interpolation functions by mesh free interpolation methods using such number of atoms. The neighbor atom search should be performed prior to construction of interpolation functions, but it requires computational complexity $O(N^2)$. The number of atoms can be millions, and thus we employ a bucketing algorithm to reduce computational costs for neighbor nodes searching. Suppose we want to find neighbor nodes around a center atom (atom O in Figure 6.13a). In the ordinary neighbor node search, the distances between all nodes are

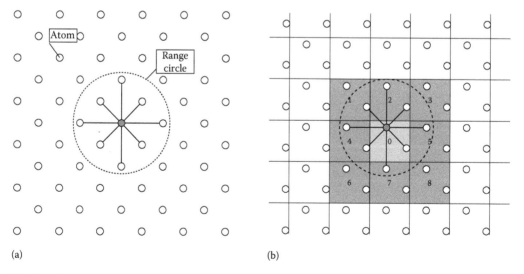

(a)   (b)

FIGURE 6.13

(a) The neighbor nodes are found based on the distance from the center atom (O) to all nodes in ordinary neighbor search procedure. (b) In the bucket search procedure, the space is partitioned into subspaces and all nodes are registered into one of the subspaces according to their position. In this case, calculations of distance with nodes belonging to regions 0–8 are enough to find neighbors.

calculated, and if the distance with an atom is less than a specified radius, the atom is one of the neighbors.

Figure 6.13b illustrates the 2D bucket search procedure. The space is divided into rectangular regions, and all atoms are registered to one of these regions according to their coordinates. In order to find neighbors, it is sufficient to calculate distances between the center atom and the other atoms belonging to the same region and neighboring regions. Suppose that neighbors that are sought for the center atom belongs to the region 0 (Figure 6.13b). Then, calculations of distance are performed only with nodes belonging to regions 0–8. This reduces neighbor search computations to the order of $O(MN)$, where M is the average number of neighbors in candidate regions. The value M does not depend on the problem size, and thus the complexity of neighbor searching depends on N using the bucket search procedure.

6.6 Comparison of Continuum Mechanics Solution with FEM

The finite element procedure for the solution of 3D anisotropic geometrically nonlinear problems has been implemented as a C++ computer code. The computer code is based on the object-oriented implementation of the finite element procedure with mesh generation as well as visualization method introduced in Ref. [104]. Also, a Java applet is created which implements the generalized plane strain solution for multilayer anisotropic structures, and it can be accessed at the URL [105]. Results of our FEM procedure and experimental data were compared in the previous study for self-positioning nanostructures with different crystal orientations [40]. Qualitative agreements are observed between the results, and we consider that differences in value may be attributed to disability of our FEM to consider imperfect crystals due to presence of dislocations. Here, results of the FEM are compared with the continuum mechanics solution for bilayer anisotropic self-positioning nanostructures. Curvature radii and strains are calculated for a bilayer self-positioning nanostructure composed of GaAs top and $In_{0.2}Ga_{0.8}As$ bottom layers for varying cubic crystal orientations (Figure 6.14). The top and the bottom layers have thickness 88 and 56 nm, respectively.

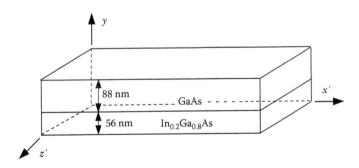

FIGURE 6.14
The bilayer system composed of GaAs top and $In_{0.2}Ga_{0.8}As$ bottom layers for curvature radius and strain estimations for varying crystal orientations.

TABLE 6.3

Constitutive Tensor Components of
GaAs and In$_{0.2}$Ga$_{0.8}$As with Unit GPa

	GaAs (Layer 2)	In$_{0.2}$Ga$_{0.8}$As
C_{11}	119.0	111.88
C_{12}	53.4	51.8
C_{44}	59.6	55.58

The material properties are shown in Table 6.3. The lattice mismatching strain ε_i^0 is estimated by using the following equation:

$$\varepsilon_i^0 = \frac{a_i - a_0}{a_0} \tag{6.123}$$

where
a_0 is an initial lattice period
a_i is the crystal lattice period of the ith layer

The lattice period of top layer (layer 2) is selected as the initial lattice period. Therefore, corresponding lattice mismatching strains are $\varepsilon_1^0 = 1.433 \times 10^{-2}$ and $\varepsilon_2^0 = 0.0$. In the FEM procedure, the initial strain ε^0 is divided into 100 increments.

Figures 6.15 and 6.16 show the final shape of the self-positioning nanostructure obtained by the FEM for orientation angle 0° with ε_x and ε_y distributions. Strain distributions are enlarged at the structure center. The structure is subjected to the displacement boundary conditions at one of the structure ends which imitates a fixed end. At another end, a massive

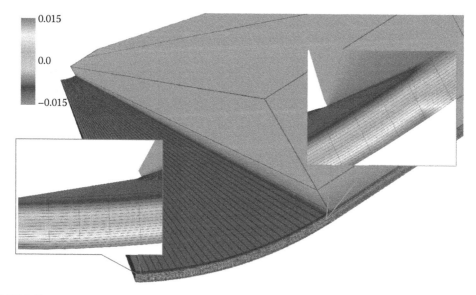

FIGURE 6.15

Final shape of self-positioning structure and distribution of ε_x for the structure with orientation angle 0°. Enlarged views are given for each end.

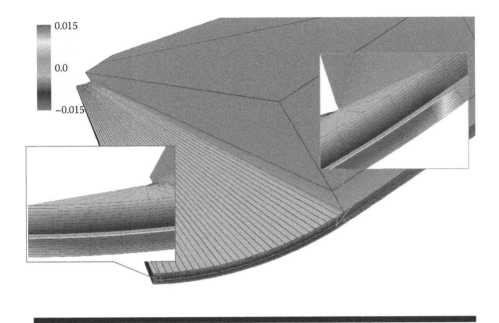

FIGURE 6.16
Final shape of self-positioning structure and distribution of ε_y for the structure with orientation angle 0°. Enlarged views are given for each end.

structure is attached to restrict curling of the film in the transverse direction. Symmetric half of the structure is created, and symmetric boundary conditions are applied at the plane of symmetry. The finite element results for the ε_x linearly depend on the structure local y coordinate as assumed in continuum mechanics solutions. The distribution of the ε_y is discontinuous at the interface of GaAs and $In_{0.2}Ga_{0.8}As$ layers due to difference in the lattice mismatching strain and material properties.

The values of curvature radius estimated by the ordinary plane strain solution (PS), generalized plane strain solution (GenPS), and FEM for varying crystal orientation angles are shown in Figure 6.17. The curvature radii vary as the sinusoidal function with 90° period, and ordinary and generalized plane strain solutions predict almost same values. Numerical and analytical solutions are in good agreement, but small differences are observed. The differences are due to ignoring components C'_{16} in the elasticity matrix (6.14) and inability of our finite element program to exactly reproduce generalized plane strain conditions. The component C'_{16} becomes zero for orientation angles 0° and 45°, and thus the continuum mechanics solutions and the FEM are in better agreement for these orientations.

Figure 6.18 shows comparison of local strain components calculated by the PS and the GenPS solutions with the FEM along local y direction. The results are shown for crystal orientation 0° and 45°. The PS solution assumes strain $\varepsilon_z = 0$, but both GenPS solution and FEM predict about 0.5% expansion in z direction. All strain components are in better agreement with the FEM using the GenPS solution. Results of the FEM and plane strain solutions were compared in this section. The continuum mechanics solutions may be more convenient for practical use, such as prediction of the self-positioning nanostructure deformation during laboratory experiments, but the FEM may be required to know detailed and more precise properties.

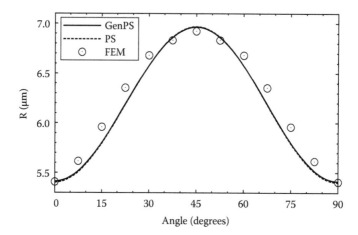

FIGURE 6.17
Curvature radius calculated by the generalized plane strain solution (GenPS), the ordinary plane strain solution (PS), and the finite element modeling (FEM) for varying orientation angle.

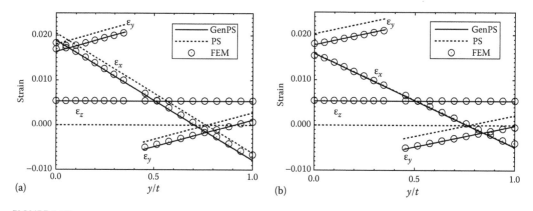

FIGURE 6.18
Comparison of strain components for orientation angles (a) 0° and (b) 45° by the generalized plane strain solution (GenPS), the ordinary plane strain solution (PS), and the finite element modeling (FEM) along local y direction. Abscissa is normalized by the total thickness t.

6.7 Comparison of Continuum Mechanics Solutions with AFEM Results

6.7.1 Atomic-Scale Modeling of Self-Positioning Nanostructures

The analytical and numerical solutions are based on continuum mechanics theory. Here, the continuum mechanics solutions are compared with the atomic-scale modeling to investigate effects of the structure size on the self-positioning. In order to model bilayer self-positioning structures consisting of GaAs top and InAs bottom layers, a problem size parameter c is defined to describe the size of atomic systems as shown in Figure 6.19. In the thickness direction, bilayer hinges are composed of c unit crystals of InAs in the bottom layer and $3c$ unit crystals of GaAs in the top layer. Thus, there are totally $4c$ unit crystals in the thickness (y) direction. In the length (x) direction, atomic layers of length $16a_0c$ are

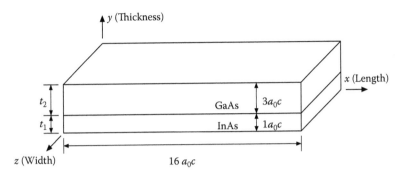

FIGURE 6.19
A schematic of the AFEM model for numerical modeling. The number of atoms in the problem is determined by the parameter c.

prepared where a_0 is an initial lattice period (Figure 6.19). The GaAs and InAs bilayer structures with the problem size parameter $c = 1, 2, 4, 8, 12, 16, 24$, and 36 are prepared to investigate size effects on equilibrium configurations. Corresponding numbers of atoms in atomic systems are 1106, 4258, 10706, 66178, 148418, 263426, 591746, and 1329986, respectively. The effect of material anisotropy is investigated for crystal orientation angles $0°, 15°, 30°, 45°, 60°, 75°$, and $90°$. For hinges with orientation angles $0°$ and $90°$, $16a_0c$ is equal to the length of $16c$ unit crystals.

In the AFEM models, periodic boundary conditions are applied to imitate atomic systems of infinite dimensions that also help to minimize number of atoms in the model. The periodic boundary condition is applied in z (width) direction for structures with orientation angles $0°, 45°$, and $90°$. Such structures consist of one complete and another incomplete crystal in the width direction, and connections across periodic boundary are created when looking for neighboring atoms. For hinges with other orientation angles ($15°, 30°, 60°$, and $75°$), enough number of atomic layers in the width direction, such as $30a_0$, are prepared and the displacement is constrained in the width direction which corresponds to the plane strain conditions ($\varepsilon_z = 0$). Also, displacement boundary conditions are applied to the model at one of the ends in the length direction in order to fix it in the space. On the plane $x = 0$, atoms are fixed in the x direction, and a few atoms near the center of the plane are fixed in the y direction to restrict translation in the vertical direction. The periodic boundaries are created across the structure width (z) direction. The model is used for investigation of curvature radius dependence on the structure thickness and crystal orientation angle. Since the crystal lattice period of InAs is larger than that of GaAs, the equilibrium configuration of the structure after self-positioning approximately corresponds to a circular arc with its center located on the positive y-axis. Since model deformation is under ordinary plane strain conditions, the results obtained by the AFEM are compared with the continuum mechanics solution under plane strain conditions for varying crystal orientation.

The thickness for structures consisting of just a few crystal layers in the thickness direction should be accurately defined when calculating curvature radius using continuum mechanics solutions. It may be appropriate to add some offset equal to a "radius" of an atom at each free surface. Adding such an offset is not critical for thick structures, but it can be important for problems with small number of unit crystals. If we adopt half of the interatom distance as an offset, then the offset is equal to $\sqrt{3}a/8$ for zincblende crystal structures, where a is a crystal lattice period. Corresponding offsets are 0.1224 nm for

GaAs and 0.1425 nm for InAs crystals. Initial strains ε_i^0 used in continuum mechanics solutions are determined by initial (a_0) and material natural lattice period (a_i) as

$$\varepsilon_i^0 = \frac{a_1 - a_0}{a_0}. \tag{6.124}$$

The initial lattice period for bilayer systems is determined using a weighted linear interpolation of GaAs and InAs natural lattice periods

$$a_0 = \frac{a_1 n_1 + a_2 n_2}{n_1 + n_2} \tag{6.125}$$

where n_1 and n_2 are number of crystals in each InAs and GaAs layer. Therefore, a_0 is assumed to be 0.57546 nm in our problems.

6.7.2 Effects of Thickness on the Curvature Radius

The first problem series calculates curvature radius for varying thickness. In the problem series, size parameter c is set to 1, 2, 4, 8, 12, 16, 24, and 36. Such c values lead to structures with thickness 2.56, 4.86, 9.46, 18.65, 27.84, 37.03, 55.41, and 82.98 nm. The largest AFEM model consists of 1,329,986 atoms that correspond to about 4 million equations. Equilibrium configurations of bilayer hinges are determined with the use of the iteration procedure (6.87). The AFEM values of the curvature radius are compared with the continuum mechanics solution under the plane strain conditions (Equation 6.44). In the continuum mechanics solution, elastic properties estimated by the AFEM on the tensile rod model are used (Table 6.2).

The computations are performed on an ordinary personal computer equipped with 2 GB RAM. In case of the smallest problem, it takes 1.2 s for assembly of the AFEM equation system, and 0.3 s for solution of the equation system per loading step. Nine iterations took 13.5 s. Solution of the largest problem requires 0.5 h for assembly, and 4.5 h for solution of the equation system per step. The number of steps was 8, and thus it took totally 40 h for the largest problem.

Figure 6.20 shows the final shape of the atomic model after self-positioning for problem size $c = 1$, which means totally four unit crystals in the thickness direction. The analysis naturally reproduces that spacing of atoms is smaller in GaAs and larger in InAs. It is also observed that the free end is not flat along vertical direction due to expansion in the bottom and compression in the top layers.

Figure 6.21 depicts the ratio of the curvature radius determined by the AFEM and by the continuum mechanics solution for varying structure thickness. Curvature radius values obtained by the AFEM modeling at the top and at the bottom of the atomic structures are calculated by using three neighbor nodes along the x direction to fit a circle. Then, the top and the bottom curvature radius values are used to calculate the curvature radius at the neutral layer by linear interpolation. The neutral layer is located at 0.54 of the thickness from the bottom of the structure in our problems. In order to estimate convergence of the curvature radius with increasing thickness, the least square fitting of the obtained eight numerical solutions is performed using power function $R(c) = \alpha_1 (\alpha_2 \, c + \alpha_3)^{-\beta} + \gamma$, where c denotes the size of the atomic system, and α_1, α_2, α_3, β, and γ are parameters found by the least square fitting. The parameter γ corresponds to a converged value of the curvature radius for infinite number of crystal layers. According to the least square fit result, the

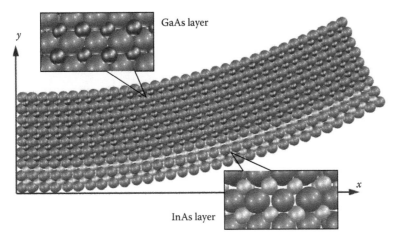

FIGURE 6.20
The final shape of atomic bilayer structure in problem size $c = 1$.

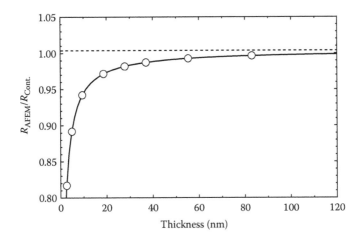

FIGURE 6.21
Ratios of the curvature radius determined by the AFEM and the continuum mechanics solution for varying structure thickness.

curvature radius by AFEM converges to 1.0037 of the plane strain solution. The difference between atomic-scale modeling and plane strain solution is −0.36% for the largest problem we investigated ($c = 36$). Therefore, the AFEM and the continuum mechanics solution agree for large thicknesses. The difference between atomic-scale and continuum mechanics curvature radius increases with reduction of the structure thickness. The difference is −18.4% for the smallest problem size $c = 1$ which corresponds to four unit crystals in the thickness direction and the thickness 2.56 nm.

6.7.3 Effects of Material Anisotropy on the Curvature Radius

The second problem series is used to determine curvature radius for varying crystal orientations. In order to investigate effect of material anisotropy on the equilibrium curvature radius, a series of AFEM solutions for problem sizes $c = 1, 2, 4$, and 8 and crystal

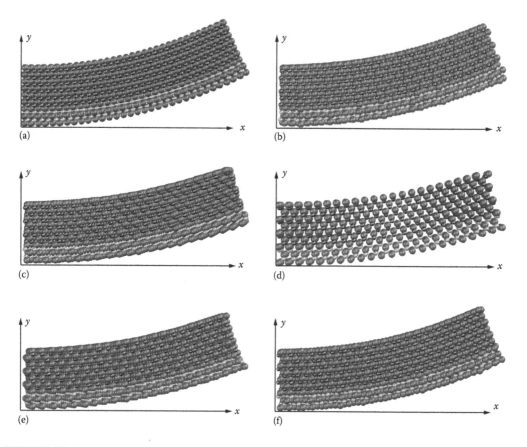

FIGURE 6.22
Final shapes of bilayer structures with the problem size $c=1$ for crystal orientation angles 0°–75°. (a) 0°, (b) 15°, (c) 30°, (d) 45°, (e) 60°, and (f) 75°.

orientation angles 0°, 15°, 30°, 45°, 60°, 75°, and 90° is prepared. Figure 6.22 shows the final shape of atomic models after self-positioning for orientation angles 0°, 15°, 30°, 45°, 60°, and 75° of problem size $c=1$.

Table 6.4 shows ratios of the curvature radius R to the thickness t obtained by the AFEM modeling for orientation angles 0°, 15°, 30°, and 45°. It also shows results calculated by the continuum mechanics solution for the orientation angle 0°.

Figure 6.23 shows dependency of the curvature radius ratio $R_{AFEM}/R_{Cont(0°)}$ on the orientation angle. The R_{AFEM} indicates the curvature radius determined by the AFEM, and $R_{Cont(0°)}$ means the continuum mechanics solution for zero orientation angle. The curvature radius value changes from minimum at orientation angles 0° and 90° to maximum at 45°. The ratio of maximal and minimal values of the curvature radius is about 1.35. This ratio is similar to experimental data and numerical FEM of GaAs and $In_{0.2}Ga_{0.8}As$ bilayer structures. Dependency of the curvature radius on the material orientation angle is close to sinusoidal function with period 90° as found by experimental investigations [40].

6.7.4 Results for Strains

Strain distributions are calculated using displacement increments during the AFEM solutions, and their contour plots are created using mesh free interpolation techniques:

TABLE 6.4

Total Thickness t (nm) and Relative Value of Curvature Radius R/t for Problem Sizes $c = 1, 2, 4$, and 8 for Varying Orientation Angle

c	t (nm)	Cont. (0°)	AFEM (0°)	AFEM (15°)	AFEM (30°)	AFEM (45°)
1	2.56	9.25	7.56	8.32	9.86	10.67
2	4.86	9.60	8.56	9.35	10.88	11.76
4	9.46	9.81	9.23	9.98	11.54	12.54
8	18.65	9.92	9.63	10.29	11.87	13.01

Note: Curvature radius value is symmetric with respect to orientation angle 45°, so results are shown up to 45°.

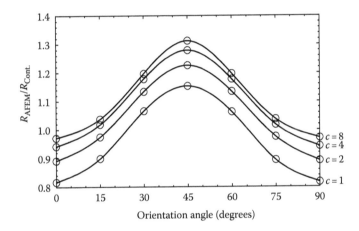

FIGURE 6.23
Dependency of curvature radius ratio $R_{AFEM}/R_{Cont(0°)}$ on orientation angle for problem sizes $c = 1$–8 (2.56–18.65 nm).

the MLS approximation, the RBF interpolation, and the Kriging interpolation. In this chapter, results of the MLS and the Kriging methods are shown since some undesirable noises are observed in the RBF interpolation using any basis functions with shape parameter selection algorithms introduced in Section 6.5.2. More detailed comparisons are found in the dissertation [106]. In the AFEM results, strain z (width) component is zero due to periodic boundary condition in the z direction. Hence, strains ε_x and ε_y components are calculated. The interpolation is constructed by taking 16 neighboring atoms around the position where strains are estimated.

6.7.4.1 Moving Least Squares Approximation

The MLS approximation produces results that are relatively smooth compared to the other interpolation methods since they do not go through exact nodal values. According to the plane strain solution, strains are linear functions of local y coordinate, and the strain ε_y is discontinuous at the interface of the bilayer due to difference in material properties and lattice mismatching strains. Figure 6.24 shows comparison of strains obtained by the MLS method and the plane strain solution along local y direction. The AFEM with the MLS approximation predicts almost same strain distributions as the plane strain

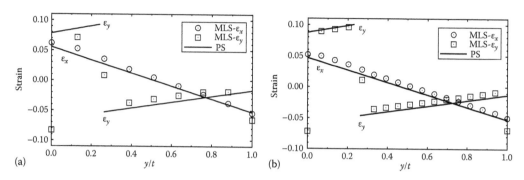

FIGURE 6.24
Strains obtained by the AFEM with MLS and the plane strain solution along local y coordinate. Abscissa is normalized by total thickness t. (a) $c=1$ and (b) $c=8$.

solution. In the AFEM results, slope of strain ε_x is slightly different at top and bottom layers due to expansion in the bottom and compression in the top layers. At $y/t=0$ and $y/t=1$, which correspond to the top and the bottom surfaces, the plane strain solution predicts ε_y strain values 0.08 and −0.002. However, strains estimated by the AFEM are about −0.084 and −0.065 for the problem size $c=1$, and about −0.074 and −0.070 for the problem size $c=8$. This indicates compression at the top and the bottom surfaces in the AFEM modeling.

Figure 6.25 depicts contour plots of the ε_x and the ε_y distribution for problem sizes $c=1$ and 8. Distribution for the problem $c=1$ looks a little rough since there are less atoms in

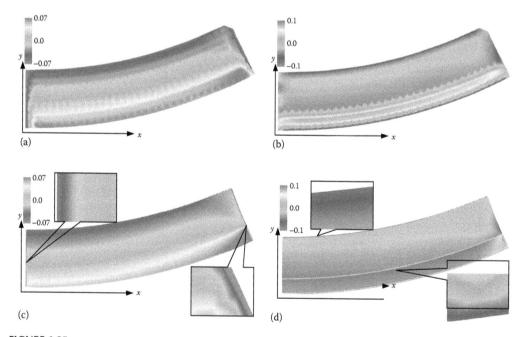

FIGURE 6.25
Contour plots of strains by the AFEM with MLS approximation for problem sizes $c=1, 8$. (a) $c=1$, ε_x, (b) $c=1$, ε_y, (c) $c=8$, ε_x, and (d) $c=8$, ε_y.

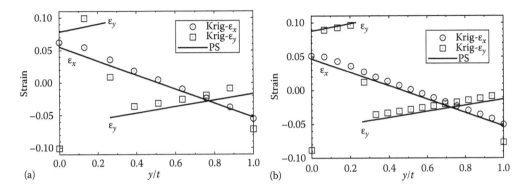

FIGURE 6.26
Strains obtained by the AFEM with Kriging interpolation and the plane strain solution along local y coordinate. (a) $c=1$ and (b) $c=8$.

the whole structure, but it can be recognized that fixed and free end effects are more significant in structures with smaller number of atoms.

6.7.4.2 Kriging Interpolation

Shape functions of the Kriging interpolation possess Kronecker delta properties. Therefore, the interpolation function is more sharp compared to the MLS interpolation. Figure 6.26 compares strains along the local y direction calculated by the plane strain solution and by the AFEM with Kriging interpolation. The strains agree well to the plane strain solution. However, at the top and the bottom surfaces, there are significant differences with plane strain solution in ε_y. These differences are also observed in the MLS approximation, but Kriging interpolation predicts slightly larger compressive strains.

Figure 6.27 presents contour plots of the ε_x and the ε_y distribution for problem sizes $c=1$ and 8. In the Kriging interpolation, sharper strain distributions at free surfaces are obtained.

Strains are calculated by the AFEM with mesh free interpolations. Significant difference of strains from the plane strain solution is observed at free surfaces. The strain component ε_y at the top and the bottom surfaces estimated by the AFEM shows the larger compressive strains than the continuum mechanics solution. The compression estimated by the continuum mechanics solution is about −0.1%, but mesh free interpolations predict about −7%. At the bottom surface, the plane strain solution predicts 10% of expansion while the AFEM with mesh free interpolation estimates compression of −9%. End effects become less significant to strain distributions as the structure size increases. Decreasing of curvature radius is observed in the atomic-scale modeling. The difference of curvature radius can be attributed to inability of continuum mechanics methods to model free surface effects. Discrepancy of ε_y at the top and at the bottom surfaces arises due to absence of atoms at each free surface. At free surfaces, atoms are attracted in a direction inside of a structure, and compression occurs (Figures 6.28 and 6.29). The AFEM with mesh free interpolation techniques reproduces such strains well.

The MLS, RBF, and the Kriging interpolations are implemented to obtain derivatives of the atomic displacements. The MLS approximation produces relatively smooth strain distributions, but some sharp features may be sometimes rounded. The RBF interpolation overcomes such difficulties, but it has another difficulty for selection of basis function or

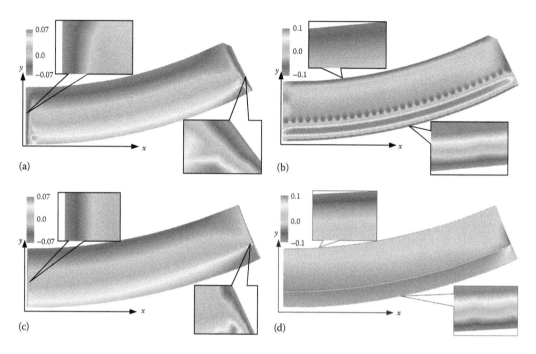

FIGURE 6.27
Contour plot of strains by the AFEM with Kriging interpolation in problem sizes $c=1, 8$. (a) $c=1$, ε_x, (b) $c=1$, ε_y, (c) $c=8$, ε_x, and (d) $c=8$, ε_y.

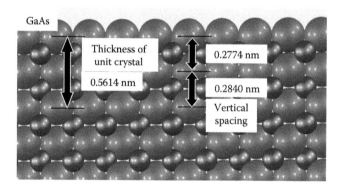

FIGURE 6.28
GaAs atomic configuration near the top surface.

suitable shape parameters. The Kriging interpolation goes through nodal values, and it is sufficient to use the quadratic basis for interpolation. We conclude that the Kriging interpolation is the most suitable method to calculate strains for the AFEM procedure. If there is some noise anticipated due to less reliable inputs, lower degrees of fitting polynomial or the MLS might be suitable for robustness.

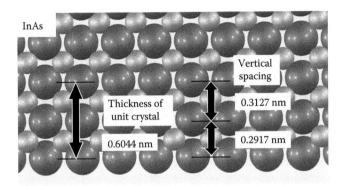

FIGURE 6.29
InAs atomic configuration near the bottom surface.

6.8 Conclusion

The self-positioning nanostructures are modeled by a continuum mechanics theory, a finite element procedure, and an atomic-scale finite element procedure taking into account material anisotropy. Continuum mechanics solutions were derived under ordinary and generalized plane strain conditions taking into account dependence of cubic crystal orientations. The generalized plane strain condition is characterized by zero total force in the structure z (width) direction. A finite element procedure was developed for modeling self-positioning structures with material anisotropy. Nonlinear deformations are controlled by using the Newton–Raphson iteration procedure with updated Lagrangian formulation. Different crystal orientations are modeled by rotating constitutive tensor using fourth-order tensor transformation law. The finite element procedure was applied to modeling of bilayer self-positioning nanostructures consisting of GaAs and $In_{0.2}Ga_{0.8}As$. In the previous study, curvature radii obtained by the FEM were compared with experimental data, and it was confirmed that the finite element results agree with the experimental data qualitatively. Comparisons of the finite element analysis and the continuum mechanics solutions show that the solution is appropriate for estimation of curvature radius and strains, especially strain in the width direction. The curvature radius is minimum at crystal orientation angles 0°, 90°, 180°, and 270°, and the largest curvature radius is observed for crystal orientation angles 45°, 135°, 225°, and 315°.

Also, an atomic-scale finite element procedure was developed for atomic-scale modeling of self-positioning nanostructures. Our AFEM procedure is formulated using the Tersoff interatomic potential function and a Nordlund parameterization for indium, gallium, and arsenide atomic systems. The deformation of self-positioning nanostructures involves large rotational and translational displacements, and thus a new iterative solution algorithm was adopted with a load relaxation to suppress nonlinear deformation at each solution step. Strain estimation procedures were also developed based on mesh free finite element interpolation techniques. The developed AFEM procedure has been used for estimation of curvature radius and strain components of bilayer self-positioning nanostructures composed of GaAs and InAs layers, and dependence of the curvature radius on varying thickness and crystal orientations are investigated. Comparisons of the AFEM results with continuum mechanics solution show that their ratio gets smaller as the structure thickness decreases. On the other hand, the curvature radius converges to the continuum mechanics solution as the structure thickness increases. Dependency of the curvature radius on the crystal orientation angle shows a periodic curve with the maximum and minimum curvature radii at angle 45°, and 0° or 90°, respectively. It was shown that hinges with different crystal orientation angles can result in curvature radii difference by 35%. Strains predicted by our AFEM procedure correspond to results of the continuum mechanics solution under plane strain conditions. However, the AFEM also predicts surface effects which are significant for thin nanostructures. The difference of curvature radius between the atomic-scale and the continuum mechanics solutions for structures of small thickness can be attributed to the inability of the continuum mechanics methods to model surface effects of neighboring atoms' absence.

References

1. V. Ya. Prinz, V. A. Seleznev, A. K. Gutakovsky, A. V. Chehovskiy, V. V. Preobrazhenskii, M. A. Putyato, and T. A. Gavrilova. Free-standing and overgrown InGaAs/GaAs nanotubes, nanohelices and their arrays. *Physica E*, 6:828–831, 2000.
2. O. G. Schmidt and K. Eberl. Nanotechnology: Thin solid films roll up into nanotubes. *Nature*, 410:168, 2001.
3. O. G. Schmidt and N. Y. Jin-Phillipp. Free-standing SiGe-based nanopipelines on Si (001) substrates. *Appl. Phys. Lett.*, 78:3310–3312, 2001.
4. O. G. Schmidt, C. Deneke, Y. M. Manz, and C. Muller. Semiconductor tubes, rods and rings of nanometer and micrometer dimension. *Physica E*, 13:969–973, 2002.
5. S. V. Golod, V. Ya. Prinz, V. I. Mashanov, and A. K. Gutakovsky. Fabrication of conducting GeSi/Si micro- and nanotubes and helical microcoils. *Semicond. Sci. Technol.*, 16:181–185, 2001.
6. P. O. Vaccaro, K. Kubota, and T. Aida. Strain-driven self-positioning of micromachined structures. *Appl. Phys. Lett.*, 78(19):2852–2854, 2001.
7. A. Vorob'ev, P. O. Vaccaro, K. Kubota, T. Aida, T. Tokuda, T. Hayashi, Y. Sakano, J. Ohta, and M. Nunoshita. SiGe/Si microtubes fabricated on a silicon-on-insulator substrate. *J. Phys. D: Appl. Phys.*, 36:L67–L69, 2003.
8. Ch. Deneke, C. Muller, N. Y. Jin-Phillipp, and O. G. Schmidt. Diameter scalability of rolled-up In(GaAs/GaAs nanotubes). *Semicond. Sci. Technol.*, 17:1278–1281, 2002.
9. B. Krause, C. Mocuta, and T. H. Metzger. Local structure of a rolled-up single crystal: An x-ray microdiffraction study of individual semiconductor nanotubes. *Phys. Rev. Lett.*, 96:165502-1–165502-4, 2006.
10. N. Y. Jin-Phillipp, J. Thomas, M. Kelsch, Ch. Deneke, R. Songmuang, and O. G. Schmidt. Electron microscopy study on structure of rolled-up semiconductor nanotubes. *Appl. Phys. Lett.*, 88:033113-1–033113-3, 2006.
11. A. Bernardi, A. R. Goñi, M. I. Alonso, F. Alsina, H. Scheel, P. O. Vaccaro, and N. Saito. Probing residual strain in InGaAs/GaAs micro-origami tubes by micro-Raman spectroscopy. *J. Appl. Phys.*, 99:063512, 2006.
12. A. Bernardi, P. D. Lacharmoise, A. R. Goñi, M. I. Alonso, P. O. Vaccaro, and N. Saito. Strain profile of the wall of semiconductor microtubes: A micro-Raman study. *Phys. Stat. Sol. (b)*, 244:380–385, 2007.
13. A. Cho. Pretty as you please, curling films turn themselves into nanodevices. *Science*, 313:164–165, 2006.

14. V. Ya. Prinz, V. A. Seleznev, A. V. Prinz, and A. V. Kopylov. 3D heterostructures and systems for novel MEMS/NEMS. *Sci. Technol. Adv. Mater.*, 10:034502, 2009.

15. H. J. In, S. Kumar, Y. Shao-Horn, and G. Barbastathis. Origami fabrication of nanostructured, three-dimensional devices: Electrochemical capacitors with carbon electrodes. *Appl. Phys. Lett.*, 88:083104-1–083104-3, 2006.

16. W. J. Arora, A. J. Nichol, H. I. Smith, and G. Barbastathis. Membrane folding to achieve three-dimensional nanostructures: Nanopatterned silicon nitride folded with stressed chromium hinges. *Appl. Phys. Lett.*, 88:053108-1–053108-3, 2006.

17. G. T. Pickett. Self-folding origami membranes. *Europhys. Lett.*, 78:480003-1–480003-6, 2007.

18. D. J. Thurmer, Ch. Deneke, Y. Mei, and O. G. Schmidt. Process integration of microtubes for fluidic applications. *Appl. Phys. Lett.*, 89:223507, 2006.

19. R. Songmuang, A. Rastelli, S. Mendach, Ch. Deneke, and O. G. Schmidt. From rolled-up Si microtubes to SiOx/Si optical ring resonators. *Microelect. Eng.*, 84:1427–1430, 2007.

20. R. Songmuang, A. Rastelli, S. Mendach, and O. G. Schmidt. SiOx/Si radial super lattices and microtube optical ring resonators. *Appl. Phys. Lett.*, 90:091905, 2007.

21. D. J. Bell, L. X. Dong, and B. J. Nelson. Fabrication and characterization of three-dimensional InGaAs/GaAs nanosprings. *Nano Lett.*, 6:725–729, 2006.

22. A. Bernardi, S. Kiravittaya, A. Rastelli, R. Songmuang, D. J. Thurmer, M. Benyoucef, and O. G. Schmidt. On-chip Si/SiOx microtube refractometer. *Appl. Phys. Lett.*, 93:094106-1–094106-3, 2008.

23. T. G. Leong, C. L. Randall, B. R. Benson, A. M. Zarafshar, and D. H. Gracias. Self-loading lithographically structured microcontainers: 3D patterned, mobile microwells. *Lab on a Chip*, 8:1621–1624, 2008.

24. T. G. Leong, B. R. Benson, E. K. Call, and D. H. Gracias. Thin film stress driven self-folding of microstructured containers. *Small*, 4:1605–1609, 2008.

25. J. H. Cho, S. Hu, and D. H. Gracias. Self-assembly of orthogonal three-axis sensors. *Appl. Phys. Lett.*, 93:043505-1–043505-3, 2008.

26. L. Zhang, L. Dong, and B. J. Nelson. Bending and buckling of rolled-up SiGe/Si microtubes using nanorobotic manipulation. *Appl. Phys. Lett.*, 92:243102-1–243102-3, 2008.

27. V. Ya. Prinz. Precise, molecularly thin semiconductor shells: from nanotubes to nanocorrugated quantum systems. *Phys. Stat. Sol. (b)*, 243:3333–3339, 2006.

28. R. Songmuang, Ch. Deneke, and O. G. Schmidt. Rolled-up micro- and nanotubes from single-material thin films. *Appl. Phys. Lett.*, 89:223109-1–223109-3, 2006.

29. F. Cavallo, R. Songmuang, C. Ulrich, and O. G. Schmidt. Rolling up SiGe on insulator. *Appl. Phys. Lett.*, 90:193120, 2007.

30. F. Cavallo, W. Siegle, and O. G. Schmidt. Controlled fabrication of Cr/Si and Cr/SiGe tubes tethered to insulator substrates. *J. Appl. Phys.*, 103:116103, 2008.

31. C.-H. Hsueh. Modeling of elastic deformation of multilayers due to residual stresses and external bending. *J. Appl. Phys.*, 91(12):9652–9656, 2002.

32. G. P. Nikishkov. Curvature estimation for multilayer hinged structures with initial strains. *J. Appl. Phys.*, 94(8):5333–5336, 2003.

33. M. Grundmann. Nanoscroll formation from strained layer heterostructures. *Appl. Phys. Lett.*, 83:2444–2446, 2003.

34. J. Zang and F. Liu. Theory of bending of Si nanocantilevers induced by molecular adsorption: A modified Stoney formula for the calibration of nanomechanochemical sensors. *Nanotechnology*, 18:405501, 2008.

35. J. Zang and F. Liu. Modified Timoshenko formula for bending of ultrathin strained bilayer films. *Appl. Phys. Lett.*, 92:021905-1–021905-3, 2008.

36. G. G. Stoney. *Proc. R. Soc. Lond.*, A82:172, 1909.

37. S. Timoshenko. Analysis of bi-metal thermostats. *J. Opt. Soc. Am.*, 11:233, 1925.

38. Y. Nishidate and G. P. Nikishkov. Generalized plane strain deformation of multilayer structures with initial strains. *J. Appl. Phys.*, 100:113518-1–113518-4, 2006.

39. G. P. Nikishkov, I. Khmyrova, and V. Ryzhii. Finite element analysis of self-positioning microstructures and nanostructures. *Nanotechnology*, 14:820–823, 2003.

40. G. P. Nikishkov, Y. Nishidate, T. Ohnishi, and P. O. Vaccaro. Effect of material anisotropy on the self-positioning of nanostructures. *Nanotechnology*, 17:1128–1133, 2006.

41. G. Fitzgerald, G. Goldbeck-Wood, P. Kung, M. Petersen, L. Subramanian, and J. Wescott. Materials modeling from quantum mechanics to the mesoscale. *Comput. Model. Eng. Sci.*, 24(3):169–183, 2008.

42. B. Liu, Y. Huang, H. Jiang, S. Qu, and K. C. Hwang. The atomic-scale finite element method. *Comput. Methods Appl. Mech. Eng.*, 193:1849–1864, 2004.

43. B. Liu, H. Jiang, Y. Huang, S. Qu, M.-F. Yu, and K. C. Hwang. Atomic-scale finite element method in multiscale computations with applications to carbon nanotubes. *Phys. Rev. B*, 72:035435-1–035435-8, 2005.

44. G. Wei, Y. Shouwen, and H. Ganyun. Finite element characterization of the size-dependent mechanical behaviour of nanosystems. *Nanotechnology*, 17:1118–1122, 2006.

45. T. C. Theodosiou and D. A. Saravanos. Molecular mechanics based finite element for carbon nanotube modeling. *Comput. Model. Eng. Sci.*, 19(2):121–134, 2007.

46. S. U. Chirputkar and D. Qian. Coupled atomistic/continuum simulation based on extended space-time finite element method. *Comput. Model. Eng. Sci.*, 24(3):185–202, 2008.

47. J. Zang, M. Huang, and F. Liu. Mechanism for nanotube formation from self-bending nanofilms driven by atomic-scale surface-stress imbalance. *Phys. Rev. Lett.*, 98:146102-1–146102-4, 2007.

48. Y. Nishidate and G. P. Nikishkov. Effect of thickness on the self-positioning of nanostructures. *J. Appl. Phys.*, 102:083501-1–083501-5, 2007.

49. Y. Nishidate and G. P. Nikishkov. Atomic-scale modeling of self-positioning nanostructures. *Comput. Model. Eng. Sci.*, 26:91–106, 2008.

50. Y. Nishidate and G. P. Nikishkov. Atomic-scale analysis of self-positioning nanostructures. *e-J. Surf. Sci. Nanotech.*, 6:301–306, 2008.

51. Y. Nishidate and G. P. Nikishkov. Curvature estimate for multilayer rolled-up nanostructures with cubic crystal anisotropy under initial strains. *J. Appl. Phys.*, 105:093536, 2009.

52. K. S. Surana. Geometrically non-linear formulation for the three dimensional solid-shell transition finite elements. *Comput. Struct.*, 15:549–566, 1982.

53. I. Pillinger, P. Hartley, C. E. N. Sturgess, and G. W. Rowe. A new linearized expression for strain increment in finite-element analyses of deformations involving finite rotation. *Int. J. Mech. Sci.*, 28:253–262, 1986.

54. S. H. Lo. Geometrically nonlinear formulation of 3D finite strain beam element with large rotations. *Comput. Struct.*, 44:147–157, 1992.

55. H. A. Elkaranshawy and M. A. Dokainish. Corotational finite element analysis of planar flexible multibody systems. *Comput. Mech.*, 54:881–890, 1995.

56. K.-J. Bathe. *Finite Element Procedures*. Prentice-Hall, Englewood Cliffs, NJ, 1996.

57. M. A. Criesfield. *Non-Linear Finite Element Analysis of Solids and Structures*. John Wiley & Sons Inc., West Sussex, U.K., 1996.

58. T. Belytschko, W. K. Liu, and B. Moran. *Nonlinear Finite Elements for Continua and Structures*. Willey, West Sussex, U.K., 2001.

59. A. Rodriguez-Ferran, A. Perez-Foguet, and A. Huerta. Arbitrary lagrangian-eulerian (ALE) formulation for hyperelastoplasticity. *Int. J. Numer. Meth. Eng.*, 53:1831–1851, 2001.

60. J. N. Reddy. *An Introduction to Nonlinear Finite Element Analysis*. Oxford University Press, Oxford, U.K., 2004.

61. O. C. Zienkiewicz and R. L. Taylor. *The Finite Element Method: Solid and Fluid Mechanics Dynamics and Non-Linearity*, Vol. 2., 6th edn., Butterworth-Heinemann, Oxford, U.K., 2005.

62. O. C. Zienkiewicz, R. L. Taylor, and J. M. Too. Reduced integration techniques in general analysis of plates and shells. *Int. J. Numer. Meth. Eng.*, 3:275–290, 1971.

63. S. Okamoto and Y. Omura. Finite-element nonlinear dynamics of flexible structures in three dimensions. *Compu. Model. Eng. Sci.*, 4:287–299, 2003.

64. S. Adachi. *GaAs and Related Materials*. World Scientific, Singapore, 1994.

65. J. E. Lennard-Jones. *Proc. Roy. Soc (Lond.)*, A106:463, 1924.

66. P. M. Morse. Diatomic molecules according to the wave mechanics. II. Vibrational levels. *Phys. Rev.*, 34:57–64, 1929.

67. N. Bernardes. Theory of solid Ne, A, Kr, and Xe at 0_K. *Phys. Rev.*, 112:1534–1539, 1958.

68. F. H. Stillinger and T. A. Weber. Computer simulation of local order in condensed phases of silicon. *Phys. Rev. B*, 31:5262–5271, 1985.

69. J. Tersoff. New empirical model for the structural properties of silicon. *Phys. Rev. Lett.*, 56:632–635, 1986.

70. J. Tersoff. New empirical approach for the structure and energy of covalent systems. *Phys. Rev. B*, 37:6991–7000, 1988.

71. J. Tersoff. Modeling solid-state chemistry: Interatomic potentials for multicomponent systems. *Phys. Rev. B*, 39:5566–5568, 1989; 41:3248, 1990 (Errata).

72. D. W. Brenner. Empirical potential for hydrocarbons for use in simulating the chemical vapor deposition of diamond films. *Phys. Rev. B*, 42:9458–9471, 1990.

73. D. W. Brenner, O. A. Shenderova, J. A. Harrison, S. J. Stuart, B. Ni, and S. B. Sinnott. A second-generation reactive empirical bond order (REBO) potential energy expression for hydrocarbons. *J. Phys.: Condens. Matter.*, 14:783–802, 2002.

74. A. Y. T. Leung, X. Guo, X. Q. He, H. Jiang, and Y. Huang. Postbuckling of carbon nanotubes by atomic-scale finite element. *J. Appl. Phys.*, 99:124308-1–124308-5, 2006.

75. P. A. Ashu, J. H. Jefferson, A. G. Cullis, W. E. Hagston, and C. R. Whitehouse. Molecular dynamics simulation of (100)InGaAs/GaAs strained-layer relaxation processes. *J. Cryst. Growth*, 150:176–179, 1995.

76. K. Nordlund, J. Nord, J. Frantz, and J. Keinonen. Strain-induced Kirkendall mixing at semiconductor interfaces. *Comput. Mater. Sci.*, 18:283–294, 2000.

77. P. Bhattacharya. *Properties of Lattice-Matched and Strained Indium Gallium Arsenide*. Institution of Electrical Engineers, London, U.K., 1993. Chapter 1.2.

78. T. Yamada and G. Yagawa. Finite element analysis with conjugate gradient method preconditioned by a norm scaling. *Transactions of JSCES*, 2001:20010033, 2001.

79. T. Belytschko, Y. Y. Lu, and L. Gu. Element-free Galerkin methods. *Int. J. Numer. Meth. Eng.*, 37:229–256, 1994.

80. W. K. Liu, S. Jun, and Y. F. Zhang. Reproducing kernel particle methods. *Int. J. Numer. Meth. Fluids*, 20:1081–1106, 1995.

81. W. K. Liu, S. Jun, S. Li, J. Adee, and T. Belytschko. Reproducing kernel particle methods for structural dynamics. *Int. J. Numer. Meth. Eng.*, 38:1655–1679, 1995.

82. S. N. Atluri and T. Zhu. A new meshless local Petrov-Galerkin (MLPG) approach in computational mechanics. *Comput. Mech.*, 22:117–127, 1998.

83. X. Zhang, K. Z. Song, M. W. Lu, and X. Liu. Meshless methods based on collocation with radial basis functions. *Comput. Mech.*, 26:333–343, 2000.

84. N. R. Aluru and G. Li. Finite cloud method: A true meshless technique based on a fixed reproducing kernel approximation. *Int. J. Numer. Meth. Eng.*, 50:2373–2410, 2001.

85. J. G. Wang. A point interpolation meshless method based on radial basis functions. *Int. J. Numer. Meth. Eng.*, 54:1623–1648, 2002.

86. S. N. Atluri and S. Shen. *The Meshless Local Petrov-Galerkin (MLPG) Method*. Tech. Science Press, Encino, CA, 2002.

87. Z. D. Han and S. N. Atluri. Meshless local Petrov-Galerkin (MLPG) approaches for solving 3D problems in elasto-statics. *Comput. Model. Eng. Sci.*, 6:169–188, 2004.

88. X. Jin, G. Li, and N. R. Aluru. On the equivalence between least-squares and kernel approximations in meshless methods. *Comput. Model. Eng. Sci.*, 2:447–462, 2001.

89. R. L. Hardy. Multiquadric equations of topography and other irregular surfaces. *J. Geophys. Res.*, 76:1905–1915, 1971.

90. E. J. Kansa. Multiquadrics—A scattered data approximation scheme with applications to computational fluid-dynamics–I: Surface approximations and partial derivative estimates. *Comp. Math. Appl.*, 19:127–145, 1990.

91. E. J. Kansa and R. E. Carlson. Improved accuracy of multiquadric interpolation using variable shape parameters. *Comp. Math. Appl.*, 24:99–120, 1992.

92. S. Rippa. An algorithm for selecting a good value for the parameter c in radial basis function. *Adv. Comput. Math.*, 11:193–210, 1999.

93. H. Wendland. Piecewise polynomial, positive definite and compactly supported radial functions of minimal degree. *Adv. Comput. Math.*, 4:389–396, 1995.

94. Z.Wu. Compactly supported, positive definite and compactly supported radial basis functions of minimal degree. *Adv. Comput. Math.*, 4:283–292, 1995.

95. M. D. Buhmann. Radial functions on compact support. *Proc. Edinburgh Math. Soc.*, 41:33–46, 1998.

96. G. R. Liu, J. Zhang, H. Li, K. Y. Lam, and Bernard B. T. Kee. Radial point interpolation based finite difference method for mechanics problems. *Int. J. Numer. Meth. Eng.*, 68:728–754, 2006.

97. R. Franke. Scattered data interpolation: Tests of some methods. *Math. Comput.*, 38:181–200, 1982.

98. K. Y. Dai, G. R. Liu, K. M. Lim, and Y. T. Gu. Comparison between the radial point interpolation and Kriging interpolation used in meshfree methods. *Comput. Mech.*, 32:60–70, 2003.

99. L. Gu. Moving kriging interpolation and the element-free Galerkin method. *Int. J. Numer. Meth. Eng.*, 56:1–11, 2003.

100. F. T. Wong and W. Kanok-Nukulchai. Kriging-based finite element method: Element-by element Kriging interpolation. *Civil Eng. Dimens.*, 11:15–22, 2009.

101. E. J. Sellountos and A. Sequeira. A hybrid multi-region BEM/LBIE-RBF velocity-vorticity scheme for the two-dimensional Navier-Stokes equations. *Comput. Model. Eng. Sci.*, 23:127–147, 2008.

102. G. Kosec and B. Sarler. Local RBF collocation method for Darcy flow. *Comput. Model. Eng. Sci.*, 23:197, 2008.

103. W. Kanok-Nukulchai and F. T. Wong. A finite element method using node-based interpolation. In *Proceedings of Third Asian-Pacific Congress on Computational Mechanics (APCOM'07)*, Kyoto, Japan, pp. PL3–PL2, 2007. (CD-ROM).

104. G. P. Nikishkov. *Programming Finite Elements in Java™*, Springer-Verlag, London, U.K., 2010.

105. http://www.u-aizu.ac.jp/labs/sw-cg/a-genps/

106. Y. Nishidate. PhD Dissertation, University of Aizu, 2009.

7
Application of Finite Element Method for the Design of Nanocomposites

Ufana Riaz

Materials Research Laboratory, Department of Chemistry, New Delhi, India

S.M. Ashraf

Materials Research Laboratory, Department of Chemistry, New Delhi, India

CONTENTS

7.1 Introduction 241
    7.1.1 Nanomaterials and Their Properties 243
7.2 Finite Element Analysis 245
    7.2.1 Fundamental Concepts in Finite Element Analysis 245
7.3 Finite Element Modeling of Nanocomposites 249
7.4 Computing the Mechanical Properties of Nanocomposites Using FEM 249
    7.4.1 Unit Cell Modeling 255
    7.4.2 Object-Oriented Modeling 256
    7.4.3 Multiscale RVE Modeling 258
7.5 Computing the Thermal Conductivity of Nanocomposites Using FEM 260
7.6 Computing the Optical Properties of Nanocomposites Using FEM 269
7.7 Computing the Magnetic Properties of Nanocomposites Using FEM 274
7.8 Computing Nanocomposites Damage Using FEM 278
7.9 Application of FEM to Biological Systems 279
7.10 Conclusion 283
Acknowledgment 284
References 285

7.1 Introduction

Computational modeling techniques are widely employed in materials chemistry as they enable rapid testing of theoretical predictions and understanding of complex experimental data. Computational modeling is a well-established technique in different areas of materials science, including polymers [1], ceramics [2], semiconductors [3], metals [4], pharmaceutical materials science [5], nanotechnology and engineering [6], biomimetic materials [7], and solid-state ionics [8], etc. Researchers generally consider computer modeling to be a theoretical technique, since the algorithms and methods involved are often expressed in highly mathematical terms, and also because theoretical groups tend to make heavy use of numerical simulations. However, this is somewhat of a misconception, since modeling

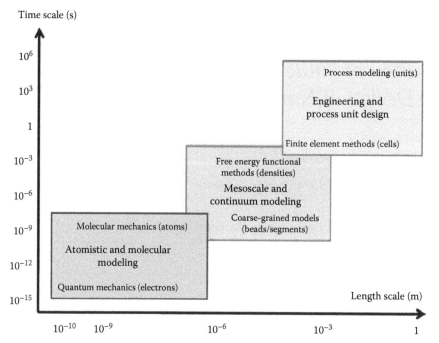

FIGURE 7.1
Hierarchy of multiscale modeling techniques. (From Elliott, J.A., *Int. Mater. Rev.*, 56, 209, 2011.)

may better be considered as a form of numerical experimentation. Numerical experiments cannot themselves explain the occurrence of a physical phenomenon, as every real experiment requires theoretical interpretation of the results. Therefore, modeling, theory, and physical experimentation have a rather more complex and subtle interdependence in the full elucidation of any scientific problem.

Multiscale modeling involves the application of modeling techniques at two or more different scales, which are often, dissimilar in their theoretical character due to the change in scale. A distinction is made between the hierarchical approach, [9–11], which involves running separate models with some sort of parametric coupling, and the hybrid approach, [12–15] in which models are run concurrently over different spatial regions of a simulation. The relationships between different categories of methods commonly used in the multiscale modeling hierarchy are shown in Figure 7.1. Although some techniques have been known for a long time and are now widely used, such as molecular dynamics (MD) and Monte Carlo (MC) methods, others such as mesoscale modeling and some more advanced methods of atomistic simulations are not as common. We have therefore included the technical summary for the benefit of nonspecialists.

The finite element analysis (FEA) is commonly used to study the thermal, mechanical, optical, morphological, and conductivity behavior of materials. The model serves as a link between computational chemistry and solid mechanics by substituting discrete molecular structures with an equivalent-continuum model.

FEA originated in the field of structural analysis and was widely explored during the 1950s and 1960s. This technique is well established in civil and aeronautical engineering and is used by mechanical engineers, for the analysis of stress in solid components. But problems in fluid mechanics and heat transfer are less commonly solved by FEAs due to the complex nature of the physical processes involved. All FEMs involve dividing the physical systems, such as structures, solid or fluid continua, into small elements. Each element is essentially a simple unit, the behavior of which can be readily analyzed. The complexities of the overall systems are adjusted by using large numbers of elements, rather than resorting to the sophisticated mathematics required by many analytical solutions. One of the main attractions of the FEA is the ease with which it can be applied to problems involving geometrically complicated systems.

7.1.1 Nanomaterials and Their Properties

Nanomaterials are generally considered as the materials that have a characteristic dimension smaller than 100 nm. Nanomaterials can be metallic, polymeric, ceramic, electronic, or composite. On the basis of geometry, nanomaterials can be classified into three categories, as shown in Figure 7.2 [16,17].

1. *Nanoparticles*: When the three dimensions of particulates are in the order of nanometers, they are called as isodimensional nanoparticles/nanocrystals.

2. *Nanotubes*: When two dimensions are in the nanometer scale forming an elongated structure, they are referred to as nanotubes/nanofibers/whiskers/nanorods.

3. *Nanolayers*: Nanolayers/nanoclays/nanosheets/nanoplatelets are characterized by only one dimension in nanometer scale and are present in the form of sheets of one to a few nanometer thick to hundreds to thousands nanometers.

Nanomaterials can also be classified as natural, incidental, and engineered nanomaterials depending on their pathway [18]. Natural nanomaterials occur in the environment (e.g., volcanic dust, lunar dust, magnetotactic bacteria, minerals, etc.). Incidental nanomaterials are created by man-made industrial processes (e.g., coal combustion, welding fumes, etc.). Engineered nanomaterials are obtained by lithographic etching of a large sample or by assembling smaller subunits through crystal growth or chemical synthesis to produce nanomaterials of desired size and dimension. Engineered nanomaterials most often have regular shapes, such as tubes, spheres, rings, etc.

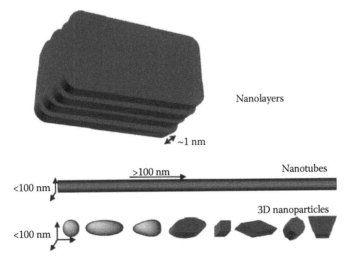

FIGURE 7.2
Classification of nanoscale materials. (From Kumar, A.P. et al., *Prog. Polym. Sci.*, 34, 479, 2009.)

Nanostructured materials are shown to possess unusual electrical, mechanical, and other physical properties as compared with their bulky counterparts. One of the reasons for unique properties of materials at nanoscale is the difference in physics at nanoscale as compared to macroscale. The fundamental laws of physics remain the same. However, their relative significance changes at the nanoscale. Gravitational and inertial forces are volume forces, which are dominant only at the macroscale and become almost negligible at the nanoscale. Frictional force that is a volume force at macroscale becomes surface force at nanoscale because adhesive forces between atoms and molecules become considerable at nanoscale [19–21]. Electrostatic and van der Waals forces are two major forces that become dominant at nanoscale. Electrostatic forces, which can be either repulsive or attractive, are very strong and act at a length scale of 1–100 nm. Van der Waals forces are attractive and act at distances less than 2 nm. There are three types of van der Waals forces:

1. Dipole–dipole forces (Keesom force) that occur between polar molecules such as water.
2. Dipole-induced dipole forces (induction of Debye force) arise when a polar molecule polarizes a nearby nonpolar molecule.
3. Induced dipole–induced dipole forces (dispersion of London force) that act on all atoms and molecules and is the most important van der Waals force [22,23].

Another difference between macroscale and nanoscale is the quantum mechanics. Quantum mechanics, instead of classical mechanics, describes the motion and energy at the nanoscale. Quantum mechanics considers the wave-particle duality of electrons. A material can exhibit entirely new properties with only a reduction in size because of the wave-particle duality of electrons. For example, gold at macroscale is yellow, inert, and a nonmagnetic metal but at 10 nm gold appears red, exhibits catalytic activity, and is magnetic [24]. Another distinctive effect that becomes dominant at the nanoscale is the Brownian motion. Brownian motion arises as the atoms are in a state of constant motion [25].

Higher aspect (As) to volume (Vs) ratio: As to Vs ratio is an indication of the quantity of interfacial region as compared to the bulk region. The interfacial region controls formation of new structural arrangements on the molecular scale. The As to Vs ratio for a spherical particle with radius r is given as [25]:

$$\frac{As}{Vs} = \frac{4\pi r^2}{4\pi r^3} = \frac{3}{r} \tag{7.1}$$

Nanoparticles have higher As to Vs ratio because of their small size (1–100 nm). Higher As to Vs ratio results in greater interfacial region, causing increased interaction between the nanoparticles. This increased interaction improves the properties of nanomaterials. Other than size, shape is an important factor in determining As to Vs ratio [25,26]. The high As to Vs ratio also makes nanoparticles more reactive as catalysts in chemical reactions. For nanoscale particles, a very small volume fraction is sufficient to achieve average distances between particles of the same order of magnitude as the radius of gyration of the macromolecules. Thus, the polymer molecule can be confined between two nanoscale particles. This is known as the confinement effect. Confinement effect reduces the number of conformations of the polymer molecules. Confinement effect is also responsible for reducing gas permeability value by providing tortuous paths for a gas molecule to diffuse through the nanocomposite [27].

Nanocomposites are composite materials in which the matrix material is reinforced by one or more separate nanomaterials in order to improve performance properties. The most common materials used as matrix in nanocomposites are polymers (e.g., epoxy, nylon, polyepoxide, polyetherimide), ceramics (e.g., alumina, glass, porcelain), and metals (e.g., iron, titanium, magnesium). As compared to conventional microcomposites, nanocomposites greatly improve the physical and mechanical properties. The nanoscale reinforcements over traditional fillers have the following advantages [28]:

1. Low-percolation threshold ($\sim$0.1–2 vol%)
2. Large number density of particles per particle volume (10^6–10^8 particles/μm^3)
3. Extensive interfacial area per volume of particles (10^3–10^4 m^2/mL)
4. Short distances between particles (10–50 nm at $\sim$1–8 vol%)

7.2 Finite Element Analysis

Finite Element Analysis (FEA) is a general numerical method for obtaining approximate solutions in space to initial-value and boundary-value problems including time-dependent processes. It employs preprocessed mesh generation, which enables the model to fully capture the spatial discontinuities of highly inhomogeneous materials. It also allows complex, nonlinear tensile relationships to be incorporated into the analysis. Thus, it has been widely used in mechanical, biological, and geological systems. In FEA, the entire domain is discretized into an assembly of simple-shaped subdomains (e.g., hexahedra or tetrahedral in three dimensions (3D), and rectangles or triangles in 2D) without gaps/overlaps. The subdomains are interconnected at the nodes. The implementation of FEM includes some important steps shown in Figure 7.3.

7.2.1 Fundamental Concepts in Finite Element Analysis

A 3D body having a volume V and a surface area S is shown in Figure 7.4.

Part of the boundary S_u is displacement constrained (the displacements are specified). Attraction T is applied on the remaining boundary S_t. The displacement due to the force applied in three directions is given by

$$u = [u \ v \ w]^T \tag{7.2}$$

The body force per unit volume is given by

$$f = \left[f_x f_y f_z \right]^T \tag{7.3}$$

The surface traction is given by

$$T = \left[T_x T_y T_z \right]^T \tag{7.4}$$

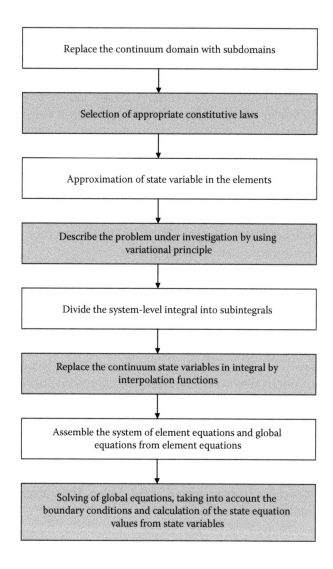

FIGURE 7.3
Steps in the FEA.

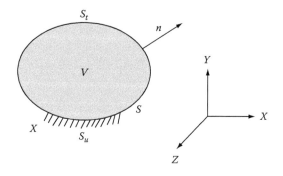

FIGURE 7.4
Three-dimensional body showing finite volume and surface. (From http://etd.ohiolink.edu/view.cgi?acc_num=ncin1122555750)

The load components are given by

$$P = \left[P_x P_y P_z \right]^{\mathrm{T}} \qquad (7.5)$$

The stress acting on an elemental volume dV is given by

$$\sigma = [\sigma_x \sigma_y \sigma_z \tau_{xy} \tau_{yz} \tau_{xz}]^{\mathrm{T}} \qquad (7.6)$$

σ,σ,σ are the normal stresses and $\tau_{xy}\tau_{yz}\tau_{xz}$ are the shear stresses
   Force equilibrium implies

$$\Sigma F_x = 0$$

$$\Sigma F_y = 0 \qquad (7.7)$$

$$\Sigma F_z = 0$$

The equilibrium equations are given by

$$\frac{do_x}{dx} + \frac{d\tau_{xy}}{dy} + \frac{d\tau_{xz}}{dz} + f_x = 0$$

$$\frac{d\tau_{xy}}{dx} + \frac{do_y}{dy} + \frac{d\tau_{yz}}{dz} + f_y = 0 \qquad (7.8)$$

$$\frac{d\tau_{xz}}{dx} + \frac{d\tau_{yz}}{dy} + \frac{do_z}{dz} + f_z = 0$$

Boundary conditions: The displacement boundary conditions are given by $u = 0$ on S_u
   n is a unit vector normal to a given surface area dA whose components are given by

$$n = [n_x n_y n_z]^{\mathrm{T}}$$

Considering the equilibrium along three perpendicular axes gives

$$\sigma_x n_x + \tau_{xy} n_y + \tau_{xz} n_z = T_x$$

$$\tau_{xy} n_x + \sigma_y n_y + \tau_{yz} n_z = T_y \qquad (7.9)$$

$$\tau_{xz} n_x + \tau_{yz} n_y + \sigma_z n_z = T_z$$

Strain–displacement relations: The strains are represented by

$$\varepsilon = \left[\varepsilon_x \ \varepsilon_y \ \varepsilon_z \ \gamma_{xy} \ \gamma_{yz} \ \gamma_{xz} \right]^{\mathrm{T}} \qquad (7.10)$$

where
   $\varepsilon_x, \varepsilon_y, \varepsilon_z$ are the normal strains
   $\gamma_{xy} \gamma_{yz} \gamma_{xz}$ are the engineering shear strains

After considering all the faces the strain equation can be written as follows:

$$\varepsilon = \left[\frac{du}{dx}, \frac{dv}{dy}, \frac{dw}{dz}, \frac{du}{dy} + \frac{dv}{dx}, \frac{dv}{dz} + \frac{dw}{dy}, \frac{du}{dz} + \frac{dw}{dx} \right] \tag{7.11}$$

For linear elastic isotropic materials, the stress–strain relations are given by Hooke's law

$$\varepsilon_x = \frac{\sigma_x}{E} - v\frac{\sigma_y}{E} - v\frac{\sigma_z}{E}$$

$$\varepsilon_y = -v\frac{\sigma_x}{E} + \frac{\sigma_y}{E} - v\frac{\sigma_z}{E} \tag{7.12}$$

$$\varepsilon_z = -v\frac{\sigma_x}{E} - v\frac{\sigma_y}{E} + \frac{\sigma_z}{E}$$

$$\gamma_{xy} = \frac{\tau_{xy}}{G}$$

$$\gamma_{yz} = \frac{\tau_{yz}}{G} \tag{7.13}$$

$$\gamma_{xz} = \frac{\tau_{xz}}{G}$$

The shear modulus is given by

$$G = \frac{E}{2(1+v)} \tag{7.14}$$

Hence, from Hooke's law

$$\varepsilon_x + \varepsilon_y + \varepsilon_z = \frac{(1-2v)}{E}(\sigma_x + \sigma_y + \sigma_z) \tag{7.15}$$

Substituting for stresses in terms of strains, we get $\sigma = D\varepsilon$
   D is a 6×6 symmetric matrix given by

$$D = \frac{E}{(1+v)(1-2v)} \begin{bmatrix} 1-v & v & v & 0 & 0 & 0 \\ v & 1-v & v & 0 & 0 & 0 \\ v & v & 1-v & 0 & 0 & 0 \\ 0 & 0 & 0 & 0.5-v & 0 & 0 \\ 0 & 0 & 0 & 0 & 0.5-v & 0 \\ 0 & 0 & 0 & 0 & 0 & 0.5-v \end{bmatrix} \tag{7.16}$$

7.3 Finite Element Modeling of Nanocomposites

There are different ways to experimentally characterize nanocomposites. The type of analysis depends on many factors such as model size, material properties, boundary conditions, and symmetry of the material. We can broadly classify the analysis into three types of analysis:

1. *2D Analysis*: This is mainly used to conserve the computational resources though less time consuming, it is less accurate in general.

2. *3D Analysis*: The main drawback of creating a 3D geometry of a component is the time required to analyze the body. Though relatively accurate than 2D analysis, it consumes higher system and computation resources.

3. *Symmetric analysis*: For some of the symmetric bodies, it is enough if we can create a quarter or half the model and analyze. The most important thing in an axis symmetric analysis lies in providing the symmetry boundary conditions. The models created in the following simulations are 3D in nature because of the anisotropy of the nanocomposites matrix.

7.4 Computing the Mechanical Properties of Nanocomposites Using FEM

Mechanical properties of nanostructured materials such as tensile and flexural tests, impact tests [29–33], and microcompression tests [34,35] can be determined by computational methods. Nanoindentation test is one of the most effective and widely used methods to measure the mechanical properties of materials. This technique uses the same principle as microindentation, but with much smaller probe and loads, so as to produce indentations from less than a hundred nanometers to a few micrometers in size. These modeling methods span a wide range of length and time scales, as shown in Figure 7.5. For the smallest length and time scales, computational chemistry techniques are primarily used to predict atomic structure using first-principles theory. For the largest length and time scales, computational mechanics is used to predict the mechanical behavior of materials and engineering structures. Modeling methods are based on well-established principles that have been developed in science and engineering. However, the intermediate length and time scales do not have general modeling methods that are as well developed as those on the smallest and largest time and length scales. Therefore, multiscale modeling techniques are employed, which take advantage of computational chemistry and mechanics methods simultaneously for the prediction of the structure and properties of materials.

Each modeling method has broad classes of relevant modeling tools (Figure 7.6). The quantum mechanical and nanomechanical modeling tools consider a discrete molecular structure of matter while micromechanics deal with the presence of a continuous material structure. Figure 7.6 shows the details of the relationship of specific modeling techniques in computational mechanics. The continuum-based methods primarily include techniques such as the FEM, the boundary element method (BEM), and the micromechanics approach developed for composite materials. Specific micromechanical techniques include the Eshelby method, the Mori–Tanaka method, and the Halpin–Tsai method.

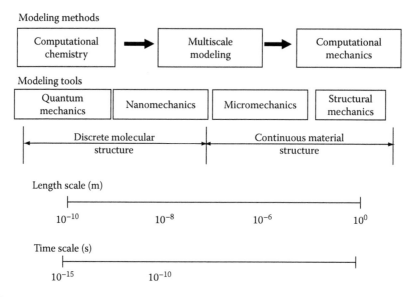

FIGURE 7.5
Material modeling techniques based on length and time scales. (From Valavala, P.K. and Odegard G.M., *Rev. Adv. Mater. Sci.* (RAMS), 9, 34, 2005.)

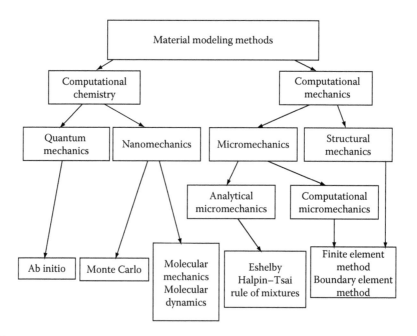

FIGURE 7.6
Computational modeling of nanomaterials using computational chemistry and mechanics. (From Valavala, P. K. and Odegard G.M., *Rev. Adv. Mater. Sci.* (RAMS), 9, 34, 2005.)

The energy in FEM is taken from the theory of linear elasticity and thus the input parameters are simply the elastic moduli and the density of the material. Since these parameters are in agreement with the values computed by MD, the simulation is consistent across the scales. More specifically, the total elastic energy in the absence of tractions and body forces within the continuum model is given by

$$U = U_v + U_k \qquad (7.17)$$

$$U_y = \frac{1}{2} \int dr \sum_{\mu,v,\lambda,\sigma}^{3} \varepsilon_{\mu v}(r) C_{\mu,v,\lambda,\sigma} \varepsilon_{\lambda,\sigma}(r) \qquad (7.18)$$

$$U_k = \int dr \rho(r) |u(r)|^2 \qquad (7.19)$$

where U_v is the Hookian potential energy term which is quadratic in the symmetric strain tensor e, contracted with the elastic constant tensor C. The Greek indices μ, v, λ, σ denote Cartesian directions and the mass density (r). The kinetic energy U_v involves the time rate of change of the displacement field μ, and the mass density r. The strains are related to the displacements according to

$$\varepsilon_{\mu v} = \frac{\delta u_\mu}{\delta r_v} + \frac{\delta u_v}{\delta r_\mu} \qquad (7.20)$$

These are fields defined throughout space in the continuum theory. Thus, the total energy of the system is an integral of these quantities over the volume of the sample dv. The FEM has been incorporated in some commercial software packages and open source codes (e.g., ABAQUS, ANSYS, Palmyra, and OOF) and widely used to evaluate the mechanical properties of polymer composites. Some attempts have recently been made to apply the FEM to nanoparticle-reinforced polymer nanocomposites.

Hardness (H) and elastic modulus (E) are calculated from the load–displacement curve obtained from a nanoindentation test. A typical load–displacement curve is shown in (Figure 7.7). As the indenter penetrates into the specimen, the loading curve climbs up. At some point, the maximum load P_{max} is reached, and then followed by the unloading. If the material is perfectly elastic and has no hysteresis, the loading curve and the unloading curve will be identical. h_{max} gives a measure of the total maximum deformation, while hf represents the maximum permanent (plastic) deformation (final penetration depth).

The most commonly used method to obtain the hardness and the elastic modulus of a material by nanoindentation is the Oliver–Pharr method [36]. According to this method, the nanoindentation hardness as a function of the final penetration depth of indent can be determined by

$$H = \frac{P_{max}}{A} \qquad (7.21)$$

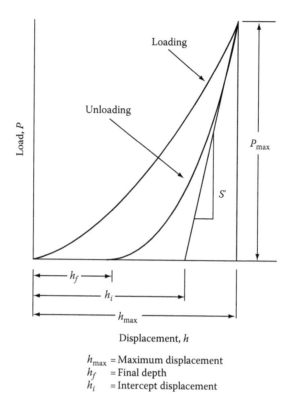

FIGURE 7.7

Typical load–displacement curve. (From Hu et al., *J. Minerals Mater. Charact. Engg.*, 9(4), 275, 2010.)

where

P_{max} is the maximum applied load measured at the maximum depth of penetration (h_{max})
A is the projected contact area between the indenter and the specimen

For a spherical indenter, $f\,A = 2\pi Rh$ (where R is the radius of the indenter), whereas for a pyramidal (Berkovich or Vickers) indenter, A can be expressed as a function of h_f as

$$A = 24.504h_f^2 + C_1h_f + C_2h_f^{1/2} + C_3h_f^{1/4} + \cdots C_8h_f^{1/128} \tag{7.22}$$

where C_1 to C_8 are constants and can be determined by standard calibration procedure. The final penetration depth, h_f, can be determined from the following expression:

$$h_f = h_{max} - \varepsilon\frac{P_{max}}{S^*} \tag{7.23}$$

where ε is a geometric constant, $\varepsilon = 0.75$ for a pyramidal indenter, and $\varepsilon = 0.72$ for a conical indenter. S^* is the contact stiffness which can be determined as the slope of the unloading curve at the maximum loading point, i.e.,

$$S^* = \left(\frac{dP}{dh}\right)_{h=h_{max}} \tag{7.24}$$

The reduced elastic modulus E_r is given by

$$E_r = \frac{S^*}{2\beta}\sqrt{\frac{\pi}{A}} \tag{7.25}$$

where β is a constant that depends on the geometry of the indenter. For both a Berkovich and a Vicker's indenter, $\beta = 1.034$, whereas for both a conical and a spherical indenter, $\beta = 1$. The specimen elastic modulus (E_s) can then be calculated as

$$\frac{1}{E_r} = \frac{1-v_s^2}{E_S} + \frac{1-v_i^2}{E_i} \tag{7.26}$$

where i s E, and i s v are the elastic modulus and Poisson's ratio, respectively, for the indenter and the specimen. For a diamond indenter, E_i is 1140 GPa and i v is 0.07. The contact stiffness, S^*, can be derived from the unloading curve which simply obeys the following power law:

$$P = B\left(h - h_f\right)^n \tag{7.27}$$

where B and n are empirical constants that can be determined by fitting the experimentally measured pairs of data (P, h) during unloading. Thus, the contact stiffness can be expressed as

$$S^* = \left(\frac{dP}{dh}\right)_{h=h_{max}} = Bn(h_{max} - h_f)^{n-1} \tag{7.28}$$

Therefore, the specimen's hardness H and elastic modulus s E will be obtained from this set of equations.

Indentation involves large plastic deformation, material nonlinearity, and contact. In order to characterize the mechanical properties for proper design of experiments, FEM is often used to simulate the nanoindentation tests [37–39]. The primary mechanical properties extracted from a nanoindentation test are the hardness and the elastic modulus. Finite element simulation could be employed to get other properties, such as yield stress and hardening [40–47]. Figure 7.8a shows the geometry of indentation of a cylindrical specimen with a conical indenter, and Figure 7.8b shows the Mises stress contour from the finite element analysis [48]. Figure 7.9 shows a 3D nanoindentation finite element mesh system. Due to symmetry, only half of the specimen volume model is shown (Figure 7.9) [39].

FEA has been used to predict the mechanical properties of composite materials since 1970 [49,50]. Various finite element models have been developed till date to characterize all kinds of composite materials. In 1991, *Sumio Iijima* discovered carbon nanotubes (CNTs) which possessed high stiffness and strength, as well as superior electrical and thermal properties [51,52], after which CNTs were used as reinforcement in developing nanocomposite materials. In the past few decades, there has been tremendous experimental [53,54], analytical [156–169], as well as finite element modeling [55,56] on developing, analyzing, and characterizing CNT-reinforced nanocomposites and other nanocomposites. In the following section, three finite element modeling (FEM) approaches have been discussed.

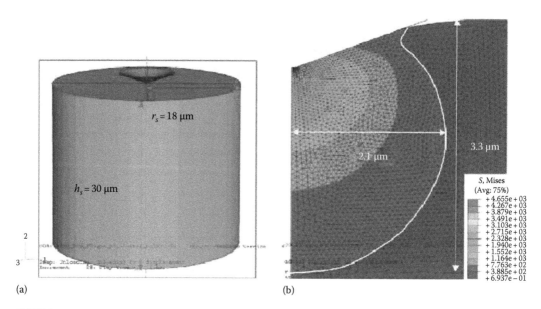

(a)   (b)

FIGURE 7.8
(a) Geometry of indentation of a cylindrical specimen with a conical indenter. (b) The Mises equivalent stress field in the specimen during indentation at $h_{max} = 600$ nm. (The stress values must be multiplied by 10^7 to respect the scale of the problem.) (From Poon, B. et al., *Int. J. Solids Struct.*, 45, 6018, 2008.)

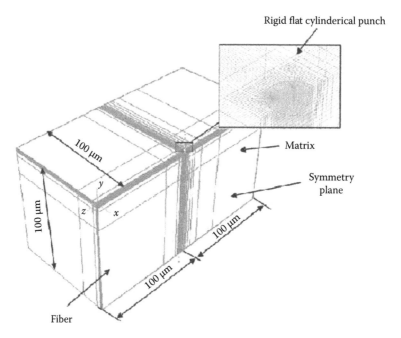

FIGURE 7.9
Illustration of a 3D nanoindentation finite element model. (From Lee, S.H. et al., *Composites: Part A*, 38, 1517, 2007.)

7.4.1 Unit Cell Modeling

The conventional unit cell concept is the same as the *representative volume element* (RVE) modeling [57]. In this case, a unit cell has a big size (usually in micrometers) and contains a significant number of fillers (usually in tens to hundreds or more). Such defined unit cell is the building block of the composite. As analytical models are difficult to establish and complicated to solve, numerical modeling and simulations are used to theoretically predict the properties of materials. The most common method used to theoretically predict the mechanical properties of nanocomposites with unit cell is the FEA. The most common method used to characterize the mechanical properties of nanocomposites with unit cell is the FEM. Hbaieb et al. [58] examined the Young's modulus of nanoclay/polymer nanocomposites with both 2D and 3D unit cells using the FEM (Figure 7.10). Four unit cells were created with 2D and 3D randomly oriented nanoclay particles models, as shown in Figure 7.10. Two kinds of boundary conditions were considered, i.e., the periodic boundary conditions and symmetrical boundary conditions. For the 2D models (both aligned and random cases), the periodic boundary conditions considered were as follows:

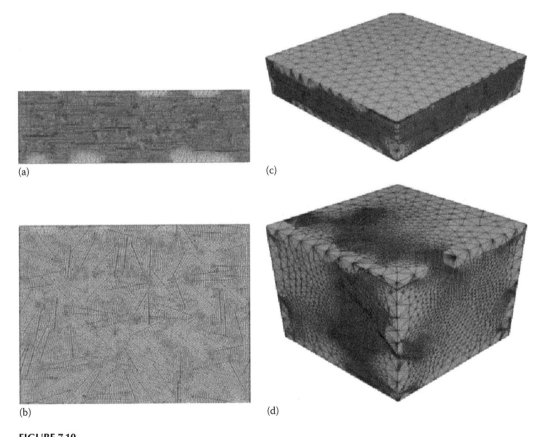

(a) (c)

(b) (d)

FIGURE 7.10
Mesh details of the model for (a) 2D aligned particle distribution, (b) 2D randomly oriented particle distribution, (c) 3D aligned particle distribution, and (d) 3D randomly oriented particle distribution. Particle volume fraction is 5%, the particle aspect ratio is 50, $E_p/E_m = 100$, $v_m = 0.35$, $v_p = 0.2$. Subscripts p and m represent particle and matrix, respectively. (From Hbaieb, K. et al., *Polymer*, 48, 901, 2007.)

$$u(\text{RE}) = u(\text{LE}) + \delta_1$$

$$v(\text{RE}) = v(\text{LE})$$

$$u(\text{TE}) = u(\text{BE})$$ (7.29)

$$v(\text{TE}) = v(\text{BE}) + \delta_2$$

where RE, LE, TE, BE, and δ_1 and δ_2 are the right, left, top, bottom edges, and the axial and transverse displacements, respectively.

The symmetrical boundary conditions considered for the 2D models were as follows:

$$u(\mathbf{LE}) = 0$$

$$v(\mathbf{BE}) = 0$$ (7.30)

$$u(\mathbf{RE}) = \delta$$

where δ is the given normal displacement in the x-direction. In addition, all edges are free of shear traction and the top edge is free of normal traction as well.

For the 3D models (both aligned and random cases), only symmetrical boundary conditions were applied as

$$u(\mathbf{LF}) = 0$$

$$v(\mathbf{BF}) = 0$$ (7.31)

$$w(\mathbf{BKF}) = 0$$

$$u(\mathbf{RF}) = \delta$$

where LF, BF, BKF, and RF denote left face, bottom face, back face, and right face. All other faces are free of any displacement or traction constraints. The numerical results indicated that 2D models do not predict the elastic modulus of clay/polymer nanocomposites accurately. However, Mori–Tanaka model [59] gave accurate predictions of the stiffness of the nanocomposites whose volume fraction was less than 5% for aligned particles. For randomly oriented particles the Wang–Pyrz model [60] overestimated the stiffness of the nanocomposites.

Lee et al. [61] used a 3D unit cell model to analyze the deformation behavior of randomly distributed $Al_{18}B_4O_{33}$ whisker-reinforced AS_{52} magnesium alloy matrix composite. The $Al_{18}B_4O_{33}$ whiskers were taken to be 10–30 μm in length and 0.5–1.0 μm in diameter. The dimensions of the unit cell taken were $10 \times 20 \times 20 \, \mu m^3$. The volume fraction of the whiskers was 15%. Figure 7.11 shows the typical unit cell (with the meshes of the whiskers) and an optical micrograph of the composite. For the Young's modulus and overall elastic–plastic response of the composite, the FEM results were observed to be in good agreement with the experimental results.

7.4.2 Object-Oriented Modeling

For highly irregular angular structure of fillers, the approximation of simple geometrical particles cannot take into account the complex morphology, size, and spatial distribution of the reinforcement. Object-oriented modeling captures the actual microstructure morphology of the nanocomposites accurately and predicts the overall properties.

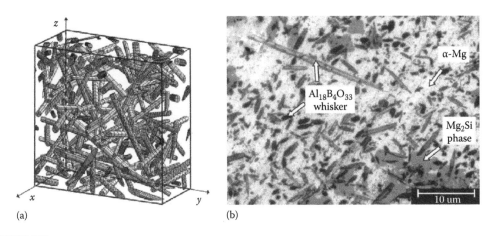

FIGURE 7.11
(a) Three-dimensional random whisker-reinforced composite model, and (b) an optical micrograph of squeeze-infiltrated $Al_{18}B_4O_{33}$/Mg random whisker composite. (From Lee, W.J. et al., *Scri. Mater.*, 61, 580, 2009.)

The object-oriented modeling incorporates the microstructure images such as scanning electron microscopy (SEM) micrographs into finite element grids. Thus the mesh produces exactly the original microstructure, inclusions size, morphology, spatial distribution, and the volume fraction of the different constituents. An object-oriented finite element code, OOF [62,63], developed by National Institute of Standards and Technology (NIST), has been extensively used in analyzing fracture mechanisms and material properties of heterogeneous materials [64,65] and mechanical properties of nanocomposites [66,67].

In RVE modeling, two basic assumptions are made:

1. Nanofillers can be idealized to simple geometries such as spheres, ellipsoids, cylinders, or cubes.

2. Nanocomposites can be reproduced by assembling a large number of such RVEs (or unit cells).

For highly irregular angular structure of fillers, the approximation of simple geometrical particles cannot take into account the complex morphology, size, and spatial distribution of the reinforcement. Therefore, the object-oriented modeling captures the actual microstructure morphology of the nanocomposites accurately and predicts the overall properties.

Dong et al. [30] studied the mechanical properties of polypropylene (PP)/organoclay nanocomposites with different clay contents ranging from 1 to 10 wt%. SEM micrographs from longitudinal loading direction of the specimen were captured and mapped onto the FEM, as shown in Figure 7.12. The actual nano-/microstructures (their size, shape, distribution, etc.) of the PP and the organoclay were used in the computational model, and each phase was attributed to the corresponding material properties. The OOF modeling results for the tensile modulus showed agreement with the experimental data and theoretical predictions. Chawala et al. [66] used 3D object-oriented FEM to evaluate the mechanical behavior of SiC particle-reinforced Al composites. For a volume of $100 \times 100 \times 20\,\mu m^3$ cell, they assumed 100 SiC particles, which lead to 20% volume fraction and compared the results of the Young's modulus and the stress–strain relations from the object-oriented (microstructure-based) model with the results of the experiment and the numerical results from simplified models (which include rectangular prism, multiparticle-ellipsoids, multiparticle spheres, etc.). Some of the results are depicted in

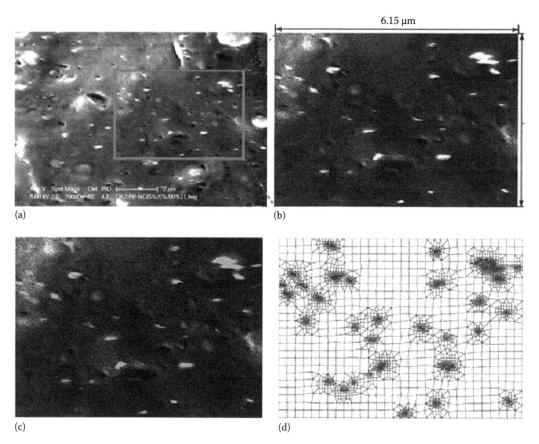

(a)   (b)

(c)   (d)

FIGURE 7.12
Typical example of creating OOF model of PP/organoclay nanocomposites (5 wt% in clay content): (a) original SEM image, (b) captured SEM image portion, (c) image segmentation using pixel selection, and (d) finite element mesh (highlighted regions contain organoclay particles and the rest are PP matrices). (From Dong, Y. et al., *Comp. Sci. Technol.*, 68, 2864, 2008.)

Figure 7.13 that indicates that 3D microstructure-based model can accurately predict the properties of particle-reinforced composites, while the simple analytical models cannot as they do not account for the microstructural factors that influence the mechanical behavior of the material. In object-oriented FEM, 2D modeling has been widely used to study the structure of nanocomposites [66,77]. There are also some works reported on 3D modeling [66]. Unfortunately, there are complex problems to be resolved in 3D modeling, especially advanced object-oriented 3D finite element codes.

7.4.3 Multiscale RVE Modeling

In multiscale RVE modeling two basic assumptions are made:

1. Nanofillers can be treated as simple shapes such as spheres, ellipsoids, cylinders, etc.
2. Nanocomposites can be reproduced by assembling a large number of such unit cells.

An RVE modeling is composed of a single or multiple nanofiller(s) with the surrounding matrix material, and proper boundary conditions to take into account the effect of the

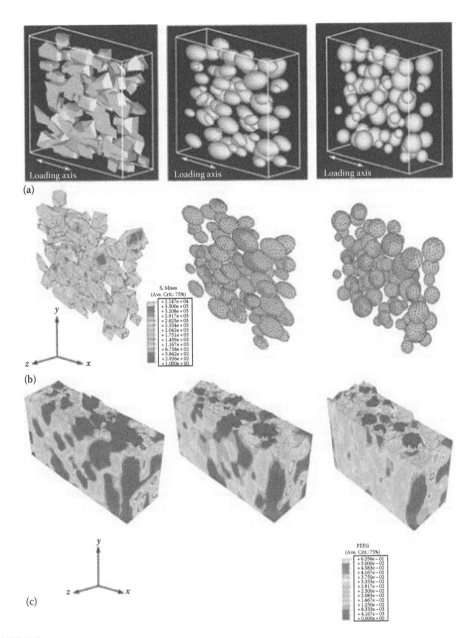

FIGURE 7.13
Comparison between 3D finite element models incorporating actual microstructure and approximation to spherical particles: (a) FEM models, (b) von Mises stress distribution in particles, and (c) plastic strain in matrix. (From Chawla, N. et al., *Acta Mater.*, 54, 1541, 2006.)

surrounding materials. Liu and Chou [68] extended the RVE concept used by Hyer [69] and Nemat-Nasser and Hori [70] for conventional fiber-reinforced composites at the microscale to nanoscale, and evaluated the effective mechanical properties of CNT-based composites by using a 3D nanoscale RVE based on elasticity theory which was solved by FEM. They modeled the nanotube at the atomistic scale and analyzed the matrix deformation using the continuum FEM. The van der Waals interactions between carbon atoms and the finite element nodes of the matrix were simulated using truss rods. Zhang et al. [71] linked continuum analysis with atomistic simulation by incorporating interatomic potential and atomic structures of CNTs directly into the constitutive law. Shi et al. [72] presented a hybrid continuum mechanics method to study the deformation and fracture behavior of CNTs embedded in composites. The method was based on a representative unit cell divided into three distinct regions which was analyzed using an atomistic potential, a continuum method based on the Cachy–Born rule and a micromechanics method. The multiscale RVE integrated nanomechanics and continuum mechanics, thus bridging the length scales from the nano- through the mesoscale. Tserpes et al. [73] proposed a multiscale RVE to investigate the tensile behavior of CNT/polymer composites. The model was based on the assumption that loaded CNTs behaved like space-frame structures. The RVE consisted of a rectangular solid whose entire volume was taken up by the matrix, and the nanotube was modeled as a 3D elastic beam. The 3D solid elements and beam elements were used to model the matrix and nanotube, respectively. The behavior of the isolated nanotube was simulated using the progressive fracture model [74]. The bonds between carbon atoms were considered as load-carrying elements while carbon atoms as joints. The nonlinear behavior of the C–C bonds was modeled by the modified Morse interatomic potential [75], and the nanotube structure was modeled by FEM. The nanotube was inserted into the matrix to form the RVE. The matrix was modeled by solid elements, and the nanotube was represented by 3D elastic beam elements created by binding the nodes of the matrix. The synthesis of the RVE is shown in Figure 7.14.

New computational tools are specially needed in the area of multiscale RVE modeling. The multiscale RVE modeling is a "local-global" approach. In order to catch the local nanocharacteristics, quantum mechanics or molecular dynamics needs to be explored. But the prediction of global macromechanical properties requires the continuum mechanics-based FEM. The transition from local to global becomes a complex issue as it involves a change of scale. Ogata et al. [12] proposed a way of combing quantum mechanics, molecular dynamics, and finite elements. In regions where the atoms obey the laws of continuum mechanics, the FEM is used. However, in critical areas such as the extremity of a fracture, molecular dynamics and quantum mechanics are required to obtain a more detailed study of the fracture process. Xiao and Belytschko [76] improved the numerical compatibility between regions modeled by molecular dynamics and those modeled using the FEM. They suggested a method introducing a broad transition region by superposing the finite element mesh of the continuum region on the atomistic structure of the molecular dynamics region. The determination of mechanical properties like effective Young's modulus along with the Poisson's ratio can be effectively and efficiently done by FEM and even with complicated analyses involving interfacial and surface properties between the matrix and filler.

7.5 Computing the Thermal Conductivity of Nanocomposites Using FEM

The thermal conductivity of a composite can be modeled using effective medium theory. According to this theory, a heterogeneous material having discontinuous properties can be replaced, by a homogeneous material that gives the same average response to a given input at the macroscopic level. This process is called homogenization. In this case, the composite is assumed to be statistically homogeneous where the fillers are uniformly dispersed within the matrix material [77]. This assumption simplifies the mathematical

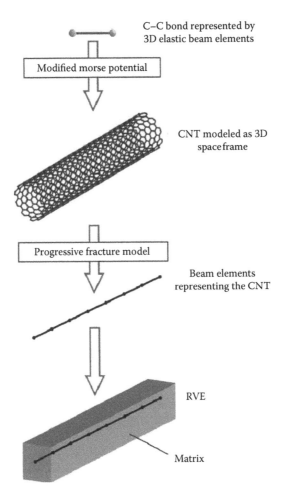

FIGURE 7.14
Synthesis of the CNT using RVE. (From Tserpes, K.I. et al., *Theoret. Appl. Fract. Mech.*, 49, 51, 2008.)

analysis to a greater extent. The steady-state heat conduction with no heat generation is then used for developing a mathematical model.

The development of a continuum model for a multiwalled nanotube (MWNT) inclusion has been carried out in two steps.

1. An equivalent continuum model of an MWNT was developed by taking into account the mechanism of heat conduction through an MWNT. The structure and properties of the nanotube were taken into account and the properties of an effective fiber were defined.

2. This effective fiber is then considered to be the inclusion phase that is embedded within the matrix material.

The composite material to be analyzed is effectively an aligned short fiber composite, where the effective fiber constitutes phase that is embedded within a polymer matrix. A mathematical solution for calculating thermal conductivity of a CNT composite in the longitudinal direction using effective medium theory was reported by Bagachi and Nomura (Figure 7.15) [78].

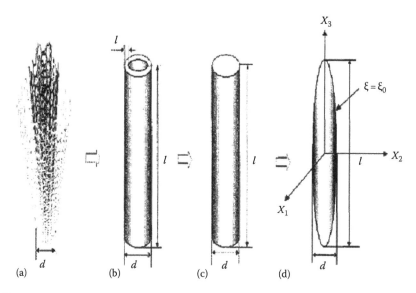

FIGURE 7.15
Development of a continuum model for an MWNT. (a) Schematic diagram of an MWNT showing concentric graphene layers, (b) equivalent continuum model, (c) effective solid fiber, and (d) a prolate spheroidal inclusion. (From Bagchi, A. and Nomura, S., *Comp. Sci. Technol.*, 66(11–12), 1703, 2006.)

A hollow cylinder having the same length and diameter as that of the nanotube was considered to represent an equivalent continuum model of the nanotube. The thickness of the cylinder wall was the same as that of outer nanotube layer (0.34 nm) and was considered to be made up of homogeneous and isotropic material that has the same physical properties as that of the nanotube. The heat carrying capacity of this hollow cylinder was applied to its entire cross section and the properties of an effective solid fiber were then investigated [78]. Effective fiber can be defined as a solid fiber that has the same length and diameter as that of the hollow cylinder and has an identical temperature gradient across its length when the same amount of heat is flowing through it. This effective fiber thus retains the geometrical properties of the nanotube while providing us with a continuum model of the nanotube structure that is suitable for mathematical analysis.

The expression for the conductivity of the effective solid fiber in the longitudinal direction is given by following equation:

$$k^{(2)} = \frac{4t}{d} k_{NT} \quad (t/d < 0.25) \tag{7.32}$$

where
  d represents nanotube diameter
  t represents the thickness of the outer wall of the nanotube

Scant studies have been carried out on the theoretical and experimental determination of conductivity of nanotubes in the transverse directions. For simplicity, the thermal conductivity of the effective fiber is assumed to be isotropic in nature. The expression for the effective thermal conductivity of a CNT composite having isotropic cylindrical short fibers as the filler material where the conductivity of the filler material is given by equation [78] and has a contact resistance at the interface is

$$k_{33}^{*} = k^{(1)}[1 + v_2(1 + \lambda B_1)f(\xi_0) \tag{7.33}$$

where v_2 is the volume fraction of the nanotube phase in B and λ are defined as

$$f(\xi_0) = \left[\frac{1}{2}\xi_0(\xi_0^2 - 1)\ln\frac{\xi_0 + 1}{\xi_0 - 1} - \xi_0^2\right]^{-1} \tag{7.34}$$

$$\lambda = \frac{k^{(2)}}{k^{(1)}} \tag{7.35}$$

The quantity ξ is the inverse of the eccentricity of the spheroid and is given by

$$\xi_0 = \left[1 - \frac{a_1^2}{a_3^2}\right]^{-1/2} \tag{7.36}$$

$$c = (a_3^2 - a_1^2)^{1/2}$$

The constant B_1 can be obtained as a solution to the following linear simultaneous equations:

$$\delta(n) + B_{2n+1}\left[1 - (1-\lambda)(\xi_0^2 - 1)P_{2n+1}(\xi_0)Q_{2n+1}(\xi_0)\right] = \left(\frac{\lambda}{\beta}\right)\sum_{m=0}^{\infty} B_{2m+1}\chi_{nm}(\xi_0) \tag{7.37}$$

where the coefficients interfacial conductance β and χ_{nm} are given by

$$\chi_{nm}(\xi_0) = \left(\frac{4n+3}{2}\right)(\xi_0^2 - 1)Q_{2n+1}(\xi_0)P_{2m+1}(\xi_0)\int_{-1}^{+1}\left(\frac{\xi_0^2 - 1}{\xi_0^2 - \mu^2}\right)^{1/2}P_{2n+1}(\mu)P_{2m+1}(\mu)d\mu \tag{7.38}$$

$$\bar{\beta} = \frac{\beta c}{k^{(1)}} \tag{7.39}$$

P_n and Q_n are known as Legendre polynomials of first kind and second kind, respectively, where

$$P_n(z) = \frac{1}{2^n n!}\frac{d}{dz^n}(z^2 - 1)^n$$

$$Q_n(z) = \frac{1}{2}P_n(z)\ln\frac{z+1}{z-1} - W_{n-1}(z) \tag{7.40}$$

where

$$W_{n-1}(z) = \frac{1}{2}P_0(z)P_{n-1}(z) + \frac{1}{n-1}P_1(z)P_{n-2}(z) + \ldots\ldots + P_{n-1}(z)P_0(z)$$

δ (n) is defined as

$$\begin{pmatrix} \delta(n) = 1 \\ if \\ n = 0 \\ otherwise = 0 \end{pmatrix}$$

To investigate the responses of individual CNTs or CNT-based nanocomposites, such as deformations, load and heat transfer mechanisms, and effective stiffness, the continuum mechanics approach has been applied. CNTs behave as a shell with approximate thickness of 0.34 nm. This is very difficult to model for various reasons like difficulty in meshing large amount of CNTs, and compatibility of shell elements with the solid elements. Shell and solid elements are not compatible with each other as they have different degrees of freedom. It is difficult to couple shell models with the 3D solid model used for the matrix at the interfaces. The solid and shell models involve different types of variables and degrees of freedom. For the analysis of CNTs embedded in matrix material, solid models of the CNTs will provide better accuracy among all the continuum mechanics models. CNT composites are made up of CNTs with different sizes and forms dispersed in a matrix. They can be single-walled or multiwalled with varying geometric parameters like length, diameter, and straight, twisted, curled, etc. The CNTs can be oriented in a particular direction and orientation in the matrix or it can be dispersed randomly. All these factors make the simulations of the mechanical nanotube-based materials extremely complicated.

The concept of unit cells or RVEs has been successfully applied in conventional fiber-reinforced composites at the microscale level and many studies have been extended to investigate CNT-based composites at the nanoscale. In RVE approach, a single nanotube with surrounding matrix material is modeled, with properly applied boundary and interface conditions taking into consideration the effect of the surrounding materials (Figure 7.16). This model has been employed to study the interactions of the nanotube with the matrix and to evaluate the effective material properties of the nanocomposite.

Different types of RVEs are used in modeling a fiber-reinforced composite material. They can be classified according to the shape of the cross section, circular RVE, rectangular or square RVE, and hexagonal RVE. The square RVE models are utilized when the CNT fibers are arranged evenly in a square pattern, while the hexagonal RVE models are used when CNT fibers are in a hexagonal pattern, in the transverse directions (Figure 7.17).

Interfaces between the CNTs and matrix are sensitive regions for the functionality and reliability of CNT-based nanocomposites. The heat carrying capacity of a CNT-based nanocomposite depends on how good the heat is transferred from the matrix material to the CNT. Since all the heat must be transferred through the interfaces to CNTs, a good thermal contact between the matrix and the CNT is required as most mechanical failures in CNT-based nanocomposites occur at or around the interfaces. The interface debonding, friction/wear, instability, or matrix cracking takes place owing to difference in the stiffness and other physical or chemical properties. Modeling thermal interfaces is always a serious challenge to any simulation technique based on the continuum mechanics and it becomes

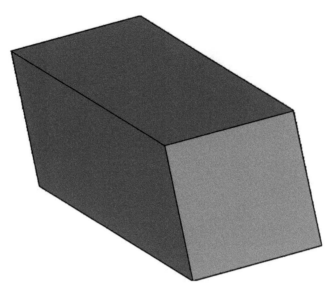

FIGURE 7.16
Representative volume element (RVE) with a single equivalent solid MWNT fiber composite inclusion. (From http://dspace.uta.edu/handle/10106/115)

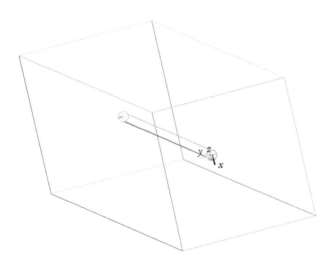

FIGURE 7.17
Wire frame model with MWNT fiber composite inclusion. (From http://dspace.uta.edu/handle/10106/115)

more difficult in the nanoscale models (Figures 7.18 and 7.19). Thus, simulation results near the interfaces require to be interpreted carefully.

Thermal conductivity of the composite is calculated using Fourier's law. According to Fourier's law, thermal conductivity is the ability of the material to transfer heat from a region of high temperature to a region of low temperature and is given by the relation

FIGURE 7.18
Meshed RVE with single effective solid fiber. (From http://dspace.uta.edu/handle/10106/115)

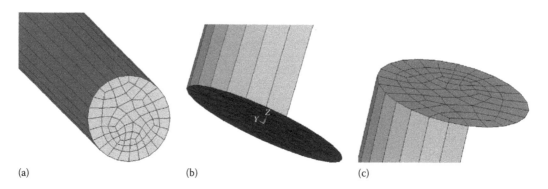

(a)   (b)   (c)

FIGURE 7.19
(a) Effective fiber meshed with solid elements, (b) partial view of effective fiber meshed with contact and target surface elements, and (c) partial view effective fiber meshed with contact surface elements. (From http://dspace. uta.edu/handle/10106/115)

$$q = - kA \, \frac{dT}{dx} \tag{7.41}$$

where
   q is the heat flow
   k is the thermal conductivity
   A is the cross-sectional area
   dT/dx is the temperature gradient in the direction of flow

Using proper tools in the analysis package, thermal heat flux and thermal gradient at each node of the meshed volume (RVE) is collected [79]. The volume average of heat flux and the thermal gradient is calculated. Effective thermal conductivity of the RVE is determined by the ratio of volume average heat flux to the volume average of temperature gradient

$$k_{eff} = \frac{\text{Volume average of heat flux}}{\text{Volume average of thermal gradient}} \qquad (7.42)$$

The volume averaging method of homogenization with the dilute assumption was considered where the composite is assumed to be statistically homogeneous, i.e., the fillers are uniformly dispersed within the matrix material. Interactions between the individual nanotubes are neglected. These two assumptions are very important factors in FEM of a CNT nanocomposite. This assumption allows us to consider the CNT nanocomposite to be made up of a number of unit cells (RVE) in a 3D space. A single unit cell (RVE) with periodic boundary conditions can be used for FEM analysis. The steady-state heat conduction with no heat generation has been used for developing the mathematical model. This assumption allows us to simplify the boundary conditions in the FEM (Figure 7.20a and b). The two faces of the RVE in the direction perpendicular to the axis of the fiber inclusion are applied with 1° temperature difference. The other four faces of the RVE are applied with adiabatic boundary conditions (Tables 7.1 through 7.4).

The aforementioned studies on the thermal conductivity in nanocomposites suggest that interphase resistances are not the prime factor for getting lower than the predicted thermal conductivities. The uniformity and consistency of certain parameters such as lengths and diameters of the CNT inclusions can greatly affect the overall thermal conductivities. Control over these parameters is vital in obtaining a nanocomposite with desired thermal properties. This approach also validates the use of FEM for the analysis of nanomaterials based on continuum mechanics approach.

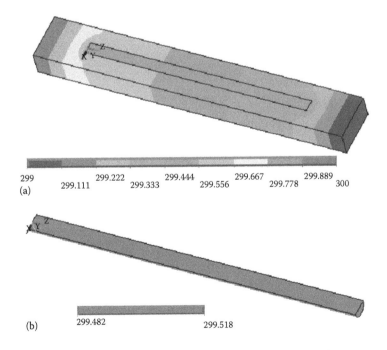

299   299.222   299.444   299.667   299.889
  (a)   299.111   299.333   299.556   299.778   300

  (b)   299.482   299.518

FIGURE 7.20
(a) Temperature profile in representative volume element (half model). (b) Temperature profile in representative volume element (half model). (From http://dspace.uta.edu/handle/10106/115)

TABLE 7.1

Theoretical and FEM Results of Thermal Conductivity
of the Nanocomposites with Varying Volume Fraction
of the Composite Material

Volume Fraction (%)	FEM Results (W/m-K)	Theoretical Results (W/m-K)	Percent Deviation from Theoretical Results (%)
0.6	1.3949	1.1783	−18.4
0.7	1.4810	1.3413	−10.4
0.8	1.5749	1.5044	−4.69
0.9	1.6711	1.6674	−0.22
1.0	1.8332	1.8305	−0.14

Source: http://dspace.uta.edu/handle/10106/115

TABLE 7.2

Theoretical and FEM Results for Effective Thermal Conductivity
of Nanocomposites with Varying Lengths of MWNT

Length of MWNT (μm)	FEM Results (W/m-K)	Theoretical Results (W/m-K)	Percent Deviation from Theoretical Results (%)
30	1.8244	1.8248	0.025
35	1.8267	1.8271	0.026
45	1.8319	1.8297	−0.11
50	1.8332	1.8305	−0.14
60	1.8358	1.8315	−0.23
70	1.8393	1.8321	−0.39

Source: http://dspace.uta.edu/handle/10106/115

TABLE 7.3

Theoretical and FEM Results for Effective Thermal Conductivity
of Nanocomposite with Varying Outer Diameter of MWNT
Keeping Length Constant

Diameter of MWNT (nm)	FEM Results (W/m-K)	Theoretical Results (W/m-K)	Percent Deviation from Theoretical Results (%)
29	2.242	2.238	−0.180
25	1.8306	1.8305	−0.007
30	1.5540	1.5587	0.305
35	1.3641	1.3646	0.035
40	1.2141	1.2190	0.406
45	1.1104	1.1058	−0.414

Source: http://dspace.uta.edu/handle/10106/115

TABLE 7.4

Theoretical and FEA Results for Effective Thermal Conductivity of Nanocomposite with Varying Thermal Contact Conductance (Fixed Aspect Ratio)

Thermal Contact Conductance (MW/m² K)	FEM Results (W/m-K)	Theoretical Results (W/m-K)	Percent Deviation from Theoretical Results (%)
12	1.8306	1.8305	−0.007
30	1.8306	1.8311	0.028
50	1.8306	1.8313	0.037
100	1.8306	1.8314	0.044
500	1.8306	1.8315	0.049
1000	1.8306	1.8315	0.049
10000	1.8306	1.8315	0.049

Source: http://dspace.uta.edu/handle/10106/115

7.6 Computing the Optical Properties of Nanocomposites Using FEM

Optimizing the performance of a given nanocomposite requires accurate knowledge of the effect of porosity, pore shape, and size as well as the optical properties of each phase on the overall optical properties of the nanocomposite. The *Maxwell-Garnett theory* (MGT) [80] was first developed to model the effective electric permittivity of heterogeneous media consisting of monodispersed spheres arranged in a cubic lattice structure within a continuous matrix and of diameter much smaller than the wavelength of the incident electromagnetic (EM) wave. The effective dielectric constant $\varepsilon_{r,eff}$ is expressed as

$$\varepsilon_{r,eff} = \varepsilon_{r,c}\left[1 - \frac{3\phi(\varepsilon_{r,c} - \varepsilon_{r,d})}{2\varepsilon_{r,c} + \varepsilon_{r,d} + \phi(\varepsilon_{r,c} - \varepsilon_{r,d})}\right] \tag{7.43}$$

where $\varepsilon_{r,c}$ and $\varepsilon_{r,d}$ are the dielectric constants of the continuous and dispersed phases, respectively, while ϕ is the porosity. For dispersed phase volume fractions larger than $\pi/6 = 52\%$ and polydispersed spheres the Bruggeman [81] model gives the following implicit equation for $\varepsilon_{r,eff}$:

$$1 - \phi = \frac{\dfrac{\varepsilon_{r,eff}}{\varepsilon_{r,c}} - \dfrac{\varepsilon_{r,d}}{\varepsilon_{r,c}}}{\left[\left(\dfrac{\varepsilon_{r,eff}}{\varepsilon_{r,c}}\right)^{1/3}\left(1 - \dfrac{\varepsilon_{r,d}}{\varepsilon_{r,c}}\right)\right]} \tag{7.44}$$

On the other hand, the *Lorentz–Lorenz model* gives the effective index of refraction n_{eff} as

$$\left(\frac{n_{eff}^2 - 1}{n_{eff}^2 + 2}\right) = (1 - \phi)\left(\frac{n_c^2 - 1}{n_c^2 + 2}\right) + \phi\left(\frac{n_d^2 - 1}{n_d^2 + 2}\right) \tag{7.45}$$

where n_c and n_d are the index of refraction of the continuous and dispersed phases, respectively. Alternatively, the parallel model gives the effective property ψ_{eff} as a linear function of the properties of the continuous and dispersed phases, i.e.,

$$\psi_{eff} = (1-\phi)\psi_c + \phi\psi_d \tag{7.46}$$

The series model, on the other hand, is expressed as

$$\frac{1}{\psi_{eff}} = \frac{1-\phi}{\psi_c} + \frac{\phi}{\psi_d} \tag{7.47}$$

In addition, Del Rio et al. [82] suggested the following effective model for electrical conductivity based on the reciprocity theorem:

$$\sigma_{eff} = \sigma_c \frac{1+\phi\sqrt{\sigma_c/\sigma_d}-1}{1+\phi\sqrt{\sigma_c/\sigma_d}-1} \tag{7.48}$$

Del Rio and Whitaker [83,84] applied the volume averaging theory (VAT) to Maxwell's equations for an ensemble of dispersed domains of arbitrary shape in a continuous matrix. They predicted the effective dielectric constant $\varepsilon_{r,eff}$, relative permeability $\mu_{r,eff}$, and electrical conductivity σ_{eff} of a two-phase mixture as

$$\varepsilon_{r,eff} = (1-\phi)\varepsilon_{r,c} + \phi\varepsilon_{r,d} \tag{7.49}$$

$$\frac{1}{\mu_{r,eff}} = \frac{1-\phi}{\mu_{r,c}} + \frac{\phi}{\mu_{r,c}} \tag{7.50}$$

$$\sigma_{eff} = (1-\phi)\sigma_c + \phi\sigma_d \tag{7.51}$$

The range of validity of these expressions, and a set of inequalities was developed by del Rio and Whitaker [83]. Their model has been numerically validated by Braun and Pilon [85] for the effective through-plane index of refraction of nonabsorbing nanoporous media with open and closed nanopores of various shapes and sizes having a wide range of porosity. Moreover, validation of the aforementioned models against experimental data often yields contradictory results [86]. These contradictions can be attributed to the fact that some of these models were not developed for the index of refraction but for the dielectric constant. However, they have been used for studying the optical properties [87–90]. Some of these models have also been derived by considering a unit cell containing one pore with uniform incident electromagnetic fields thus ignoring possible interference taking place between adjacent pores [80,81]. Huge experimental uncertainty exists in the measure of the porosity and the retrieval of the complex index of refraction from transmittance and reflectance measurements. The latter is very sensitive to the surface roughness of the film and to the uniformity and value of the film thickness. Unfortunately, often, neither the film thickness L nor the experimental uncertainty for both and ϕ m_{eff} are reported. Modeling both the through-plane effective index of refraction and absorption index of nanoporous thin films consists of horizontally aligned cylindrical nanopores or nanowires with different diameters and various porosities and of dielectric medium with embedded metallic nanowires. Such thin films are anisotropic and depend on properties in the direction normal to the film surface. It is limited to nonmagnetic materials for which $\mu_{r,c}=\mu_{r,d}=\mu_{r,eff}=1$. Spectral normal–normal transmittance and reflectance are obtained by numerically solving Maxwell's equations and have been used to obtain the effective index of refraction and absorption index.

In order to develop the numerical model, a surrounding environment of medium 1 where n_1, $k_1=0$ is considered from which an EM wave is incident on an absorbing thin film of medium 2 (n_2, k_2) deposited onto an absorbing dense substrate which is medium 3 (n_3, k_3). A linearly polarized plane wave in transverse electric (TE) mode is incident normal to the film top surface and propagates through the 2D thin film along the x-direction. As the wave propagates in the x–y plane, it has only one electric field component in the z direction, while the magnetic field has two components in the x–y plane such that in time-harmonic form it can be expressed as

$$\vec{E}(x,y,t) = E_z(x,y)e^{i\omega t}\vec{e}_z \tag{7.52}$$

$$\vec{H}(x,y,t) = [H_x(x,y)\vec{e}_x + H_y(x,y)\vec{e}_y]e^{i\omega t} \tag{7.53}$$

where
$\vec{E}$ is the electric field vector
$\vec{H}$ is the magnetic field vector
$\vec{e}_x, \vec{e}_y, \vec{e}_z$ are the unit vectors
$\omega = 2\pi c_0/\lambda$ is the angular frequency of the wave

For general time-varying fields in a conducting medium, Maxwell's equations can be written as

$$\nabla x\left[\frac{1}{\mu_r\mu_o}\nabla x\vec{E}(x,y,t)\right]-\omega^2\varepsilon_r^*\varepsilon_o\,\vec{E}(x,y,t)=0, \tag{7.54}$$

$$\nabla x\left[\frac{1}{\varepsilon_r^*\varepsilon_o}\nabla x\vec{H}(x,y,t)\right]-\omega^2\mu_r\mu_o\,\vec{H}(x,y,t)=0 \tag{7.55}$$

where μ_0 and μ_r are the magnetic permeability of vacuum and the relative magnetic permeability, respectively, while ε_r^* ($= n^2-k^2-i2nk$) is the complex dielectric constant. The associated boundary conditions are

$$\vec{n}(x)(\overrightarrow{H_1}-\overrightarrow{H_2})=0 \tag{7.56}$$

at the surroundings–film interface,

$$\vec{n}(x)\vec{H}=0 \tag{7.57}$$

at symmetry boundaries,

$$\sqrt{\mu_r \mu_o}(\vec{n}(x)\overrightarrow{H}) + \sqrt{\varepsilon_o \varepsilon_r^*}\overrightarrow{E} = 0 \tag{7.58}$$

at the film–substrate interface, and

$$\sqrt{\mu_r \mu_o}(\vec{n}(x)\overrightarrow{H}) + \sqrt{\varepsilon_o \varepsilon_r^*}\overrightarrow{E} = 2\sqrt{\varepsilon_o \varepsilon_r^*}\overrightarrow{E} \tag{7.59}$$

at the source surface, where $\vec{n}$ is the normal vector to the appropriate interface. The equation corresponds to the impedance boundary condition for a semi-infinite substrate while

$$\sqrt{\mu_r \mu_o}(\vec{n}(x)\overrightarrow{H}) + \sqrt{\varepsilon_o \varepsilon_r^*}\overrightarrow{E} = 0 \tag{7.60}$$

$$\sqrt{\mu_r \mu_o}(\vec{n}(x)\overrightarrow{H}) + \sqrt{\varepsilon_o \varepsilon_r^*}\overrightarrow{E} = 2\sqrt{\varepsilon_o \varepsilon_r^*}\overrightarrow{E_o} \tag{7.61}$$

is the low reflecting boundary condition to model the imaginary source surface where the incident EM wave $E_o = \vec{E}_o \vec{e}_z$ is emitted and that is transparent to the reflected waves. The Poynting vector $\vec{\pi}$ is defined as the cross product of the electric and magnetic vectors $\vec{\pi} = 1/2 \operatorname{Re}\left\{\vec{E}xH\right\}$.

Its magnitude corresponds to the energy flux carried by the propagating EM waves. Averaging the x component of the Poynting vector at location (x, y) over a period $2\pi/\omega$ of the EM wave gives

$$\left.|\pi_x|_{avg}(x,y) = \frac{1}{2}\operatorname{Re}\left\{E_z H_y^*\right\} \tag{7.62}$$

and

$$\left.|\pi_x|_{avg}(x,y) = \frac{1}{2}\operatorname{Re}\left\{E_z H_x^*\right\} \tag{7.63}$$

The incident electric field $\vec{E}_o = E_o \vec{e}_z$ and, therefore, the incident time-averaged Poynting vector $\left.|\vec{\pi}_o|\right._{avg}$ are imposed at all locations along the source surface. The values of the x component of the Poynting vector along the film–substrate interface are then calculated numerically and averaged along the boundary to yield $\left.|\vec{\pi}_{x,t}|\right._{avg}$. The transmittance of the thin film is then recovered by taking the ratio of the transmitted to the incident average Poynting vectors, i.e.,

$$T_{num} = \frac{\left.|\pi_{x,t}|\right._{avg}}{\left.|\pi_{x,0}|\right._{avg}} \tag{7.64}$$

Similarly, the magnitude of the x component of the reflected time-averaged Poynting vector $\left.|\pi_{x,t}|\right._{avg}$ is computed numerically, and the reflectance of the film is computed according to

$$R_{num} = \frac{\left|\pi_{x,r}\right|_{avg}}{\left|\pi_{x,0}\right|_{avg}} \qquad (7.65)$$

The preceding equations can be solved numerically using a commercially available finite element solver by applying the *Galerkin* FEM based on unstructured meshes. Figure 7.21 is a schematic representation of an actual model consisting of three nanopores or nanowires of diameter $D = 10\,nm$ and cell width H of 20 nm corresponding to a volume fraction $= \pi D^2/4H^2 = 0.1963$. Figure 7.21 indicates material properties of the different domains and the locations at which each of the boundary conditions are applied. The lines separating two adjacent cubic cells do not correspond to actual boundary conditions. Both the matrix and the nanodomains are treated as homogeneous and isotropic with index of refraction n and absorption index k equal to that of the bulk. The effective optical properties of the nanocomposites can be obtained by minimizing the root mean square of the relative errors for the transmittance and reflectance. Under certain conditions, the effective index of refraction or absorption index of the composite material can be smaller than that of both the continuous and dispersed phases. The same results and conclusions are expected for spherical pores and nanoparticles. These results can be used to design and optimize nanocomposite materials with tunable optical properties.

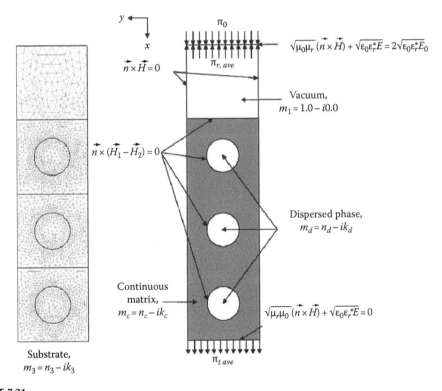

FIGURE 7.21
Physical model and the corresponding finite element grid of the absorbing nanoporous thin film along with the boundary conditions. (From Pilon et al., *J. Appl. Phy.*, 101, 014320, 2007.)

7.7 Computing the Magnetic Properties of Nanocomposites Using FEM

The magnetic behavior in magnetic elements results from size and shape of the elements, and the dynamics of domain formation. The theoretical treatment of magnetization reversal dynamics requires the solution of the Gilbert equation [91]

$$\frac{\delta J}{\delta t} = |\gamma| J x H_{eff} + \frac{\alpha}{J_S} x \frac{\delta J}{\delta t} \tag{7.66}$$

which describes the physical path of the magnetic polarization J toward equilibrium. Here, γ is the gyromagnetic ratio of the free electron spin and is the Gilbert damping constant. The effective field H_{eff}, given by the negative variational derivative of the total magnetic Gibb's free energy, is the sum of the exchange field, the magnetocrystalline anisotropy field, the demagnetizing field, and the external field. The Gilbert equation is a partial differential equation in space and time. The space discretization of the Gilbert equation integration leads to a system of ordinary differential equations. Implicit time integration schemes require solving a system of nonlinear equations at each time steps. The corresponding system matrix will be fully populated if the demagnetizing field is directly calculated from the magnetization distribution. Sparse matrix schemes can be constructed if a magnetic vector potential or a magnetic scalar potential is introduced as an additional variable and space transformations [92] or asymptotic boundary conditions [93] are used to treat the open boundary problem. However, in soft magnetic elements, the formation of domains is very sensitive to the shape of the sample. The error introduced using spatial transformations was found to influence the final domain pattern. A hybrid FEM/BEM is applied to treat the open boundary problem, which shows better convergence properties. The error in the demagnetizing field decreases rapidly as the number of finite elements is increased [94].

The demagnetizing field can be calculated from a magnetic scalar potential, $H_d = -\nabla' U$, which fulfills Poisson's equation within the magnet, Ω_m, and Laplace's equation outside the magnetic particle, Ω_e. The basic concept of the hybrid FEM/BEM method is to split the magnetic scalar potential into two parts: $U = U_1 + U_2$ where U_1 accounts for the divergence of the magnetization within the particle and U_2 is required to meet the boundary conditions. The latter also carries the magnetostatic interactions between spatially independent particles. The potential U_1 is the solution of the Poisson equation

$$\nabla^2 U_1(r) = \nabla M(r) \quad for\ r \in \Omega \tag{7.67}$$

with the natural boundary condition

$$\nabla U_1 . n = M . n \tag{7.68}$$

and U_1 outside the magnetic particles, where n is the outward normal at the boundary Γ of the magnetic region Ω_m. The potential U_2 satisfies the Laplace equation

$$\nabla^2 U_2(r) = 0$$

$$for \tag{7.69}$$

$$r \in \Omega_m \cup \Omega_e$$

and shows a jump at the boundary of the magnetic particles

$$U_2^{in} - U_2^{out} = U_1 \tag{7.70}$$

A standard FEM may be used to solve these equations which define a double-layer potential which is created by a dipole sheet at $\acute{\Gamma}$ with magnitude U_1. At the boundary U_1 the potential U_2 is given by the integral

$$U_2(r) = \frac{1}{4\pi} \int_r U_1(r') \nabla' \left(\frac{1}{|r-r'|} \right) . n' d^2 r' + \left(\frac{\Omega(r)}{4\pi} - 1 \right) U_1(r) \tag{7.71}$$

which can be evaluated using the BEM [95]. The solid angle $\Omega(r)$, subtended by Γ' at r, equals 2π for a smooth surface point r. The preceding equation contains interactions between the boundary nodes of the finite element mesh. Therefore, its discretization requires a full matrix. After discretization, the potential U_2 at the boundary nodes follows from a matrix vector multiplication

$$U_2 = BU_1 \tag{7.72}$$

The $n \times n$ matrix B results from the boundary element discretization of $U_2(r)$, where n denotes the number of nodes at the boundary. The values of U_2 in the interior of the particle follow from the solution of the Laplace equation with Dirichlet boundary conditions, which again can be solved by a standard FEM. Matrix B depends only on the geometry and the finite element mesh and thus has to be computed only once for a given finite element mesh. Since the hybrid FEM/BEM method does not introduce any approximations, the method is accurate and effective. However, the size of the full matrix B may be a problem for large magnetic particles. After discretization of the integral equation $U_2(r)$, all the information on the spatial arrangement of the particles is contained in matrix B, which relates the nodes on boundary of the particles with each other. The grains of the polycrystalline magnetic thin films are divided into tetrahedral finite elements.

Within a tetrahedron, the direction cosines of the magnetization are interpolated by linear functions, whereas quadratic functions represent the magnetic potentials. Figure 7.22 gives the corresponding mixed finite element. The use of an individual order of interpolation for the different components, popular in computational fluid dynamics, is called the mixed FEM [96]. It adapts the finite element subspaces to the physical nature of the problem.

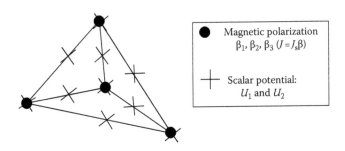

FIGURE 7.22

Mixed finite element used for the calculations. (●): Nodes for the linear interpolation of the direction cosines of the magnetic polarization J. (+): Nodes for the quadratic interpolation of the magnetic scalar potential. (From Schrefl, T., *J. Mag. Mag. Mater.*, 207, 66, 1999.)

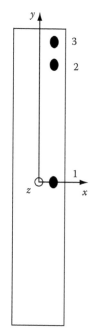

FIGURE 7.23

Points for the test of magnetostatic field calculations. (From Schrefl, T., *J. Mag. Mag. Mater.*, 207, 66, 1999.)

The demagnetizing field arising from magnetic volume charges varies linearly with the magnetic polarization. In order to avoid loss of accuracy with differentiation, the interpolation order of the scalar potential should be higher than that of the magnetic polarization.

The formation of domains is very sensitive to the accuracy of the magnetostatic field calculation.

A uniformly magnetized thin film with a thickness of 21 nm, and a lateral extension of 200×1600 nm (Figure 7.23) checks the accuracy of the magnetostatic field calculation. The use of a quadratic interpolation drastically reduces the error owing to corner singularities. Figure 7.24 shows the polycrystalline microstructure of Co nanoelement and the corresponding finite element mesh. The ill-shaped elements result from the underlying microstructure. The very short edges are well below the exchange length, which prevent any additional error in the exchange energy owing to the bad shape of some elements. The condition number of the system matrix is improved by precondition techniques before solving the equations.

Figure 7.25 shows the numerically calculated end domain in an NiFe nanoelement with one pointed end. These end domains occur only at the flat end of the elements. However, the magnetization rotates out of the long axis near the pointed end, where it becomes arranged parallel to the inclined surfaces forming the pointed ends. This configuration reduces the magnetic surface charges and requires less exchange energy than any possible end domain near the pointed end. For the calculation of the stable end domain, zero anisotropy is assumed. The end domain evolves into a small-scale zigzag domain pattern in elements with a transverse anisotropy of $Ku = 38$ kJ/m^3. Figure 7.25 gives the time evolution of the magnetic domain structure assuming a transverse anisotropy. The pointed ends suppress the formation of end domains. Magnetization reversal starts from magnetization ripple which originates near the corners of the inclined surfaces forming the pointed end. The switching field of the isolated element was found to depend on the value of the exchange constant assumed in the calculations. The exchange constant of a granular

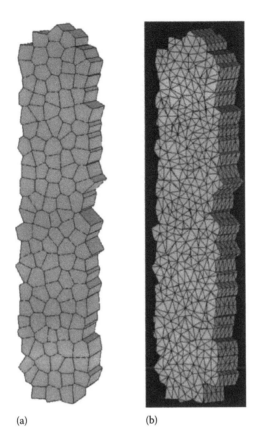

(a)　　　　　　　　(b)

FIGURE 7.24
Model of the polycrystalline microstructure (a) and the finite element mesh (b) of a 200 nm long, 40 nm wide, and 25 nm thick Co nanoelement. (From Schrefl, T., *J. Mag. Mag. Mater.*, 207, 66, 1999.)

NiFe may be reduced as compared to the bulk value [97]. Switching field comparable with the experimental measurements is obtained assuming an exchange constant $= 8 \times 10^{-12}$ J/m, a saturation polarization $Js = 1$ T, and zero magnetocrystalline anisotropy. This result suggests that the small-scale domain pattern observed in *Lorentz* imaging of NiFe has to be attributed to a transverse anisotropy, which may arise from internal strain [98]. With the aforementioned value for the transverse anistropy, the influence of the bar width on the domain pattern for zero applied field can be simulated in good agreement with the experiments. Micromagnetic modeling provides a precise understanding of domain formation and of the switching mechanism in patterned magnetic elements. Small-scale domains in the remnant state of NiFe elements can be explained by a uniaxial anisotropy parallel to the short axis. Micromagnetic simulations successfully reproduce the influence of bar width and tip shape on the domain structure. The switching field of individual elements in an array of closely packed NiFe elements strongly depends on the strength and direction of the interaction field, which is determined by the magnetization pattern of several neighboring elements. The dependence of the switching field of Co elements on the shape of the ends is attributed to the demagnetizing field at the corners causing the nucleation of reversed domains. Micromagnetic models based on the Gilbert equation of motion resolve magnetization processes in space and time. In addition to the hysteresis properties, the

FIGURE 7.25
(a) End domain at the flat end of a NiFe nanoelement ($L=2500$ nm, $W=200$ nm, $P=500$ nm, $T=25$ nm). The arrow indicates the direction of the magnetization. (b) Evolution of a zigzag domain pattern in elements with transverse anisotropy ($L=1600$ nm, $W=200$ nm, $P=500$ nm, $T=21$ nm). The gray scale maps the magnetization component parallel to the short axis of the element. (From Schrefl, T., *J. Mag. Mag. Mater.*, 207, 66, 1999.)

micromagnetic simulations predict where the reversed domains nucleate and how they expand with time. The polycrystalline microstructure of Co films and edge irregularities significantly influence the magnetization reversal mechanism. The comparison with experimental results shows that accurate numerical predictions of the switching fields of patterned magnetic elements require precise models of the microstructure.

7.8 Computing Nanocomposites Damage Using FEM

When combining highly stiff nanofibers with a matrix, the final nanocomposite material possesses higher strength than the matrix. The aim of the FEM is to investigate the stresses at both the matrix and most importantly the reinforcement. The FEM can be carried out using ANSYS software to derive various stresses at the matrix/nanofiber interfaces. Maximum principal stresses, von Mises stress, and normal stresses have been analyzed along nanofiber short interface (x-direction in this study) and along a rounded nanofiber cross section in which stresses were noticed along the circumference as well as at the highest stress points on the nanofiber. Three different scenarios have been investigated [99] as mentioned in the following.

1. Long and short nanofibers (50 and 20 nm, respectively)
2. High modulus and low modulus nanofibers (600 and 50 GPa, respectively)
3. Shaped nanofibers (rounded, star, and hexagonal cross-section shapes)

Micromechanical matrix/reinforcement interlocking, chemical bonding, and the weak van der Waals force are three main mechanisms of interfacial load transfer [100]. Star-shaped and hexagonal-shaped nanofibers with the same volume were simulated under loading using FEM. The aim was to investigate their performance in improving matrix/reinforcement stress transfer compared to rounded cross section nanofibers. The star-shaped nanofibers carry higher stresses than rounded and hexagonal-shaped nanofibers even if the nanofiber length is increased. Hexagonal-shaped nanofibers performed better than rounded nanofibers 20–50 nm long. It was found that von Mises minimum stresses did not appear as they were close to 0. The results revealed that by maintaining a strong bond between the nanofiber and the matrix, star-shaped fibers showed a good potential in producing high strength nanocomposites. However, experimental investigations are required to confirm this study since researchers have reported some discrepancies between FEM studies and their equivalent experiments [101] because inefficient shear stress transfer can lead to poor nanocomposites properties [102]. In particular, the FEM of various forms of nanofiber shapes and volume fractions revealed, as expected, that interfacial debonding is most likely the main source of nucleation damage. It is found that the nanofiber/matrix debonding can be attributed to the high stress concentrations at the nanofiber ends which can be made more severe with poor interfacial shear stress transfer.

7.9 Application of FEM to Biological Systems

Tremendous efforts have been made to develop modeling and simulation approaches for fluid–structure interaction problems. With FEM for both fluid and solid domains, the submerged structure is solved more realistically and accurately in comparison with the corresponding fiber network representation. The fluid solver is based on a stabilized equal-order finite element formulation applicable to problems involving moving boundaries [103–105]. This stabilized formulation prevents numerical oscillations without introducing excessive numerical dissipation. As the background fluid mesh does not have to follow, it is possible to assign a sufficiently refined fluid mesh within the region around the immersed, moving, deformable structures.

The immersed finite element analysis (IFEA) was developed by Zhang et al. [106] to solve complex fluid and deformable structure interaction problems. Consider an incompressible 3D deformable structure in Ω^s completely immersed in an incompressible fluid domain Ω^f. Together, the fluid and the solid occupy a domain Ω, but they do not intersect:

$$\Omega^f \cup \Omega^s = \Omega$$

$$\Omega^f \cup \Omega^s = \varnothing$$

(7.73)

The solid domain can occupy a finite volume in the fluid domain. Since we assume both fluid and solid to be incompressible, the union of two domains can be treated as one incompressible continuum with a continuous velocity field. In the computation, the fluid spans the entire domain Ω, thus an *Eulerian* fluid mesh is adopted, whereas a *Lagrangian* solid mesh is constructed on top of the *Eulerian* fluid mesh. The coexistence of fluid and solid in Ω^s requires some considerations when developing the momentum and continuity equations.

In the computational fluid domain Ω, the fluid grid is represented by the time-invariant position vector x, while the material points of the structure in the initial solid domain

$\Omega_0{}^s$ and the current solid domain Ω^s are represented by X^s and x^s (X^s, t), respectively. The superscript s is used in the solid variables to distinguish the fluid and solid domains.

In the fluid calculations, the velocity v and the pressure p are the unknown fluid field variables, whereas the solid domain involves the calculation of the nodal displacement u^s, which is defined as the difference of the current and the initial coordinates: $u^s = x^s - X^s$. The velocity v^s is the material derivative of the displacement du^s/dt.

We define the fluid–structure interaction force within the domain Ω^s as $f_i{}^{FSI}$, where FSI stands for fluid–structure interaction:

$$f_i^{FSI,s} = \left(\rho^s - \rho^f\right)\frac{dvi}{dt} + \sigma_{ij,j}{}^s - \sigma_{ij,j}{}^f + \left(\rho^s - \rho^f\right)gi$$

$$x \in \Omega^s$$

(7.74)

The interaction force $f_i{}^{FSI}$ is calculated with the *Lagrangian* description. A Dirac delta function, δ, is used to distribute the interaction force from the solid domain onto the computational fluid domain

$$f_i^{FSI}(x,t) = \int_{\Omega^s} f_i^{FSI,s}(x^s,t)\delta(x - x^s(X^s,t))d\Omega$$

(7.75)

The equation for the fluid can be derived by combining the fluid terms and the interaction force as

$$\rho^f\frac{dvi}{dt} = \sigma^f{}_{ij,j} + f_i^{FSI}; \quad x \in \Omega$$

(7.76)

Since the entire domain Ω is incompressible, we need to apply the incompressibility constraint once in the entire domain Ω:

$$v_{i,i} = 0$$

To delineate the *Lagrangian* description for the solid and the *Eulerian* description for the fluid, we introduce different velocity field variables v_i^s and v_i to represent the motions of the solid in the domain Ω^s and the fluid within the entire domain Ω. The coupling of both velocity fields is accomplished with the Dirac delta function.

$$v_i^s(X^s,t) = \int_{\Omega} v_i(x,t)\delta(x - x^s(X^s,t))d\Omega$$

(7.77)

Assuming that there is no traction applied on the fluid boundary, $\int_{\Gamma_{hi}} \delta v_i h_i d\Gamma = 0$ applying integration by parts and the divergence theorem, we can get the final weak form (with stabilization terms):

$$\int_{\Omega} (\delta v_i + \tau^m v_k \delta v_{i,k} + \tau^c \delta p_i)[\rho^f(v_{i,t} + v_j v_{i,j}) - f_i^{FSI}]d\Omega + \int_{\Omega} (\delta v_{ij}\sigma^f_{ij,}d\Omega$$

$$- \sum_e \int_{\Omega_e} (\tau^m v_k \delta v_{ik} + \tau^c \delta p_i)\sigma^f_{ij,j}d\Omega + \int_{\Omega} (\delta p + \tau^c \delta v_{i,i})v_{i,j}d\Omega = 0$$

(7.78)

The nonlinear systems are solved with the *Newton–Raphson method*. Moreover, to improve computational efficiency, we also employ the iterative algorithm and compute the residuals based on matrix-free techniques [107–108]. The transformation of the weak form from the updated Lagrangian to the total Lagrangian description is to change the integration domain from Ω^s to Ω_0^s. Since we consider an incompressible fluid and solid, and the *Jacobian* determinant is one in the solid domain, the transformation of the weak form to total Lagrangian description yields

$$\int_{\Omega_0^s} \delta u_i [(\rho^s - \rho^f) u_i^s - \frac{\delta P_{ji}}{\delta X_j^s} - (\rho^s - \rho^f) gi + f_i^{FSI,s}] d\Omega_0^s = 0 \tag{7.79}$$

where the first Piola–Kirchhoff stress P_{ij} is defined as $P_{ij} = JF_{ik}^{-1}\sigma_{kj}^s$ and the deformation gradient $F_{ij} = \delta x_i^s / \delta X_j^s$. Using integration by parts and the divergence theorem, we can rewrite

$$\int_{\Omega_0^s} \delta u_i [(\rho^s - \rho^f) u_i^s d\Omega_0^s + \int_{\Omega_0^s} \delta u_{ji} P_{ji} d\Omega_0^s - \int_{\Omega_0^s} \delta u_i (\rho^s - \rho^f) gid\Omega_0^s + \int_{\Omega_0^s} \delta u_i f_i^{FSI,s} d\Omega_0^s = 0 \tag{7.80}$$

For structures with large displacements and deformations, the second Piola–Kirchhoff stress S_{ij} and the Green–Lagrangian strain E_{ij} are used in the total Lagrangian formulation:

$$S_{ij} = \frac{\delta W}{\delta E_{ij}}$$

and

$$E_{ij} = \frac{1}{2}(C_{ij} - \delta_{ij}) \tag{7.81}$$

where the first Piola–Kirchhoff stress P_{ij} can be obtained from the second Piola–Kirchhoff stress as $P_{ij} = S_{ik}F_{jk}$.

Finally, in the interpolation process from the fluid onto the solid grid, the discretized form can be written as

$$f_{ij}^{FSI} = \sum_I f_{il}^{FSI,s}(X^s, t)\phi_I(x_J - x_I^s), \quad x_I^s \in \Omega_{\phi J} \tag{7.82}$$

The solid velocity v_i^s at node I can be calculated by gathering the velocities at fluid nodes within the influence domain $\Omega_{\phi J}$. A dual procedure takes place in the distribution process from the solid onto the fluid grid. The discretized form is expressed as

$$v_{il}^s = \sum_J v_{ij}(t)\phi_J(x_J - x_I^s), \quad x_J \in \Omega_{\phi I} \tag{7.83}$$

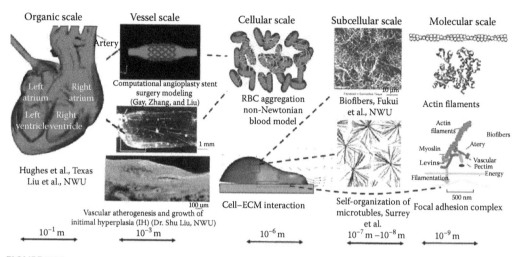

FIGURE 7.26
Modeling of biological processes using a 3D multiscale technique. (From Liu et al., *Comput. Method Appl. Mech. Engg.*, 195, 1722, 2006.)

This ensures the no-slip boundary condition on the surface of the solid, and prevents the fluid from penetrating the solid, provided the solid mesh is at least twice as dense as the surrounding fluid mesh.

The ultimate goal of the 3D multiple scale modeling is to better understand biological phenomena that span the five scales depicted in Figure 7.26. The measurements of traction forces and the simultaneous imaging of the fibrous structure of the cell will provide critical input to the simulation of cell motility.

$$f_{il}^{FSI,s} = f_{il}^{inert} - f_{il}^{int} + f_{il}^{ext}, in\Omega^s,$$

$$f_{il}^{FSI} = \sum_I f_{il}^{FSI,s}(X^s,t)\phi_I(x_J - x_I^s), x_I^s \in \Omega_{\phi_J},$$

$$\rho^f(v_{i,t} + v_j v_{ij}) = \sigma_{ijj} + \rho g_i + f_{il}^{FSI}, in\Omega, \tag{7.84}$$

$$v_{ij} = 0, in\Omega,$$

$$v_{il}^s = \sum_J v_{ij}(t)\phi_J(x_J - x_I^s), x_J \in \Omega_{\phi_I}.$$

Using the IFEM formulation, a simple case of blood flow through a capillary vessel can be studied, where the middle quarter of the vessel is designated to be injured and capable of activating platelets. The capillary vessel has a diameter of $50\,\mu m$ and a length of $100\,\mu m$ [109]. The adhesion between the platelet and injured vessel wall is described by an attractive force combined with an elastic link which provides the resistance of the platelet to shear after bonding to the blood vessel. The activation of the platelet is described by dynamically updating the array, which stores information for the activated platelet. Due to the small-scale of the platelet (the diameter is around $2\,lm$), it is treated as a rigid particle immersed in plasma. The density of the platelet is very close to that of the plasma, thus $\rho_s = \rho_f$.

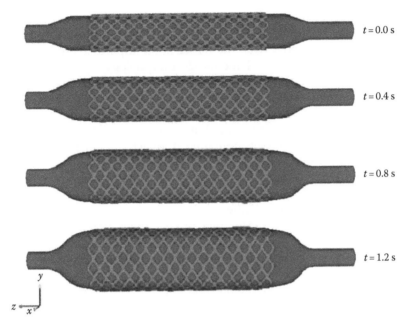

FIGURE 7.27
Deployment of the stent through the inflation of the balloon at different time steps. (From Liu et al., *Comput. Method Appl. Mech. Engg.*, 195, 1722, 2006.)

Coronary stents physically open the channel of constricted arterial segments by fatty deposits or calcium accumulations. During stenting, a balloon deploys the stent which is kept inflated for 30s and is then deflated. At the end of the process, the expanded stent is embedded into the wall of the diseased artery and holds it open. By using the IFEM method, we can study the flow pattern during the deployment of a stent, its deformation, and stress distribution. The results show that this computational method provides a useful tool for future stent designs (Figure 7.27).

Furthermore, the IFEM has been combined with electrokinetics to study the mechanisms of bio-nanoelectromechanical (NEMS) devices [109]. In the 3D dynamic simulations, the fluid flow and solid deformation/motion are reasonably captured. Using novel NEMS devices, it will be able to simultaneously visualize the cellular scale structures and measure the cellular traction forces and adhesion forces.

7.10 Conclusion

A main goal of computational materials science is the rapid and accurate prediction of new materials and their new properties and features, which is very difficult to achieve with traditional modeling and simulation methods at a single length and time scale with the current computer power. Therefore, it is expected to use the FEM to bridge the models and simulation techniques across a broad range of length and time scales to address the macroscopic or mesoscopic behaviors of materials from a detailed molecular description.

The development of polymer nanocomposites relies largely on our understanding of the structure–property relationship of the materials which requires a multiscale model to predict the material properties from the information of particle properties, molecular structure, molecular interaction, and mesoscale morphology. The current research in modeling and simulation of polymer nanocomposites are largely limited to individual length and time scale. However, it should be noted that some efforts have recently been made to develop multiscale strategies for predicting the multiscale level of structure, properties, and processing performance of polymer nanocomposites based on nanoparticle reinforcement (e.g., nanosphere, nanotube, nanorod, and clay platelet). The development of polymer nanocomposites necessitates a comprehensive understanding of the phenomena at different time and length scales. The challenge for FEM is to move, as seamlessly as possible, from one scale to another so that the calculated parameters, properties, and numerical information can be efficiently transferred across scales. In case of polymer nanocomposites, it is required to accurately predict their hierarchical structures and behaviors and to capture the phenomena on length scales that span typically 5–6 orders of magnitude and time scales that can span a dozen orders of magnitude. For example, a clay particle with a diameter of 0.5 mm and 100 layers would have about 85 million atoms. If such a particle is dispersed into polymer matrix to form polymer nanocomposites containing 5% of clay in weight, the system would then have about 3 billions of atoms. In the past decade or so, this need has significantly stimulated the development of computer modeling and simulation, either as a complementary or alternative technique to experimentation. These techniques indeed represent approaches at various time and length scales from molecular scale (e.g., atoms), to microscale (e.g., coarse-grains, particles, monomers) and then to macroscale (e.g., domains), and have shown success to various degrees in addressing many aspects of polymer nanocomposites. Despite the progress over the past years, there are a number of challenges in computer modeling and simulation of nanocomposites. There is a need to develop new and improved simulation techniques at individual time and length scales. It is also important to integrate the developed methods at wider range of time and length scales, spanning from quantum mechanical domain (a few atoms) to molecular domain (many atoms), to mesoscopic domain (many monomers or chains), and finally to macroscopic domain (many domains or structures), to form a useful tool for exploring the structural, dynamic, and mechanical properties, as well as optimizing design and processing control of polymer nanocomposites. Developing such a multiscale method is very challenging but indeed represents the future of computer simulation and modeling, not only in polymer nanocomposites but also other fields. New concepts, theories, and computational tools should be developed in the future to make truly seamless multiscale modeling a reality. Such development is crucial in order to achieve the longstanding goal of predicting particle–structure–property relationships in material design and optimization.

Acknowledgment

The corresponding author Dr Ufana Riaz wishes to acknowledge the Department of Science and Technology (DST), Science and Engineering Research Council (SERC), India, for granting the corresponding author Fast Track Project vide sanction no SR/FT/CS. 012/2008.

References

1. Baschnagel, J., Binder, K., Doruker, P., Gusev, A.A., Hahn, O., Kremer, K., Mattice, W.L., Müller-Plathe, F., Murat, M., Paul, W., Santos, S., Suter U.W. and Tries V. 2000. Bridging the gap between atomistic and coarse-grained models of polymers: Status and perspectives. *Adv. Polym. Sci.*, 152: 41–156.

2. Rudd, R.E. and Broughton, J.Q. 2000. Concurrent coupling of length scales in solid state systems. *Phys. Status Solidi.* (b), 217(1): 251–291.

3. Smith, G.S., Tadmor, E.B., and Kaxiras, E. 2000. Multiscale simulation of loading and electrical resistance in silicon nanoindentation. *Phys. Rev. Lett.*, 84(6): 1260–1263.

4. Flewitt, P.E.J. 2004. The use of multiscale materials modelling within the UK nuclear industry. *Mater. Sci. Eng. A*, A365(1–2): 257–266.

5. Elliott, J.A. and Hancock, B.C. 2006. Pharmaceutical materials science: An active new frontier in materials research. *Mater. Res. Soc. Bull.*, 31(11): 869–873.

6. Liu, W.K., Karpov, E.G., Zhang, S., and Park, H.S. 2004. An introduction to computational nanomechanics and materials. *Comput. Methods. Appl. Mech. Eng.*, 193(17–20): 1529–1578.

7. Stoneham, A.M. 2003. The challenges of nanostructures for theory. *Mater. Sci. Eng. C*, C23(1–2): 235–241.

8. Kim, S., Yamaguchi, S., and Elliott, J.A. 2009. Solid-state ionics in the 21st century: Current status and future prospects. *Mater. Res Soc. Bull.*, 34(12): 900–906.

9. Nieminen, R.M. 2002. From atomistic simulation towards multiscale modelling of materials. *J. Phys. Condens. Matter.*, 14(11): 2859–2876.

10. Berendsen, H.J.C. 2005. *Simulating the Physical World: Hierarchical Modeling from Quantum Mechanics to Fluid Dynamics*. Cambridge, U.K., Cambridge University Press.

11. Groh, S. and Zbib, H.M. 2009. Advances in discrete dislocations dynamics and multiscale modeling. *J. Eng. Mater. Technol. ASME*, 131(4): 041209–041219.

12. Ogata, S., Lidorikis, E., Shimojo, F., Nakano, A., Vashishta, P., and Kalia, R.K. 2001. Hybrid finite element/molecular-dynamics/electronic-density-functional approach to materials simulations on parallel computers. *Comp. Phys. Comm.*, 138(2): 143–154.

13. Miller, R.E. and Tadmor, E.B. 2007. Hybrid continuum mechanics and atomistic methods for simulating materials deformation and failure. *Mater. Res. Soc. Bull.*, 32(11): 920–926.

14. Miller, R.E. and Tadmor, E.B. 2009. A unified framework and performance benchmark of fourteen multiscale atomistic/continuum coupling methods. *Model. Simul. Mater. Sci. Eng.*, 17(5): 053001–053052.

15. Bernstein, N., Kermode, J.R., and Csanyi, G. 2009. Hybrid atomistic simulation methods for materials systems. *Rep. Progr. Phys.*, 72(2): 026501–026526.

16. Braun, T., Schubert, A., and Zsindely, S. 1997. Nanoscience and nanotechnology on the balance. *Scientometrics*, 38: 321–325.

17. Alexandre, M. and Dubois, P. 2000. Polymer-layered silicate nanocomposites: Preparation, properties and uses of a new class of materials. *Mater. Sci. Eng.*, 28: 1–63.

18. Kumar, A.P., Depan, D., Tomer, N.S., and Singh, R.P. 2009. Nanoscale particles for polymer degradation and stabilization-Trends and future perspectives. *Prog. Polym. Sci.*, 34: 479–515.

19. Ramakrishnan, N. and Arunachalam, V.S. 1997. Finite element methods for materials modelling. *Progr. Mater. Sci.*, 42(1–4): 253–261.

20. Zienkiewicz, O.C., Taylor, R.L., and Zhu, J.Z. 2005. *The Finite Element Method: Its Basics and Fundamentals*. Oxford, U.K., Butterworth- Heinemann.

21. Rogers, B., Pennathur, S., and Adams, J. 2008. *Nanotechnology Understanding Small Systems*. Boca Raton, FL, CRC Press, p. 398.

22. Li, T.D., Gao, J., Szoszkiewicz, R., Landman, U., and Riedo, E. 2007. Structured and viscous water in sub nanometer gaps. *Phys. Rev. B*, 75(11): 115415–115421.
23. Eijkel, J.C.T. and Berg, V.D.A. 2005. Nanofluidics: What is it and what can we expect from it. *Microfluid Nanofluidics*, 1: 249–267.
24. Roduner, E. 2006. Size matters: Why nanomaterials are different. *Chem. Soc. Rev.*, 35: 583–592.
25. Jones, R.A.L. 2004. *Soft Machines: Nanotechnology and Life*. Oxford, U.K., Oxford University Press, p. 229.
26. Crosby, A.J. and Lee, J.Y. 2007. Polymer nanocomposites: The "nano" effect on mechanical properties. *Polym. Rev.*, 47: 217–229.
27. Damme, H.V. 2008. Nanocomposites: The end of compromise. In: Brechignac, C., Houdy, P., and Lahmani, M., eds., *Nanomaterials and Nanochemistry*. New York, Springer, pp. 348–380.
28. Vaia, R.A. and Wagner, H.D. 2004. Framework for nanocomposites. *Mater. Today*, 7: 32–37.
29. Thilly, L., Petegem, S.V., Renault, P.O., Lecouturier, F., Vidal, V., Schmitt, B., and Swygenhoven, V. 2009. A new criterion for elastio-plastic transition in nanomaterials: Application to size and composite effects on Cu-Nb nanocomposite wires. *Acta Materialia*, 57: 3157–3169.
30. Dong, Y., Bhattacharyya, D., and Hunter, P.J. 2008. Experimental characterization and object-oriented finite element modeling of polypropylene/organoclay nanocomposites. *Comp. Sci. Technol.*, 68: 2864–2875.
31. Chen, Q., Chasiotis, I., Chen, C., and Roy, A. 2008. Nanoscale and effective mechanical behavior and fracture of silica nanocomposites. *Comp. Sci. Technol.*, 68: 3137–3144.
32. Chen, W.H., Cheng, H.C., Hsu, Y.C., Uang, R.H., and Hsu, J.S. 2008. Mechanical material characterization of Co nanowires and their nanocomposite. *Comp. Sci. Technol.*, 68: 3388–3395.
33. Song, S.Y. and Youn, J.R. 2006. Modelling of effective elastic properties of polymer based carbon nanotube composites. *Polymer*, 47: 1741–1748.
34. Schuster, B.E., Wei, Q., Ervin, M.H., Hruszkewycz, S.O., Miller, M.K., Hufnagel, T.C., and Ramesh, K.T. 2007. Bulk and microscale compressive properties of a pd-based metallic glass. *Scripta Materialia*, 57(6): 517–520.
35. Zhang, H., Schuster, B.E., Wei, Q., and Ramesh, K.T. 2006. The design of accurate micro-compression experiments. *Scripta Materialia*, 54(2): 181–186.
36. Oliver, W.C. and Pharr, G.M. 1992. An improved technique for determining hardness and elastic modulus using load and displacement sensing indentation experiments. *J. Mater. Res.*, 7(6): 1564–1583.
37. Han, C.F. and Lin, J.F. 2008. Modeling to evaluate the contact areas of hard materials during the nanoindentation tests. *Sens. Actuat. A*, 147(1): 229–241.
38. Ling, Z. and Hou, J. 2007. A nanoindentation analysis of the effects of microstructures on elastic properties of Al_2O_3/SiC composites. *Comp. Sci. Technol.*, 67: 3121–3129.
39. Lee, S.H., Wang, S., Pharr, G.M., and Xu, H. 2007. Evaluation of interphase properties in a cellulose fiber reinforced polypropylene composite by nanoindentation and finite element analysis. *Composites: Part A*, 38: 1517–1524.
40. Toparli, M. and Koksal, N.S. 2005. Hardness and yield strength of dentin from simulated nanoindentation tests. *Comp. Meth. Progr. Biomedi.*, 77: 253–257.
41. Maritza, G.J., Veprek, H., Ratko, G., Veprek, R.G., Argon, A.S., Parks, D.M., and Veprek, S. 2009. Non-linear finite element constitutive modeling of indentation into super- and ultra hard materials: The plastic deformation of the diamond tip and the ratio of hardness to tensile yield strength of super- and ultra hard nanocomposites. *Surf. Coat. Technol.*, 203: 3385–3391.
42. Jiang, W. and Batra, R.C. 2009. Identification of elastic constants of FCC metals from 2D load indentation curves. *Comp. Mater. Sci.*, 45(2): 511–515.
43. Harsono, E., Swaddiwudhipong, S., and Liu, Z.S. 2009. Material characterization based on simulated spherical-Berkovich indentation tests. *Scrip. Mater.*, 60(11): 972–975.
44. Geng, K., Yang, F., and Grulke, E.A. 2008. Nanoindentation of submicron polymeric coating systems. *Mater. Sci. Eng. A.*, 479(1/2): 157–163.

45. Farrissey, L.M. and McHugh, P.E. 2005. Determination of elastic and plastic material properties using indentation: Development of method and application to a thin surface coating. *Mater. Sci. Eng. A*, 399: 254–266.
46. Zeng, K., Soderlund, E., Giannakopoulos, A.E., and Rowcliffe, D.J. 1996.Controlled indentations: A general approach to determine mechanical properties of brittle materials. *Acta Mater.*, 44:1127–1141.
47. Yoshino, M., Aoki, T., Chandrasekaran, N., Shirakashi, T., and Komanduri, R. 2001. Finite element simulation of plane strain plastic–elastic indentation on single-crystal silicon. *Int. J. Mech. Sci.*, 43(2): 313–333.
48. Poon, B., Rittel, D., and Ravichandran, D. 2008. An analysis of nanoindentation in linearly elastic solids. *Int. J. Solids Struct.*, 45: 6018–6033.
49. Broutman, L.J. and Panizza, G. 1971. Micromechanics studies of rubber-reinforced glassy polymers. *Int. J. Polym. Mater.*, 1: 95–109.
50. Agarwal, B.D. and Broutman, L.J. 1974. Three-dimensional finite element analysis of spherical particle composites. *Fibre. Sci. Tech.*, 7: 63–77.
51. Iijima, S. 1991. Helical microtubules of graphitic carbon. *Nature*, 354: 56–58.
52. Treacy, M.M.J., Ebbesen, T.W., and Gibson, T.M. 1996. Exceptionally high Young's modulus observed for individual carbon nanotubes. *Nature*, 381: 680–687.
53. Uddin, M.F. and Sun, C.T. 2008. Strength of unidirectional glass/epoxy composite with silica nanoparticle-enhanced matrix. *Compos. Sci. Technol.*, 68: 1637–1643.
54. Zhang, J., Wu, T., Wang, L., Jiang, W., and Chen, L. 2008. Microstructure and properties of Ti_3SiC_2/SiC nanocomposites fabricated by spark plasma sintering. *Compos. Sci. Technol.*, 68: 499–505.
55. Boutaleb, S., Zairi, F., Mesbah, A., Nait-Abdelaziz, M., Gloaguen, J.M., Boukharouba, T., and Lefebvre, J.M. 2009. Micromechanics-based modeling of stiffness and yield stress for silica/polymer nanocomposites. *Int. J. Solids. Struct.*, 46: 1716–1726.
56. Ashrafi, B. and Hubert, P. 2006. Modeling the elastic properties of carbon nanotube array/polymer composites. *Compos. Sci. Technol.*, 66: 387–396.
57. Steglich, D., Siegmund, T., and Brocks, W. 1999. Micromechanical modeling of damage due to particle cracking in reinforced metals. *Comput. Mater. Sci.*, 16: 404–413.
58. Hbaieb, K., Wang, Q.X., Chia, Y.H.J., and Cotterell, B. 2007. Modeling stiffness of polymer/clay nanocomposites. *Polymer*, 48: 901–909.
59. Tandon, G.P. and Weng, G.J. 1984. The effect of aspect ratio of inclusions on the elastic properties of unidirectionally aligned composites. *Polym. Compos.*, 5: 327–333.
60. Wang, J. and Pyrz, R. 2004. Prediction of the overall moduli of layered silicate-reinforced nanocomposites-part I: Basic theory and formulas. *Compos. Sci. Technol.*, 64: 925–934.
61. Lee, W.J., Son, J.H., Kang, N.H., Park, I.M., and Park, Y.H. 2009. Finite-element analysis of deformation behaviors in random-whisker-reinforced composite. *Scripta Mater.*, 61: 580–583.
62. Langer, S.A., Fuller, E.R., and Carter, W.C. 2001. OOF: An image-based finite element analysis of material microstructures. *Comput. Sci. Eng.*, 3: 15–23.
63. Langer, S.A., Reid, A.C.E., Haan, S.I., and Garcia, R.E. The OOF2 manual: Revision 3.2 for OOF2 Version 2.0 beta 8, NIST, Gaithersburg, MD. Online: http://www.ctcms.nist.gov/~langer/oof2man/index.html.
64. Ganesh, V.V. and Chawla, N. 2005. Effect of particle orientation anisotropy on the tensile behavior of metal matrix composites: Experiments and microstructure-based simulation. *Mater. Sci. Eng. A*, 391: 342–353.
65. Cannillo, V., Manfredini, T., Montorsi, M., Boccaccini, A.R. 2004. Use of numerical approaches to predict mechanical properties of brittle bodies containing controlled porosity. *J. Mater. Sci.*, 39: 4335–4337.
66. Chawla, N., Sidhu, R.S., and Ganesh, V.V. 2006. Three-dimensional visualization and microstructure based modeling of deformation in particle-reinforced composites. *Acta Mater.*, 54: 1541–1548.

67. Cannillo, V., Bondioli, F., Lusvarghi, L., Montorsi, M., Avella, M., Errico, M.E., and Malinconico, M. 2006. Modeling of ceramic particles filled polymer-matrix nanocomposites. *Compos. Sci. Technol.*, 66: 1031–1037.

68. Li, C. and Chou, T.W. 2006. Multiscale modeling of compressive behavior of carbon nanotube/polymer composites. *Compos. Sci. Technol.*, 66: 2409–2414.

69. Hyer, M.W. 1998. *Stress Analysis of Fiber-Reinforced Composite Materials*. Boston, MA, McGraw-Hill.

70. Nemat-Nasser, S. and Hori, M. 1999. Micromechanics: Overall properties of heterogeneous materials. Amsterdam, the Netherland, Elsevier.

71. Zhang, P., Huang, Y., Geubelle, P.H., Klein, P.A., and Hwang, K.C. 2002. The elastic modulus of single wall carbon nanotubes: A continuum analysis incorporating interatomic potential. *Int. J. Solids Struct.*, 39: 3893–3906.

72. Shi, D., Feng, X., Jiang, H., Huang, Y.Y., and Hwang, K. 2005. Multiscale analysis of fracture of carbon nanotubes embedded in composites. *Int. J. Fract.*, 134: 369–386.

73. Tserpes, K.I., Papanikos, P., Labeas, G., and Pantelakis, Sp.G. 2008. Multi-scale modeling of tensile behavior of carbon nanotube-reinforced composites. *Theoret. Appl. Fract. Mech.*, 49: 51–60.

74. Tserpes, K.I. and Papanikos, P. 2006. A progressive fracture model for carbon nanotubes. *Compos. Part B*, 37: 662–669.

75. Belytschko, T., Xiao, S., Schatz, G., and Ruoff, R. 2002. Atomistic simulations of nanotube fracture. *Phys. Rev. B*, 65: 235430–235442.

76. Xiao, S.P. and Belytschko, T. 2004. A bridge domain method for coupling continua with molecular dynamics. *Comput. Methods. Appl. Mech. Eng.*, 193: 1645–1669.

77. Avella, M., Bondioli, F., Cannillo, V., Errico, M.E., Ferrari, A.M., Focher, B., Malinconico, M., Manfredini, T., and Montorsi, M. 2004. Preparation, characterization and computational study of poly (ε-caprolactone) based nanocomposites. *Mater. Sci. Technol.*, 20: 1340–1344.

78. Bagchi, A. and Nomura, S. 2006. On the effective thermal conductivity of carbon nanotube reinforced polymer composites. *Comp. Sci. Technol.*, 66(11–12): 1703–1712.

79. Liu, Y.J. and Chen, X.L. 2003. Continuum models of carbon nanotube-based composites using the boundary element method. *Elect. J. Bound. Elem.*, 1: 316–335.

80. Garnett, J.C.M. 1904. Colours in metal glasses and in metallic films. *Philos. Trans. R. Soc. Lond. A*, 203: 385–420.

81. Bruggeman, D.A.G. 1935. Berechnung verschiedener physikalischer Konstanten von heterogenen Substanzen. I. Dielektrizitätskonstanten und Leitfähigkeiten der Mischkörper aus isotropen Substanzen. *Ann. Phys. Leipzig,* 416(7): 636–664.

82. del Río, J.A., Zimmerman, R.W., and Dawe, R.A. 1998. Formula for the conductivity of a two-component material based on the reciprocity theorem. *Solid. State. Commun.*, 106(4): 183–186.

83. del Río, J.A. and Whitaker, S. 2000. Maxwell's equations in two-phase systems I: Local electrodynamic equilibrium. *Transp. Porous. Media*, 39: 159–186.

84. del Río, J.A. and Whitaker, S. 2000. Maxwell's equations in two-phase systems II: Two-equation model. *Transp. Porous. Media*, 39: 259–287.

85. Braun, M.M. and Pilon, L. 2006. Effective optical properties of non-absorbing nanoporous thin films. *Thin Solid Films*, 496: 505–514.

86. Braun, M.M. 2004. Effective optical properties of nanoporous thin-films. MS thesis, Mechanical and Aerospace Engineering Department, Los Angeles, CA, University of California.

87. Sanchez, M.I., Hedrick, J.L., and Russell, T. P. 1995. Nanofoam porosity by infrared spectroscopy. *J. Polym. Sci., Part B: Polym. Phys.*, 33: 253–257.

88. Sanchez, M.I., Hedrick, J.L., and Russell, T.P. 1996. In: *Microporous and Macroporous Materials Research Society Symposium Proceedings Materials Research Society*. Pittsburgh, PA, Vol. 431, pp. 475–480.

89. Loni, A., Canham, L.T., Berger, M.G., Arens-Fischer, R., Munder, H., Luth, H., Arrand, H.F., and Benson, T.M. 1996. Porous silicon multilayer optical waveguides. *Thin Solid Films*, 276: 143–146.

90. Postava, K. and Yamaguchi, T. 2001. Optical functions of low-k materials for interlayer dielectrics. *J. Appl. Phys.*, 89: 2189–2193.

91. Gilbert, T.L. 1955. A Lagrangian formulation of gyromagnetic equation of the magnetization field. *Phys. Rev.* 100: 1243–1250.
92. Brunotte, X., Meunier, G., and Imho, J.F. 1992. Finite element modeling of unbounded problems using transformations: A rigorous, powerful and easy solution. *IEEE Trans. Magn.*, 28: 1663–1666.
93. Khebir, A., Kouki, A., and Mittra, R. 1990. Asymptotic boundary conditions for finite element analysis of three dimensional transmission line discontinuities. *IEEE Trans. Magn.*, 38: 1427–1432.
94. Schreff, T., Fidler, J., Kirk, K.J., and Chapman, J.N. 1997. A higher order FEM-BEM method for the calculation of domain processes in magnetic nano-elements. *IEEE Trans. Magn.*, 33: 4182–4184.
95. Banerjee, P.K. 1994. *The Boundary Element Methods in Engineering*. London, U.K., McGraw-Hill.
96. Brezzi, F. and Fortin, M. 1991. *Mixed and Hybrid Finite Element Methods*. New York, Springer.
97. Victoria, R.H. 1987. Quantitative theory for hysteretic phenomena in CoNi magnetic thin films. *Phys. Rev. Lett.*, 58: 1788–1791.
98. Hubert, A. and Ruhrig, M. 1991. Micromagnetic analysis of thin film elements (invited). *J. Appl. Phys.*, 69: 6072–6077.
99. Bourchak, M., Kada, B., Alharbi, M., and Aljuhany, K. 1999. Nanocomposites damage characterization using finite element analysis. *Int. J. Nanoparticles*, 2(1–6): 467–475.
100. Schadler, L.S., Giannaris, S.C., and Ajayan, P.M. 1998. Load transfer in carbon nanotube epoxy composites. *Appl. Phy. Lett.*, 73: 38–42.
101. Spanos, P.D. and Kontsos, A. 2008. A multiscale Monte Carlo finite element method for determining mechanical properties of polymer nanocomposites. *Probab. Eng. Mech.*, 23(4): 456–470.
102. Xu, L.R., Li, L., Lukehart, C.M., and Kuai, H. 2007. Mechanical characterization of nanofibre-reinforced composite adhesives. *J. Nanosci. Nanotechnol.*, 7: 1–3.
103. Tezduyar, T.E. 1992. Stabilized finite element formulations for incompressible-flow computations. *Adv. Appl. Mech.*, 28: 1–44.
104. Tezduyar, T.E. 2001. Finite element methods for flow problems with moving boundaries and interfaces. *Arch. Comput. Methods. Engg.*, 8(2): 83–130.
105. Hughes, T.J.R., Franca, L.P., and Balestra, M. 1986. A new finite element formulation for computational fluid dynamics: V. Circumventing the Babuska–Brezzi condition: A stable Petrov–Galerkin formulation of the Stokes problem accommodating equal-order interpolations. *Comput. Methods Appl. Mech. Engg.*, 59: 85–99.
106. Zhang, L., Gerstenberger, A., Wang, X., and Liu, W.K. 2004. Immersed finite element method. *Comput. Methods Appl. Mech. Engg.*, 193(21–22): 2051–2067.
107. Saad, Y. and Schultz, M.H. 1986. GMRES: A generalized minimal residual algorithm for solving nonsymmetric linear systems. *SIAM. J. Sci. Stat. Comput.*, 7(3): 856–869.
108. Zhang, L., Wagner, G.J., and Liu, W.K. 2002. A parallelized mesh free method with boundary enrichment for large-scale CFD. *J. Comput. Phys.*, 176: 483–506.
109. Liu, W.K., Liu, Y., Farrell, D., Lucy, Z., Wang, X.S., Fukui, Y., Patankar, N., Zhang, Y., Bajaj, C., Lee, J., Hong, J., Chen, X., and Hsu, H. 2006. Immersed finite element method and its applications to biological systems. *Comput. Methods Appl. Mech. Engg.*, 195: 1722–1749.

8

Finite Element Modeling of Carbon Nanotubes and Their Composites

Mahmoud Nadim Nahas

King Abdulaziz University, Jeddah, Saudi Arabia

CONTENTS

8.1 Introduction...291
8.2 Literature Survey 292
8.3 Fundamental...296
8.4 Applications..296
8.5 Analysis of Graphene Sheet 296
8.6 Analysis of Carbon Nanotubes..301
8.7 Future Work..305
8.8 Conclusions..306
Acknowledgments 306
References...306

8.1 Introduction

Right from their discovery in 1991 by Iijima (1991), carbon nanotubes (CNTs) have spurred considerable interest among scientists and engineers because of their atypical physical properties.

CNTs are finding increasing application in composites because of their high modulus and electrical conduction. CNTs are of molecular dimensions and consist of one large molecule, so molecular mechanics should be an attractive modeling technique. Mechanical properties can be better evaluated using finite element analysis (FEA). Single-walled CNT (SWCNT) and multi-walled CNT (MWCNT) varieties are available.

The use of CNTs as reinforcing materials in nano-composites has made it necessary to predict their mechanical properties and assess their deformation under loading. This characterization is more complex than that of conventional materials due to the fact that their mechanical properties depend on nanostructure. Direct experimental measurements are difficult due to the nanosize of CNTs. Hence, theoretical approaches seem to be an excellent alternative.

The theoretical approaches are of two main types: the atomistic methods and the continuum mechanics. The former includes classical molecular dynamics as discussed by Iijima et al. (1996) and Yakobson et al. (1997), tight-binding molecular dynamics of Hernandez et al. (1998), and density functional theory of Sanchez-Portal et al. (1999).

The continuum mechanics methods mainly involve classical continuum mechanics, see for instance Li and Chou (2003) and Tersoff (1992), or continuum shell modeling, see Yakobson et al. (1996) and Ru (2000).

In this chapter, the finite model analysis is used to model graphene sheet using truss elements. The developed model is then used to model CNTs. The finite element (FE) model is used to find the equivalent stiffness of the nano-structured materials.

8.2 Literature Survey

Mazahery and Shabani (2012) conducted a FE investigation on the effects of the increase in the contents of nano-silicon carbides in an aluminum alloy on its hardness, porosity, elongation, yield strength, and ultimate tensile strength. Their results show that as the content of nanoparticles increase, the composite is strengthened but its ductility is retained.

Yan et al. (2011) constructed an axisymmetric FE model to predict the elastic modulus of a nanoparticle embedded in a composite matrix for two cases, a stiff particle in a soft matrix and for a soft particle in a stiff matrix. They showed that the Olive-Phar indentation method can be used with excellent accuracy if the indentation depth lies within the particle dominated depth.

Shabani and Mazahery (2012) used FEM along with artificial neural network (ANN) to determine the mechanical properties of the composite and genetic algorithms (GA) to determine the optimal stir casting conditions of Al–Si aluminum alloy reinforced with alumina nanoparticles. The results showed that the proposed model can help in finding the optimal processing condition for stir casting and that the presence of nanoparticles significantly can help in improving the mechanical properties of the composite.

Deb et al. (2011) performed an explicit FEA comparison between a plain polypropylene (PP), which is thermoplastic polymer used in vehicles and nanoclay–polypropylene composite on a vehicle's A-pillar trim. Results show that the nanocomposite improved the head impact safety of the A-pillar trim when compared to plain PP.

Sun et al. (2011) constructed an analytical model to calculate the effective mechanical properties of nanoparticle reinforced composite. Then FEA was used to study the fracture of a compact tension sample. Their results are in good agreement with published data.

Qin and Ma (2011) conducted a FEA to study the dynamic performance of zinc–aluminum alloys reinforced with TiC nanoparticles in bearing bushings. Although plain aluminum alloys have limitations to operating temperature and speed, they have high strength and friction characteristics.

Mortazavi et al. (2011) used molecular dynamics (MD) to obtain the mechanical properties of glass silica nanoparticle reinforced epoxy polymer. Then they incorporated the obtained data in a FE model to construct a representative volume element (RVE).

Wang and Zhao (2010) modeled a cylindrical RVE with a central carbon nanofiber to investigate the temperature distribution. The results show that the nanofibers maintained constant temperature during heat transfer process due to the large difference in thermal properties between the fiber and the matrix.

Motamedi et al. (2012) compared a rubber matrix with a CNT reinforced rubber by constructing an RVE. The addition of CNTs helped in increasing the mechanical properties of the composite. They also investigated the effect of CNTs waviness on their mechanical properties compared to straight CNTs.

Lapeyronnie et al. (2011) developed a FE model to estimate the mechanical properties of a 3D layer to layer angle interlock fabrics. The obtained mechanical properties were implemented in a heterogeneous model and validated with experimental results.

Ghosh et al. (2011) investigated the rolling resistance of nano-composite (organoclay carbon back or organoclay silica fillers) passenger car tire. A FE model was used to obtain the elastic strain energy in a steady state rolling simulation, which showed that the nano-composites improved the rolling resistance of the wheel.

Zhu et al. (2012) investigated the bending and free vibration of a SWCNT reinforced composite plate using FEM. CNTs were uniaxially aligned and were distributed either uniformly or by three kinds of functionally graded distribution. Different boundary conditions were applied and different concentrations of CNTs by volume were used to obtain the response of the plates and natural frequencies and mode shapes.

Gao et al. (2011) and Wu et al. (2011) constructed a FE model of a (PP)/nano-TiO_2 composite to investigate the stress and strain response of the composite. It is found that at low volume concentrations the particles have less influence on each other local stress field. It is also found that the stress level rises at the interaction region between the particle and the matrix.

Jingxin et al. (2008) constructed an RVE composing of two nano-composite ceramics to study its mechanical behavior and damage in it. They showed that the stress distribution is uneven and that damage occurs in the elements near the nano-ceramic particles. Also, the damage route was evaluated.

Ding et al. (2007) constructed a FE model to investigate the stress field in a ZrO2/Cu composite. Experimental results show two types of crystals, tetragonal and monoclinic zirconia particles. High stress concentration at the tip of a monoclinic zirconia was evident, which can lead to initiation of cracks compared to tetragonal zirconia.

Jung et al. (2006) used glass fiber reinforced nanocomposite as a radar absorbing structure (RAS). The composite consists of three phases, glass fiber, epoxy, and nano-carbon materials. The composite showed excellent RAS efficiency at X-band frequency range (8–12 GHz) but thermal spring back occurred due to the drop from curing temperature to room temperature. To overcome this spring back, two composite shells were modeled, carbon/epoxy and glass/epoxy. The spring back was predicted using FEM.

Luo et al. (2002) studied the wear resistance and interfacial failure in a TiN-reinforced TiNi/TiC alloy. FEM was used to investigate the behavior of the composite when TiN powder was added to the TiNi matrix.

Takano et al. (2008) modeled the interface between natural rubber and carbon black (average diameter 30–122 nm). It is shown that the ratio of the layer between carbon black and the rubber (sticky hard layer) to carbon black diameter influences the properties of the composite.

Giannopoulos et al. (2010) used FEM to estimate the effective Young's modulus of SWCNT-reinforced composites at different volume fractions. A cylindrical RVE is used to model the matrix as a continuum media. The reinforcing CNTs were modeled according to their atomistic microstructure using spring elements. Joint elements of changeable stiffness are used to simulate the interfacial region.

Wang et al. (2005) studied the effect of the change in CNTs diameter on the bending elastic modulus. A nonlinear FEA was conducted using ABAQUS FE solver to show that the formation of ripples on cantilevered CNTs inner arc can lead to a substantial decrease in the effective bending modulus of the tube.

Tserpes et al. (2006) conducted a FEM to predict the failure stress and strains in CNTs with missing carbon atoms and 10% weakening in C–C bonds. The tubes were subjected to axial tension. The results obtained correlate with the results from the literature.

Liu and Chen et al. (2003) developed a nanoscale RVE to evaluate the effective mechanical properties of CNT-based composite based on continuum mechanics and FEM. The results show that the addition of 2% and 5% volume fraction increases the stiffness of the composite 0.7 and 9.5 times.

Georgantzinos et al. (2009) used FEM to investigate the stress–strain behavior of SWCNT-reinforced rubber composites at different volume fractions. A cylindrical RVE is used to model the matrix as continuum solid elements. The reinforcing CNTs were modeled according to their atomistic microstructure using nonlinear spring elements. Joint elements of changeable stiffness are used to simulate the interfacial region.

Tserpes et al. (2008) modeled a RVE to study the tensile behavior of CNT reinforced polymer composites and the effect of interfacial shear strength. FEM was used to model the RVE. A perfect bond was assumed between the matrix and the CNT until the interfacial shear stress exceeded the strength. The debonded region in the CNTs is then prohibited from any load transfer to simulate the failure in this region. The stiffness remains unaffected but the tensile strength decreases significantly.

Shokrieh and Rafiee (2010b) studied the tensile behavior of short CNTs embedded in polymer matrix with the effects of van der Waals bonds. A FEA was carried to investigate the effect of CNTs' length on the mechanical properties. For lengths less than 100 nm, the effect of CNTs becomes negligible. The improvement comes to saturation at a length of 10 μm and more. Shokrieh and Rafiee (2010c) used a nonlinear FEA on a 3D model consisting of CNT, nonbonded interphase region, and the surrounding polymer matrix. Van der Waals interaction was used for the bond between the CNT and the surrounding polymer. The results show a nonlinear relation between the stress and strain. The length of the CNTs affect the efficiency of the reinforcement. Shokrieh and Rafiee (2010a) used an equivalent long fiber model to conduct a FEA to investigate the mechanical properties of CNT reinforced polymer matrix. The FE model consists of the CNT, interphase (treated as Van der Waals interaction), and polymer matrix. The results show that the rule of mixture overestimates the properties of the investigated model.

Kulkarni et al. (2010) conducted a FE and experimental investigation on the reinforcement of the laminated composites by growing CNTs on the surface of the fiber filaments, which increase the effective diameter of the fibers and the interface area for the polymeric matrix on the fiber. The elastic, shear modulus, and Poisson's ratio were obtained.

Kuronuma et al. (2011) investigated the effect of CNTs addition on the crack growth of CNT/polycarbonate composites at room temperature and liquid nitrogen at 77 K. The growth of the crack during fatigue loading was modeled using elastic-plastic FEA to investigate the effects of CNTs addition.

Li and Chou (2006) investigated the compressive behavior of CNT reinforced polymer composite. The tubes are modeled at an atomistic scale and the matrix is bonded as a continuum medium. The tubes and matrix are bonded by Van der Waals interaction. The buckling force at different CNT lengths was calculated and the results show that the reinforcement of CNT can increase the buckling resistance of the composite.

Yas and Heshmati (2012) investigated the dynamic behavior of glass fiber nanobeams reinforced with SWCNTs under different loading conditions. The FEM was used to discretize the model and obtain numerical results of the motion equation. Parameters discussed in the work are the load velocity, shear deformation, slenderness ratio, and boundary conditions.

Shindo et al. (2012) investigated the change in the mechanical and electrical response of cracked CNT reinforced polymer composites. A single edge cracked specimen was subjected to tensile loading at room temperature and liquid nitrogen temperature of 77 K. An analytical model was developed to predict the change in the resistance resulting from the crack propagation. The results from the analytical model and experimental model were compared. Also the fracture properties were assessed using FEM in terms of J-integral.

Haque and Ramasetty (2005) investigated the load transfer in SWCNT reinforced polymer matrix. An analytical model and a 2D FE model were constructed to predict the axial stress and interfacial shear in the CNT. The effect of the CNT length, aspect ratio, matrix modulus, CNT volumetric ratio on the axial stress, and interfacial shear stress was investigated.

Chen et al. (2010) used atomistic simulation, shear-lag theory, and fracture mechanics to investigate the fracture toughness of CNT-reinforced composites. The results show that neither longer CNT nor stronger CTN and matrix interface can lead to a better toughness. FEM was used to study the fracture toughness of CNT-reinforced composites at different interfacial bond densities. The results show that for a (6, 6) CNTs, a 5%–10% interfacial chemical bond density is the optimum and that increasing the length of CNTs over 100 nm would not improve fracture toughness, but will lead to clustering and self-folding of the tubes.

Ashrafi and Hubert (2006) investigated the elastic properties of SWCNT-reinforced composites. FEM was used to predict the properties of twisted SWCNT. These properties are then used in a twisted CNT reinforced polymer composite. The effect of CNT volume fraction and aspect ratio for the composite are then calculated for aligned and randomly reinforced composites to get the elastic properties.

Rokni et al. (2012) studied a 2D optimum distribution of MWCNTs in the longitudinal and thickness direction of a polymer composite micro to achieve maximum natural frequency. Different weight ratios of MWCNT-reinforced thermosetting polyester epoxy/amine were used. A user code written in Python was used to generate a 3D model in ABAQUS of the beam and to evaluate the optimum distribution under various boundary conditions. The distribution of the CNTs is made by dividing the beam in the longitudinal and thickness direction. The same authors (Rokni et al., 2011) studied the optimum distribution of MWCNTs in a microbeam to obtain maximum natural frequency. The beam was divided into 10 segments of thermosetting polyester epoxy/amine resin reinforced and different weight ratio of CNTs were used. A computer program was written in Python to model the beam in ABAQUS and to calculate the optimum distribution of CNTs at different boundary conditions. The results show that the optimum distribution depends on the vibration mode shape.

Ayatollahi et al. (2011) investigated the effect of MWCNT-reinforcement on the fracture toughness of the CNT/epoxy composite. FEM was conducted to simulate the response of the single edge notch bend specimen under shear loading and calculate the fracture toughness. The effect of MWCNT reinforcement shows an increase in fracture toughness (more in shear loading compared to normal loading).

Selmi et al. (2007) investigated the elastic properties of SWCNT-reinforced composites using several micromechanical models and a comparative study between them. Validation of the results is done by experimental work or by using the FEM based on a 2D periodic cell or (RVE). The results show that there are agreements between the models in most cases but some fail for some composite materials and for some properties.

Formica et al. (2010) conducted FEA to investigate the effect of CNTs alignment and volume fraction on the elastic properties of the composite. Three types of isotropic matrices were used (epoxy, rubber, and concrete). The results show an increase of up to 500% in the lowest natural frequency of the composite.

Fisher et al. (2003) investigated the effect of wavy CNT reinforcement on the effective reinforcing modulus of the CNT/polymer composite. The results show that even the slightest curvature in the tube can lead to a significant decrease in the effective modulus when compared to straight reinforcing tubes. These results show that the curvature can be limiting to the enhancement of the composite modulus.

Nouri et al. (2012) investigated the mechanical properties of CNT reinforced aluminum nanocomposite (CNRAN). FEM was used to predict the hardness and elastic modulus of an indentation test. The results of the model show good agreement with experimental results.

8.3 Fundamental

A CNT is a cylinder of graphitic carbon atoms bonded on the surface in a hexagonal array consistent with a space-frame structure. The hexagonal covalent bonding has been used as truss elements for a mesh for FEA using different software. The contributions of wall thickness, tube diameter, and bonding chirality on the elastic modulus of SWCNT and MWCNT are investigated here. The space-frame model of CNT merges the real-world scale functions of FEA with those of molecular mechanics at the atomistic scale, equating the modulus of the frame with the force field of the covalent bonds.

The forces between individual atoms in nanostructure are characterized by a force field as discussed by Abd-Rabou and Nahas (2009), where the total molecular potential energy is the sum of the energies associated with bond stretching, angle variation, torsion, and inversion the energies associated with bond stretching, angle variation, torsion, and inversion, and the nonbonded interaction energies of van der Waals and electrostatic terms. Various functional forms may be used for these energy terms depending on the particular material and loading conditions. Since experimental data are unavailable, FEA is used as a source of information for defining the force field.

8.4 Applications

Two applications are presented. The first is the analysis of a graphene sheet and the second one is the analysis of SWCNT and MWCNT.

8.5 Analysis of Graphene Sheet

A truss model is used here to represent the forces between two atoms as shown in Figure 8.1.

Continuum representation of the molecules is used in the FE modeling to simulate the molecular behavior of the carbon atoms in the graphene sheet. The chemical bonds between the atoms exert a force that helps in keeping the atoms in place, which are viewed as hanging mass being held by these bonds. This force can be represented as an elastic spring. A linear truss T3D2 element in ABAQUS (the FE package employed in this work) is used to model these bonding and nonbonding chemical bonds to simulate the displacement in the atoms. This model allows the mechanical behavior of the nano-structured system to be modeled accurately in terms of displacements of the atoms.

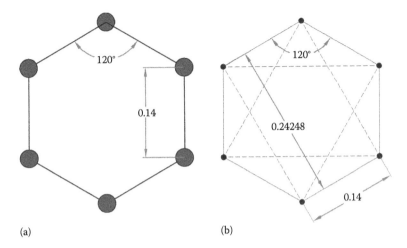

(a) (b)

FIGURE 8.1
(a) Molecular representation of the graphene sheet molecules and (b) FE representation (dimensions are in nanometers).

The modulus of elasticity used in the FEA model is 6.52×10^{-5} (N/nm²) for the bonded elements and 3.352×10^{-6} (N/nm²) for the nonbonded elements, with Poisson's ratio 0.3 for both. The lengths of the element are shown in Figure 8.1, while the cross-sectional area is 0.0014 (nm²) for both elements.

Two load cases are analyzed. The first one is when the graphene sheet is pulled in the vertical direction (y-direction), while the second load case is when the sheet is pulled in the horizontal direction (x-direction), see Figure 8.2. Due to symmetry, the figure shows only one-fourth of the graphene sheet.

(a) (b)

FIGURE 8.2
Boundary conditions with (a) first load case and (b) second load case.

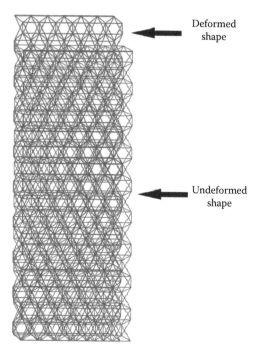

FIGURE 8.3
Deformed and undeformed shapes for the first load case.

A displacement of 0.001 nm was given in the y-direction (in the first load case) and in the x-direction (in the second load case). Since the value of the displacement is known, the stiffness can be calculated by finding the reaction forces in the model. To prevent rigid body motion, the nodes on one side are prevented from moving in the x-direction, while the nodes in the bottom are prevented from moving in the y-direction, as shown in Figure 8.2.

The deformed and the undeformed shapes of both cases are shown in Figures 8.3 and 8.4.

To calculate the stiffness of the sheet in both directions, the reaction forces are first found by ABAQUS. The stiffness of the graphene sheet, K, can be calculated from the equation:

$$\sum_{i}^{n} F = K\delta \qquad (8.1)$$

where
  F represents the reaction force of the element i
  n is the total number of elements
  δ is the displacement of the sheet

The stiffness of the graphene sheet in the vertical and horizontal directions is calculated using the reaction forces as given in Figures 8.5 and 8.6, respectively.

To complete the picture, Figures 8.7 and 8.8 show the generated Von Mises stresses in both load cases. For the first load case the maximum Von Mises stress was 1.85 GPa, while for the second load case, it was 3.8 GPa. The difference in the two values is due to the change in the orientation of the truss elements in each load case.

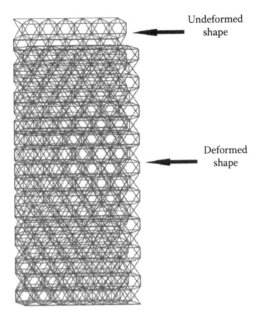

FIGURE 8.4
Deformed and undeformed shapes for the second load case.

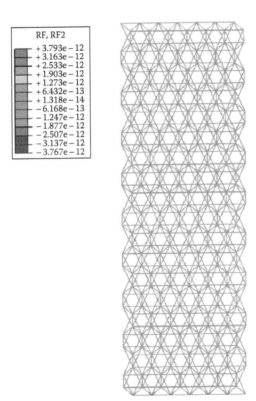

FIGURE 8.5
Reaction forces in first load case.

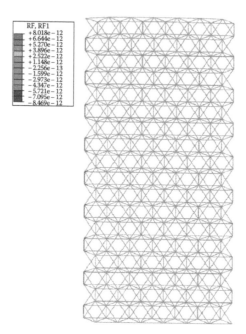

FIGURE 8.6
Reaction forces in the horizontal direction.

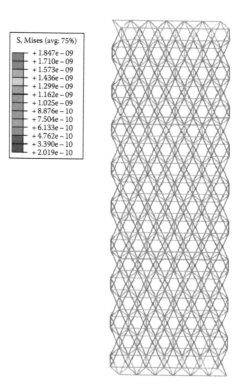

FIGURE 8.7
Von Mises stresses in the vertical direction.

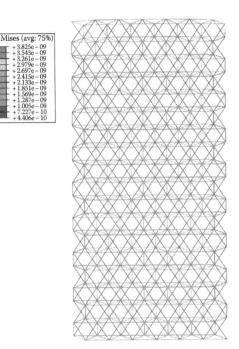

FIGURE 8.8
Von Mises stresses in the horizontal direction.

8.6 Analysis of Carbon Nanotubes

The CNT is modeled by bending the previously developed graphene sheet around its vertical edge as discussed by Nahas and Abd-Rabou (2010a,b). Two-dimensional truss elements of graphene sheet are replaced with 3D truss elements, which have three degrees of freedom per node. Figure 8.9 shows two types of CNTs that were modeled, a (12, 0) SWCNT and a (12, 0), (11, 0), (13, 0) MWCNT. To complete the cylindrical shape using the graphene sheet, more elements are developed to connect the free nodes in the vertical direction. The same mechanical and geometrical properties of molecular mechanics model of the graphene sheet are used here.

The two models were studied under two load cases, axial loading, Figure 8.10, to calculate the axial stiffness, and transverse loading, Figure 8.11, to calculate the bending stiffness. The axial loading is simulated by applying a known axial displacement (0.001 nm) at the top end of the tube while keeping the other end clamped to prevent the rigid body motion. Using a known displacement, the stiffness of the tube can be calculated from knowing the reaction forces generated in the model. For the transverse load case, the load is simulated by applying a known transverse displacement (0.001 nm) to the top end and fixing the lower end of the tube.

The reaction forces of the FE analysis for the axial loading of the tubes are shown in Figure 8.12. The stiffness of the tube K can be calculated from Equation 8.1. The reactions for the transverse loading are shown in Figure 8.13. The same procedure is used to calculate the bending stiffness of the tube.

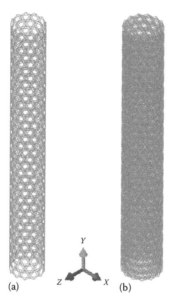

FIGURE 8.9
(a) SWCNT finite elements mesh and (b) MWCNT finite elements mesh.

FIGURE 8.10
Axial loading and boundary conditions: (a) SWCNT and (b) MWCNT.

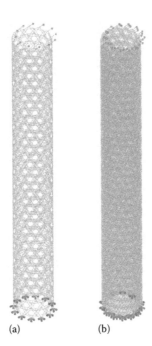

(a)   (b)

FIGURE 8.11
Transverse loading and boundary conditions: (a) SWCNT and (b) MWCNT.

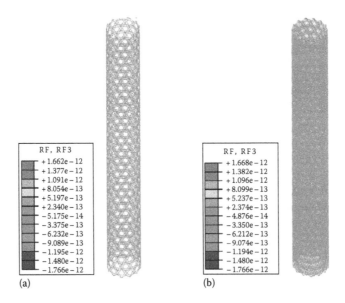

RF, RF3
+ 1.662e − 12
+ 1.377e − 12
+ 1.091e − 12
+ 8.054e − 13
+ 5.197e − 13
+ 2.340e − 13
− 5.175e − 14
− 3.375e − 13
− 6.232e − 13
− 9.089e − 13
− 1.195e − 12
− 1.480e − 12
− 1.766e − 12

(a)

RF, RF3
+ 1.668e − 12
+ 1.382e − 12
+ 1.096e − 12
+ 8.099e − 13
+ 5.237e − 13
+ 2.374e − 13
− 4.876e − 14
− 3.350e − 13
− 6.212e − 13
− 9.074e − 13
− 1.194e − 12
− 1.480e − 12
− 1.766e − 12

(b)

FIGURE 8.12
Reaction forces generated in axial loading: (a) SWCNT and (b) MWCNT.

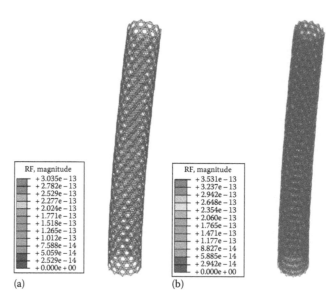

FIGURE 8.13
Reaction forces generated in transverse loading: (a) SWCNT and (b) MWCNT.

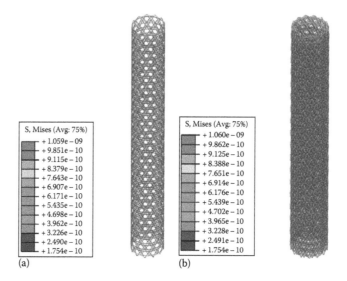

FIGURE 8.14
Von Mises stresses generated in axial loading: (a) SWCNT and (b) MWCNT.

The resulting Von Mises stresses are shown in Figures 8.14 and 8.15. In axial loading, the generated Von Mises stresses were 1.059×10^{-9} for SWCNT and 1.06×10^{-9} for MWCNT. For the transverse loading case, the generated Von Mises Stresses were 15.6×10^{-9} (N/nm^2) in SWCNT and 18.2×10^{-9} (N/nm^2) in MWCNT.

The modulus of elasticity of the tube is related to the axial stiffness by the relation shown in Equation 8.2:

$$K = \frac{EA}{L} \qquad (8.2)$$

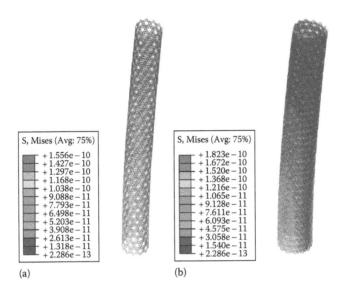

FIGURE 8.15
Von Mises stresses generated in transverse loading: (a) SWCNT and (b) MWCNT.

where
 A is the cross-sectional area of the tube
 L is the tube length

The cross-sectional area of the tube can be found by calculating the perimeter of the tube, as in Equation 8.3:

$$A = \pi D t \tag{8.3}$$

where
 D is the tube diameter
 t is the wall thickness of the tube

Young's modulus can then be calculated by rewriting Equation 8.2 as follows:

$$E = \frac{KL}{\pi D t} \tag{8.4}$$

The calculated Young's modulus for SWCNT was 1.192 TPa and for MWCNT it was 1.197 TPa.

8.7 Future Work

The author is now working on some other related problems. Some have been already submitted for publications. One paper studies the effect of inclusion of monolayer graphene sheets on the mechanical properties of epoxy resin. Another paper studies the effect of addition of CNTs on the mechanical properties of epoxy resin. In both cases, numerical analysis using FEM is conducted. Other papers are using FEA in vibration problems and fatigue problems resulting from cyclic loadings.

8.8 Conclusions

FEA has been used to model structure–property relationships of nano-structured materials. The model links computational chemistry (used to predict molecular properties) and solid mechanics (used to describe macroscopic mechanical behavior). Discrete molecular structures are modeled as equivalent truss element by equating the molecular potential energy of nano-structured materials with the mechanical strain energy of truss element model. This modeling method has been applied to a graphene sheet. The FE model performs very well and gives good results. As the FE model comprises a small number of elements, it performs under minimal computational time.

The model has also been used to investigate the properties of SWCNTs and MWCNTs.

The obtained values of Young's modulus agree very well with the corresponding theoretical results and experimental measurements that are available in the literature.

The results demonstrate that the proposed FE model may provide a valuable tool for studying the mechanical behavior of CNTs and nanocomposites based on them.

Acknowledgments

The author is thankful to his colleagues Professor Mahmoud Abd-Rabou and Engineer Mahmoud Al-Zahrani for their valuable help. The author is also grateful to the Deanship of Scientific Research of King Abdulaziz University for funding some of his projects which led to writing this chapter.

References

Abd-Rabou, M. and M. N. Nahas. 2009. Finite element modeling of nano-structured materials. *International Journal of Nano and Biomaterials*, 2 (1/2/3/4/5):263–272.

Ashrafi, B. and P. Hubert. 2006. Modeling the elastic properties of carbon nanotube array/ polymer composites. *Composites Science and Technology*, 66 (3–4):387–396. doi: 10.1016/j. compscitech.2005.07.020.

Ayatollahi, M. R., S. Shadlou, and M. M. Shokrieh. 2011. Fracture toughness of epoxy/multi-walled carbon nanotube nano-composites under bending and shear loading conditions. *Materials and Design*, 32 (4):2115–2124. doi: 10.1016/j.matdes.2010.11.034.

Chen, Y. L., B. Liu, X. Q. He, Y. Huang, and K. C. Hwang. 2010. Failure analysis and the optimal toughness design of carbon nanotube-reinforced composites. *Composites Science and Technology*, 70 (9):1360–1367. doi: 10.1016/j.compscitech.2010.04.015.

Deb, A., G. S. Venkatesh, A. Karmarkar, and B. Gurumoorthy. 2011. Application of a nanoclay-polypropylene composite to efficient vehicle occupant safety countermeasure design. Paper read at *Nanotech 2011*, Boston, MA.

Ding, J., N. Zhao, C. Shi, and J. Li. 2007. Effect of nano ZrO_2 crystal structure on copper matrix composite interface. *Beijing Keji Daxue Xuebao/Journal of University of Science and Technology Beijing*, 29 (2):119–124.

Fisher, F. T., R. D. Bradshaw, and L. C. Brinson. 2003. Fiber waviness in nanotube-reinforced polymer composites—I: Modulus predictions using effective nanotube properties. *Composites Science and Technology*, 63 (11):1689–1703. doi: 10.1016/s0266-3538(03)00069-1.

Formica, G., W. Lacarbonara, and R. Alessi. 2010. Vibrations of carbon nanotube-reinforced composites. *Journal of Sound and Vibration*, 329 (10):1875–1889. doi: 10.1016/j.jsv.2009.11.020.

Gao, M., Y. Wu, Y. Chen, J. Li, and S. Ju. 2011. Numerical simulation of tensile behaviors in the elastic region for $PP/nano-TiO_2$ composites. *Gaofenzi Cailiao Kexue Yu Gongcheng/Polymeric Materials Science and Engineering*, 27 (3):101–104.

Georgantzinos, S. K., G. I. Giannopoulos, and N. K. Anifantis. 2009. Investigation of stress–strain behavior of single walled carbon nanotube/rubber composites by a multi-scale finite element method. *Theoretical and Applied Fracture Mechanics*, 52 (3):158–164. doi: 10.1016/j.tafmec.2009.09.005.

Ghosh, S., R. A. Sengupta, and G. Heinrich. 2011. Investigations on rolling resistance of nanocomposite based passenger car radial tyre tread compounds using simulation technique. *Tire Science and Technology*, 39 (3):210–222.

Giannopoulos, G. I., S. K. Georgantzinos, and N. K. Anifantis. 2010. A semi-continuum finite element approach to evaluate the Young's modulus of single-walled carbon nanotube reinforced composites. *Composites Part B: Engineering*, 41 (8):594–601. doi: 10.1016/j.compositesb.2010.09.023.

Haque, A. and A. Ramasetty. 2005. Theoretical study of stress transfer in carbon nanotube reinforced polymer matrix composites. *Composite Structures*, 71 (1):68–77. doi: 10.1016/j.compstruct.2004.09.029.

Hernandez, E., C. Goze, P. Bernier, and A. Rubio. 1998. Elastic properties of C and BxCyNz composite nanotubes. *Physical Review Letters*, 80:4502–4505.

Iijima, S. 1991. Helical microtubules of graphitic carbon. *Nature*, 354 (6348):56–58.

Iijima, S., C. Brabec, A. Maiti, and J. Bernholc. 1996. Structural flexibility of carbon nanotubes. *Journal of Chemical Physics*, 104:2089–2092.

Jingxin, K., N. Xinhua, Z. Jan, and Z. Yongling. 2008. Damage finite element analysis of nano-composite ceramics. *Key Engineering Materials*, 368–372 (February):1657–1658. doi: 10.4028/www.scientific.net/KEM.368-372.1657.

Jung, W. K., B. Kim, M. S. Won, and S. H. Ahn. 2006. Fabrication of radar absorbing structure (RAS) using GFR-nano composite and spring-back compensation of hybrid composite RAS shells. *Composite Structures*, 75 (1–4):571–576.

Kulkarni, M., D. Carnahan, K. Kulkarni, D. Qian, and J. L. Abot. 2010. Elastic response of a carbon nanotube fiber reinforced polymeric composite: A numerical and experimental study. *Composites Part B: Engineering*, 41 (5):414–421. doi: 10.1016/j.compositesb.2009.09.003.

Kuronuma, Y., Y. Shindo, T. Takeda, and F. Narita. 2011. Crack growth characteristics of carbon nanotube-based polymer composites subjected to cyclic loading. *Engineering Fracture Mechanics*, 78 (17):3102–3110. doi: 10.1016/j.engfracmech.2011.09.006.

Lapeyronnie, P., P. Le Grognec, C. Binétruy, and F. Boussu. 2011. Homogenization of the elastic behavior of a layer-to-layer angle-interlock composite. *Composite Structures*, 93 (11):2795–2807.

Li, C. and T.-W. Chou. 2003. A structural mechanics approach for the analysis of carbon nanotubes. *International Journal of Solids Structures*, 40:2487–2499.

Li, C. and T.-W. Chou. 2006. Multiscale modeling of compressive behavior of carbon nanotube/polymer composites. *Composites Science and Technology*, 66 (14):2409–2414. doi: 10.1016/j.compscitech.2006.01.013.

Liu, Y. J. and X. L. Chen. 2003. Evaluations of the effective material properties of carbon nanotube-based composites using a nanoscale representative volume element. *Mechanics of Materials*, 35 (1–2):69–81. doi: 10.1016/s0167-6636(02)00200-4.

Luo, Y. C., R. Liu, and D. Y. Li. 2002. Investigation of the mechanism for the improvement in wear resistance of nano-TiN/TiC/TiNi composite: A study combining experiment and FEM analysis. *Materials Science and Engineering A*, 329–331:768–773.

Mazahery, A. and M. O. Shabani. 2012. Nano-sized silicon carbide reinforced commercial casting aluminum alloy matrix: Experimental and novel modeling evaluation. *Powder Technology*, 217 (February):558–565.

Mortazavi, B., S. Ahzi, J. Bardon, A. Laachachi, and D. Ruch. 2011. The Minerals, Metals and Materials Society (TMS). Investigation of mechanical properties of silica/epoxy nano-composites by molecular dynamics and finite element modeling. *Supplemental Proceedings: Materials Fabrication, Properties, Characterization, and Modeling*, Vol 2, John Wiley & Sons, Hoboken, NJ.

Motamedi, M., M. Eskandari, and M. Yeganeh. 2012. Effect of straight and wavy carbon nanotube on the reinforcement modulus in nonlinear elastic matrix nanocomposites. *Materials and Design*, 34:603–608.

Nahas, M. N. and M. Abd-Rabou. 2010a. Finite Element Modeling of Carbon Nanotubes. *International Journal of Mechanical and Mechatronics Engineering*, 10 (3):19–24.

Nahas, M. N. and M. Abd-Rabou. 2010b. Finite element modeling of multi-walled carbon nanotubes. *International Journal of Engineering and Technology*, 10 (4):71–91.

Nouri, N., S. Ziaei-Rad, S. Adibi, and F. Karimzadeh. 2012. Fabrication and mechanical property prediction of carbon nanotube reinforced aluminum nano-composites. *Materials and Design*, 34 (0):1–14. doi: 10.1016/j.matdes.2011.07.047.

Qin, Z. and K. Ma. 2011. Study on bearing dynamic properties of nano particle reinforced ZA27 alloy composite material. *Advanced Materials Research*, 328–330 (Sept.):1467–1470. doi: 10.4028/www.scientific.net/AMR.328-330.1467.

Rokni, H., A. S. Milani, and R. J. Seethaler. 2011. Maximum natural frequencies of polymer composite micro-beams by optimum distribution of carbon nanotubes. *Materials and Design*, 32 (6):3389–3398. doi: 10.1016/j.matdes.2011.02.008.

Rokni, H., A. S. Milani, and R. J. Seethaler. 2012. 2D optimum distribution of carbon nanotubes to maximize fundamental natural frequency of polymer composite micro-beams. *Composites Part B: Engineering*, 43(2):779–785. doi: 10.1016/j.compositesb.2011.07.012.

Ru, C. Q. 2000. Effective bending stiffness of carbon nanotubes. *Physical Reviews B*, 62:9973–9976.

Sanchez-Portal, D., E. Artacho, J. M. Soler, A. Rubio, and P. Ordejon. 1999. Ab initio structural elastic and vibrational properties of carbon nanotubes. *Physical Reviews B*, 59:12678–12688.

Selmi, A., C. Friebel, I. Doghri, and H. Hassis. 2007. Prediction of the elastic properties of single walled carbon nanotube reinforced polymers: A comparative study of several micromechanical models. *Composites Science and Technology*, 67 (10):2071–2084. doi: 10.1016/j.compscitech.2006.11.016.

Shabani, M. O. and A. Mazahery. 2012. The GA optimization performance in the microstructure and mechanical properties of MMNCs. *Transactions of the Indian Institute of Metals*, 65(1):77–83. doi: 10.1007/s12666-011-0110-9.

Shindo, Y., Y. Kuronuma, T. Takeda, F. Narita, and S.-Y. Fu. 2012. Electrical resistance change and crack behavior in carbon nanotube/polymer composites under tensile loading. *Composites Part B: Engineering*, 43 (1):39–43. doi: 10.1016/j.compositesb.2011.04.028.

Shokrieh, M. M. and R. Rafiee. 2010a. Investigation of nanotube length effect on the reinforcement efficiency in carbon nanotube based composites. *Composite Structures*, 92 (10):2415–2420. doi: 10.1016/j.compstruct.2010.02.018.

Shokrieh, M. M. and R. Rafiee. 2010b. On the tensile behavior of an embedded carbon nanotube in polymer matrix with non-bonded interphase region. *Composite Structures*, 92 (3):647–652. doi: 10.1016/j.compstruct.2009.09.033.

Shokrieh, M. M. and R. Rafiee. 2010c. Prediction of mechanical properties of an embedded carbon nanotube in polymer matrix based on developing an equivalent long fiber. *Mechanics Research Communications*, 37 (2):235–240. doi: 10.1016/j.mechrescom.2009.12.002.

Sun, L., R. F. Gibson, and F. Gordaninejad. 2011. Multiscale analysis of stiffness and fracture of nanoparticle-reinforced composites using micromechanics and global-local finite element models. *Engineering Fracture Mechanics*, 78 (15):2645–2662.

Takano, N., M. Asai, T. Kouda, and K. Hashimoto. 2008. Three-dimensional morphology analysis and finite element modeling of nano particle dispersed materials. *Zairyo/Journal of the Society of Materials Science, Japan*, 57 (5):423–429.

Tersoff, J. 1992. Energies of fullerenes. *Physical Reviews B*, 46:15546–15549.

Tserpes, K. I., P. Papanikos, G. Labeas, and Sp. G. Pantelakis. 2008. Multi-scale modeling of tensile behavior of carbon nanotube-reinforced composites. *Theoretical and Applied Fracture Mechanics*, 49 (1):51–60. doi: 10.1016/j.tafmec.2007.10.004.

Tserpes, K. I., P. Papanikos, and S. A. Tsirkas. 2006. A progressive fracture model for carbon nanotubes. *Composites Part B: Engineering*, 37 (7–8):662–669. doi: 10.1016/j.compositesb.2006.02.024.

Wang, X., X. Y. Wang, and J. Xiao. 2005. A non-linear analysis of the bending modulus of carbon nanotubes with rippling deformations. *Composite Structures*, 69 (3):315–321. doi: 10.1016/j.compstruct.2004.07.009.

Wang, H. and X. Zhao. 2010. Hybrid finite element formulation for temperature prediction in carbon nanofiber based composites. *Advanced Materials Research*, 139–141 (October):39–42. doi: 10.4028/www.scientific.net/AMR.139-141.39.

Wu, Y., M. Gao, Y. Chen, J. Li, and S. Ju. 2011. Finite element simulation of tensile behaviors in the elastic region for PP/nano-TiO_2 composites. *Advanced Materials Research*, 150–151 (October):1819–1823. doi: 10.4028/www.scientific.net/AMR.150-151.1819.

Yakobson, B. I., C. J. Brabec, and J. Bernholc. 1996. Nanomechanics of carbon tubes: Instabilities beyond linear range. *Physical Review Letters*, 76:2511–2514.

Yakobson, B. I., M. P. Campbell, C. J. Brabec, and J. Bernholc. 1997. High strain rate fracture and C-chain unraveling in carbon nanotubes. *Computational Materials Science*, 8:341–348.

Yan, W., C. L. Pun, Z. Wu, and G. P. Simon. 2011. Some issues on nanoindentation method to measure the elastic modulus of particles in composites. *Composites Part B: Engineering*, 42 (8):2093–2097.

Yas, M. H. and M. Heshmati. 2012. Dynamic analysis of functionally graded nanocomposite beams reinforced by randomly oriented carbon nanotube under the action of moving load. *Applied Mathematical Modelling*, 36 (4):1371–1394. doi: 10.1016/j.apm.2011.08.037.

Zhu, P., Z. X. Lei, and K. M. Liew. 2012. Static and free vibration analyses of carbon nanotube-reinforced composite plates using finite element method with first order shear deformation plate theory. *Composite Structures*, 94(4):1450–1460. doi: 10.1016/j.compstruct.2011.11.00.

9

Finite Element–Aided Electric Field Analysis of Needleless Electrospinning

Haitao Niu
Deakin University, Geelong, Victoria, Australia

Xungai Wang
Deakin University, Geelong, Victoria, Australia

Tong Lin
Deakin University, Geelong, Victoria, Australia

CONTENTS

9.1 Introduction 311
9.2 Electrospinning and Electrospun Nanofibers 312
9.3 Electric Field in Electrospinning 314
9.4 FEM Analysis of Rotary Needleless Spinnerets 317
9.5 Effect of Electrospinning Parameters 323
9.6 FEM Analysis of Other Needleless Spinnerets 324
9.7 Conclusion 326
9.8 Summary 326
References 327

9.1 Introduction

Electrospinning has become an efficient technology to produce nano-fibrous materials. It conventionally uses a needle-like nozzle to process polymer fluids into fibrous materials. However, needle electrospinning has very limited fiber production ability, and this considerably restricts its wide applications in practice. Needleless electrospinning appeared 45 years later after Anton Formhals (Anton 1934) invented electrospinning in 1934. The first needleless electrospinning system used a ring spinneret for the production of fiber fleece (Simm, 1976). However, needleless electrospinning did not exhibit its potential in the mass production of nanofibers until the invention of roller electrospinning (Jirsak et al., 2005) in 2005. Since then, many needleless electrospinning systems have been reported. Needleless electrospinning is highly determined by the fiber generators that have a considerable influence on the electric field profiles. Because of the difficulty in direct measurement of high electric field, finite element methods (FEMs) have become the main technique to understand electric field profiles in electrospinning zone and fiber generator surface.

This chapter reviews the recent development in the analysis of electric field in needleless spinnerets using FEM. The influences of spinneret shape and electrospinning conditions on the electric field profile are presented, and the relationship between electric field and electrospinning performance is introduced.

9.2 Electrospinning and Electrospun Nanofibers

The basic electrospinning setup comprises a high-voltage power supply, a needle-like nozzle, and a counter-electrode collector, as shown in Figure 9.1. During electrospinning, charged by a high electric voltage, the solution droplet at the needle tip is electrified (Li and Xia, 2004b). Under the action of electric force, the solution droplet is drawn toward the opposite electrode and as a result deforms into a conical shape (also known as "Taylor cone" (Taylor, 1969)). When the electric force overcomes the surface tension of the polymer solution, the polymer solution can be ejected off the "Taylor cone," forming a solution jet. This charged jet is further stretched into a fine filament due to its strong interaction with the external electric field and the charge repulsion within the filament. The rapid evaporation of solvent from the filaments results in dry fibers that deposit randomly on the collector, forming a nonwoven nanofiber web in most of the cases.

Electrospinning technology has many advantages such as good versatility in processing different materials (e.g., polymers, chemicals, biomaterials, and nanoparticles), low cost, simplicity, and high efficiency. Electrospun nanofibers exhibit many unique properties, including high aspect ratio, excellent pore interconnectivity, and high porosity. Depending on materials, operating parameters, and nozzle configuration, electrospun nanofibers can have different morphologies. Beaded fibers (Fong et al., 1999), porous fibers (Bognitzki et al., 2001; Chen et al., 2011) or grooved (Huang et al., 2011) surface, ribbons (Koombhongse et al., 2001), hollow fibers (Li and Xia, 2004a) can be prepared using a normal needle nozzle, while side-by-side (Lin et al., 2005), helical (Lin et al., 2005; Chang and Shen, 2011), and core-sheath (Sun et al., 2003) nanofibers are normally prepared using a special spinneret. Although electrospun nanofibers are generally collected in the form of nonwoven mat,

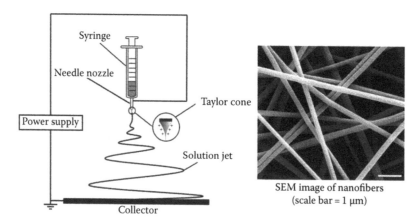

FIGURE 9.1
Schematic illustration of a basic electrospinning setup, Taylor cone, and a SEM image of electrospun nanofibers.

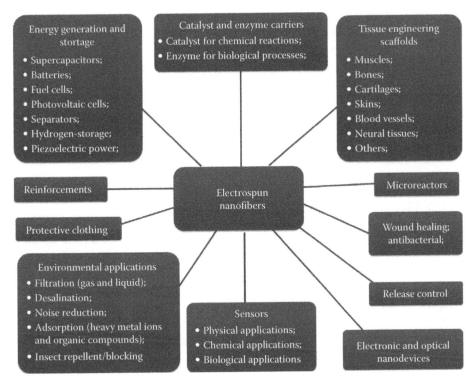

FIGURE 9.2
Applications of electrospun nanofibers.

aligned nanofibers (Theron et al., 2001; Li et al., 2003, 2004; Katta et al., 2004), 3D nano-fiber structure (Zhang and Chang, 2008), and nanofiber yarns (Dalton et al., 2005; Smita et al., 2005; Huan et al., 2006; Ali et al., 2012) can be obtained using special fiber-collecting techniques. These characteristics make electrospun nanofibers have wide applications in diversified areas.

Figure 9.2 lists the most important applications of electrospun nanofibers. They include environmental protection, energy generation and storage, tissue engineering scaffolds, release control, catalyst and enzyme carriers, and sensors (Fang et al., 2008; Thavasi et al., 2008; Lu et al., 2009), and many others.

Recently, needleless electrospinning has emerged with the ability of producing nanofibers on large scales. Unlike needle electrospinning that often uses a needle nozzle as the fiber generator, needleless electrospinning uses a spinneret that contains no needles, and nanofibers are generated from an open liquid surface. Among all the spinnerets reported for needleless electrospinning, rotating spinnerets are the most promising, because they can produce quality nanofibers continuously and massively. Several different rotating needleless electrospinning setups have already been invented by different research groups. In 2005, a roller electrospinning was commercialized by Elmarco Co. with the brand name "NanospiderTM" (Jirsak et al., 2005).

The jet formation in rotating needleless electrospinning can be divided into a few stages. First, a thin polymer solution layer is applied onto the spinneret surface as a result of partial immersion in the solution and the rotation of the spinneret. The rotation also causes perturbations of the solution layer on the spinneret, thus inducing the formation of conical spikes on the surface. When a high voltage is applied, the spikes concentrate electric forces, intensifying the perturbations to form "Taylor cones." Jets are finally stretched out from the "Taylor cones," resulting in nanofibers.

It has been reported that the rotating needleless electrospinning setup is simple, efficient, and easy to maintain, but only occupies a small space. The rotation of spinnerets ensures conveying the polymer solution to electrospinning sites for continuous nanofiber production. When nanofibers are electrospun upward, this process effectively prevents nonelectrospun solution droplets from contaminating the nanofiber products.

9.3 Electric Field in Electrospinning

In electrospinning, the critical voltage for initiating an electrospinning process has been proposed to follow the equation (Taylor 1969)

$$V_c^2 = 4\ln(2h/R)(1.3\pi R\gamma)0.09 \tag{9.1}$$

where
  h is the distance from the needle tip to the collector
  R is the needle outer radius
  γ is the surface tension

The factor 0.09 is inserted to predict the voltage. Although voltage is used in this equation, electric force (F) is the actual driving force to jet formation, which is determined by the electric field intensity (E) and charge (q).

$$F = E \cdot q \tag{9.2}$$

For needleless electrospinning, although jets are initiated from an open polymer solution surface, there is a critical voltage or critical electric field intensity that it still needs to overcome. Lukas et al. (2008) have indicated that the self-organization of jets takes place on free liquid surface during needleless electrospinning, and the critical electric field intensity (E_c) for needleless electrospinning is predicted as follows:

$$E_c = \sqrt[4]{\frac{4\gamma\rho g}{\varepsilon^2}} \tag{9.3}$$

where
  ρ is liquid mass density
  g is gravity acceleration
  γ is surface tension
  ε is the permittivity

This equation suggests that jet initiation in needleless electrospinning is only determined by solution properties.

Since the electric force is the driving force for both needle and needleless electrospinning, fibers should be easier to generate from a spinneret that has a larger electric field regardless of the electrospinning mode. In other words, nanofibers should form earlier from an area where the electric field intensity is higher than the critical value required for initiating an electrospinning process. This is important to needleless electrospinning because the electric field intensity along the entire fiber-generating surface on the spinneret is not the same. When the voltage is above the critical value, the higher electric field enables more solution jets to be generated and fiber stretching is also faster, leading to increase in the fiber productivity.

It is practically difficult to directly measure the electric field intensity and its profile along the spinneret due to the very high voltage involved. Calculations based on classic physical principles become an important alternative method to gain understanding on the electric field intensity in a high eclectic field.

Partial differential equations (PDEs) as typical differential equations are often used to describe the propagation of sound, heat, electrostatics, electrodynamics, fluid flow, or elasticity. Just as ordinary differential equations often modeling dynamical systems, PDEs can model multidimensional systems. The electric field (E) due to a static charge density $\rho(x, y, z)$ distribution is determined by Gauss' Law:

$$\nabla \cdot E = \frac{\rho(x,y,z)}{\varepsilon_0} \tag{9.4}$$

where ε_0 is the permittivity constant. A static electric field can be derived from a scalar potential $\psi\,(x, y, z)$

$$E = -\nabla \psi \tag{9.5}$$

which obeys Poisson's equation:

$$\nabla^2 \psi = \left(\frac{\partial^2}{\partial x^2} + \frac{\partial^2}{\partial y^2} + \frac{\partial^2}{\partial z^2} \right) \psi = -\frac{\rho(x,y,z)}{\varepsilon_0} \tag{9.6}$$

These equations indicate that even under the same applied voltage, the electric field formed by different spinnerets may have different intensities and profiles depending on their geometric shapes.

In practice, the arbitrary shapes of spinnerets in a three-dimensional (3D) object make it almost impossible to find the solution to these PDEs if classical analytical methods are employed. FEM is a numerical technique for finding approximate solutions of PDEs, and it has been used to solve a wide range of physical and engineering PDE problems. Finite element analysis was first developed in 1943 by R. Courant (Courant and Hilbert, 1943), who used the Ritz method of numerical analysis and minimization of variational calculus to obtain approximate solutions to vibration systems. Attributed to the rapid decline in the cost of computers and the enhancement in the computing power, FEM has been developed greatly and able to produce accurate results for PDEs. At present, solving of PDE can be simply realized by commercial FEM programs such as COMSOL, Maxwell SV 2D, or Maxwell 3D.

The basic approach of solving PDE by FEM includes defining the physical problem based on which to develop a model, formulating the governing equations (initial conditions and/or boundary conditions), discretizing equations, solving the discrete system of equations, and result interpretation and error analysis. Since electrostatics is a physical phenomenon,

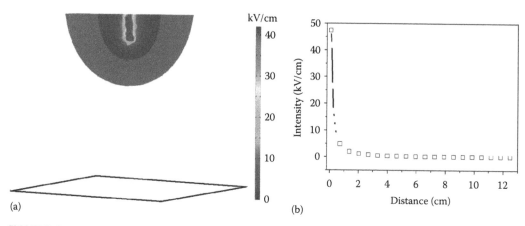

FIGURE 9.3
(a) Electric field profile of a needle nozzle and (b) electric field intensity from nozzle tip to the collector.

which can be described in terms of PDEs, FEM can be an attractive and practical method to analyze the electric field in electrospinning and design spinnerets for needleless electrospinning. The calculated electric field and visualized results can assist in understanding the relationship between the electric field and spinneret geometry shape, and the influence of spinneret shape on electrospinning performance.

To calculate the electric field intensity of an electrospinning setup, the physical objects in the electrospinning setup (e.g., spinneret, solution container, polymer solution, and collector) are setup first in the program according to their actual dimensions, location, and relative permittivity. A high voltage is set according to the practical connection. The meshing and solving of equations are conducted by the computer program. Finally, the electric field intensity and distribution profile can be obtained and visualized.

The electric field profile of a needle nozzle is shown in Figure 9.3a. An intensified electric field is typically formed near the needle tip, which declines rapidly toward the collector. Based on this 3D profile, the electric field intensity change along any specific direction or surface can be easily acquired. Figure 9.3b is an example showing the change of electric field intensity from nozzle tip to the collector. The collectors in needle electrospinning have far more important influences on the electric field formed and nanofiber collection, since the fiber generator in needle electrospinning remains the same (needle).

FEM has been successfully used for analyzing the collections of aligned nanofibers (Li et al., 2003; Liu and Dzenis, 2008; Park and Yang, 2011), and patterned nanofibers (Salim et al., 2008; Ding et al., 2009). Recently, it was also used to understand jet repulsion in multijet electrospinning (Angammana and Jayaram, 2011).

In needleless electrospinning, fiber generators play a key role in determining the electric field intensity and profile. Niu et al. (2009) first used FEM to analyze the electric field intensity of two rotary needleless spinnerets, disk, and cylinder. They found that a disk spinneret formed a higher intensity electric field with a narrower distribution than a cylinder spinneret under the same applied voltage. They also experimentally proved the better electrospinning performance of the disk spinneret than the cylinder spinneret. The result that a spinneret containing surface with a larger curvature forms an electric field of higher intensity, thus having a better electrospinning performance was also reported by other researchers (Wang et al., 2009; Lin et al., 2010; Thoppey et al., 2010, 2011; Niu et al., 2011).

This chapter focuses on the recent progress in FEM electric field analysis of needleless electrospinning and the influence of spinnerets on electrospinning productivity and fiber quality. In the following section, rotating spinnerets will be employed as a main example to elucidate the electric field profile and its influence on the electrospinning process.

9.4 FEM Analysis of Rotary Needleless Spinnerets

In rotary needleless electrospinning, when a high voltage is applied, the surface fluctuations of solution layer on the spinneret are amplified, causing the formation of conical spikes on the surface, which can be stretched into fine solution jets. A good understanding of the electric field in needleless electrospinning can facilitate the optimization of fiber generators. Figure 9.4 schematically shows three rotating spinnerets for needleless electrospinning. They all partially immersed into a solution container underneath, and a drum collector is set right upward. Based on the physical mode, material, and the operating parameters, FEM electric field analysis can be performed. Figures 9.5 through 9.9 show the result of electric field analysis for cylinder, ball, and disk electrospinning systems.

Figure 9.5 shows the electric field profile of cylinder electrospinning under different dimensions. These figures only provide the profile around the spinnerets since there

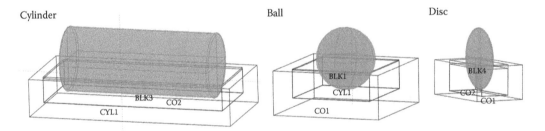

FIGURE 9.4
Schematic illustration of rotating spinnerets.

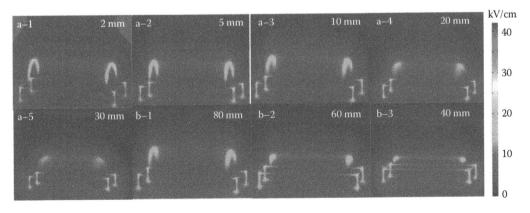

FIGURE 9.5
Electric field profiles of cylinder spinnerets with (a) different rim radii (length 200 mm, and diameter 80 mm) and (b) different diameters (length 200 mm, and rim radius 10 mm). (From Niu, H. et al., *J. Text. Inst.*, 1, 2011.)

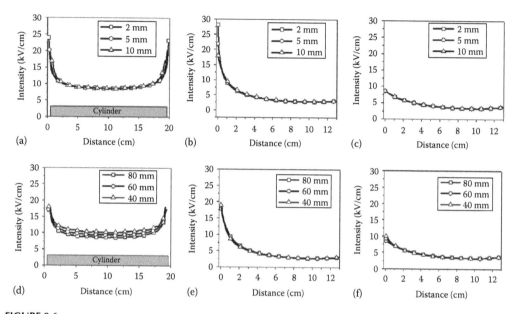

FIGURE 9.6

(a–c) Electric field intensities of cylinder spinnerets with the same diameter but different rim radii, (a) along cylinder length, (b) from cylinder end to the collector, and (c) from cylinder middle to the collector; (d–f) electric field intensities of cylinder spinnerets with the same rim radius but different cylinder diameters, (d) along cylinder length, (e) from cylinder end to the collector, and (f) from cylinder middle to the collector. (From Niu, H. et al., *J. Text. Inst.*, 1, 2011.)

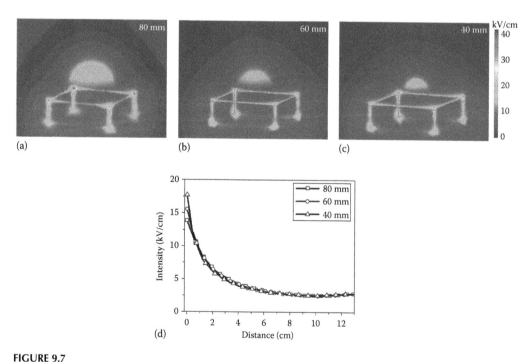

FIGURE 9.7

(a–c) Electric field profiles of ball spinnerets with different diameters. (From Niu, H. et al., *J. Text. Inst.*, 1, 2011.)
(d) Electric field intensities of ball spinnerets with different diameters.

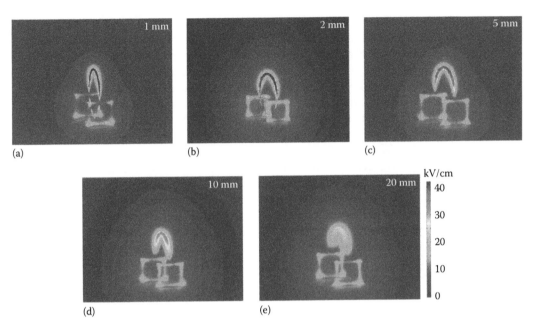

FIGURE 9.8
Electric field profiles of disk spinnerets with different thicknesses ((a) 2 mm, (b) 5 mm, (c) 10 mm, (d) 20 mm, and disk diameter 80 mm). (From Niu, H. et al., *J. Text. Inst.*, 1, 2011.)

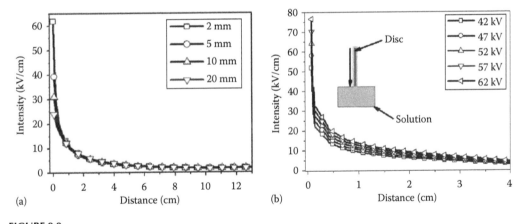

FIGURE 9.9
(a) Electric field intensities of disk spinnerets with different thicknesses. (From Niu, H. et al., *J. Text. Inst.*, 1, 2011.)
(b) Electric field intensifies from disk top to the solution surface (diameter 80 mm.)

is little difference in the collector area. The cylinder top ends always formed higher electric field intensity than the middle surface. The electric field at the cylinder ends varied in intensity depending on the rim radius of the cylinder end. With an increase in the rim radius, the electric field intensity at the cylinder ends declined. The intensity discrepancy between cylinder end and middle areas was reduced but not eliminated completely (Figure 9.5a). When cylinders were kept in the same length and rim radius, decreasing the cylinder diameter led to higher strength of electric fields in the middle area (Figure 9.5b).

Figure 9.6 shows the electric field profile along cylinder length and from cylinder surface to the collector. When increasing the rim radius, the electric field intensity at cylinder ends reduced, and the area that had high-intensity value expanded slightly toward the cylinder middle (Figure 9.6a and b). The intensity from cylinder middle area to the collector was little affected by the rim radius (Figure 9.6c). Reducing the cylinder diameter slightly increased the electric field intensity on the cylinder middle surface (Figure 9.6d), and the electric field profile from the spinneret surface to the collector remained almost unchanged (Figure 9.6e and f).

Figure 9.7 shows the electric field analysis result of ball electrospinning in the spinneret area. The electric field was evenly distributed on the top half. However, the intensity was lower than that at cylinder ends. When the ball diameter reduced from 8 to 4 cm, electric field intensity on the ball surface increased; however, the intensity profile from 1 cm off the ball top to the collector remained unchanged (Figure 9.7d).

Disk spinneret formed highly intensified electric fields narrowly distributed at the top circumference (Figure 9.8). The electric field intensity was much larger than that of cylinder and ball spinnerets. With a decrease in the disk thickness, the electric field intensity on the disk surface increased.

Figure 9.9a shows the influence of disk thickness on the electric field profile from disk top to the collector. The change of disk thickness had little effect on the profile, except that the electric field intensity on the disk surface changed greatly. The electric field intensity on the 20 mm thick disk generator was 24 kV/cm. The intensity increased to 62 kV/cm, when the thickness reduced to 2 mm. Further reducing the thickness from 2 mm to 1 mm increased the electric field intensity to 200 kV/cm. Figure 9.9b shows that the electric field intensity decayed rapidly away from the disk top area to the solution surface.

By comparing the electric field profiles of these three spinnerets, it is easy to conclude that the geometric shape and dimension of a spinneret have significant effects on the intensity and profile of electric field on the spinneret surface, which should greatly influence the electrospinning process, fiber quality, and productivity. To verify this, cylinder, ball, and disk electrospinning was performed at the same electrospinning parameters using the same polymer solution. Figure 9.10 shows the electrospinning processes of three needleless electrospinning systems.

Table 9.1 lists the fiber diameter and production rate of polyvinyl alcohol (PVA) nanofibers electrospun using different spinnerets. The differences in these experiment results

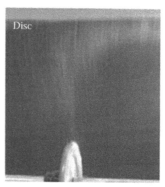

FIGURE 9.10
Photographs of cylinder, disk, and ball electrospinning processes (Polyvinyl alcohol (PVA) solution, applied voltage 57 kV). (From Niu, H. et al., *J. Appl. Polym. Sci.*, 114(6), 3524, 2009.)

TABLE 9.1

Spinnerets and Their Electrospinning Productivities

Cylinder (L = 200 mm)						
Cylinder	R (mm)	2	5	10	20	30
(Φ = 80 mm)	Fiber diameter (nm)	334 ± 118	357 ± 127	366 ± 116	×	×
	Productivity (g/h)	6.3	8.6	5.8	×	×
Cylinder	Φ (mm)	80	60		40	
(R = 10 mm)	Fiber diameter (nm)	366 ± 116	325 ± 150		289 ± 132	
	Productivity (g/h)	5.8	7.2		8.4	
Disk (Φ = 80 mm)						
Disk	L (mm)	1	2	5	10	20
	Fiber diameter (nm)	×	257 ± 77	277 ± 92	287 ± 91	321 ± 108
	Productivity (g/h)	×	6.2	5.2	5.6	5.2
Ball						
Ball	Φ (mm)	80		60		40
	Fiber diameter (nm)	344 ± 105		×		×
	Productivity (g/h)	3.1		×		×

Source: Niu, H. et al., *J. Text. Inst.*, 1, 2011.
Note: ×, electrospinning failure; R, rim radius; Φ, diameter; L, thickness.

can be explained by the FEM electric field analysis. Due to the uneven distributed electric field intensity along the cylinder surface, nanofibers were unevenly produced from cylinder surface. Fibers were easily generated from the cylinder ends in comparison with the middle area. Cylinders produced nanofibers from the whole top surface only when the applied voltage was over 57 kV/cm.

When the cylinder rim radius was 5 mm, lower electric field intensity was formed at the end area compared to that with a rim radius of 2 mm. However, the electric field intensity in this area was still large enough to initiate electrospinning. Because of the increased high electric field surface, more solution jets were generated from the cylinder surface, resulting in an improved productivity of 8.6 g/h. The reduced electric field intensity due to increase of cylinder rim radius could also be used to explain the failure of cylinder electrospinning when the rim radius was larger than 20 mm. With such a large rim radius, the electric field intensity at the cylinder ends was lower than the critical value required for initiating an electrospinning process. The change of cylinder diameter also led to changes in fiber productivity, and the increase of electric field intensity on the middle surface of cylinders resulted in an increased fiber production rate. Based on the FEM calculation and experiment results, the critical electric field intensity to initiate an electrospinning process on the needleless electrospinning nozzle should be around 18 kV/cm.

The ball spinneret had a low production rate of 3.1 g/h due to the low electric field intensity, and it required a high voltage (57 kV/cm) to initiate the electrospinning process. Even though the fact that smaller balls generated stronger electric fields than the threshold value for electrospinning (14 kV/cm, based on the ball diameter 80 mm), they are unable to generate jets.

For a thin disk of 2 mm in thickness, electrospinning can be performed at an applied voltage as low as 42 kV, and the fiber production rate was 6.2 g/h when the applied voltage was 57 kV/cm. The productivity of disk electrospinning increased with the decreased disk

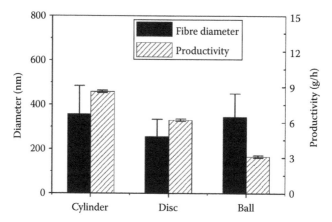

FIGURE 9.11

Comparison among cylinder, disk, and ball spinnerets (applied voltage 57 kV, spinning distance 13 cm, cylinder diameter 80 mm, cylinder rim radius 5 mm, disk diameter 80 mm, disk thickness 2 mm, ball diameter 80 mm). (From Niu, H. et al., *J. Text. Inst.*, 1, 2011.)

thickness. When the applied voltage was 57 kV, the electric field intensity of disk (thickness 1 mm) was so high that a corona discharge happened, preventing electrospinning from working properly, although the disk is covered by a thin solution layer.

The relationship between electrospinning performance and spinneret shape is shown in Figure 9.11. Under the best working conditions, the disk produced the finest nanofibers. The narrowly distributed electric field also made jets travel in a similar rate through the electrospinning zone, resulting in nanofibers with a narrow diameter distribution. Among the cylinder, disk, and ball electrospinning systems, cylinder electrospinning produced coarse nanofibers with the largest fiber productivity. Compared to the disk, the ball spinneret produced coarser nanofibers with a lower productivity.

Lin et al. (2010) designed a novel needleless electrospinning using a spiral coil as spinneret. FEM analysis of the electrospinning system and photograph of electrospinning process are shown in Figure 9.12. An intensified electric field was formed on the surface of

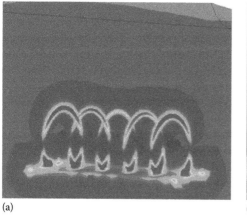

(a)

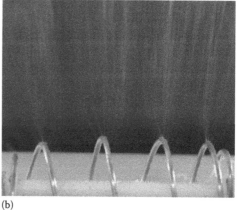

(b)

FIGURE 9.12

(a) Electric field profile of coils spinneret and (b) photograph of coil electrospinning process.

spiral coil because of the small diameter of the coil wire, and the high electric field intensity was just formed narrowly on the coil surface. When a high voltage was applied, fine and uniform nanofibers were generated from the coil surface with high fiber productivity (Lin et al., 2010).

9.5 Effect of Electrospinning Parameters

The applied voltage and collecting distance are two important factors that affect the electric field. The influences of applied voltage on the electric field of cylinder, ball, and disk spinnerets are shown in Figure 9.13. The electric field profiles were geometry-specific. The electric field intensity was increased at a high applied voltage. However, the intensity distribution did not change much. The increase of intensity near the spinneret was more obvious than that close to the collector. With increasing the applied voltage, the most significant improvement in the electric field intensity took place on the disk top, followed by the cylinder ends, then the ball top, and the cylinder middle area. In addition, reducing the spinning distance also improved the electric field intensity in the electrospinning zone.

The influences of applied voltage on the electrospinning productivity and fiber diameter have been reported by Niu et al. (2009). Due to the unevenly distributed electric field on the cylinder surface, nanofibers produced by a cylinder spinneret always had a large diameter distribution (Figure 9.14a). The change of applied voltage had a little influence on the average diameter of cylinder electrospun nanofibers. However, for the disk spinneret,

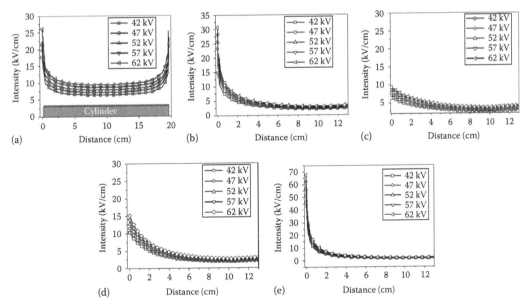

FIGURE 9.13
Electric field intensities under different applied voltages: (a) along cylinder length, (b) from cylinder end to the collector, (c) from cylinder middle to the collector, (d) from ball top to the collector, and (e) from disk top to the collector (cylinder diameter 80 mm, rim radius 2 mm, ball diameter 80 mm, disk thickness 2 mm, applied voltage 57 kV, and spinning distance 13 cm).

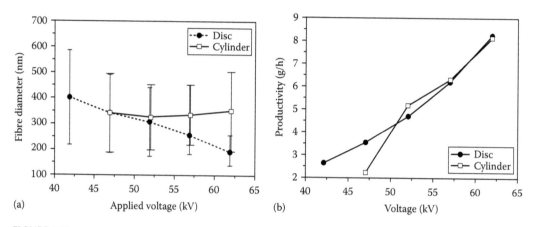

FIGURE 9.14

Influences of applied voltage on (a) fiber diameter and (b) fiber productivity of disk and cylinder electrospinning systems (collecting distance 13 cm, PVA concentration 9 wt%, cylinder diameter 80 mm, cylinder rim radius 2 mm, disk diameter 80 mm, disk thickness 2 mm). (From Niu, H. et al., *J. Appl. Polym. Sci.*, 114(6), 3524, 2009.)

increasing the voltage resulted in a reduced fiber diameter with a narrowed diameter distribution. Figure 9.14b shows nanofiber production rates of disk and cylinder electrospinning systems. The productivity increased with increasing the applied voltage from 47 to 62 kV for both systems. The influence of applied voltage on the electrospinning process, fiber diameter, and productivity for other needleless electrospinning systems is similar to those of disk or cylinder electrospinning.

9.6 FEM Analysis of Other Needleless Spinnerets

FEM electric field analysis was also used to assist in understanding the electrospinning processes of other needleless electrospinning systems including conical coil (Wang et al., 2009), plate (Thoppey et al., 2010), and bowel edge (Thoppey et al., 2011).

Wang et al. (2009) reported a conical coil spinneret to electrospin nanofibers in a downward manner. FEM analysis showed that an intensified electric field was formed on the wire surface and it had higher intensity on the wires near the coil ends (Figure 9.15a and b). In comparison, a needle nozzle formed an intensified electric field at its tip, but with a lower intensity due to the lower applied voltage (Figure 9.15c). From spinneret to the collector, the electric fields formed by both spinnerets decayed rapidly and then stabilized near to the collector (Figure 9.15d). When the conical coil was loaded with polymer solution, jets were stretched out from both the wire surface and the wire gap under the strong electric field (Figure 9.15e). The fiber production rate of electrospun PVA nanofibers increased with increasing the applied voltage (45–60 kV).

Figure 9.16a and b show the electric fields of a needle and a plate electrospinning system reported by Thoppey et al. (2010). Both systems showed a similar electric field profile. The plate formed an intensified electric field at the plate edge and a low electric field near the collector (Figure 9.16b). During electrospinning, numerous solution jets can be easily

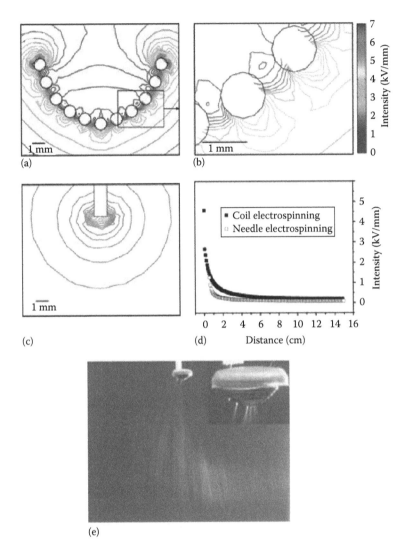

FIGURE 9.15
Cross-sectional view of electric field profiles of (a) and (b) conical coil spinneret (applied voltage 60 kV), (c) needle spinneret (applied voltage 22 kV), (d) electric field profiles along the electrospinning direction, and (e) photograph of conical-coil electrospinning process. (From Wang, X. et al., *Polym. Eng. Sci.*, 49(8), 1582, 2009.)

ejected off from the plate edge, resulting in an improved fiber production rate compared with the needle electrospinning. When a stack of plates was used, the electric field intensity at the plate edges reduced (Figure 9.16c). However, it was still strong enough to drive the jet generation and fiber production.

When a bowel was used to electrospin nanofibers, FEM analysis showed the formation of an intensified electric field at the bowl edge, the profile of which was similar to that of needle electrospinning (Figure 9.17a and b) (Thoppey et al., 2011). The high electric field forming area and the fiber-generating area overlapped at the bowl edge. Figure 9.17c shows that jets are ejected off from the bowl edge when a high voltage is applied. The narrowly distributed intensified electric field along the narrow spinning sites (bowl edge) enabled

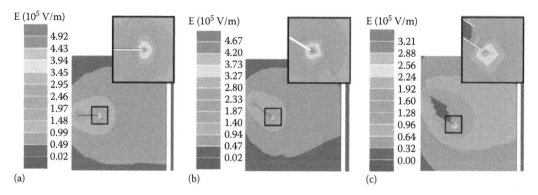

FIGURE 9.16
Electric field intensity distributions of (a) conventional needle electrospinning, (b) edge-plate electrospinning, and (c) waterfall electrospinning (collecting distance 15 cm, applied voltage 15 kV; insets are the magnifications of the indicated square areas in each figure). (From Thoppey, N.M. et al., *Polymer*, 51(21), 4928, 2010.)

jets to be equally stretched. As a result, the collected nanofibers have a similar quality as those from needle electrospinning under the same electrospinning conditions.

9.7 Conclusion

The electric field in needleless electrospinning is highly dependent on the geometric shape of spinnerets, which considerably affects the electrospinning process and productivity. FEM is an effective method to calculate electric fields and intensity profiles in needleless electrospinning. It provides a visualized electric field profile, which greatly assists in correlating with electrospinning experiments. Although several works have been reported on FEM analysis of electric field in needleless electrospinning, the work in this area is mainly based on static analysis without considering dynamic influences of spinning solution and Taylor cones. It can be expected that more studies on electric field analysis using FEM will enable to design and optimize needleless spinnerets, analyze the interaction between the charged solution jet and the electric field, and predict deposition structure of nanofibers.

9.8 Summary

Needleless electrospinning is advantageous in producing polymeric nanofibers on a large scale. It has become a promising way to provide nanofibers for diversely different applications in practice. The fiber generator in needleless electrospinning has a considerable influence on electric field intensity and profile, which ultimately determine the electrospinning performance, fiber quality, and fiber productivity. This chapter summarizes the recent progress in using FEM to analyze the electric field formed in needleless electrospinning and the relationship between the electric field and needleless electrospinning performance.

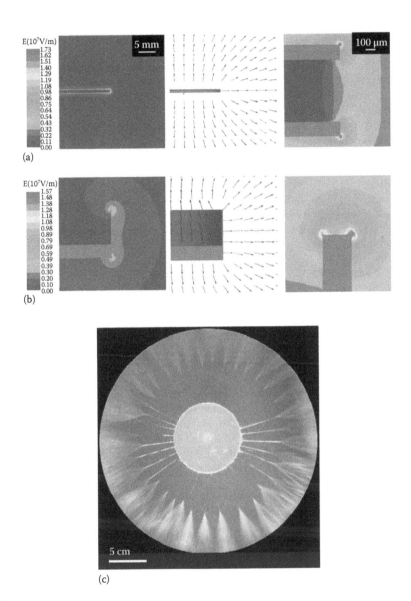

FIGURE 9.17
Electric field distributions: (a) conventional needle electrospinning (collecting distance 15 cm, applied voltage 11 kV), (b) bowl electrospinning (collecting distance 15 cm, applied voltage 55 kV), and (c) photograph of bowl electrospinning process (top-view). (From Thoppey, N.M. et al., *Nanotechnology*, 22(34), 345301, 2011.)

References

Ali, U., Y. Zhou, X. Wang, and T. Lin. 2012. Direct electrospinning of highly twisted, continuous nanofiber yarns. *Journal of the Textile Institute* 103(1):80–88.

Angammana, C.J. and S.H. Jayaram. 2011. The effects of electric field on the multijet electrospinning process and fiber morphology. *IEEE Transactions on Industry Applications* 47(2):1028–1035.

Anton, F. 1934. Process and apparatus for preparing artificial threads. United States: Richard, Schreiber Gastell, Anton, Formhals.

Bognitzki, M., W. Czado, T. Frese et al. 2001. Nanostructured fibers via electrospinning. *Advanced Materials* 13(1):70–72.

Chang, G. and J. Shen. 2011. Helical nanoribbons fabricated by electrospinning. *Macromolecular Materials and Engineering* 296(12):1071–1074.

Chen, H., J. Di, N. Wang et al. 2011. Fabrication of hierarchically porous inorganic nanofibers by a general microemulsion electrospinning approach. *Small* 7(13):1779–1783.

Courant, R. and D. Hilbert. 1943. *Methoden der mathematischen Physik,* Vol. 1, New York: Interscience.

Dalton, P.D., D. Klee, and M. Möller. 2005. Electrospinning with dual collection rings. *Polymer* 46(3):611–614.

Ding, Z., A. Salim, and B. Ziaie. 2009. Selective nanofiber deposition through field-enhanced electrospinning. *Langmuir* 25(17):9648–9652.

Fang, J., H. Niu, T. Lin, and X. Wang. 2008. Applications of electrospun nanofibers. *Chinese Science Bulletin* 53(15):2265–2286.

Fong, H., I. Chun, and D.H. Reneker. 1999. Beaded nanofibers formed during electrospinning. *Polymer* 40(16):4585–4592.

Huan, P., L. Li, L. Hu, and X. Cui. 2006. Continuous aligned polymer fibers produced by a modified electrospinning method. *Polymer* 47:4901–4904.

Huang, C., Y. Tang, X. Liu et al. 2011. Electrospinning of nanofibres with parallel line surface texture for improvement of nerve cell growth. *Soft Matter* 7(22):10812–10817.

Jirsak, O., F. Sanetrnik, D. Lukas, V. Kotek, L. Martinova, and J. Chaloupek. 2005. A method of nanofibres production from a polymer solution using electrostatic spinning and a device for carrying out the method. WO 2005/024101 A1.

Katta, P., M. Alessandro, R.D. Ramsier, and G.G. Chase. 2004. Continuous electrospinning of aligned polymer nanofibers onto a wire drum collector. *Nano Letters* 4(11):2215–2218.

Koombhongse, S., W. Liu, and D.H. Reneker. 2001. Flat polymer ribbons and other shapes by electrospinning. *Journal of Polymer Science Part B: Polymer Physics* 39(21):2598–2606.

Li, D., Y. Wang, and Y. Xia. 2003. Electrospinning of polymeric and ceramic nanofibers as uniaxially aligned arrays. *Nano Letters* 3(8):1167–1171.

Li, D., Y. Wang, and Y. Xia. 2004. Electrospinning nanofibers as uniaxially aligned arrays and layer-by-layer stacked films. *Advanced Materials* 16(4):361–366.

Li, D. and Y. Xia. 2004a. Electrospinning of nanofibers: Reinventing the wheel? *Advanced Materials* 16(14):1151–1170.

Li, D. and Y. Xia. 2004b. Direct fabrication of composite and ceramic hollow nanofibers by electrospinning. *Nano Letters* 4(5):933–938.

Lin, T., H. Wang, and X. Wang. 2005. Self-crimping bicomponent nanofibers electrospun from polyacrylonitrile and elastomeric polyurethane. *Advanced Materials* 17(22):2699–2703.

Lin, T., X. Wang, X. Wang, and H. Niu. 2010. Electrostatic spinning assembly. WO/2010/043002.

Liu, L. and Y.A. Dzenis. 2008. Analysis of the effects of the residual charge and gap size on electrospun nanofiber alignment in a gap method. *Nanotechnology* 19(35):355307.

Lu, X., C. Wang, and Y. Wei. 2009. One-dimensional composite nanomaterials: Synthesis by electrospinning and their applications. *Small* 5(21):2349–2370.

Lukas, D., A. Sarkar, and P. Pokorny. 2008. Self-organization of jets in electrospinning from free liquid surface: A generalized approach. *Journal of Applied Physics* 103:084309.

Niu, H., T. Lin, and X. Wang. 2009. Needleless electrospinning. I. A comparison of cylinder and disk nozzles. *Journal of Applied Polymer Science* 114(6):3524–3530.

Niu, H., X. Wang, and T. Lin. 2011. Needleless electrospinning: Influences of fibre generator geometry. *Journal of the Textile Institute*:1–8.

Park, S.H. and D.-Y. Yang. 2011. Fabrication of aligned electrospun nanofibers by inclined gap method. *Journal of Applied Polymer Science* 120(3):1800–1807.

Salim, A., C. Son, and B. Ziaie. 2008. Selective nanofiber deposition via electrodynamic focusing. *Nanotechnology* 19(37):375303.

Simm, W. 1976. Apparatus for the production of filters by electrostatic fiber spinning. United States: Bayer Aktiengesellschaft (Leverkusen, DT).

Smita, E., U. Buttnerb, and R.D. Sanderson. 2005. Continuous yarns from electrospun fibers. *Polymer* 46(8):2419–2423.

Sun, Z., E. Zussman, A.L. Yarin, J.H. Wendorff, and A. Greiner. 2003. Compound core–shell polymer nanofibers by co-electrospinning. *Advanced Materials* 15(22):1929–1932.

Taylor, G. 1969. Electrically driven jets. *Proceedings of the Royal Society of London. Series A, Mathematical and Physical Sciences* 313(1515):453–475.

Thavasi, V., G. Singh, and S. Ramakrishna. 2008. Electrospun nanofibers in energy and environmental applications. *Energy & Environmental Science* 1(2):205–221.

Theron, A., E. Zussman, and A.L. Yarin. 2001. Electrostatic field-assisted alignment of electrospun nanofibres. *Nanotechnology* 12(3):384–390.

Thoppey, N.M., J.R. Bochinski, L.I. Clarke, and R.E. Gorga. 2010. Unconfined fluid electrospun into high quality nanofibers from a plate edge. *Polymer* 51(21):4928–4936.

Thoppey, N.M., J.R. Bochinski, L.I. Clarke, and R.E. Gorga. 2011. Edge electrospinning for high throughput production of quality nanofibers. *Nanotechnology* 22(34):345301.

Wang, X., H. Niu, T. Lin, and X. Wang. 2009. Needleless electrospinning of nanofibers with a conical wire coil. *Polymer Engineering & Science* 49(8):1582–1586.

Zhang, D. and J. Chang. 2008. Electrospinning of three-dimensional nanofibrous tubes with controllable architectures. *Nano Letters* 8(10):3283–3287.

10

Molecular Dynamic Finite Element Method (MDFEM)*

Lutz Nasdala

Offenburg University of Applied Sciences, Offenburg, Germany

Andreas Kempe

Leibniz Universität at Hannover, Hannover, Germany

Raimund Rolfes

Leibniz Universität at Hannover, Hannover, Germany

CONTENTS

10.1 Introduction 332
10.2 Chemical and Physical Bonds 335
    10.2.1 Covalent Bonding 335
    10.2.2 Ionic Bonding 336
    10.2.3 Metallic Bonding 337
    10.2.4 Van der Waals Bonding 337
    10.2.5 Hydrogen Bonding 338
10.3 From Quantum to Molecular Mechanics 339
    10.3.1 Schrödinger Wave Equation 339
    10.3.2 Ab Initio Molecular Dynamics 341
        10.3.2.1 Born–Oppenheimer Approximation 341
        10.3.2.2 Hartree–Fock Theory 341
        10.3.2.3 Density Functional Theory 341
    10.3.3 Semiempirical Models 342
    10.3.4 Classical Molecular Dynamics 342
        10.3.4.1 Newton's Equation of Motion 342
        10.3.4.2 Lagrange's Equation of Motion 343
        10.3.4.3 Hamilton's Equations of Motion 343
10.4 Molecular Dynamic Finite Element Method 344
    10.4.1 Requirements for the Finite Element Method 344
        10.4.1.1 Force Fields 344
        10.4.1.2 Short- and Long-Range Force Field Potentials 344
        10.4.1.3 Multi-Body Force Field Potentials 346
        10.4.1.4 Natural Lengths and Angles 347
        10.4.1.5 Linear and Nonlinear Analysis 347
        10.4.1.6 Material Parameters 348
        10.4.1.7 Finite Elements without Rotational Degrees of Freedom ... 348

* Reprinted with the permission of Tech Science Press from *CMC: Computers, Materials & Continua*, 19(1), 2010.

           10.4.1.8 Time Integration Schemes 349

           10.4.1.9 Relaxation Step to Determine the Reference Configuration............. 351

       10.4.2 User Elements... 352

           10.4.2.1 Two-Node Element for Bond Stretch..353

           10.4.2.2 Three-Node Element for Bending 354

           10.4.2.3 Four-Node Element for Torsion 355

       10.4.3 Inversion.. 356

           10.4.3.1 Spacial Structures 357

           10.4.3.2 Planar Structures 357

           10.4.3.3 Force Field Potentials.. 358

           10.4.3.4 Modeling Inversion by Improper Torsion 359

10.5 Numerical Examples 360

       10.5.1 Torsion of Single-Walled Carbon Nanotube.. 360

       10.5.2 Inelastic Behavior of Elastomeric Material... 361

           10.5.2.1 Relaxation Step 361

           10.5.2.2 Loading Step 363

10.6 Conclusions... 364

Acknowledgment.. 365

Appendix 10A: Newton–Raphson Method.. 365

Appendix 10B: Stiffness Matrices of MDFEM Elements 367

References.. 370

10.1 Introduction

A common question that arises at the beginning of a numerical simulation is how to capture relevant effects while keeping the numerical effort as low as possible: Can the material be treated as homogeneous and isotropic, or is it necessary to consider each individual atom? Cracks in materials which might be observed as a result of mechanical exposure or aging are usually treated at another analysis level than buckling phenomena of thin-walled engineering structures like cooling towers.

A lot of constitutive material models have been developed so far to describe softening behavior, hysteresis loops, friction, or fatigue. Though many of these models are physically motivated, they are often enriched by rheological, that is, phenomenological components like dashpot or friction elements in order to account for inelastic effects. Rheological models are suited for efficient simulations. However, in order to provide an explanation for the physical background of material inelasticity, the aforementioned phenomena have to be deduced from the formulation and breaking of chemical and physical bonds.

Figure 10.1 gives an overview of different simulation levels starting with the quantum scale up to macroscale. The most complex and detailed level is the quantum scale which is based on the Schrödinger equation that describes the interactions between electrons, neutrons, and protons. The second level is the nanoscale which can be reached by neglecting quantum effects (step 1). Force fields treat atoms as point masses connected by spring elements that represent different bonding types. The mechanical behavior observed in these classical molecular dynamic (MD) simulations can be used as motivation for models on microscale (step 2). For larger structures such as tires, the results on microscale must be transferred to macroscale which is usually done by homogenization (step 3). Certain problems require the introduction of further simulation levels. For instance, if cracks have to be

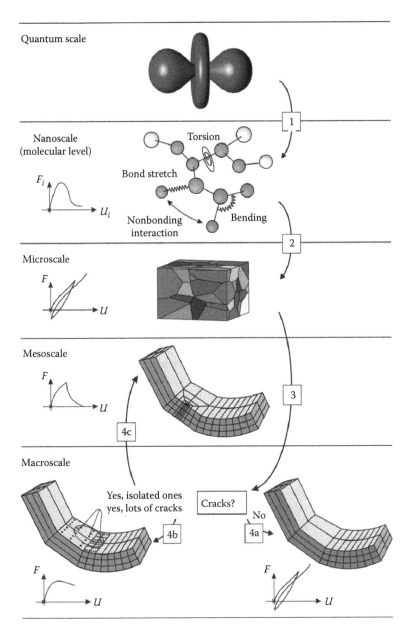

FIGURE 10.1
Simulation models at different length scales.

taken into account, they are often considered by mesoscale models (step 4c). If crack propagation can be neglected (step 4a) or if there are lots of microcracks which can be smeared (step 4b), macroscale models will suffice.

Regardless of the length scale, numerical models can be classified by the simulation technique used, namely, particle mechanics or continuum mechanics. In particle mechanics, the physical system is treated in a "discrete" way using particles and particle interactions. From a mathematical point of view, particles do not have to be atoms but can be seen in a more general way so that even whole galaxies can be simulated by means of particle mechanics, which is based on Newton's second law: The force acting on a body equals the product of mass and acceleration. Particle mechanics is based on force fields that use potentials to describe the interactions between two or more bodies. The most simple example is Hooke's law for linear springs which is derived from a so-called harmonic potential, a quadratic polynomial. Other well-known examples include the potential for gravitational attraction between objects with mass, the Coulomb potential for electrostatic forces, or van der Waals bonds and hydrogen bonds that are often analyzed by using the potential of Lennard-Jones (1929), a universal approach that considers both attraction and repulsion forces.

Simulations based on particle models, especially MD simulations, are usually very expensive in terms of computational time. Typical models to study crystal crack propagation comprise up to several million particles which at first glance seems to be a lot. However, compared to one cubic meter gas at a temperature of $273.15\,\mathrm{K}$ and a hydrostatic pressure of $p_\mathrm{n} = 101.325\,\mathrm{kPa}$ which contains $N_\mathrm{L} = 2.686763 \cdot 10^{25}\,\mathrm{m}^{-3}$ molecules (the Loschmidt constant) or compared to $12\,\mathrm{g}$ of the carbon isotope $\mathbf{C}_{12}$ which consists of $N_\mathrm{A} = 6.0221367 \cdot 10^{23}\,\mathrm{mol}^{-1}$ atoms (the Avogadro constant) or compared to the 200 billion stars comprising the Milky Way, it is obvious that not all problems can be handled by particle mechanics.

For homogeneous structures, continuum mechanical approaches can be applied which are capable of simulating engineering structures like bridges, airplanes, and cars. However, a simple question of how to explain crack propagation exceeds the capabilities of both particle and continuum mechanics, so that concurrent multiscale simulations are necessary.

A common approach to link a discrete atomic structure to a continuum region is to apply the Cauchy–Born hypothesis which goes back to Born and Huang (1954) and Ericksen (1984): The bond distance vector $\mathbf{r} = \mathbf{F}\mathbf{r}_0$ in the deformed configuration can be mapped from the bond distance vector $\mathbf{r}_0$ in the undeformed configuration by the deformation gradient tensor $\mathbf{F}$. $\mathbf{F}$ can be decomposed into a stretch and a rotation part, so that the bond length $|\mathbf{r}|$ can be given in terms of the right Cauchy–Green strain tensor. It should be noted that the Cauchy–Born rule involves an approximation because according to the continuum mechanics framework, line elements mapped by $\mathbf{F}$ from the undeformed to the deformed configuration must be infinitesimally small. Bond vectors, however, are of finite length. In order to overcome stability problems, unrealistic wave reflections and other numerical problems resulting from the transition region, various alternative formulations can be found in the literature like the exponential Cauchy–Born rule proposed by Arroyo and Belytschko (2002).

Among different solution techniques that can be applied to a continuum region, the finite element method (FEM) is the most versatile and widespread approach. Due to its very broad range of applications, it has become the dominating engineering simulation method. This leads to the desire to also integrate MD simulations to the well-established FEM framework. As shown by Nasdala and Ernst (2005) by the example of the Dreiding force field proposed by Mayo et al. (1990), the "molecular dynamic finite element method" (MDFEM) requires special finite elements if multi-body potentials have to be considered.

This chapter is addressed to users of finite element codes who want to learn more about MDFEM, its background in molecular mechanics, the new class of finite elements needed to obtain the same results as traditional MD software, applicable time integration schemes, and other implementation issues.

Two examples are provided. In order to demonstrate the robustness and reliability of MDFEM even for geometrically highly nonlinear problems, the first example is about carbon nanotubes with Stone–Wales defects that buckle when subjected to torsional loads. The second example illustrates the benefits for computational material scientists. It is shown that inelastic material behavior can be deduced from the rearrangement of bonds which makes the use of rheological elements obsolete.

10.2 Chemical and Physical Bonds

For a better understanding of MDFEM, particularly concerning the numerical effort, a brief overview of the different bond types is given in this chapter. There are two main classes:

- Chemical bonds, the so-called strong bonds:
  The three types of chemical bonds are ionic bonds, covalent bonds, and metallic bonds. The electrons of an atom's outermost orbital are called valence electrons. With the exception of the noble gases whose atoms' outer electron shells are completely filled, each atom tries to obtain a noble gas configuration by either gaining or losing valence electrons, depending on its electronegativity. The electronegativity difference between two atoms of a covalent bond is usually less than 1.7 eV. When the difference is 1.7 eV or greater, the bond is predominantly ionic. In a metallic bond, valence electrons are free to move through the crystalline lattice.
- Physical bonds or interactions, which are often referred to as weak bonds:
  The range of weak bondings comprises dipole–dipole interactions, dipole–ion interactions, van der Waals interactions, and hydrogen bondings. The distinction between these categories is ambiguous, for example, hydrogen bondings are often considered to be a special case of van der Waals bonds.

The term "orbital" refers to a mathematical function that describes the probability of an electron's position. The first and simplest one is a sphere and is denoted as s-orbital. It is followed by the p-orbital which has the shape of a dumbbell. Further examples include the d-orbital and the f-orbital.

From an energetic point of view, the overlap of the orbitals must be as large as possible. As shown in Figure 10.2, one s-orbital can mix with one, two, or three p-orbitals to form so-called hybrid orbitals. While the two sp^1-hybrid orbitals are located on a straight line, the three sp^2-hybrid orbitals are in a plane at an included angle of 120°. Between the four energetically equivalent sp^3-hybrid orbitals, which are also referred to as q-orbitals, there is a tetrahedral angle of 109.5°. For the chemical element carbon, all three types are possible: The two carbon atoms of acetylene are sp^1 hybridized, the carbon atoms of graphite have sp^2-hybrid orbitals, and to achieve the high strength of diamond, carbon must be sp^3 hybridized.

10.2.1 Covalent Bonding

A covalent bond can be formed by two atoms of the same kind as well as by different types of atoms. In contrast to ionic bonds, the electrons are not transferred completely but shared pairwise between the participating atoms. The overlapping atomic orbitals form joint electron clouds, the so-called molecular orbitals. Since they provide more space for the electrons, according to Heisenberg's uncertainty relation, molecular orbitals have a smaller impulse and therefore also a lower kinetic energy than the individual atomic orbitals.

If two s-orbitals or an s-orbital and a hybrid orbital are involved, we speak of a σ-bond. Each atom contributes one of its valence electrons to the common bond cloud. Diamond owes its high hardness to the fact that each of its carbon atoms forms four σ-bonds which are very strong due to the low kinetic energy. The interaction between two aligned p-orbitals is called π-bond.

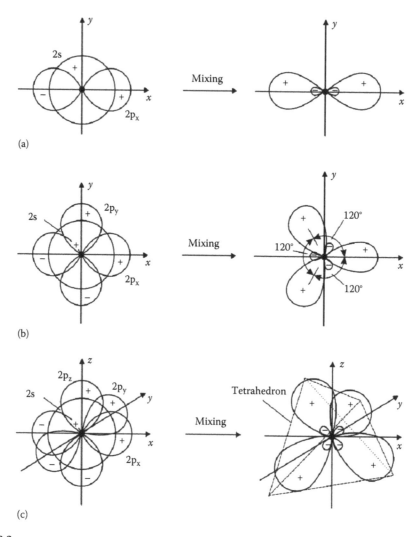

FIGURE 10.2
Orbital hybridizations of the second shell. (a) One s-orbital and one p-orbital combined to two sp^1-hybrid orbitals. (b) One s-orbital and two p-orbitals combined to three sp^2-hybrid orbitals. (c) One s-orbital and three p-orbitals combined to four sp^3-hybrid orbitals.

While single bonds such as the hydrogen–carbon bonds are σ-bonds, the valence electrons of double and triple bonds are also connected by one or two π-bonds. Figure 10.3 shows ethene (ethylene), a main component of many elastomeric materials, which has a double bond, and ethyne where each atom has to share three of its valence electrons. Multiple bonds are not very stable.

10.2.2 Ionic Bonding

When atoms have a large difference in electronegativity, one or more electrons can be completely transferred from the atom with distinct metallic properties to the atom with nonmetallic properties. This process results in a positively charged ion and a negatively

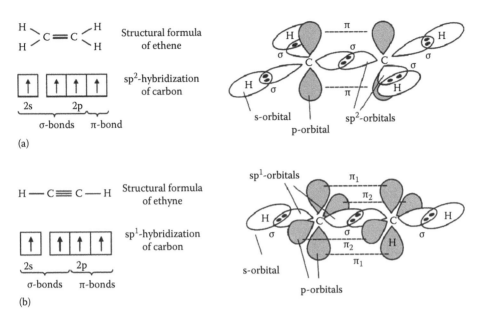

(a)

(b)

FIGURE 10.3
Double and triple bonds composed of σ- and π-bonds: (a) Ethene, C_2H_4. (b) Ethyne, C_2H_2.

charged ion being attracted to each other by electrostatic forces. A typical feature of ionic bonds is the formation of regular ionic crystals, for example, sodium chloride (NaCl), commonly known as table salt.

10.2.3 Metallic Bonding

In a metallic bond, the atoms are organized in a regular crystalline structure. In contrast to the previously introduced valence bonds, the electrons are no longer assigned to specific atoms. Instead, valence electrons separate from the atoms, which then become positively charged metallic ions, while the electrons can move freely and randomly through the crystalline lattice. This "electron gas" accounts for the high electrical and thermal conductivity of metals.

10.2.4 Van der Waals Bonding

The type of bonding that is named after the physicist Johannes Diderik van der Waals (1837–1923) describes the electrostatic interaction between dipoles. As shown in Figure 10.4, dipoles are caused by an uneven distribution of electron density within atoms or molecules. As a consequence, there is an offset between the center of mass of all electrons and the protons' center of mass, so that partial charges evolve. Opposite partial charges attract each other whereas equal partial charges repel. Depending on the electronegativity, some compounds have a permanent dipole while other molecules are initially nonpolar. When such nonpolar molecules approach a dipole, they become dipoles themselves, so-called induced dipoles.

Even if two nonpolar atoms or molecules approach one another, they can induce dipoles. The constant fluctuation of the electrons in a nonpolar molecule evokes a momentary

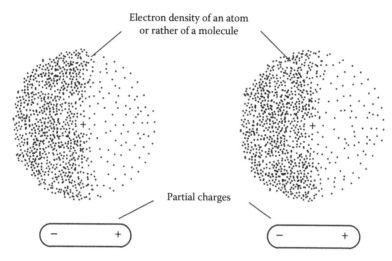

FIGURE 10.4
Van der Waals interaction (dipole–dipole interaction).

dipole for a short amount of time which may then polarize a neighboring molecule. These types of van der Waals bondings are often referred to as London dispersion forces and temporary dipoles as the electrical charges are oscillating back and forth.

Compared to an intramolecular force of a covalent or ionic bond, each van der Waals interaction is very weak and only significant when two atoms come close to one another. However, as an intermolecular force, van der Waals bondings can act simultaneously in various directions. Long hydrocarbon chains which can be found in fats and oils as well as in elastomers and polymers owe their elasticity and strength to the collective action of lots of van der Waals forces. They are responsible also for many other chemical properties such as the high boiling point of fats or the fact that even noble gas atoms unite as noble gas crystals—but only at very low temperatures. Van der Waals interactions are the only linkage between the different walls of a multi-walled carbon nanotube, see e.g., Xie et al. (2007) and Xie and Long (2006).

When the kinetic energy increases, the probability that atoms approach one another decreases. Hence, van der Waals bondings tend to break with rising temperatures. This kind of bond degradation often comes along with a phase transition from the solid to the liquid state or from the liquid to the gaseous state.

10.2.5 Hydrogen Bonding

Hydrogen bonding can be regarded as a special and important case of van der Waals bonding. When a hydrogen atom is covalently bonded to a strong electronegative atom, such as fluorine, oxygen, or nitrogen, very strong permanent dipoles arise because each hydrogen atom contains only one proton and therefore cannot attract the joint electron pair very well. As the hydrogen atom has a large positive partial charge, it can now form hydrogen bonds to other molecules with negative partial charges.

The part which provides the hydrogen atom is called donor while the other part is the acceptor. Examples of H-donors include –OH, –NH$_2$, or –COOH. Typical H-acceptors are compounds that contain oxygen or nitrogen atoms. Hydrogen bonds are considerably strong physical bondings. For this reason water is a very stable compound.

10.3 From Quantum to Molecular Mechanics

In the last section, different types of bonds have been introduced, however, without quantifying how strong they really are or how the equilibrium distances and angles between the individual atoms can be predicted. In theory, the exact behavior of all chemical and physical bonds can be computed with the help of the famous Schrödinger equation formulated by Erwin Schrödinger (1926). It is the central equation of quantum mechanics, as it describes the interactions between the electrons and the atomic nuclei accurately. An exact solution is available as of today only for the hydrogen atom. Therefore, a variety of approximation procedures has been developed. The range extends from ab initio methods over semi-empirical models up to classical MDs. A large body of literature exists on this topic (see Levine 1991, Pauling and Wilson Jr. 1985, and Shen and Atluri 2004). The most important methods will be discussed subsequently.

10.3.1 Schrödinger Wave Equation

A time-dependent and a time-independent version of the Schrödinger wave equation have been derived. For a system that consists of K nuclei and N electrons, the general time-dependent Schrödinger equation reads

$$\hat{E}\Psi(\mathbf{R}, \mathbf{r}, t) = \hat{H}\Psi(\mathbf{R}, \mathbf{r}, t) \tag{10.1}$$

with the energy operator

$$\hat{E} = i\hbar \frac{\partial}{\partial t} \tag{10.2}$$

and the Hamilton operator, the Hamiltonian

$$\hat{H}(\mathbf{R}, \mathbf{r}) = \underbrace{\frac{e^2}{4\pi\varepsilon_0} \sum_{k=j+1}^{N} \sum_{j=1}^{N} \frac{1}{|\mathbf{r}_k - \mathbf{r}_j|}}_{=\hat{V}_e} - \underbrace{\frac{e^2}{4\pi\varepsilon_0} \sum_{k=1}^{K} \sum_{j=1}^{N} \frac{Z_k}{|\mathbf{R}_k - \mathbf{r}_j|}}_{=\hat{V}_{eK}} +$$

$$+ \underbrace{\frac{e^2}{4\pi\varepsilon_0} \sum_{k=j+1}^{K} \sum_{j=1}^{K} \frac{Z_k Z_j}{|\mathbf{R}_k - \mathbf{R}_j|}}_{=\hat{V}_K} - \underbrace{\frac{\hbar^2}{2m_e} \sum_{j=1}^{N} \Delta_{\mathbf{r}_j}}_{=\hat{T}_e} - \underbrace{\frac{\hbar^2}{2} \sum_{k=1}^{K} \frac{1}{M_j} \Delta_{\mathbf{R}_j}}_{=\hat{T}_K}. \tag{10.3}$$

As solution of the general time-dependent Schrödinger equation, the state function Ψ $(\mathbf{R}, \mathbf{r}, t)$ depends on the coordinate vectors $\mathbf{R} = (\mathbf{R}_1, \mathbf{R}_2, ..., \mathbf{R}_K)$ and $\mathbf{r} = (\mathbf{r}_1, \mathbf{r}_2, ..., \mathbf{r}_N)$ of nuclei and electrons as well as on the time t and yields, according to Heisenberg's uncertainty principle, probability distributions of the location and the impulse of these particles. In terms of its structure, the Schrödinger equation can be seen in analogy to an electromagnetic wave as a wave equation. Hence, Ψ is often denoted as wave equation. On neglecting the spin, the external forces acting on the system, and relativistic interactions, the Hamiltonian $\hat{H}$ comprises the operators of the Coulomb potentials $\hat{V}_e$, $\hat{V}_{eK}$, and $\hat{V}_K$ as well as the operators of the kinetic energies $\hat{T}_e$ and $\hat{T}_K$. M_j and Z_j are the masses and charge numbers of the nuclei, m_e is the mass of an electron, ε_0 the electric field constant, $i = \sqrt{-1}$ the imaginary unit, and $\hbar = h/2\pi$ a natural constant based on Planck's quantum of action h. The Laplace operators $\Delta_{\mathbf{R}_j} = \dfrac{\partial^2}{\partial R_{j,x}^2} + \dfrac{\partial^2}{\partial R_{j,y}^2} + \dfrac{\partial^2}{\partial R_{j,z}^2}$ and $\Delta_{\mathbf{r}_j} = \dfrac{\partial^2}{\partial r_{j,x}^2} + \dfrac{\partial^2}{\partial r_{j,y}^2} + \dfrac{\partial^2}{\partial r_{j,z}^2}$ indicate that the kinetic energies depend on the curvature of the wave equation.

For reasons of simplicity, the Hamiltonian $\hat{H}$ in Equation 10.3 has been assumed to be independent of time t. In this case, a decoupling of the state function

$$\Psi(\mathbf{R}, \mathbf{r}, t) = \psi(\mathbf{R}, \mathbf{r}) \cdot f(t) \tag{10.4}$$

into a time-independent part $\psi(\mathbf{R}, \mathbf{r})$ and a time-dependent part $f(t)$ is possible. Substitution into Equation 10.1 yields

$$\psi(\mathbf{R}, \mathbf{r}) \cdot \hat{E} f(t) = f(t) \cdot \hat{H} \psi(\mathbf{R}, \mathbf{r}). \tag{10.5}$$

After division by $f(t) \neq 0$ and using

$$E = \frac{\hat{E} f(t)}{f(t)} \tag{10.6}$$

we obtain the stationary Schrödinger equation

$$\hat{H} \psi(\mathbf{R}, \mathbf{r}) = E \psi(\mathbf{R}, \mathbf{r}) \tag{10.7}$$

which is easier to solve. E in contrast to $\hat{E}$ is not an operator but a simple number. The total energy of the solution

$$E_{\text{tot}} = i\hbar \frac{1}{f(t)} \frac{\partial f(t)}{\partial t} \tag{10.8}$$

is both space- and time-dependent.

Although the stationary Schrödinger equation given in Equation 10.7 is not time-dependent, this does not imply that only static problems can be solved. It just means that the time dependence of the solutions is known. After multiplication with $f(t)$, Equation 10.8 represents a first-order differential equation. Its solution

$$f(t) = c \, \exp\left(-\frac{iE_{\text{tot}}t}{\hbar}\right) \tag{10.9}$$

yields the temporal evolution of the wave equation, simple sine or cosine oscillations.

An important feature of the Schrödinger equation is that new solutions can be obtained by superposition, which is a typical property of wave equations. For the time-dependent Schrödinger equation, Equation 10.1, the general solution approach for ψ^* and its corresponding complex conjugate ψ^* is given by

$$\Psi(\mathbf{R}, \mathbf{r}, t) = \sum_n c_n \exp\left(-\frac{iE_n t}{\hbar}\right) \psi_n(\mathbf{R}, \mathbf{r}) \tag{10.10}$$

with

$$c_n = \int \psi_n^*(\mathbf{R}, \mathbf{r}) \Psi(\mathbf{R}, \mathbf{r}, 0) d\mathbf{R} \, d\mathbf{r}. \tag{10.11}$$

To obtain a unique solution, initial conditions have to be defined.

10.3.2 Ab Initio Molecular Dynamics

Between both extremes, pure ab initio methods on the one hand, which can be used to solve the Schrödinger equation for hydrogen exactly, and classical MDs on the other hand, where atoms are treated as point masses, a variety of different approaches exists. In order to distinguish themselves from "empirical" classical MDs, many methods claim to be an "ab initio" approach. As different notations and ratings can be found in the literature, for nonphysicists, an evaluation of these methods can be quite challenging. The same approach can be denoted as "quantum mechanical method" by some authors, while others prefer to write "semiempirical method." Additional information on "ab initio methods" can be found in Marx and Hutter (2000).

10.3.2.1 Born–Oppenheimer Approximation

A basic simplification of the Schrödinger equation which is named after Max Born and Julius Robert Oppenheimer is to separate the equations of motion with regard to the electrons and the nuclei. It exploits the fact that electrons have a much lower mass and move faster than the nuclei, which allows for an immediate adaption to new nuclei positions. The Schrödinger equation of the nuclei is replaced by Newton's equation of motion. The remaining part is called the electronic Schrödinger equation.

Hence, with the help of the Born–Oppenheimer approximation, the Schrödinger equation can be reduced to a many-electron problem. To solve this problem, further approximations such as the Hartree–Fock approach (HF) or the density functional theory (DFT) are necessary.

10.3.2.2 Hartree–Fock Theory

The approach developed by Hartree (1928, 1932) and Fock (1930) reduces the many-electron problem to coupled one-electron problems. The electron–electron interaction terms are replaced by an averaged potential for each electron considering the average potential of the other electrons. On the other hand, the neglect of electron correlation effects is the main source of error of the HF method.

The HF method is comparable to the Ritz approach. In order to determine the Hamiltonian's ground state, trial functions within the framework of a variational principle are introduced. The so-called slater determinants of one-electron wave equations can be used as test functions, see Slater et al. (1969).

10.3.2.3 Density Functional Theory

The density functional theory is based on a publication by Hohenberg and Kohn (1964). They show that the electron energy in the ground state is not only well defined by a functional of the wave function but also by a functional of the electron density. On this basis, Kohn and Sham (1965) present the so-called local-density approximation (LDA) which can be used to approximate the electron ground-state energy. In combination with a variational principle for the density functional, this approach finally leads to the one-electron Schrödinger equations, also known as Kohn–Sham equations.

Compared to pure ab initio methods with high numerical effort of order $\mathcal{O}(N^4)$, a reduction to order $\mathcal{O}(N^3)$ can be achieved. Methods based on the DFT approach with a reduced effort of $\mathcal{O}(N^2)$ are Car–Parrinello MDs and the conjugate gradient method (CG) whereas the latter is considered to be a bit more efficient than the method developed by Car and Parrinello (1985).

10.3.3 Semiempirical Models

The time-dependent self-consistent field (TDSCF) approximation introduced by Dirac (1930) is based on an estimation of the electron distribution, which is used to determine the potential for each single electron with respect to the remaining electrons. Atomic nuclei are assumed to move according to the rules of classical mechanics. The electron's effective potential is also known as Ehrenfest potential, see Ehrenfest (1927) for details. The computed electron distribution acts as a starting point for a new iteration step, which is repeated until the atomic orbitals are determined with sufficient accuracy.

Originally introduced by Bloch (1928) and revised by Slater and Koster (1954), the tight-binding method is another typical representative of the semiempirical methods. A linear combination of atomic orbitals (LCAO) allows for a parametrization of the Hamiltonian such that the total energy and the eigenvalues of the electronic Schrödinger equation can be determined. Interatomic forces are computed by means of the Hellmann–Feynman theorem.

An alternative to semiempirical methods is the so-called hybrid QM/MD methods. As in a typical multiscale model, only a small portion of the system is treated quantum mechanically, while for the rest of the system MD simulations based on force fields are carried out.

10.3.4 Classical Molecular Dynamics

In classical MDs, the parameterized potential functions only depend on the nuclei positions. Not having to bother with the electrons, it is possible to simulate the interactions between many thousands or even many millions of atoms. In the following, the equations of motion are given in three different forms, which can be easily transformed into each other.

10.3.4.1 Newton's Equation of Motion

For a system of N atoms, the equations of motion, also known as Newton's second law, are given as

$$\mathbf{F}_j = \dot{\mathbf{I}}_j = M_j\ddot{\mathbf{R}}_j = -\nabla_{\mathbf{R}_j}V \quad \text{with } j = 1,\ldots,N \tag{10.12}$$

with the total empirical potential $V = V(\mathbf{R}_1, \mathbf{R}_2, \ldots, \mathbf{R}_j, \ldots, \mathbf{R}_N)$. $\mathbf{F}_j$, M_j, and $\mathbf{R}_j$ denote the inner forces, masses, and coordinates of atom j, respectively. $\mathbf{I}_j = M_j\dot{\mathbf{R}}_j$ is the momentum vector.

10.3.4.2 Lagrange's Equation of Motion

Using the Lagrangian

$$L = E_{\text{kin}} - V, \tag{10.13}$$

the Lagrange form of the equations of motion reads

$$\frac{d}{dt}\frac{\partial L}{\partial \dot{\mathbf{R}}_j} - \frac{\partial L}{\partial \mathbf{R}_j} = 0 \quad \text{with } j = 1, \ldots, N. \tag{10.14}$$

Utilizing the generalized coordinates $(q_1, q_2, q_3, q_4, \ldots) = (R_{1,x}, R_{1,y}, R_{1,z}, R_{2,x}, \ldots)$, it can be written as

$$\frac{d}{dt}\frac{\partial L}{\partial \dot{q}_i} - \frac{\partial L}{\partial q_i} = 0 \quad \text{with } i = 1, \ldots, 3N \tag{10.15}$$

and the kinetic energy

$$E_{\text{kin}} = \sum_{j=1}^{N} M_j \frac{\dot{\mathbf{R}}_j^2}{2} = \sum_{i=1}^{3N} M_i \frac{\dot{q}_i^2}{2}. \tag{10.16}$$

10.3.4.3 Hamilton's Equations of Motion

Hamilton's equations of motion comprise two first-order differential equations

$$\dot{p}_i = -\frac{\partial H}{\partial q_i} \quad \text{and} \quad \dot{q}_i = \frac{\partial H}{\partial p_i} \tag{10.17}$$

with the Hamilton's principal function

$$H = E_{\text{kin}} + V = -L + \sum_{i=1}^{3N} p_i \dot{q}_i \tag{10.18}$$

and the generalized momentum

$$p_i = \frac{\partial L}{\partial \dot{q}_i} \tag{10.19}$$

with $(p_1, p_2, p_3, p_4, \ldots) = (I_{1,x}, I_{1,y}, I_{1,z}, I_{2,x}, \ldots)$. Like the Hamiltonian (Equation 10.3), Hamilton's principal function (Equation 10.18) consists of both a kinetic energy E_{kin} contribution and a potential energy V contribution.

10.4 Molecular Dynamic Finite Element Method

10.4.1 Requirements for the Finite Element Method

Since MD simulations do not belong to the standard applications of FEM codes, the user has to make the necessary adjustments. In this section, all aspects of the MDFEM, in particular the differences to traditional MD software, are discussed from a FEM user's point of view.

10.4.1.1 Force Fields

In MDs, the potential energy which, in the previous chapter, is denoted by the variable V as it is usual in quantum mechanics to avoid confusion with the energy operator is denoted by the variable E and called force field. Force fields can be derived in two complementary ways (see Ercolessi and Adams 1994):

1. From quantum mechanical calculations
2. From experimental work

Well-known force field approaches include AMBER (Assisted Model Building with Energy Refinement, see Weiner and Kollman 1981), CHARMM (Chemistry at HARvard Molecular Mechanics, see Brooks et al. 1983), the "Molecular Mechanics" force fields MM1, MM2, MM3, and MM4 (Allinger and Chen 1996), ECEPP (Roterman et al. 1989), and UFF (Universal Force Field, see Rappé et al. 1992).

The main application of classical MD simulations is a calculation of equilibrium configurations within a so-called conformational analysis, without considering bond forming and bond rupture reactions, cf. Schlick (2002). Force field potentials are often extended and adapted in order to account for specific material aspects. The potential by Stillinger and Weber (1985) for instance is well suited and efficient for the approximation of crystalline silicon as it includes terms to enforce a diamond-like tetrahedral local structure which results in a more stable and realistic than compact structure. For the treatment of nonbonded interactions, the popular potential by Lennard-Jones (1929) is often preferred due to its efficiency, while, for example, the Buckingham (1938) potential introduces an exponential function of distance to capture the exchange repulsion stemming from the Pauli exclusion principle more precisely. Some approaches make use of the concept of local environment, with bond strengths depending on the bonding environment for example, the many-body potentials by Tersoff (1988) and Brenner (1990). ReaxFF by van Duin et al. (2001) replaces explicit bonds with bond orders to account for continuous formation and breaking of bonds.

In general, MDFEM can be applied to all kinds of force fields. However, it should be noted that many force fields are designed for a limited range of chemical elements or rather very special substances like proteins or peptides and cannot be used to examine other structures.

10.4.1.2 Short- and Long-Range Force Field Potentials

Most chemical and physical bonds can be described using pair potentials. From a numerical point of view, a distinction has to be made between short-range (bonded) and long-range (nonbonded) interactions. In the following, the implications for the FEM are discussed by the example of the Dreiding force field (Mayo et al. 1990).

The bond stretch between the atoms I and J can be expressed by a linear approach

$$E_B^{lin} = \frac{1}{2} k_e (R_{IJ} - R_e)^2 \tag{10.20}$$

(parabolic potential) that is well suited for small deformations. It depends on the natural bond length R_e, the interatomic distance R_{IJ} and the stiffness k_e. To account for nonlinear bending at large deformations, the function derived by Morse (1929)

$$E_B = D_e \left[\exp(-\alpha n (R_{IJ} - R_e)) - 1 \right]^2 \quad \text{with } \alpha = \frac{1}{n} \sqrt{\frac{k_e}{2D_e}} \tag{10.21}$$

is applied. The parameter D_e denotes the fracture energy. n is the bond order, for example, $n = 1$ for the σ-bonds of diamond, $n = 1.333$ for graphene, $n = 1.5$ for benzene, $n = 2$ for the π-bond of ethene or $n = 3$ for the π-bond of ethyne. The nonlinear Morse approach corresponds to the linear one if deformations are small $R_{IJ} \rightarrow R_e$.

Bond stretch can be seen as a quite stable connection between an atom and its closest neighbors. For instance, hydrogen has one neighbor, the carbon atoms of ethyne have two neighboring atoms, and the carbon atoms of graphene are neighbored by three atoms. The number of neighboring atoms is limited to four which is the case for diamond.

An approach widely used to describe long-range van der Waals interactions is given by Lennard-Jones (1929). The general form reads

$$E_{m,n}^{LJ}(R_{IJ}) = \frac{\varepsilon}{n-m} \left(\frac{n^n}{m^m} \right)^{\frac{1}{n-m}} \left[\left(\frac{\sigma}{R_{IJ}} \right)^n - \left(\frac{\sigma}{R_{IJ}} \right)^m \right] \quad \text{with } m < n. \tag{10.22}$$

While repulsion forces are expressed by the first term, the second term accounts for attraction forces. ε denotes the size of the forces and σ refers to the zero-crossing of the potential. Often the form $(m,n) = (6,12)$ is applied, whereas $m = 6$ reflects the actual van der Waals force for large distances, and $n = 12$ is chosen for simplicity reasons without a physical motivation. The resulting large repulsion forces avoid that two atoms come too close.

Substituting the parameters ε and $\sigma = \sqrt[6]{\frac{1}{2}} R_{e,vdW}$ with the fracture energy $D_{e,vdW}$ and the natural length $R_{e,vdW}$ leads to the (6,12)-form of the Lennard-Jones (LJ) potential

$$E_{vdW}^{LJ}(R_{IJ}) = D_{e,vdW} \left[\left(\frac{R_{IJ}}{R_{e,vdW}} \right)^{-12} - 2 \left(\frac{R_{IJ}}{R_{e,vdW}} \right)^{-6} \right]. \tag{10.23}$$

Hydrogen bondings are characterized with the help of a modified LJ approach. The general equation (10.22) is altered by substituting $(m,n) = (10,12)$ and multiplying with $\cos^4 \Theta_{DHA}$. The potential function then reads

$$E_{hb}(R_{DA}, \Theta_{DHA}) = D_{hb} \left[5 \left(\frac{R_{DA}}{R_{hb}} \right)^{-12} - 6 \left(\frac{R_{DA}}{R_{hb}} \right)^{-10} \right] \cos^4 \Theta_{DHA}. \tag{10.24}$$

Θ_{DHA} denotes the angle between the donor D, hydrogen H, and acceptor A. R_{DA} is the distance between the donor and the acceptor. Both material parameters, the modified fracture energy D_{hb} and the natural equilibrium distance R_{hb}, depend on the hydrogen bridge type and hence have to be fitted to experimental data.

Other force fields often use a simple (10,12) LJ approach

$$E_{hb}^{LJ}(R_{DA}) = D_{hb}\left[5\left(\frac{R_{DA}}{R_{hb}}\right)^{-12} - 6\left(\frac{R_{DA}}{R_{hb}}\right)^{-10}\right] \qquad (10.25)$$

to describe hydrogen bonding.

Coulomb or electrostatic interactions can be simulated by means of the Coulomb potential

$$E_Q(R_{IJ}) = \frac{1}{4\pi\varepsilon_0}\frac{q_I q_J}{R_{IJ}} \qquad (10.26)$$

for two point loads q_I and q_J with a distance R_{IJ}. ε_0 denotes the dielectric constant. Charges with equal algebraic signs repel whereas opposite charges attract each other.

In contrast to bond stretch, the long-range van der Waals, hydrogen, and Coulomb interactions connect an atom to a multitude of atoms. In theory, each atom has a direct connection to all atoms of the same structure. For practical applications, it is recommended to introduce cutoff distances or special algorithms in order to reduce the numerical effort of an MDFEM analysis:

1. Since interaction forces approach zero for large distances, bondings that exceed a certain range, for example, three times the natural length, can be neglected.

2. For a geometric nonlinear analysis, the selection should be revised in regular intervals as the interatomic distances are supposed to change.

3. For very large structures and long-range interactions that decay slowly with distance, it is necessary to bundle long-range interactions using special algorithms. For instance, the Ewald summation has been derived for the fast treatment of electrostatic interactions and is often used in biomolecular systems such as proteins and enzymes in a crystalline state, whereas methods based on multipole expansions are often used for nonperiodic systems, such as an enzyme in solution. For details, see Schlick (2002).

For static and implicit dynamic procedures, long-range potentials lead to a very large bandwidth of the stiffness matrix. Therefore, it should be considered to use an explicit time-integration scheme, even if the problem is quasi-static.

10.4.1.3 Multi-Body Force Field Potentials

Sophisticated force field approaches also include multi-body potentials. The simplest example is a three-body potential that considers an energy increase if the angle Θ_{IJK} between two bonds IJ and JK differs from the natural angle Θ_J^0. The resulting deformation mode is called bending. In analogy to the natural lengths introduced in the previous section, the natural angle yields the energy-free equilibrium configuration with regard to bending. For small deformations, the Dreiding force field approach for bending reads

$$E_A^{\text{lin}} = \frac{1}{2} K_{IJK} \left[\Theta_{IJK} - \Theta_j^0 \right]^2. \tag{10.27}$$

For large deformations, an extended cosine approach

$$E_A = \begin{cases} \dfrac{1}{2} C_{IJK} \left[\cos \Theta_{IJK} - \cos \Theta_j^0 \right]^2 & \text{for } \Theta_j^0 \neq 180° \\[2ex] K_{IJK} [1 + \cos \Theta_{IJK}] & \text{for } \Theta_j^0 = 180° \end{cases} \tag{10.28}$$

can be used which accounts for additional natural angles.

The Dreiding torsion energy

$$E_T = E_{IJKL} = \frac{1}{2} V_{IJKL} \left[1 - \cos \left[n_{JK} \left(\varphi - \varphi_{JK}^0 \right) \right] \right] \tag{10.29}$$

is a function of the dihedral angle $\varphi = \varphi_{IJKL}$ that is defined by the two planes IJK and JKL, the natural angle φ_{JK}^0, and the periodicity n_{JK}. A linearization of Equation 10.29 can be carried out in case of small deformations $\varphi \rightarrow \varphi_{JK}^0$. This leads to the torsion energy

$$E_T^{\text{lin}} = \frac{1}{4} V_{IJKL} n_{JK}^2 \left(\varphi - \varphi_{JK}^0 \right)^2. \tag{10.30}$$

As shown in Section 10.4.2, special user elements are required to consider multi-body potentials within the MDFEM framework.

10.4.1.4 Natural Lengths and Angles

Instead of studying the behavior of molecules under external loading, MD simulations usually aim to determine minimum energy conformations. In most cases, molecules can be seen as statically overdetermined systems and as such the equilibrium distances and angles of the overall structure usually differ from the natural lengths and angles which are also referred to as constitutive lengths and angles.

In order to utilize force fields in finite element simulations, MD finite elements must be able to "memorize" the natural distances and angles with the help of intrinsic material parameters. As a consequence, in contrast to standard elements, the coordinates of the initial configuration only have to be given approximately.

10.4.1.5 Linear and Nonlinear Analysis

It is obvious that the finite element code must be able to consider both geometrical and physical nonlinearities since interatomic bonds and interactions can undergo large motions and rotations leading to bond breakage or rather a rearrangement of bonds.

Nevertheless, it is often desirable to also have the opportunity to perform a linear analysis. As already mentioned and shown in Section 10.4.1.9 in more detail, the equilibrium configuration is generally unknown, even for an unloaded structure. A linear analysis using the harmonic force field potentials can improve the initial configuration at low numerical cost since no iteration is required. Then, the linear force field potentials are exchanged by their nonlinear counterparts and a nonlinear simulation step is performed to finish the conformational analysis.

10.4.1.6 Material Parameters

In structural mechanics, usually comprehensive tests are required to obtain material parameters for sophisticated material models such as a viscoplastic damage model. MDFEM users, however, do not have to worry about expensive and time-consuming testing issues because all the parameters are already provided by the force fields. The different force field approaches can be seen as large databases from which the required parameters can be extracted.

It is recommended to start with a so-called universal force field which usually includes the complete range of chemical elements. Force fields developed for special applications such as the analysis of proteins or DNAs may give better results if similar structures are to be examined, but are often limited to a small subset of the periodic table of the chemical elements. Hence, before applying a specialized force field, the user has to check whether all chemical elements of his structure are involved.

To evaluate the convergence of the numerical solution, FE solvers use error tolerances which are based either on relative or on absolute values for the displacements and forces. The latter case is inadmissible if SI units are used because, among other reasons, the tolerances would be too high so that false results could be accepted as true. Therefore, a general recommendation is to replace the SI units m, N, kg, and s by the following unit system: $nm = 10^{-9}\,m = 10\,\text{Å}$, $nN = 10^{-9}\,N$, $akg = 1$ atto $kg = 10^{-18}\,kg$, and $ns = 10^{-9}\,s$.

10.4.1.7 Finite Elements without Rotational Degrees of Freedom

In MDs, atoms are treated as point masses which possess only translational degrees of freedom, implying that only finite elements without rotational degrees of freedom should be used. However, a common approach is to describe the atomic interactions by means of standard beam elements which have rotational degrees of freedom, cf. Wang and Wang (2004), Tserpes and Papanikos (2005), Harik (2002), and Li and Chou (2004). At this point, it must be stressed that beam models can only be regarded as a work-around, being used for simplicity reasons.

As shown in Section 10.4.2, special user elements with the following characteristics have to be introduced:

- To be able to decouple bending and torsion energies, the finite elements must not have rotational degrees of freedom.
- The force field parameters can be applied directly.
- Compared to beam models that use rotational degrees of freedom, only half the number of degrees of freedom is necessary which is much more efficient.

Note that for a beam element model, the force field parameters have to be transformed to normal, bending, and torsional stiffnesses and strengths. And what is more: It can be shown that the specification of the bending and torsion parameters is ambiguous because even for the pure torsion mode depicted in Figure 10.6e some beam elements, namely, the cantilever beams, are subjected to bending. That means, torsion between the planes of two adjacent atom groups *IJK* and *JKL* can be considered both by torsion of the beam *JK* and by bending of the beams *IJ* and *KL*.

10.4.1.8 Time Integration Schemes

The equation of motion can be written in the general form

$$\mathbf{R} = \mathbf{M}\ddot{\mathbf{u}} + \mathbf{I} - \mathbf{P} = 0 \tag{10.31}$$

with the residual vector $\mathbf{R}$, the mass matrix $\mathbf{M}$, the acceleration vector $\ddot{\mathbf{u}}$, and the vectors of the internal and external forces $\mathbf{I} = \mathbf{I}(\mathbf{u},\dot{\mathbf{u}})$ and $\mathbf{P}(\mathbf{u},\dot{\mathbf{u}})$.

In MDs, often explicit time integration schemes are applied. Verlet (1967) developed one of the most popular algorithms, which can be derived using a double Taylor series with two time points $t_0 - \Delta t$ and $t_0 + \Delta t$, that is truncated after the cubic term. However, the Verlet algorithm is not available in most finite element codes. In addition, it is unusual to apply multi-step algorithms such as the higher order Runge–Kutta methods. Considering this, we concentrate on the three most important time integration schemes available in commercial finite element codes:

1. *The implicit HHT method*: The HHT method is an extension of the implicit time integration scheme developed by Newmark (1959) that is based on a Taylor series expansion of the displacements

$$\mathbf{u}_{n+1} = \mathbf{u}_n + \Delta t \dot{\mathbf{u}}_n + \Delta t^2 \left[\left(\frac{1}{2} - \beta \right) \ddot{\mathbf{u}}_n + \beta \ddot{\mathbf{u}}_{n+1} \right] \tag{10.32}$$

and the velocities

$$\dot{\mathbf{u}}_{n+1} = \dot{\mathbf{u}}_n + \Delta t [(1 - \gamma)\ddot{\mathbf{u}}_n + \gamma \ddot{\mathbf{u}}_{n+1}] \tag{10.33}$$

where quadratic and higher order terms are approximated by a quadrature rule. After rearranging the equations, we obtain the displacements at time t_{n+1}

$$\dot{\mathbf{u}}_{n+1} = \frac{\gamma}{\beta \Delta t}(\mathbf{u}_{n+1} - \mathbf{u}_n) + \left(1 + \frac{\gamma}{\beta} \right) \dot{\mathbf{u}}_n + \Delta t \left(1 - \frac{\gamma}{2\beta} \right) \ddot{\mathbf{u}}_n \tag{10.34}$$

and the accelerations at time t_{n+1}

$$\ddot{\mathbf{u}}_{n+1} = \frac{1}{\beta(\Delta t)^2}(\mathbf{u}_{n+1} - \mathbf{u}_n) - \frac{1}{\beta \Delta t}\dot{\mathbf{u}}_n + \left(1 - \frac{1}{2\beta} \right) \ddot{\mathbf{u}}_n. \tag{10.35}$$

The accuracy and stability depends on the choice of the quadrature parameters $\beta \in]0,1]$ and $\gamma \in]0,1]$. For example, if $\beta = \frac{1}{4}$ and $\gamma = \frac{1}{2}$ (constant acceleration) or $\beta = \frac{1}{6}$ and $\gamma = \frac{1}{2}$ (linear acceleration), the Newmark method is unconditionally stable, that is, its robustness is independent of Δt.

Hilber et al. (1978) extended Newmark's method by introducing the parameter α

$$\mathbf{M}\ddot{\mathbf{u}}\big|_{t_{n+1}} + (1+\alpha)(\mathbf{I}-\mathbf{P})\big|_{t_{n+1}} - \alpha(\mathbf{I}-\mathbf{P})\big|_{t_n} = 0 \tag{10.36}$$

which shifts the force vector term from the new time point t_{n+1} to an intermediate time point $t_{n+1+\alpha}$ with $\alpha \in \left[-\frac{1}{3}, 0\right]$. This damps high frequencies and stabilizes the time integration scheme. A typical value is $\alpha = -\frac{1}{20}$. For $\alpha=0$, the HHT algorithm is equivalent to Newmark's method.

2. *The implicit Euler backward method*: Like the HHT method, the Euler backward method is an implicit time integration scheme, which implies that displacements

$$\mathbf{u}_{n+1} = \mathbf{u}_n + \Delta t\,\dot{\mathbf{u}}_{n+1} \tag{10.37}$$

and velocities

$$\dot{\mathbf{u}}_{n+1} = \dot{\mathbf{u}}_n + \Delta t\,\ddot{\mathbf{u}}_{n+1} \tag{10.38}$$

have to be updated iteratively. The Euler backward method is unconditionally stable.

When time increments become large, the dynamic response is affected by numerical damping, which is more distinct compared to the HHT method. Hence, the HHT integration scheme is a good choice, when dynamic effects are important, while the Euler backward method should be preferred over HHT for a quasi-static analysis.

3. *The explicit midpoint method*: No iterations are required for the explicit time integration scheme. Velocities

$$\dot{\mathbf{u}}_{n+1/2} = \dot{\mathbf{u}}_{n-1/2} + \frac{\Delta t_{n+1} + \Delta t_n}{2}\ddot{\mathbf{u}}_n \tag{10.39}$$

and subsequently the displacements

$$\mathbf{u}_{n+1} = \mathbf{u}_n + \Delta t_{n+1}\dot{\mathbf{u}}_{n+1/2} \tag{10.40}$$

are updated by means of the explicit midpoint rule. Multiplying the equation of motion, Equation 10.31, with the inverse mass matrix $\mathbf{M}^{-1}$ yields the accelerations

$$\ddot{\mathbf{u}}_{n+1} = \mathbf{M}^{-1}(\mathbf{P}_{n+1} - \mathbf{I}_{n+1}) \tag{10.41}$$

as a function of the internal forces $\mathbf{I}_{n+1} = \mathbf{I}(\mathbf{u}_{n+1}, \dot{\mathbf{u}}_{n+1/2})$ and the external forces $\mathbf{P}_{n+1} = \mathbf{P}(\mathbf{u}_{n+1}, \dot{\mathbf{u}}_{n+1/2})$. Since atoms are regarded as point masses, $\mathbf{M}$ is a diagonal mass matrix and thus can be easily inverted.

The stability and accuracy depends on the time increment Δt. For linear problems, if the harmonic force field potentials are chosen, the optimal time increment size

$$\Delta t = \frac{2}{\omega_{max}} \tag{10.42}$$

is limited by the highest eigenfrequency ω_{max}. Since for systems including a large number of atoms, it is not feasible to determine ω_{max}, the highest element eigenfrequency or rather "bond eigenfrequency" ω_{max}^{elem} can be used instead. In structural mechanics, this corresponds to a time increment size of

$$\Delta t = \frac{L_{min}}{c} \tag{10.43}$$

where
L_{min} is the smallest characteristic element dimension
c is the dilatational wave speed c of the material

Based on the angular frequency $\omega = \sqrt{k/m}$ of the simple harmonic oscillator, we propose to update the time increment for each increment in accordance with Equation 10.42

$$\Delta t = 2\eta\sqrt{\frac{m_{IJ}}{|k_{IJ}|}} \quad \text{with } 0 < \eta \le 1 \tag{10.44}$$

where k_{IJ} and m_{IJ} denote the current bond stiffness, which may also be negative, and the mass of the two atoms I and J. The parameter η controls accuracy and efficiency of the explicit time integration scheme, for example, $\eta = 0.5$.

10.4.1.9 Relaxation Step to Determine the Reference Configuration

Usually, the equilibrium configuration of an atomic structure is unknown because the interatomic distances and angles differ from the natural bond distances and angles. From a numerical point of view, damping can help to determine the equilibrium state. The conformational analysis then becomes a "relaxation step." This technique can be compared to a conventional finite element analysis where initial stresses or bolt pretensions are applied in a first analysis step. In general, equilibrium configurations can be obtained in three different ways:

1. For statically determinate systems or symmetric structures such as graphene, it is possible to specify the coordinates of the atoms exactly.

2. If the structure is statically indeterminate, but the equilibrium configuration can be given approximately, a simple static analysis is sufficient. As an example, consider a (10,10) armchair carbon nanotube with a Stone–Wales defect. The initial configuration is shown in Figure 10.5, top left. Using the nonlinear force field potentials, within a single static analysis step the equilibrium state shown top right can be reached. If more than one equilibrium configuration exists, a two-step static conformational analysis is recommended starting with the harmonic potentials.

3. For complex structures with multiple equilibrium states, convergence of a pure static analysis is quite unlikely. Even a dynamic analysis usually faces numerical problems because the initial potential energies of the bonds can be fairly high if the distance between two atoms is too close which often cannot be avoided. The high potential energy then is transformed to kinetic energy resulting in enormous oscillations or rather "explosions."

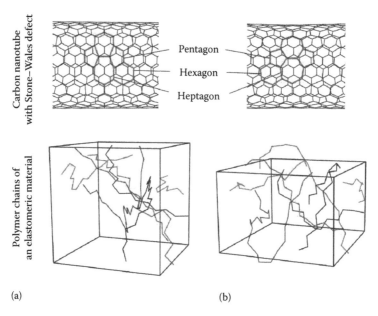

FIGURE 10.5
Reorganization of bonds during conformational analysis: (a) generated initial meshes and (b) equilibrium states after relaxation step.

An example where numerical damping has to be introduced to dissipate the energy is elastomeric material. Figure 10.5, bottom left, shows seven polymer chains which are interconnected by van der Waals forces. While the chemical bonds between the 107 atoms can be modeled in accordance with the natural lengths and angles, the van der Waals bonding lengths are generated more or less randomly. The equilibrium state shown bottom right is simulated using the Euler backward method in combination with an automatic time incrementation algorithm. Even though no additional damping is introduced, the quasi-static solution can be achieved in "only" about 250 time increments.

For static analyses and dynamic analyses using a different time integration scheme, it is recommended to start with large values for the damping parameters, for example, for a Rayleigh damping model. During the relaxation step, damping should be reduced in accordance with the bonds' energies until a valid quasi-static equilibrium state is reached. The simulation times are comparable with the Euler backward scheme.

10.4.2 User Elements

Since conventional finite elements do not meet the requirements listed in the previous section, in the following, a new class of finite elements is introduced that can be used for MD simulations within the FEM framework. As pointed out in Section 10.4.1.7, the developed finite elements which are presented in Figure 10.6 only use translational degrees of freedom.

Bending and torsional moments are applied by means of force couples. The lever arms depend on the bond lengths R_{IJ}^0, R_{JK}^0 and R_{KL}^0 and angles Θ_J^0 and Θ_K^0 shown in Figure 10.6a. The force directions are described using the unit vectors depicted in Figure 10.6b. The two-, three-, and four-node elements for bond-stretch, bending, and torsion given in Figure 10.6c through e, have been implemented in the finite element codes Feap and Abaqus using a superposition technique.

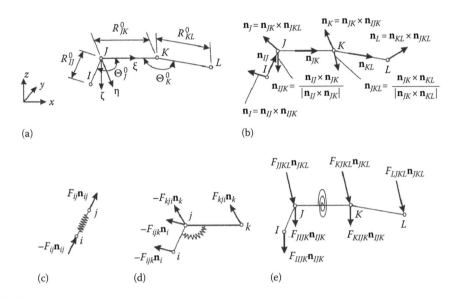

FIGURE 10.6
Molecular dynamic finite elements without rotational degrees of freedom: (a) coordinate systems and dimensions, (b) unit vectors, (c) two-node spring element, (d) three-node bending element, and (e) four-node torsion element.

In this section, the force vectors, the so-called right-hand side vectors are given. They are obtained by derivatives of the force field potentials with respect to the atomic coordinates and have to be defined for all analysis types. In contrast to the explicit time integration scheme which does not use the Newton–Raphson method given in Appendix A, user elements for static and implicit dynamic analyses also require stiffness matrices. They can be found in Appendix 10B.

10.4.2.1 Two-Node Element for Bond Stretch

The element force vector of the two-node element includes the node vectors of the atoms i and j

$$\mathbf{F}_i = -F_{ij}\mathbf{n}_{ij}, \quad \mathbf{F}_j = -\mathbf{F}_i \tag{10.45}$$

and can be used for both chemical and physical bonds. It is defined by its unit direction vector $\mathbf{n}_{ij}$ and the magnitude F_{ij} which depends on the force field potential. The derivative of Equation 10.20 yields

$$F_{ij}^{\text{lin}} = k_{ij}\left(|\mathbf{x}_j - \mathbf{x}_i| - R_{ij}^0\right). \tag{10.46}$$

From the Morse potential (Equation 10.21) we get

$$F_{ij}^{\text{Morse}} = -2\alpha_{ij}n_{ij}D_{ij}\left[\exp\left(-2\alpha_{ij}n_{ij}\left(|\mathbf{x}_j - \mathbf{x}_i| - R_{ij}^0\right)\right)\right.$$
$$\left. - \exp\left(-\alpha_{ij}n_{ij}\left(|\mathbf{x}_j - \mathbf{x}_i| - R_{ij}^0\right)\right)\right]. \tag{10.47}$$

For the LJ approach Equation 10.23, the force magnitude is given as

$$F_{ij}^{LJ} = 12 \frac{D_{ij,\text{vdW}}}{R_{ij,\text{vdW}}^0} \left[-\left(\frac{|\mathbf{x}_j - \mathbf{x}_i|}{R_{ij,\text{vdW}}^0} \right)^{-13} + \left(\frac{|\mathbf{x}_j - \mathbf{x}_i|}{R_{ij,\text{vdW}}^0} \right)^{-7} \right] \qquad (10.48)$$

and the electrostatic force derived from Equation 10.26 reads

$$F_Q(R_{IJ}) = -\frac{1}{4\pi\varepsilon_0} \frac{q_I q_J}{R_{IJ}^2}. \qquad (10.49)$$

Note that Coulomb forces can be neglected for the examples given in Section 10.5.

10.4.2.2 Three-Node Element for Bending

The force vector of the three-node bending element consists of three components:

$$\mathbf{F}_i = F_{ijk}\mathbf{n}_i, \quad \mathbf{F}_k = F_{kji}\mathbf{n}_k, \quad \mathbf{F}_j = -\mathbf{F}_i - \mathbf{F}_k. \qquad (10.50)$$

The magnitude F_{ijk} results from the derivative of the harmonic potential (Equation 10.27)

$$F_{ijk}^{\text{lin}} = \frac{K_{ijk}}{R_{ij}^0} \left[\Theta_{ijk} - \Theta_j^0 \right] \qquad (10.51)$$

or the extended cosine approach (Equation 10.28)

$$F_{ijk} = \begin{cases} -\dfrac{C_{ijk}}{R_{ij}^0} \left[\cos\Theta_{ijk} - \cos\Theta_j^0 \right] \sin\Theta_{ijk} & \text{for } \Theta_j^0 \neq 180° \\[2ex] -\dfrac{K_{ijk}}{R_{ij}^0} \sin\Theta_{ijk} & \text{for } \Theta_j^0 = 180° \end{cases} \qquad (10.52)$$

with respect to the unknown bending angle

$$\Theta_{ijk} = \arccos(-\mathbf{n}_{ij} \cdot \mathbf{n}_{jk}) \in [0°,\ 180°]. \qquad (10.53)$$

The unit vectors $\mathbf{n}_{ij}$ and $\mathbf{n}_{jk}$ defining Θ_{ijk} can be taken from Figure 10.7. Considering that the lever arms are generally different, if follows for the magnitude

$$F_{kji} = F_{ijk} \frac{R_{ij}^0}{R_{jk}^0}. \qquad (10.54)$$

At this point it should be noted that angles between 180° and 360° can be omitted. For symmetry reasons, they are covered by the torsion potential with its 0° - and 180°-equilibrium angles. For example, the combination $\Theta_{ijk} = 190°$ and $\varphi_{IJKL} = 4°$ is equivalent to $\Theta_{ijk} = 170°$ and $\varphi_{IJKL} = 184°$.

Instead of Equation 10.53 for the computation of the bending angle, many authors suggest to make use of the formula $\Theta_{ijk} = 180° - \arcsin |\mathbf{n}_{ij} \times \mathbf{n}_{jk}|$. However, it is necessary to

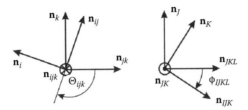

FIGURE 10.7
Bending and torsion angles derived from unit vectors.

point out that this approach is not suitable for large deformations because it only covers values between 90° and 180°.

10.4.2.3 Four-Node Element for Torsion

As shown in Figure 10.6e, there are six concentrated forces acting on the four-node torsion element whose magnitudes generally differ from each other. The element vector

$$\mathbf{F}_I = F_{IIJK}\mathbf{n}_{IJK}$$

$$\mathbf{F}_L = F_{LJKL}\mathbf{n}_{JKL}$$

$$\mathbf{F}_J = F_{JIJK}\mathbf{n}_{IJK} + F_{JJKL}\mathbf{n}_{JKL} = \alpha_{JIJK}\,\mathbf{F}_I + \frac{\alpha_{JJKL}}{\alpha_{LJKL}}\,\mathbf{F}_L \tag{10.55}$$

$$\mathbf{F}_K = F_{KIJK}\mathbf{n}_{IJK} + F_{KJKL}\mathbf{n}_{JKL} = \alpha_{KIJK}\,\mathbf{F}_I + \frac{\alpha_{KJKL}}{\alpha_{LJKL}}\,\mathbf{F}_L$$

contains four node vectors. While the first index of the force magnitudes denotes the node the force is acting on, indices 2–4 refer to one of the two planes *IJK* and *JKL*, represented by the unit vectors $\mathbf{n}_{IJK}$ and $\mathbf{n}_{JKL}$, see Figure 10.7.

Considering the lever arm $R_{IJ}^0 \sin\Theta_J^0$, the magnitude F_{IIJK} follows from the derivative of Equation 10.30

$$F_{IIJK}^{\text{lin}} = -\frac{V_{IJKL}\, n_{JK}^2}{2R_{IJ}^0 \sin\Theta_J^0}\left(\varphi_{IJKL} - \varphi_{JK}^0\right) \tag{10.56}$$

or Equation 10.29

$$F_{IIJK} = -\frac{V_{IJKL}\, n_{JK}}{2R_{IJ}^0 \sin\Theta_J^0}\sin\left[n_{JK}\left(\varphi_{IJKL} - \varphi_{JK}^0\right)\right] \tag{10.57}$$

with respect to the dihedral angle φ_{IJKL}. For the determination of φ_{IJKL}, a case distinction is required depending on the number of natural angles φ_{JK}^0; for example, for graphene $\varphi_{JK}^0 = 0°$ if $\mathbf{n}_{IJK} \cdot \mathbf{n}_{JKL} \geq 0$ or $\varphi_{JK}^0 = 180°$ if $\mathbf{n}_{IJK} \cdot \mathbf{n}_{JKL} < 0$, and thus:

$$\varphi_{IJKL} = \begin{cases} +\arcsin[(\mathbf{n}_{IJK} \times \mathbf{n}_{JKL})\cdot\mathbf{n}_{JK}] \in [-90°, +90°] & \text{for } \mathbf{n}_{IJK}\cdot\mathbf{n}_{JKL} \geq 0 \\ 180° - \arcsin[(\mathbf{n}_{IJK} \times \mathbf{n}_{JKL})\cdot\mathbf{n}_{JK}] \in [+90°,+270°] & \text{for } \mathbf{n}_{IJK}\cdot\mathbf{n}_{JKL} < 0. \end{cases} \tag{10.58}$$

Furthermore, it is necessary to project the cross product $\mathbf{n}_{IJK} \times \mathbf{n}_{JKL}$ on the associated unit vector $\mathbf{n}_{JK}$ in order to determine the sign of φ_{IJKL} for $\varphi_{JK}^0 = 0°$ or rather if φ_{IJKL} is larger or smaller than 180° for $\varphi_{JK}^0 = 180°$.

From simple geometric considerations, we get the magnitudes

$$F_{KIJK} - \underbrace{\frac{R_{IJ}^0 \cos\Theta_J^0}{R_{JK}^0}}_{= \alpha_{KIJK}} F_{IIJK}, \qquad F_{JIJK} = -F_{IIJK} - F_{KIJK} = \underbrace{(-1 - \alpha_{KIJK})}_{= \alpha_{JIJK}} F_{IIJK} \qquad (10.59)$$

associated with plane *IJK* and the forces

$$F_{LJKL} = -\underbrace{\frac{R_{IJ}^0 \sin\Theta_J^0}{R_{KL}^0 \sin\Theta_K^0}}_{= \alpha_{LJKL}} F_{IIJK}$$

$$F_{JJKL} = -\underbrace{\frac{R_{KL}^0 \cos\Theta_K^0}{R_{JK}^0}}_{} F_{LJKL} = \underbrace{\left(-\alpha_{LJKL} \frac{R_{KL}^0 \cos\Theta_K^0}{R_{JK}^0}\right)}_{= \alpha_{JJKL}} F_{IIJK} \qquad (10.60)$$

$$F_{KJKL} = -F_{LJKL} - F_{JJKL} = \underbrace{(-\alpha_{LJKL} - \alpha_{JJKL})}_{= \alpha_{KJKL}} F_{IIJK}$$

which act normal to plane *JKL*.

To avoid a coupling of the different energy potentials, changes of the lever arms due to changes of the equilibrium distances are neglected.

10.4.3 Inversion

Some force fields such as the Dreiding approach also include potential energies for inversion. In geometry, inversion in a point or point reflection is equivalent to a 180°-rotation and a reflection at a plane, as illustrated in Figure 10.8. In MDs, inversion is only relevant when at least four atoms are involved which must be arranged in a specific manner. In contrast to torsion, where the four atoms form a chain-like structure, three atoms form a triangle through which the fourth may pass. Hence, for some structures such as elastomeric material consisting of several polymer chains, inversion is irrelevant.

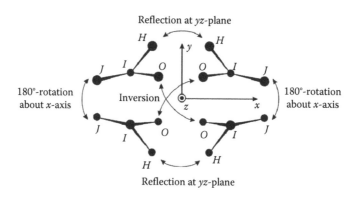

FIGURE 10.8
Inversion as a combination of rotation and reflection.

It is not trivial to set up an appropriate potential since inversion can be separated from the other energy forms only in very rare cases such as for ammonia, NH_3. For most molecules, interactions between inversion and torsion energy and, for spacial structures, also bending energy has to be considered. Therefore, some approaches such as the CHARMM force field represent inversion by "improper" torsion, that is, inversion is incorporated by means of torsion energy.

10.4.3.1 Spacial Structures

From a mechanical point of view, inversion of spacial structures such as an ammonia molecule, NH_3, is a snap-through problem, as illustrated in Figure 10.9. As opposed to planar structures, spatial structures have two equilibrium states or even more if multiple local inversions can occur. Ammonia has two natural bending angles: $\Theta_j^0 = 106.7°$ and $2 \cdot 120° - 106.7° = 133.3°$.

10.4.3.2 Planar Structures

For planar structures such as graphene, inversion refers to the motion of an atom out of the plane defined by its three neighbors, cf. Figure 10.10a. The out-of-plane displacement w of node I is expressed by the inversion angle

$$\Psi = \Psi_1 + \Psi_r = \arctan\left(\frac{2w}{R_e}\right) + \arctan\left(\frac{w}{R_e}\right) \tag{10.61}$$

which corresponds to the angle between plane *HIO* and bond *IJ*. Due to the structure's symmetry, the same angle can be found between plane *IJO* and bond *HI* and between plane *IJH* and bond *IO*. As depicted in Figure 10.10b, the deformation leads to so-called improper torsion, for the atoms *I*, *J*, *K*, and *L*.

Figure 10.11 illustrates that a total of 24 torsion angles are affected when atom I is moved out of plane *HJO*. For small deformations, the following torsion angles emerge:

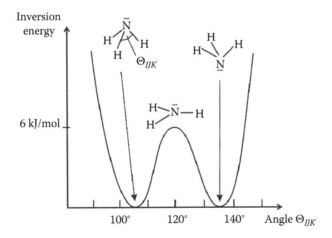

FIGURE 10.9
Energy profile during snap-through (inversion) of a NH_3-molecule.

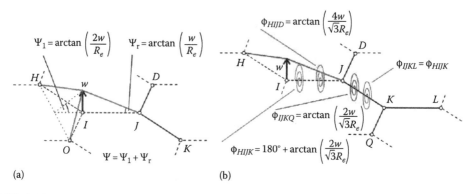

(a)   (b)

FIGURE 10.10
Improper torsion resulting from inversion for the example of graphene: (a) inversion and (b) additional torsion.

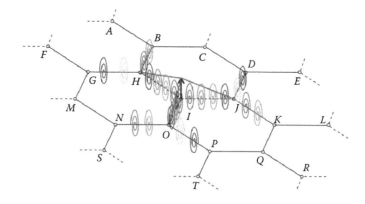

FIGURE 10.11
Torsion angles affected by deflection of atom I out of plane defined by atoms H, J, and O according to Equation 10.62.

$$\varphi_{HIJD} = \varphi_{OIHG} = \varphi_{JIOP} = +\frac{4w}{\sqrt{3}R_e}, \quad \varphi_{OIJK} = \varphi_{JIHB} = \varphi_{HION} = -\frac{4w}{\sqrt{3}R_e}$$

$$\varphi_{IJKQ} = \varphi_{IHBC} = \varphi_{IONM} = +\frac{2w}{\sqrt{3}R_e}, \quad \varphi_{IJDC} = \varphi_{IHGM} = \varphi_{IOPQ} = -\frac{2w}{\sqrt{3}R_e}$$

$$\varphi_{HIJK} = \varphi_{OIHB} = \varphi_{JION} = \varphi_{IJKL} = \varphi_{IHBA} = \varphi_{IONS} = 180° + \frac{2w}{\sqrt{3}R_e}$$

$$\varphi_{OIJD} = \varphi_{JIHG} = \varphi_{HIOP} = \varphi_{IJDE} = \varphi_{IHGF} = \varphi_{IOPT} = 180° - \frac{2w}{\sqrt{3}R_e}.$$

(10.62)

10.4.3.3 Force Field Potentials

In spectroscopy, the calculation of the inversion energy is often carried out with the help of the linear approach

$$E_I^{lin} = \frac{1}{2}K_{inv}(\Psi - \Psi_0)^2.$$

(10.63)

To address the fact that the derivative of the inversion energy has to vanish, $dE_I/d\Psi = 0$, when reaching the snap-through point $\Psi = 0°$, Mayo et al. (1990) suggest the following formulation:

$$E_I = \begin{cases} \dfrac{1}{2}C_{inv}(\cos\Psi - \cos\Psi_0)^2 & \text{for} \quad \Psi_0 \neq 0° \\[2mm] K_{inv}(1 - \cos\Psi) & \text{for} \quad \Psi_0 = 0° \end{cases} \tag{10.64}$$

with the stiffness $K_{inv} = \sin^2\Psi_0 C_{inv}$; for example, $K_{inv} = 40 \dfrac{\text{kcal}}{\text{mol rad}^2}$ for graphene. The approach distinguishes between spatial structures $\Psi_0 \neq 0°$ and planar structures $\Psi_0 = 0°$. For spatial structures, the maximum energy at the snap-through point $\Psi = 0°$ is

$$E_I^{bar} = 2C_{inv}\sin^2\left(\frac{1}{2}\Psi_0\right), \tag{10.65}$$

for example, $E_I^{bar} = 6\dfrac{\text{kcal}}{\text{mol}}$ for ammonia.

10.4.3.4 Modeling Inversion by Improper Torsion

For planar structures, the inversion energy can be taken into account using additional torsion energy. For demonstration purposes, we consider the example of graphene shown in Figures 10.10 and 10.11. The deflection w of atom I leads to an inversion energy with the direct neighbors H, J, and O. As a consequence, atom J moves out of its plane IDK by $-w/3$, etc.

With the inversion and torsion angles given in Equations 10.61 and 10.62, the ratio of inversion energy to torsion energy for small deformations can be derived as

$$\frac{E_{I,ges}^{lin}}{E_{T,ges}^{lin}} = \frac{E_I^{lin}(\Psi_1 + \Psi_r) + 3E_I^{lin}(\Psi_r)}{6E_T^{lin}(\varphi_{HIJD}) + 18E_T^{lin}(\varphi_{IJKQ})} =$$

$$= \frac{\dfrac{1}{2}K_{inv}\left(\dfrac{3w}{R_e}\right)^2 + 3\cdot\dfrac{1}{2}K_{inv}\left(\dfrac{w}{R_e}\right)^2}{6\cdot\dfrac{1}{4}V_{IJKL}n_{JK}^2\left(\dfrac{4w}{\sqrt{3}R_e}\right)^2 + 18\cdot\dfrac{1}{4}V_{IJKL}n_{JK}^2\left(\dfrac{2w}{\sqrt{3}R_e}\right)^2} =$$

$$= \frac{6K_{inv}}{14V_{IJKL}n_{JK}^2} = \frac{6.40}{14\cdot6.25\cdot2^2} = 68.6\%. \tag{10.66}$$

Hence, torsional stiffness has to be increased by 68.6% while, in turn, the inversion energy can be neglected. This way, we can obtain identical results for planar symmetric structures like graphene if deformations are small. For large deformations, a small error has to be accepted. Nevertheless, this approach is recommended as it is favorable from a mathematical and mechanical point of view when all energy forms are independent of one another.

10.5 Numerical Examples

In order to demonstrate the capabilities of MDFEM and to verify the robustness and reliability of the finite elements introduced in Section 10.4.2, two different examples are presented. The first one deals with single-walled carbon nanotubes as for such structures chemical bonds are dominant. Neglecting physical interactions, even with an implicit time integration scheme, structures with more than 1 million atoms can be simulated.

For the second example, a model of elastomeric material, physical bondings have to be considered. As a consequence, static and implicit dynamic analyses are limited to a few thousand atoms because the long-range potentials lead to a very large bandwidth of the stiffness matrix. To overcome this limitation, the explicit midpoint method is used.

10.5.1 Torsion of Single-Walled Carbon Nanotube

Carbon nanotubes are subject of many numerical investigations in literature, as their mechanical and electrical properties are remarkable, but very hard to determine experimentally. Defects stemming from the manufacturing process may lead to drastic changes in the material behavior.

A systematic investigation of the effects of atomistic defects on the nanomechanical properties and fracture behavior of single-walled carbon nanotubes using MD simulation is provided by Cheng et al. (2009). Their results show that the properties highly depend on the defect rate but also on the distribution pattern and that the failure of the nanotubes can be regarded as brittle whereas the cracks propagate along the areas with high tensile stress concentration.

The effects of the nanotube helicity, the nanotube diameter, and the percentage of vacancy defects on the bond length, bond angle, and tensile strength of zigzag and armchair single-walled carbon nanotubes is subject of a study by Jeng et al. (2009). A good agreement of the stress–strain response between MDs and molecular statics simulations is observed.

The MDFEM simulation in Figure 10.12 shows the failure mode of a (10,10)-armchair carbon nanotube with two Stone–Wales defects at a torsion angle of about 50°. The cross section, which is initially circular adopts an elliptic shape that propagates helically over the entire length of the tube until the opposite walls come close to each other. A similar behavior was observed by Rochefort et al. (1999) and Chakrabarty and Cagin (2008).

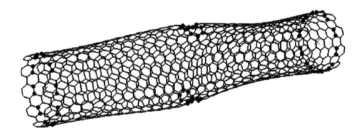

FIGURE 10.12
Buckling of carbon nanotube with two Stone–Wales defects at the top and at the bottom (SW3) at about 50°-torsion.

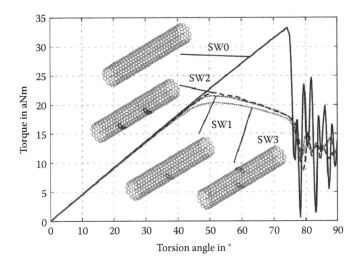

FIGURE 10.13
Influence of Stone–Wales defects on torsion load resistance.

The torque–rotation curves in Figure 10.13 demonstrate the influence of the Stone–Wales (SW) defects. While for a nanotube without defects (SW0) a bifurcation can be observed, defects transform the bifurcation problem into a snap-through problem. Instead of an instantaneous collapse the shape of the nanotube changes slowly. The maximum torsional moment of 33 aN-m (SW0) decreases to 22 aN-m for the nanotube with two defects at the bottom site (SW2), 21.5 aN-m for the nanotube with only one defect (SW1), and 20 aN-m for the nanotube in Figure 10.12 with two opposing defects (SW3).

At this point, it should be stressed that the results are identical to a classical MD simulation, given that the same force field potentials are applied. Here, we used the Dreiding approach proposed by Mayo et al. (1990). The reader who is interested in different load cases, namely, when the nanotubes are subjected to tension, compression, or bending loads, is referred to the publication by Nasdala et al. (2005).

10.5.2 Inelastic Behavior of Elastomeric Material

As pointed out in Section 10.4.1.9, MDFEM simulations usually start with a relaxation step. Elastomeric material is a good example where it is impossible to guess a valid equilibrium state. In addition to the challenging conformational analysis, this example also provides explanations for inelastic material behavior. Using elastic force field potentials, it is possible to simulate softening and hysteresis effects.

10.5.2.1 Relaxation Step

For demonstration purposes, we start with the small example shown in Figure 10.5, bottom. Since only 107 atoms are involved, it is possible to apply the implicit Euler backward method for the relaxation step. This method is very efficient since large time increments lead to high numerical damping which dissipates the kinetic energy of the system.

If the implicit HHT method is used without introducing additional damping, the analysis does not converge, as demonstrated in Figure 10.14. In contrast to the equilibrium

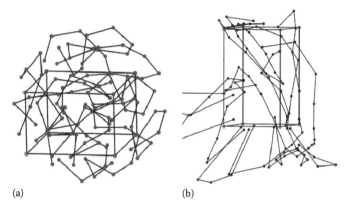

(a)   (b)

FIGURE 10.14
Convergence problems of analyses without damping: (a) implosion and (b) explosion.

state, the van der Waals bond energies are significantly higher than the bond stretch energies of the chemical bonds. This stems from the fact that the van der Waals bond energies are very high for atoms close to each other according to the LJ approach. In a dynamic analysis, the potential energy is transformed to kinetic energy but as the bond fracture energies are small compared to the initial bond potential energies, the bonds are destroyed. Depending on the size of the RVE, an "implosion" may occur first, then followed by an "explosion" or, if the RVE is small, the analysis already starts with the second part.

For larger structures, which may be more than 1000 atoms, the explicit time integration method is preferable. For the example shown in Figure 10.15 that consists of 10,051 atoms, computational costs can be efficiently reduced, when the analysis starts with a very high amount of damping, for example, α-Rayleigh damping, which then is reduced in about 10 steps, each time by an order of magnitude. This approach can be optimized by setting the velocities to zero before starting each new step.

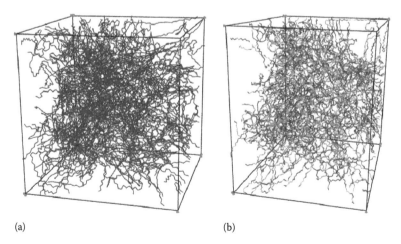

(a)   (b)

FIGURE 10.15
Conformational analysis of an elastomer with 10,051 atoms: (a) generated initial mesh and (b) equilibrium state after relaxation step.

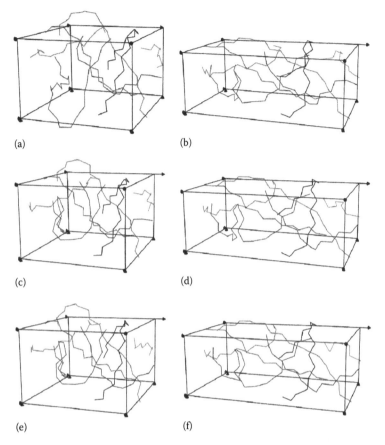

FIGURE 10.16
Continuous reorganization of bonds during cyclic loading: (a) 0% strain (initial equilibrium state), (b) 50% strain (4. cycle), (c) 0% strain (start of 5. cycle), (d) 50% strain (5. cycle), (e) 0% strain (start of 6. cycle), and (f) 50% strain (6. cycle).

10.5.2.2 Loading Step

Both models are subjected to cyclic loading at 25%, 50%, and 100% strain amplitude, 3 cycles at each amplitude. Constraints keep the volume of the RVE constant. The loading can be regarded as quasi-static. As can be seen in Figure 10.16 for cycles 4–6 of the 107-atom model, there is a continuous rearrangement of chains within the polymer network. The chemical bonds can sustain the applied load. Some van der Waals interactions, however, break when the fracture energy is exceeded while other interactions come into existence when two atoms approach each other. Note that the physical bondings as well as covalent interchain cross-linkages, built during the vulcanization process are not shown for the sake of clarity.

The 107-atom model is too small for being representative, i.e., the load–deflection curve is very jagged and similar models would lead to different results. Therefore, we shall discuss only the response of the 10,051-atom model, shown in Figure 10.17.

It is remarkable that all the main characteristics of elastomeric materials, namely,

- Mullins effect (softening)
- Curvature change from negative to positive

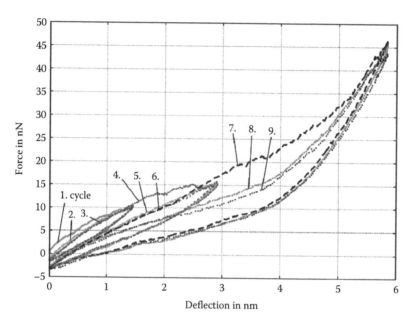

FIGURE 10.17
Load–deflection curve of an elastomer with 10,051 atoms.

- Permanent set (plasticity or viscoelasticity)
- Hysteresis loops (energy dissipation)

can be found in the simulated force–deflection curve though only (elastic) force fields potentials are used.

It should be noted that the load–deflection curve is not smooth. This is due to the fact that a rupture of bondings reduces the overall stiffness while the creation of bondings strengthens the structure.

The results prove that it is generally possible to simulate inelastic material behavior without using damping or friction elements. The underlying mechanism is the rearrangement of bondings which causes the polymer chains to vibrate. Potential energy is transformed to kinetic energy. The oscillations on the nanoscale can be observed on the macroscale as a temperature increase of the material.

10.6 Conclusions

It is possible to perform MD simulations within the framework of the finite element method. However, this is not an easy task since a special class of finite elements is required for force fields using multi-body potentials. This chapter presents the theoretical background of the MDFEM elements as well as guidelines for the implementation and usage.

Apart from mesh generation techniques, which are not covered, all important aspects of MDFEM are discussed from a FEA software user's point of view: what time integration schemes are usually available and when to use which, what is the difference between natural and equilibrium bond lengths and angles, how to obtain an equilibrium configuration, or when inversion energy is important and how it can be transformed to torsion energy. Two examples demonstrate the accuracy and efficiency of the introduced MDFEM elements.

MDFEM provides a framework that is more than performing simple MD simulations. Conventionally, MD programs are used in chemistry and physics to perform conformational studies based on force fields. The goal is to determine equilibrium states rather than to study the response of atomic structures under mechanical loading.

The main benefit of MDFEM is that concurrent multiscale simulations, that is, a combination of continuum and atomistic regions, are feasible. Complex models can be developed to predict the properties of composites containing nanoparticles which determine the macroscopic material behavior. For such models, parametric studies in terms of computer-aided material design can be carried out to analyze the influence of changes in the atomic structure, namely, the particle size, distribution, or the particle–matrix interface. The results can then be used to identify the basic mechanisms that lead to the enhancement of characteristic values of such composites and subsequently exploited to improve the manufacturing processes.

Acknowledgment

The authors acknowledge funding by the Helmholtz Association of German Research Centres. The Institute of Structural Analysis (ISD) is member of its virtual institute "Nanotechnology in Polymer Composites." ISD is also member of the Laboratory of Nano and Quantum Engineering (LNQE), the support of which is gratefully acknowledged.

Appendix 10A: Newton–Raphson Method

In statics and implicit dynamics, the Newton–Raphson method, also known as Newton's method, is the predominant technique for solving the system of nonlinear equations

$$\mathbf{R}(\bar{\mathbf{u}}) = \mathbf{0}. \tag{10.67}$$

As a direct calculation of the solution vector

$$\bar{\mathbf{u}} = \mathbf{u} + \mathbf{c} \tag{10.68}$$

or the vector $\mathbf{c}$, which has to be added to a previously determined approximation $\mathbf{u}$, is not feasible, a Taylor series expansion

$$\mathbf{R}(\bar{\mathbf{u}}) = \mathbf{R}(\mathbf{u}) + D\mathbf{R}(\mathbf{u}) \cdot \mathbf{u} + \cdots \tag{10.69}$$

is performed which then is truncated after the linear term $D\mathbf{R}(\mathbf{u}) \cdot \mathbf{u}$:

$$D\mathbf{R}(\mathbf{u}) = \operatorname{grad}\mathbf{R} = \frac{d\mathbf{R}}{d\mathbf{x}} = \frac{d\mathbf{R}}{d\mathbf{u}} = \mathbf{K}_{\mathrm{T}}. \tag{10.70}$$

In structural mechanics, $\mathbf{R}$ denotes the vector of the residual forces and $\mathbf{K}_T$ the tangential stiffness matrix

$$K_{\mathrm{T},ij}(\mathbf{u}^k) = \left.\frac{\partial R_i}{\partial u_j}\right|_{\mathbf{u}^k} \tag{10.71}$$

with the degrees of freedom i and j. The solution vector $\mathbf{c}$ of the system of linear equations

$$\mathbf{K}_T(\mathbf{u}^k)\cdot\mathbf{c} = -\mathbf{R}(\mathbf{u}^k) \tag{10.72}$$

has to be added to the approximation of the previous iteration step k

$$\mathbf{u}^{k+1} = \mathbf{u}^k + \mathbf{c} \tag{10.73}$$

until increment $\mathbf{c}$ and residuum $\mathbf{R}$ are sufficiently small. Alternatively, an "energy norm" $\mathbf{R}\cdot\mathbf{c}$ can be computed, which may also be used to verify the quadratic convergence rate of the Newton–Raphson method

$$\frac{(\mathbf{R}\cdot\mathbf{c})^{k+2}}{(\mathbf{R}\cdot\mathbf{c})^{k+1}} \approx \left[\frac{(\mathbf{R}\cdot\mathbf{c})^{k+1}}{(\mathbf{R}\cdot\mathbf{c})^{k}}\right]^2 \tag{10.74}$$

in the vicinity of the solution.

To illustrate the Newton–Raphson method, Figure 10.18 shows a system of two non-linear equations $\mathbf{R}=\mathbf{0}$ or $(R_1,R_2)=(0,0)$ with the degrees of freedom u_1 and u_2. The solution $\bar{\mathbf{u}}=(\bar{u}_1,\bar{u}_2)$ can be interpreted geometrically as the intersection point of the two planes

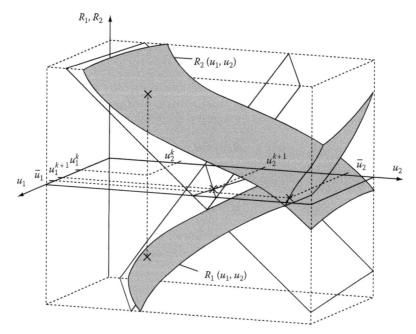

FIGURE 10.18
Illustration of Newton–Raphson method.

$R_1 = R_1(u_1,u_2)$ and $R_2 = R_2(u_1,u_2)$ with the u_1–u_2-plane. For the solution, at first two tangential planes are constructed using the points $R_1\left(u_1^k,u_2^k\right)$ and $R_2\left(u_1^k,u_2^k\right)$ defined by the approximation $\mathbf{u}^k = \left(u_1^k,u_2^k\right)$ of the time increment k. The intersection point of the two planes with the $u_1 - u_2$-plane yields a better approximation $\mathbf{u}^{k+1} = \left(u_1^{k+1},u_2^{k+1}\right)$ which then is used to start the next iteration step.

Appendix 10B: Stiffness Matrices of MDFEM Elements

A consistent linearization of the MDFEM elements introduced in Section 10.4.2 is required to obtain a quadratic convergence rate of the Newton–Raphson method. For the two-node element, the derivative of the nodal forces $\mathbf{F}_i$ with respect to the displacements $\mathbf{u}_j$ is obtained by means of the product rule

$$\frac{\partial \mathbf{F}_i}{\partial \mathbf{u}_i} = -\frac{\partial \mathbf{F}_i}{\partial \mathbf{u}_j} = -\frac{\partial \mathbf{F}_j}{\partial \mathbf{u}_i} = \frac{\partial \mathbf{F}_j}{\partial \mathbf{u}_j} = -\mathbf{n}_{ij} \otimes \frac{\partial F_{ij}}{\partial \mathbf{u}_i} - F_{ij} \frac{\partial \mathbf{n}_{ij}}{\partial \mathbf{u}_i} \tag{10.75}$$

with the derivatives of the unit vectors

$$\frac{\partial \mathbf{n}_{ij}}{\partial \mathbf{u}_i} = -\frac{\partial \mathbf{n}_{ij}}{\partial \mathbf{u}_j} = \frac{\mathbf{n}_{ij} \otimes \mathbf{n}_{ij} - 1}{|\mathbf{x}_j - \mathbf{x}_i|} \tag{10.76}$$

and the derivatives of the force magnitudes. Depending on the chosen approach, we get

$$\frac{\partial F_{ij}^{\text{lin}}}{\partial \mathbf{u}_i} = -\frac{\partial F_{ij}^{\text{lin}}}{\partial \mathbf{u}_j} = -k_{ij}\mathbf{n}_{ij} \tag{10.77}$$

or

$$\frac{\partial F_{ij}^{\text{Morse}}}{\partial \mathbf{u}_i} = -\frac{\partial F_{ij}^{\text{Morse}}}{\partial \mathbf{u}_j} = -(2\alpha_{ij}n_{ij})^2 D_{ij}\left[\exp\left(-2\alpha_{ij}n_{ij}\left(|\mathbf{x}_j - \mathbf{x}_i| - R_{ij}^0\right)\right)\right.$$
$$\left. -0.5\exp\left(-\alpha_{ij}n_{ij}\left(|\mathbf{x}_j - \mathbf{x}_i| - R_{ij}^0\right)\right)\right]\mathbf{n}_{ij} \tag{10.78}$$

for the chemical bonds and

$$\frac{\partial F_{ij}^{\text{LJ}}}{\partial \mathbf{u}_i} = -\frac{\partial F_{ij}^{\text{LJ}}}{\partial \mathbf{u}_j} = 12\frac{D_{ij,\text{vdW}}}{\left(R_{ij,\text{vdW}}^0\right)^2}\left[-13\left(\frac{|\mathbf{x}_j - \mathbf{x}_i|}{R_{ij,\text{vdW}}^0}\right)^{-14} + 7\left(\frac{|\mathbf{x}_j - \mathbf{x}_i|}{R_{ij,\text{vdW}}^0}\right)^{-8}\right]\mathbf{n}_{ij} \tag{10.79}$$

for van der Waals bondings.

In case of the three-node element, we obtain

$$\frac{\partial \mathbf{F}_i}{\partial \mathbf{u}_a} = \mathbf{n}_i \otimes \frac{\partial F_{ijk}}{\partial \mathbf{u}_a} + F_{ijk}\frac{\partial \mathbf{n}_i}{\partial \mathbf{u}_a}$$

$$\frac{\partial \mathbf{F}_k}{\partial \mathbf{u}_a} = \mathbf{n}_k \otimes \frac{\partial F_{kji}}{\partial \mathbf{u}_a} + F_{kji}\frac{\partial \mathbf{n}_k}{\partial \mathbf{u}_a} \tag{10.80}$$

$$\frac{\partial \mathbf{F}_j}{\partial \mathbf{u}_a} = -\frac{\partial \mathbf{F}_i}{\partial \mathbf{u}_a} - \frac{\partial \mathbf{F}_k}{\partial \mathbf{u}_a} \quad \text{with } a = i, j, k$$

as a function of

$$\frac{\partial \mathbf{n}_i}{\partial \mathbf{u}_a} = \mathbf{n}_{ij} \times \frac{\partial \mathbf{n}_{ijk}}{\partial \mathbf{u}_a} - \mathbf{n}_{ijk} \times \frac{\partial \mathbf{n}_{ij}}{\partial \mathbf{u}_a}$$

$$\frac{\partial \mathbf{n}_k}{\partial \mathbf{u}_a} = \mathbf{n}_{jk} \times \frac{\partial \mathbf{n}_{ijk}}{\partial \mathbf{u}_a} - \mathbf{n}_{ijk} \times \frac{\partial \mathbf{n}_{jk}}{\partial \mathbf{u}_a} \quad \text{with } a = i, j, k \tag{10.81}$$

with

$$\frac{\partial \mathbf{n}_{ijk}}{\partial \mathbf{u}_a} = \frac{\dfrac{\partial(\mathbf{n}_{ij} \times \mathbf{n}_{jk})}{\partial \mathbf{u}_a}|\mathbf{n}_{ij} \times \mathbf{n}_{jk}| - (\mathbf{n}_{ij} \times \mathbf{n}_{jk}) \otimes \dfrac{\partial|\mathbf{n}_{ij} \times \mathbf{n}_{jk}|}{\partial \mathbf{u}_a}}{|\mathbf{n}_{ij} \times \mathbf{n}_{jk}|^2}$$

$$= \frac{\left(\mathbf{n}_{ij} \times \dfrac{\partial \mathbf{n}_{jk}}{\partial \mathbf{u}_a} - \mathbf{n}_{jk} \times \dfrac{\partial \mathbf{n}_{ij}}{\partial \mathbf{u}_a}\right) - \mathbf{n}_{ijk} \otimes \mathbf{n}_{ijk}\left(\mathbf{n}_{ij} \times \dfrac{\partial \mathbf{n}_{jk}}{\partial \mathbf{u}_a} - \mathbf{n}_{jk} \times \dfrac{\partial \mathbf{n}_{ij}}{\partial \mathbf{u}_a}\right)}{|\mathbf{n}_{ij} \times \mathbf{n}_{jk}|}$$

$$= \frac{(\mathbf{1} - \mathbf{n}_{ijk} \otimes \mathbf{n}_{ijk})\left(\mathbf{n}_{ij} \times \dfrac{\partial \mathbf{n}_{jk}}{\partial \mathbf{u}_a} - \mathbf{n}_{jk} \times \dfrac{\partial \mathbf{n}_{ij}}{\partial \mathbf{u}_a}\right)}{|\mathbf{n}_{ij} \times \mathbf{n}_{jk}|} \quad \text{with } a = i, j, k \tag{10.82}$$

and the derivatives of the magnitudes

$$\frac{\partial F_{ijk}^{\text{lin}}}{\partial \mathbf{u}_a} = \frac{K_{ijk}}{R_{ij}^0}\frac{\partial \Theta_{ijk}}{\partial \mathbf{u}_a} \quad \text{with } a = i, j, k \tag{10.83}$$

or

$$\frac{\partial F_{ijk}}{\partial \mathbf{u}_a} = \begin{cases} \dfrac{C_{ijk}}{R_{ij}^0}\left[\cos\Theta_j^0 \cos\Theta_{ijk} - \cos(2\Theta_{ijk})\right]\dfrac{\partial \Theta_{ijk}}{\partial \mathbf{u}_a} & \text{for } \Theta_j^0 \neq 180° \\[2em] -\dfrac{K_{ijk}}{R_{ij}^0}\cos\Theta_{ijk}\dfrac{\partial \Theta_{ijk}}{\partial \mathbf{u}_a} & \text{for } \Theta_j^0 = 180° \text{ with } a = i, j, k \end{cases} \tag{10.84}$$

with

$$\frac{\partial \Theta_{ijk}}{\partial \mathbf{u}_a} = \frac{\mathbf{n}_{ij} \cdot \dfrac{\partial \mathbf{n}_{jk}}{\partial \mathbf{u}_a} + \mathbf{n}_{jk} \cdot \dfrac{\partial \mathbf{n}_{ij}}{\partial \mathbf{u}_a}}{\sqrt{1 - (\mathbf{n}_{ij} \cdot \mathbf{n}_{jk})^2}}. \tag{10.85}$$

The derivatives $\partial \mathbf{n}_{ij}/\partial \mathbf{u}_a$ are given in Equation 10.76. Note that the following derivatives vanish: $\partial \mathbf{n}_{ij}/\partial \mathbf{u}_k = \mathbf{0}$, if $k \neq i \wedge k \neq j$. The derivatives of the load magnitudes F_{ijk} and F_{kji} with respect to the displacements $\mathbf{u}_a$ depend on the lever arms and are related as follows:

$$\frac{\partial F_{kji}}{\partial \mathbf{u}_a} = \frac{\partial F_{ijk}}{\partial \mathbf{u}_a} \frac{R_{ij}^0}{R_{jk}^0}. \tag{10.86}$$

For the four-node torsion element, the computation of

$$\frac{\partial \mathbf{F}_I}{\partial \mathbf{u}_a} = \mathbf{n}_{IJK} \otimes \frac{\partial F_{IIJK}}{\partial \mathbf{u}_a} + F_{IIJK} \frac{\partial \mathbf{n}_{IJK}}{\partial \mathbf{u}_a}$$

$$\frac{\partial \mathbf{F}_L}{\partial \mathbf{u}_a} = \mathbf{n}_{JKL} \otimes \frac{\partial F_{LJKL}}{\partial \mathbf{u}_a} + F_{LJKL} \frac{\partial \mathbf{n}_{JKL}}{\partial \mathbf{u}_a}$$

$$\frac{\partial \mathbf{F}_J}{\partial \mathbf{u}_a} = \alpha_{JIJK} \frac{\partial \mathbf{F}_I}{\partial \mathbf{u}_a} + \frac{\alpha_{JJKL}}{\alpha_{LJKL}} \frac{\partial \mathbf{F}_L}{\partial \mathbf{u}_a} \tag{10.87}$$

$$\frac{\partial \mathbf{F}_K}{\partial \mathbf{u}_a} = \alpha_{KIJK} \frac{\partial \mathbf{F}_I}{\partial \mathbf{u}_a} + \frac{\alpha_{KJKL}}{\alpha_{LJKL}} \frac{\partial \mathbf{F}_L}{\partial \mathbf{u}_a} \quad \text{with } a = I, J, K, L$$

requires the derivatives $\partial \mathbf{n}_{ijk}/\partial \mathbf{u}_a$ already given in Equation 10.82. Note that $\partial \mathbf{n}_{IJK}/\partial \mathbf{u}_L = \mathbf{0}$ and $\partial \mathbf{n}_{JKL}/\partial \mathbf{u}_I = \mathbf{0}$. The derivatives of the nodal force magnitudes are given as

$$\frac{\partial F_{IIJK}^{\text{lin}}}{\partial \mathbf{u}_a} = -\frac{V_{IJKL} n_{JK}^2}{2 R_{IJ}^0 \sin \Theta_J^0} \frac{\partial \varphi_{IJKL}}{\partial \mathbf{u}_a} \quad \text{with } a = I, J, K, L \tag{10.88}$$

or

$$\frac{\partial F_{IIJK}}{\partial \mathbf{u}_a} = -\frac{V_{IJKL} n_{JK}^2}{2 R_{IJ}^0 \sin \Theta_J^0} \cos \left[n_{JK} (\varphi_{IJKL} - \varphi_{JK}^0) \right] \frac{\partial \varphi_{IJKL}}{\partial \mathbf{u}_a} \quad \text{with } a = I, J, K, L \tag{10.89}$$

with

$$
\frac{\partial \varphi_{IJKL}}{\partial \mathbf{u}_a} =
\begin{cases}
+\dfrac{(\mathbf{n}_{IJK} \times \mathbf{n}_{JKL})\dfrac{\partial \mathbf{n}_{JK}}{\partial \mathbf{u}_a} + \mathbf{n}_{JK}\left(\mathbf{n}_{IJK} \times \dfrac{\partial \mathbf{n}_{JKL}}{\partial \mathbf{u}_a} - \mathbf{n}_{JKL} \times \dfrac{\partial \mathbf{n}_{IJK}}{\partial \mathbf{u}_a}\right)}{\sqrt{1 - \left[(\mathbf{n}_{IJK} \times \mathbf{n}_{JKL})\cdot \mathbf{n}_{JK}\right]^2}} \\[4pt]
\quad \text{for } (\mathbf{n}_{IJK} \cdot \mathbf{n}_{JKL}) \geq 0 \\[10pt]
-\dfrac{(\mathbf{n}_{IJK} \times \mathbf{n}_{JKL})\dfrac{\partial \mathbf{n}_{JK}}{\partial \mathbf{u}_a} + \mathbf{n}_{JK}\left(\mathbf{n}_{JK} \times \dfrac{\partial \mathbf{n}_{JKL}}{\partial \mathbf{u}_a} - \mathbf{n}_{JKL} \times \dfrac{\partial \mathbf{n}_{IJK}}{\partial \mathbf{u}_a}\right)}{\sqrt{1 - \left[(\mathbf{n}_{IJK} \times \mathbf{n}_{JKL})\cdot \mathbf{n}_{JK}\right]^2}} \\[4pt]
\quad \text{for } (\mathbf{n}_{IJK} \cdot \mathbf{n}_{JKL}) < 0.
\end{cases}
\tag{10.90}
$$

Note that the torsion angle $\varphi_{IJKL} = \varphi_{IJKL}(\mathbf{u}_I, \mathbf{u}_J, \mathbf{u}_K, \mathbf{u}_L)$ depends on the displacements of all four atoms involved.

References

Allinger, N.L.; Chen, K. (1996) Hyperconjugative effects on carbon-carbon bond lengths in molecular mechanics (MM4). *Journal of Computational Chemistry*, 17, 747–755.

Arroyo, M.; Belytschko, T. (2002) An atomistic-based finite deformation membrane for single layer crystalline films. *Journal of the Mechanics and Physics of Solids*, 50, 9, 1941–1977.

Bloch, F. (1928) Über die Quantenmechanik der Elektronen in Kristallgittern. *Zeitschrift für Physik*, 52, 555–600.

Born, M.; Huang, K. (1954) *Dynamical Theory of Crystal Lattices*. Oxford University Press, Oxford, U.K.

Brenner, D.W. (1990) Empirical potential for hydrocarbons for use in simulating the chemical vapor deposition of diamond films. *Physical Review B*, 42, 15, 9458–9471.

Brooks, B.R.; Bruccoleri, R.E.; Olafson, B.D.; States, D.J.; Swaminathan, S.; Karplus, M. (1983) CHARMM: A program for macromolecular energy, minimization, and dynamics calculations. *Journal of Computational Chemistry*, 4, 187–217.

Buckingham, R.A. (1938) The classical equation of state of gaseous helium, neon and argon. *Proceedings of the Royal Society of London. Series A, Mathematical and Physical Sciences*, 168, 933, 264–283.

Car, R.; Parrinello, M. (1985) Unified approach for molecular dynamics and density functional theory. *Physical Review Letters*, 55, 22, 2471–2474.

Chakrabarty, A.; Cagin, T. (2008) Computational studies on mechanical and thermal properties of carbon nanotube based nanostructures. *Computers, Materials & Continua*, 7, 3, 167–189.

Cheng, H.-C.; Hsu, Y.-C.; Chen, W.-H. (2009) The influence of structural defect on mechanical properties and fracture behaviors of carbon nanotubes. *Computers, Materials & Continua*, 11, 2, 127–146.

Dirac, P. (1930) Note on exchange phenomena in the Thomas atom. *Proceedings of the Cambridge Philosophical Society*, 26, 376–385.

van Duin, A.C.T.; Dasgupta, S.; Lorant, F.; Goddard (III), W.A. (2001) Reaxff: A reactive force field for hydrocarbons. *The Journal of Physical Chemistry A*, 105, 41, 9396–9409.

Ehrenfest, P. (1927) Bemerkung über die angenäherte Gültigkeit der klassischen Mechanik innerhalb der Quantenmechanik. *Zeitschrift für Physik*, 45, 455–457.

Ercolessi, F.; Adams, J. (1994) Interatomic potentials from first-principles calculations: The force-matching method. *Europhysics Letters*, 26, 583–588.

Ericksen, J.L. (1984) The Cauchy and Born hypotheses for crystals. In Gurtin, M.E. (ed.), *Phase Transformations and Material Instabilities in Solids*, pp. 61–77. Academic Press, Inc., New York.

Fock, V. (1930) Näherungsmethode zur Lösung des quantenmechanischen Mehrkörperproblems. *Zeitschrift für Physik*, 61, 126–148.

Harik, V.M. (2002) Mechanics of carbon nanotubes: Applicability of the continuum-beam models. *Computational Materials Science*, 24, 328–342.

Hartree, D.R. (1928) The wave mechanics of an atom with a noncoulomb central field. Part I: Theory and methods. *Proceedings of the Cambridge Philosophical Society*, 24, 89.

Hartree, D.R. (1932) A practical method for the numerical solution of differential equations. *Memoirs and Proceedings of the Manchester Literary and Philosophical Society*, 77, 91.

Hilber, H.M.; Hughes, T.J.R.; Taylor, R.L. (1978) Collocation, dissipation and 'overshoot' for time integration schemes in structural dynamics. *Earthquake Engineering and Structural Dynamics*, 6, 99–117.

Hohenberg, P.; Kohn, W. (1964) Inhomogeneous electron gas. *Physical Review*, 136, 3B, 864–871.

Jeng, Y.-R.; Tsai, P.-C.; Huang, G.-Z.; Chang, I.-L. (2009) An investigation into the mechanical behavior of single-walled carbon nanotubes under uniaxial tension using molecular statics and molecular dynamics simulations. *Computers, Materials & Continua*, 11, 2, 109–125.

Kohn, W.; Sham, J. (1965) Self-consistent equations including exchange and correlation effects. *Physical Review*, 140, 4A, 1133–1138.

Lennard-Jones, J.E. (1929) The electronic structure of some diatomic molecules. *Transactions of the Faraday Society*, 25, 668–686.

Levine, I.N. (1991) *Quantum Chemistry*, 4th edn. Prentice-Hall, New York.

Li, C.; Chou, T.-W. (2004) Modeling of elastic buckling of carbon nanotubes by molecular structural mechanics approach. *Mechanics of Materials*, 36, 11, 1047–1055.

Marx, D.; Hutter, J. (2000) Ab initio molecular dynamics: Theory and implementation. In Grotendorst, J. (ed.), *Modern Methods and Algorithms of Quantum Chemistry*, pp. 301–449. John von Neumann Institute for Computing, Jülich, Germany.

Mayo, S.L.; Olafson, B.D.; Goddard (III), W.A. (1990) DREIDING: A generic force field for molecular simulations. *Journal of Physical Chemistry*, 94, 8897–8909.

Morse, P.M. (1929) Diatomic molecules according to the wave mechanics II: Vibrational levels. *Physical Review*, 34, 57–64.

Nasdala, L.; Ernst, G. (2005) Development of a 4-node finite element for the computation of nano-structured materials. *Computational Materials Science*, 33, 4, 443–458.

Nasdala, L.; Ernst, G.; Lengnick, M.; Rothert, H. (2005) Finite element analysis of carbon nanotubes with Stone-Wales defects. *Computer Modeling in Engineering and Sciences*, 7, 3, 293–304.

Newmark, N.M. (1959) A method of computation for structural dynamics. *Journal of Engineering Mechanics Division*, 85, 67–94.

Pauling, L.; Wilson Jr., E.B. (1985) *Introduction to Quantum Mechanics with Applications to Chemistry*. Dover, New York.

Rappé, A.K.; Casewit, C.J.; Colwell, K.S.; Goddard (III), W.A.; Skiff, W.M. (1992) UFF, a full periodic table force field for molecular mechanics and molecular dynamics simulations. *Journal of the American Chemical Society*, 114, 10024–10035.

Rochefort, A.; Avouris, P.; Lesage, F.; Salahub, D.R. (1999) Electrical and mechanical properties of distorted carbon nanotubes. *Physical Review B*, 60, 19, 13824–13830.

Roterman, I.K.; Lambert, M.H.; Gibson, K.D.; Scheraga, H.A. (1989) A comparison of the CHARMM, AMBER and ECEPP potentials for peptides. I. Conformational predictions for the tandemly repeated peptide (Asn-Ala-Asn-Pro)$_9$. *Journal of Biomolecular Structures and Dynamics*, 7, 391–419.

Schlick, T. (2002) *Molecular Modeling and Simulation*. Springer, New York.

Schrödinger, E. (1926) Quantisierung als Eigenwertproblem. *Annalen der Physik*, 79, 361–376, 489–527.

Shen, S.; Atluri, S.N. (2004) Computational nano-mechanics and multi-scale simulation. *Computers, Materials & Continua*, 1, 1, 59–90.

Slater, J.C.; Koster, G.F. (1954) Simplified LCAO method for the periodic potential problem. *Physical Review*, 94, 6, 1498–1524.

Slater, J.C.; Wilson, T.M.; Wood, J.H. (1969) Comparison of several exchange potentials for electrons in the Cu+ ion. *Physical Review*, 179, 1, 28–38.

Stillinger, F.H.; Weber, T.A. (1985) Computer simulation of local order in condensed phases of silicon. *Physical Review B*, 31, 8, 5262–5271.

Tersoff, J. (1988) Empirical interatomic potential for carbon, with applications to amorphous carbon. *Physical Review Letters*, 61, 25, 2879–2882.

Tserpes, K.I.; Papanikos, P. (2005) Finite element modeling of single-walled carbon nanotubes. *Composites Part B: Engineering*, 36, 468–477.

Verlet, L. (1967) Computer experiments on classical fluids. I. Thermodynamical properties of Lennard-Jones molecules. *Physical Review*, 159, 1, 98.

Wang, X.Y.; Wang, X. (2004) Numerical simulation for bending modulus of carbon and some explanations for experiment. *Composites Part B: Engineering*, 35, 79–86.

Weiner, P.K.; Kollman, P.A. (1981) Description of the AMBER program for molecular mechanics. *Journal of Computational Chemistry*, 2, 287–303.

Xie, G.Q.; Long, S.Y. (2006) Elastic vibration behaviors of carbon nanotubes based on micropolar mechanics. *Computers, Materials & Continua*, 4, 1, 11–19.

Xie, G.Q.; X. Han; Long, S.Y. (2007) Characteristic of waves in a multi-walled carbon nanotube. *Computers, Materials & Continua*, 6, 1, 1–11.

11

Application of Biomaterials and Finite Element Analysis (FEA) in Nanomedicine and Nanodentistry

Andy H. Choi
The University of Hong Kong, Pokfulam, Hong Kong, People's Republic of China

Jukka P. Matinlinna
The University of Hong Kong, Pokfulam, Hong Kong, People's Republic of China

Richard C. Conway
University of Technology, Sydney, New South Wales, Australia

Besim Ben-Nissan
University of Technology, Sydney, New South Wales, Australia

CONTENTS

11.1 Introduction 373
11.2 FEA in Nanomedicine and Dentistry 377
11.3 Coating of Dental and Medical Implants: Introduction and Processes 381
11.4 Nanocrystalline Coatings 381
    11.4.1 Physical Vapor Deposition 382
    11.4.2 Chemical Vapor Deposition 382
    11.4.3 Sol–Gel Process 382
        11.4.3.1 Sol–Gel Synthesis of Nanohydroxyapatite 383
11.5 Diamond-Like Carbon 384
11.6 Thin Film Mechanical Testing: Introduction and Methods 384
    11.6.1 Adhesion and Wear Test Methods 385
    11.6.2 Instrumented Nanoindentation 385
    11.6.3 Micro- and Nanoindentation in Medicine and Dentistry 387
    11.6.4 Micro- and Nanocoatings and FEA: Past and Present 391
11.7 Conclusion 394
References 395

11.1 Introduction

"Nanostructured materials" refer to certain materials that have delicate structures and sizes that fall within the range of 1–100 nm. As a consequence of this size, an extensive development of nanotechnology has taken place in the fields of materials science and engineering in the past decade. Yet, such developments have not come as a surprise, when it is appreciated that these nanostructured materials have the ability to be adapted and integrated into biomedical devices. This is possible because most biological systems, including viruses, membranes and protein complexes, exhibit natural nanostructures.

A biomaterial, by definition, is a nondrug substance that is suitable for inclusion in systems that augment or replace the function of bodily tissues or organs. A century ago, artificial devices made from materials as diverse as gold and wood were developed to a point where they could replace the various components of the human body. These materials were capable of being in contact with bodily fluids and tissues for prolonged periods of time, while eliciting little, if any, adverse reactions.

When these synthetic materials are placed within the human body, the tissues react toward the implant in a number of ways. The mechanism of tissue interaction at a nanoscale level is dependent on the response to the implant surface, and as such three terms which describe a biomaterial, with respect to the tissues' responses, have been defined, namely, bioinert, bioresorbable, and bioactive (Figure 11.1):

- Bioactive—refers to a material which interacts with the surrounding bone and, in some cases, even soft tissue, upon being placed within the human body (e.g., hydroxyapatite [HAp]).

- Bioinert—refers to any material that, once placed within the human body, has a minimal interaction with its surrounding tissue; examples include stainless steel, titanium, alumina, partially stabilized zirconia, and ultra-high-molecular weight polyethylene.

- Bioresorbable—refers to a material that, upon placement within the human body, begins to dissolve or to be resorbed and slowly replaced by the advancing tissues (e.g., bone, tricalcium phosphate, bioglass).

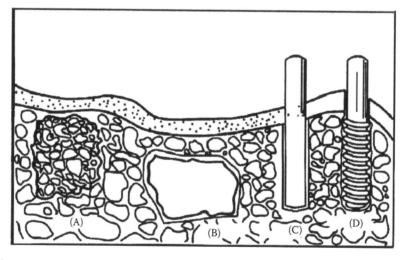

FIGURE 11.1
Classification of bioceramics according to their bioactivity: (A) Bioresorbable tricalcium phosphate implant $[Ca_3(PO_4)_2]$; (B) surface-active, bioglass or A-W glass; (C) bioactive, hydroxyapatite $(Ca_{10}(PO_4)_2(OH)_2)$ coating on a metallic dental implant; and (D) bioinert (alumina dental implant). (Adapted from Ben-Nissan, B. and Choi, A.H.: Nanoceramics for medical applications. In *Advanced Nanomaterials*. 2010. Weinheim, Germany. Copyright Wiley-VCH Verlag GmbH & Co. KgaA.)

In the early 1970s, bioceramics were employed as implants to perform singular, biologically inert roles. The limitations of these synthetic materials as tissue substitutes were highlighted with the increasing realization that the cells and tissues of the body perform many other vital regulatory and metabolic roles. The demands of bioceramics have since changed, from maintaining an essentially physical function without eliciting a host response, to providing a more positive interaction with the host. This has been accompanied by increasing demands on medical devices that they not only improve the quality of life but also extend its duration. Most importantly, nanobioceramics—at least potentially—can be used as body interactive materials, helping the body to heal, or promoting the regeneration of tissues, thus restoring physiological functions.

The main factors in the clinical success of any biomaterial are its biofunctionality and biocompatibility, both of which are related directly to interactions at the tissue and implant interface. This approach is currently being explored in the development of a new generation of nanobioceramics with a widened range of dental and biomedical applications. The improvement of interface bonding by nanoscale coatings, based on biomimetics, has been of worldwide interest in the past decade, and today several companies are in early commercialization stages of new-generation, nanoscale-modified implants for orthopedic, ocular, and maxillofacial surgery, as well as for hard- and soft-tissue engineering. Modeling and analysis of these nanoscale structures are current interests in both engineering and clinical science. Tissue–implant interactions are generated in nano- and mesoscale, and mathematical analyses of these also current interests.

The properties and microstructure of nanostructured materials depend in an extreme manner on their processing route as well as on their synthesis method. As a result, it is extremely important to select the most appropriate technique when preparing nanomaterials with desired properties and property combinations. The most commonly used synthesis technique for the production of advanced ceramics include pressing, as well as wet chemical processing techniques such as co-precipitation and sol–gel, all of which have been used to produce nanocoatings, nanoparticles, and nanostructured solid blocks and shapes.

The bone mineral is composed of nanocrystals, or more accurately, nanoplatelets originally described as HAp and similar to the mineral dahllite. It is now agreed that bone apatite can be better described as carbonate hydroxyapatite (CHA) and approximated by the formula $(Ca,Mg,Na)_{10}(PO_4CO_3)_6(OH)_2$. The composition of commercial CHA is similar to that of bone mineral apatite. Bone pore sizes range from 1 to 100 nm in normal cortical bone and from 150 to 400 μm in trabecular bone tissue, and the pores are interconnected.

Calcium phosphates are classed according to particular solubilities, for example, when bonding to the surrounding tissue, and their ability to degrade and be replaced by advancing bone growth. The solubilities of various calcium phosphate compounds can be shown as [2] follows:

Amorphous calcium phosphate (ACP) > dibasic calcium phosphate (DCP) > tetracalcium phosphate (TTCP) > α-tricalcium phosphate (α-TCP) > β-TCP > HAp.

It has long been established that porous bulk HAp cannot be used for load-bearing applications due to its unfavorable mechanical properties. As a result, HAp has been used instead as a coating in orthopedic surgery on metallic alloys, metals giving the support required.

In the past 30 years, four general conventional industrial coating methods have been proposed for the production of bioactive coatings for clinical applications. The first method, developed by Hench, and later by Ducheyne, and their colleagues utilizes spray coating method that uses relatively thick calcium phosphate coatings (100 μm–2 mm) for bone ingrowth [3]; for the second, Hench and colleagues developed thick bioglass coatings for surface bioactivity [4]. The third method, developed in the early 1990s by Kokubo and colleagues, was based on self-assembly by precipitation in a simulated body fluid (SBF) solution [5]. Although thick bioglass coatings and coatings based on self-assembly are effective, spray coating is the only coating method that has been applied commercially to orthopedic implants. A fourth, newer, and very promising method involves dipping in sol–gel derived HAp solutions to produce strong nanocoatings, which was developed by Ben-Nissan and coworkers [1,6].

Currently, the most common materials in clinical use are those selected from a handful of well-characterized and available biocompatible ceramics, metals, polymers, and their combinations as composites or hybrids. These unique production techniques, together with the development of new enabling technologies such as microscale, nanoscale, bioinspired fabrication (biomimetics) and surface modification methods, have the potential to drive at an unprecedented rate the design and development of new nanomaterials useful for medical and dental applications.

The current focus is on the manufacture of new nanoceramics that are relevant to a wide range of applications, including: modeling and finite element analysis (FEA), implantable surface-modified dental and medical devices for better hard- and soft-tissue attachment; imaging; materials for minimally invasive surgery; increased bioactivity for tissue regeneration and engineering; treatment of bacterial and viral infections; drug and gene delivery; cancer treatment; and delivery of oxygen to damaged tissues. A more futuristic view, which could in fact become reality within two decades, includes nanorobotics, nanobiosensors, and micronanodevices for a wide range of biomedical applications.

Biomimetic processing is based on the notion that biological systems process and store information at the molecular level, and the extension of this concept to the processing of nanocomposites for biomedical devices and tissue engineering, such as scaffolds for bone regeneration, has been brought out in the past decade [7]. A number of research groups have reported through a self-assembly process, the synthesis of novel bone nanocomposites of HAp and collagen, gelatin, or chondroitin sulfate. These self-assembled experimental bone nanocomposites have been reported to exhibit similarities to natural bone in not only their structure but also their physiological properties [8].

The term nanocomposite can be defined as a heterogeneous combination of two or more materials, in which at least one of those materials should be on a nanoscale. It is possible to manipulate, using the composite approach, the mechanical properties of the composites, such as strength and modulus, closer to those of natural bone, with the aid of secondary substitution phases.

The manufacture of a nanocomposite can be accomplished by physically mixing or introducing a new component into an existing nanosized material, which allows for property modifications of the nanostructured materials. This may also offer new material functions. For example, some biopolymers and biomolecules, such as poly(lactic acid) (PLA), poly(lactic-co-glycolic acid) (PLGA), polyamide, collagen, silk fibrin, chitosan, and alginate have been reported to mix into nanohydroxyapatite (nanoHAp) systems [7–10].

Another form of nanocomposite developed for biomedical applications is the gel system. For this, nanostructured materials can be entrapped in to a gel, which is a three-dimensional (3D) network immersed in a fluid, such that the properties of the nanomaterials can be improved and tailored to suit the specific needs of certain biomedical devices. A nanogel, which is a nanosized, flexible hydrophilic polymer gel [9], is an example of a gel that can be used in drug delivery carriers.

These nanogels can spontaneously bind and encapsulate, through ionic interactions, any type of negatively charged oligonucleotide drug. A key advantage of nanogels is that they allow for a high "payload" of macromolecules (up to 50 wt%), a value which normally cannot be approached with conventional nanodrug carriers [10]. Recently, a novel intracellular biosensor has been fabricated by entrapping indicator dyes into an acrylamide hydrogel [11,12], while a carbon nanotube (CNT) aqueous gel has been developed as an enzyme-friendly platform for use in enzyme-based biosensors [13].

Although there has been a large amount of development in nanobiomaterial processing, synthesis and properties, modeling and analysis by FEA has been restricted to a few dedicated groups involved in modeling of bone–implant interfaces, analyzing bone resorption, modified implant surfaces and nanolayer coatings under nanoindentation techniques.

The aim of this chapter is to provide a brief background on the current applications of FEA in nanomedicine and dentistry. A brief examination of the processes used for the production of nanocoatings will be followed by its nanomechanical properties using nanoindentation method, modeling, and their analysis with the finite element approach.

11.2 FEA in Nanomedicine and Dentistry

The science and engineering behind the design, production, characterization, and application of materials and processes whose smallest functional organization is on the nanometer scale is referred to as nanotechnology.

In the last two decades, there has been a major increase in interest in nanostructured material in advanced technologies, such as medical technology. Nanostructured materials are associated with a variety of uses within the dental and medical fields, for example, restorative dentistry is based on the nanostructured natural materials such as enamel and dentin, white spot lesions (WSL) is generated through nanoscale interactions, nanoparticles in drug-delivery systems, nanocoatings for dental implants and prostheses, in biomaterials science and diagnostic systems and in regenerative medicine [14,15].

The finite element method (FEM) was first introduced in 1956 and was extensively used in the fields of engineering and in 1970s in orthopedic biomechanics to evaluate stresses in human bones during functional loadings, and implant design and analysis and in dentistry related to the deformations under functional loadings. Since then, this method has widely been accepted in engineering and in biomedical systems and applied with increasing frequency for stress analyses of bone and bone-prosthesis structures, dental implants and devices, fracture fixation devices and various kinds of tissues other than bone (Figure 11.2). More recently, FEM was used to investigate the interactions and properties of nanofibers and nanoparticles within composite structures. More importantly, FEM has also been accepted to nanoindentation and nanomechanical testing to evaluate the biomechanical properties of nanocoatings such as zirconia and HAp on metallic implants and devices.

A quadratic remodeling formula was utilized by Lin [16] to evaluate bone resorption due to occlusal overload. A 2D FEM with a single unit implant in the mesial-distal section is considered in this study. The model was constructed from a computed tomography (CT) scan, together with a single unit titanium implant and a ceramic crown, embedded in the cortical bone with thickness of approximately 2 mm surrounding the cancellous bone. A highly dense finite element mesh was generated, to better capture bone resorption

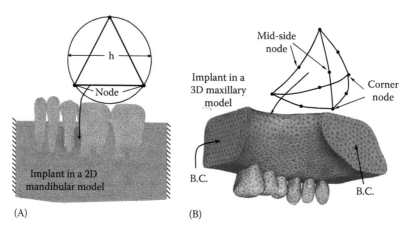

FIGURE 11.2
Characteristics of typical finite elements in bone remodeling simulation: (A) the circumcircle of 2D triangular element with element size in diameter "h" and (B) a quadratic 3D tetrahedral element with mid-side node to capture curved geometry. (Adapted from Lin, C.F., A computational protocol to the prediction of dental implant induced bone remodelling and its material design: A novel approach, PhD dissertation, University of Sydney, New South Wales, Australia, 2010.)

progressing in detail. The finite element model comprises in total 62,884 3-node triangular elements, featuring 60,170 3-node triangular elements in the cortical regions. The loading applied to the implant represents a higher range of magnitude of 402 N at an inclination angle of 1.32° to simulate the scenario of occlusal overload. ABAQUS was employed for finite element calculation and a user-subroutine in Python script was used for the bone remodeling calculations.

Xing et al. [17] measured the adhesion forces between nanofibers and a colloidal polystyrene (PS) AFM probe. The Classical JKR model and 3D FEA were used to simulate the interaction between nanofiber and nanoparticle. Nanofibers were assumed to be cylindrical in shape and nanoparticles spherical. ANSYS (version 10.0) was used to simulate the interactions between a nanoparticle and a nanofiber or the interactions between a nanoparticle and a planar film. The element type solid 185 was used. To simulate the pull-off process, the "death and birth" algorithm of the cable element was activated to mimic the failure when the element had reached its limit; the failed element was given a zero stiffness matrix. Two processes were simulated: (1) the probe attaches to the fiber, simulated by giving a displacement loading (negative loading indicating compression) to the cantilever or the nanoparticle, where the initial gap between fiber and particle was 0.35 nm; (2) the probe detaches from the fiber, simulated by giving the same displacement loading as in the first process (positive loading indicating a pulling force). Thirty substeps were set to complete the loading processes. The reaction forces at the bottom line of the nanofiber or the bottom surface of the nanofilm were calculated from the solution of the substeps. According to the authors, this study is expected to provide approaches and information useful in the design of nanomedicine and scaffold based on nanofibers for tissue engineering and regenerative medicine.

The impact of intravenously injected gold nanoparticles (GNPs) on interstitially delivered laser-induced thermal therapy (LITT) in the liver was examined by Elliot and coworkers [18]. Three-dimensional finite element modeling, ex vivo canine liver tissue containing GNPs absorbing at 800 nm, and agar gel phantoms were used to simulate the presence of nanoparticles in the liver during LITT. The model was implemented using COMSOL

Multiphysics (Heat Transfer Module). The mesh generated for the finite element calculations was a 2D triangular mesh with an average triangle size of $0.02\,cm^2$. The density of this mesh was increased by a factor of 2 in the regions of greatest temperature increase, this increased computational accuracy without a large increase in computation time.

Lin et al. [19] developed a systematic protocol to assess mandibular bone remodeling induced by dental implantation, which extends the remodeling algorithms established for the long bones into dental settings. A 3D FEA model was developed, representing a segment of the human mandible with a number of adjacent teeth. The mandibular bone protrudes slightly beyond the central incisor, aiming to simulate the far field boundary conditions in the FEA model. The model was constructed from in vivo CT scan images, and processed in Rhinoceros 3.0. The FEA mesh was generated in 10-node tetrahedral elements using MSC PATRAN 2005, comprising totally 109,020 elements after a convergence test including 58,885 elements assigned to the cancellous bone and 20,747 elements to the cortical bone. The cortical bone has a thickness around 2 mm, representing Class 2 bone. In their work, a period of 48 months was chosen as the remodeling duration, in which the changes in the bone densities, occlusal displacement, and natural frequencies are compared to explore the effects of remodeling.

Using FEA, Sassaroli et al. [20] have modeled the process of heating of a spherical GNP by nanosecond laser pulses and of heat transfer between the particle and the surrounding medium, with no mass transfer. In their analysis, they assume that the temperature of the GNP remains below melting temperature. In addition, the particle is considered to be spherical, rigid, and larger than 20 nm so that quantum mechanical and surface effects can be neglected. Under these conditions, the process of heating of a GNP subject to laser pulses of at least nanosecond duration. For easy implementation and reliability of the solution, the thermal equations have been solved using a commercial finite element package Comsol Multiphysics 3.5.

Verhulp et al. [21] investigated effects of element size, order, and material models on the results of post-yield trabecular bone simulations. A compression experiment of a trabecular bone specimen to failure was simulated using different micro-FE meshes and material models based on cortical bone. Experimental and simulated results were compared both at the apparent level, by comparing load–displacement curves, and at the local level, by comparing microscopic deformations. Four different FE meshes were created from the first three-dimensional scan by simply converting bone voxels to hexahedral elements using a fixed global threshold value. In order to achieve both sufficient numerical accuracy and to limit the computational costs, three meshes were created using linear hexahedral elements of 40, 60, and 80 µm, with 40 µm being slightly less than one-fourth of the mean trabecular thickness. This resulted in high-resolution FE meshes with 539,793, 156,007, and 63,933 elements, and 772,328, 254,539, and 117,387 nodes, respectively, with the corresponding volume fractions equal to 27%, 26.5%, and 24.4%. Three different isotropic material models were used to describe trabecular-tissue yield and post-yield behavior. All the micro-FE analyses were performed with the FE package MSC.Marc, incorporating large deformations (geometrical nonlinearity). Linear analyses were performed to determine the tissue effective elastic modulus for each FE mesh and to determine the loading mode (tension or compression) of each element.

In 2008, Verhulp and coworkers [22] also tested cortical and trabecular stress transfer in the proximal femur of two bones, one normal and one osteoporotic, using micro-FE analysis. The micro-FE meshes were created from high-resolution CT images of the proximal 10 cm of a healthy and a severely osteoporotic femur. The bone voxels were directly converted to equally sized 80-µm brick elements, rendering micro-FE meshes of 97 and 72 million elements and 130 and 100 million nodes for the healthy and osteoporotic femurs, respectively. In each femur, cortical and trabecular bone tissue were identified based on the number of elements in a fixed neighborhood. The linear-elastic micro-FE models were solved using an iterative element-by-element solver. Tissue stresses and strains were used to compute strain-energy densities (SED), effective strains and maximal principal strains.

A study on the fracture mechanisms in dental nanocomposites were conducted by Chan et al. [23]. Finite-element method was utilized to analyze the growth of an interface crack around an elastic particle embedded within an elastic/plastic matrix. A hard particle embedded in an elastic matrix separated by an interface was generated, which contains 50% particle and 50% matrix by area, and was subjected to principal stresses applied along the x-axis and y-axis. Two boundary layers of elements of uniform size were specified along the interface between the matrix and the particle for avoiding any artifact that may result from nonuniform element size. A small interface crack was placed between the two boundary layers by releasing the appropriate nodes at the apex of the circular particle. For a given set of the applied stresses, the stress field around the interface crack was computed.

Isaksson et al. [24] compare various mechano-regulation algorithms' abilities to describe normal fracture healing in one computational model. Additionally, it was hypothesized that tissue differentiation during normal fracture healing could be equally well regulated by the individual mechanical stimuli, for example, deviatoric strain, pore pressure, or fluid velocity. A biphasic finite element model of an ovine tibia with a 3mm fracture gap and callus was used to simulate the course of tissue differentiation during normal fracture healing. For the computational model, a mechano-regulated adaptive axisymmetric finite element model of an ovine tibia was created. The geometry involved a 3mm transverse fracture gap and an external callus. The external surface of the callus, the ends of the cortical bone, and the intramedullary canal were assumed to be covered by fascia and impermeable. The loads were applied to the cortical bone at the top of the model. The callus consisted of 779, the marrow of 1060, and the cortical bone of 540 elements, which were all 8-noded biquadratic displacement, bilinear pore-pressure elements. The finite element solver used was ABAQUS and the adaptive process of fracture healing was implemented in MATLAB®. The load applied was regulated in a biofeedback loop, where the load magnitude was determined by the interfragmentary movement in the fracture gap.

Taylor and coworkers [25] developed a technique to simulate the tensile fatigue behavior of human cortical bone. A combined continuum damage mechanics (CDM) and FEA approach was used to predict the number of cycles to failure, modulus degradation and accumulation of permanent strain of human cortical bone specimens. The simulation of fatigue testing of eight dumb-bell specimens of cortical bone was performed and the predictions compared with existing experimental data. A three-dimensional finite element model of a dumb-bell specimen of human cortical bone was generated. Only one quarter of the specimen was modeled due to symmetry and the appropriate symmetry constraints were applied. The model consisted of 599 8-noded reduced integration elements and 910 nodes. The cortical bone was assumed to be isotropic, homogeneous, and linear elastic, with a Poisson's ratio of 0.35. No attempt was made to simulate the cyclic viscoelastic properties of cortical bone. All analyses were performed using Marc version 7.3.

Crestal bone loss is observed around various designs of dental implants. A possible cause of this bone loss is related to the stresses acting on periimplant bone. Vaillancourt et al. [26] investigated the relationship between stress state and bone loss and analyzed two-dimensional finite element models corresponding to bucco-lingual and mesio-distal sections of canine mandibles with one of two designs of porous-coated dental implants.

A fully porous-coated design consisting of a solid Ti6A14V core had a porous coating over the entire outer surface of the implant component, while a partially porous-coated design had the porous coating over the apical two-thirds of the implant surface only. Occlusal forces with axial and transverse components were assumed to act on the implant with interface bonding and effective force transfer at all porous coat–bone interfaces and no bonding for the nonporous-coated regions. The results of the analysis indicated that at most implant aspects (buccal, lingual, mesial, and distal), the equivalent stresses in crestal bone adjacent to the coronal-most, nonporous-coated zone of the partially porous-coated implants were lower than around the most coronal region of the fully porous-coated implants. The region of lower stress around the partially porous-coated implants corresponded to observed areas of crestal bone loss in animal studies, suggesting that crestal bone loss in this case was due to bone disuse atrophy.

11.3 Coating of Dental and Medical Implants: Introduction and Processes

Coating is a technique for modifying the surface of the base material in order to improve the mechanical and/or physical performances of implants and devices. Coatings offer the possibility of modifying the surface properties of dental and surgical-grade materials to achieve improvements in biocompatibility, clinical reliability, and performance. Techniques such as chemical vapor deposition (CVD), physical vapor deposition (PVD), electrochemical vapor deposition, metal–organic chemical vapor deposition (MOCVD), thermal or diffusion conversion, and sol–gel processing have been used to produce coatings on both the micro- and nanoscale.

In the context of biomedical applications, the definition of macro-, micro-, thin film, and nanothickness, or more generally, thin films, have been used interchangeably and/or wrongly. The authors of this chapter believe that coatings greater than 1000 μm should be considered thick or macrocoatings, 1–1000 μm should be considered thin-film coatings or microcoatings, and below 1 μm should be considered nanocoatings.

The term "thermal-spraying" covers processes, for example, plasma spraying, high-velocity oxygen flame-spraying, flame-spraying, and detonation gun-spraying. Plasma-spraying can be conducted in an ambient atmosphere (atmospheric plasma-spraying [APS]), under vacuum (vacuum plasma-spraying [VPS]) or under controlled atmospheres, such as nitrogen. Plasma-spraying uses a direct current arc to produce gas plasma, although other sources, such as radiofrequency-generated plasmas, can be employed. Each is capable of producing coating thicknesses from a few microns to a few millimeters [27].

Plasma-spraying is the only widely used coating process to produce medical implants on a commercial scale. It uses an electrical discharge to convert a carrier gas, for example, argon, into plasma. Rapid gas expansion induces speeds of up to 800 m/s. The plasma heats the powder to a partially liquid form and propels it toward the substrate.

11.4 Nanocrystalline Coatings

Nanocrystalline coatings with grain sizes in the nanometer range are known to exhibit superior hardness and strength [28]. The quest for nanostructured coatings is driven by the improvement in nanocoating techniques and the availability of various kinds of precursor materials and sources.

Existing PVD and CVD processes for preparing microcrystalline coatings can be utilized to fabricate nanostructured coatings by modifying the processing parameters or by using feedstock powders having nanograined structures. In addition, conventional plasma spraying and high velocity oxygen fuel thermal spraying are viable high-rate deposition techniques to produce nanocrystalline coatings using nanosized feedstock powders.

11.4.1 Physical Vapor Deposition

PVD is a versatile synthesis method and is capable of preparing thin film materials with structural control at the atomic or nanometer scale and this can be achieved by carefully monitoring the processing conditions. PVD involves the generation of vapor phase species either via evaporation, sputtering, laser ablation, or ion beam. In most PVD-based processing approaches, it is not possible to uniformly coat nonplanar substrates without sophisticated substrate translation/rotation or the use of multiple, spatially distributed sources. This arises because the vapor atoms are created in high vacuum that results in nearly collisionless vapor transport to the substrate. As a result, only regions in the line-of-sight of the vapor source are coated.

Physical vapor deposition of thin films has found widespread use in many industrial sectors. State-of-the-art magnetron sputtering processes allow the deposition of metals, alloys, ceramic, and polymer thin films onto a wide range of substrate materials. Therefore, it is used in many coating applications, including biomedical applications. There is an increasing demand for coatings with tailored and enhanced properties such as wear, corrosion resistance, high hardness, low friction, and specific optical or electrical properties as well as decorative colors, and often, complex combinations of those properties are required.

11.4.2 Chemical Vapor Deposition

CVD of coatings and films involves the chemical reactions of gaseous reactants on or near the vicinity of a heated substrate surface. This atomistic deposition method can provide highly pure materials with structural control at atomic or nanoscale levels. Furthermore, it can produce single layer, multilayer, composite, nanostructured, and functionally graded coating materials with well-controlled dimension and unique structure at low processing temperatures. In addition, the unique feature of CVD over other deposition techniques such as the nonline-of-sight-deposition capability has allowed the coating of complex-shaped biomedical prostheses and the fabrication of nanodevices and composites.

The flexibility of CVD has led to rapid growth in the use of functional devices and it has become one of the main processing methods for the deposition of thin films and coatings for a wide range of applications, including refractory ceramic materials used for hard coatings and metallic films for protective coatings [29].

11.4.3 Sol–Gel Process

Sol–gel processing is a versatile and attractive technique since it can be used to fabricate ceramic coatings from solutions by chemical means. The sol–gel process is relatively easy to perform and complex shapes can be coated, and it has also been demonstrated that the nanocrystalline grain structure of sol–gel coatings produced results in improved mechanical properties [30–33].

The sol–gel technique dates back to the genesis of chemistry. It was first discovered in 1846 as an application technology, when Ebelmen [34] observed the hydrolysis and polycondensation of tetraethylorthosilicate (TEOS). In 1939, the first sol–gel patent was published covering the preparation of SiO_2 and TiO_2 coatings [35]. In 1955, Roy and Roy [36] recognized the potential for producing high-purity glasses using methods not possible with traditional ceramic-processing techniques. In doing so, they generated the first report on the use of sol–gel technology to produce homogeneous multicomponent glasses.

In 1965, Schroeder [37] reported the first investigation conducted by Schott Glass involving sol–gel synthesized coatings. Mixed-oxides coatings were developed although they were mainly interested in single-oxide optical coatings of SiO_2 and TiO_2.

Sol–gel technology also found applications in a number of technology fields, such as biomedical applications [6,38] in the late 1980s and 1990s. A number of excellent review articles, book chapters, and books cover the science and technology of the basics of sol–gel technology for various ceramic oxide systems [4,39–46].

A sol, by definition, is a suspension of colloidal particles in a liquid [47]. A sol differs from a solution in that a sol is a two-phase, solid–liquid system, whereas a solution is a single-phase system. Colloidal particles can be in the approximate size range of 1–1000 nm; for this reason, gravitational forces on these colloidal particles are negligible and interactions are dominated by short-range forces such as van der Waals forces and surface charges. Diffusion of the colloids by Brownian motion leads to a low-energy arrangement, thus imparting stability to the system [48]. The stability of the sol particles can be modified by reducing their surface charge. If the surface charge is significantly reduced, then gelation is induced and the resultant product is able to maintain its shape without the assistance of a mould.

Gels are regarded as composites, since gels consist of a solid skeleton or network that encloses a liquid phase or excess of solvent. Depending on their chemistry, gels can be soft and have a low elastic modulus, usually obtained through controlled polymerization of the hydrolyzed starting compound. In this case, a three-dimensional network forms, resulting ultimately in a high molecular weight polymeric gel. The resultant gel can be thought of as a macroscopic molecule that extends throughout the solution. The gelation point is the time taken for the last bond in this network to form. This gelation can be used to produce a nanostructured monolith or nanosized coatings, depending on the process applied.

11.4.3.1 Sol–Gel Synthesis of Nanohydroxyapatite

Hydroxyapatite, $Ca_{10}(PO_4)_6(OH)_2$, is widely accepted as a biocompatible material, which resembles the mineral component of bone and teeth [2–4].

Since the early 1980s, calcium phosphates have been used as porous coating materials on a range of metallic implants for dental and orthopedic applications [49–52]. This was initiated to compensate for the poor mechanical properties of porous bulk calcium phosphate materials and utilize the excellent mechanical properties of the metallic substrates.

Nanocrystalline HAp can be fabricated using a number of different production methods and it can be used as nanocoatings, monolithic solid ceramic products, or nanosized powders and platelets for a number of applications.

To prepare nanocrystalline apatites, methods of hydrothermal synthesis, wet chemical precipitation, sol–gel synthesis, mechanochemical synthesis, mechanical alloying, coprecipitation, ball milling, liquid–solid–solution synthesis, vibromilling of bones, radio frequency induction plasma, flame spray pyrolysis, electrocrystallization, microwave processing, hydrolysis of other calcium orthophosphates, double step stirring, emulsion-based, or solvothermal syntheses, and several other techniques are known. Continuous preparation procedures are also available [15,27,28].

In addition, nanodimensional HAp might be manufactured by a laser-induced fragmentation of HAp microparticles in water and in solvent-containing aqueous solutions, while dense nanocrystalline HAp films might be produced by radio frequency magnetron sputtering [53]. A comparison between the sol–gel synthesis and wet chemical precipitation technique has been performed and both methods appear to be suitable for synthesis of nanocrystalline apatite.

Various sol–gel routes have been employed for the production of synthetic HAp powders since the early 1990s. Ben-Nissan and coworkers introduced alkoxide-based nanocoatings as early as in 1989 and since then a number of excellent studies have been conducted on a range of precursors to produce pure nanocrystalline apatite powders, solid products, or coatings for medical and engineering applications [15,27,54]. The major ones are calcium acetate, calcium alkoxide, calcium chloride, calcium hydroxide, calcium nitrate, and dicalcium phosphate dihydrate. It has been reported by some of the investigators that the thickness of the sol–gel derived HAp coatings produced are in the 70–100 nm range [15,27,54].

11.5 Diamond-Like Carbon

In recent years, diamond-like carbon (DLC) has been the focus of extensive research due to its potential application in many technological areas. The combination of biocompatibility, wear resistance, chemical inertness, hemocompatibility, low friction, wear resistance, and high hardness makes it ideal for a number of applications ranging from the coatings of stents, heart valves, orthopedic components, and prostheses in the biomedical industry [55,56].

DLC and doped DLC have been extensively investigated for possible biomedical applications. It has been demonstrated that doping with elements such as Si, N, F, Ca, and P improves the properties of DLC such as biocompatibility, infection resistance, and mechanical properties [57–60]. DLC films have been employed to coat the surfaces of blood-interfacing prostheses.

11.6 Thin Film Mechanical Testing: Introduction and Methods

With the ever increasing demands imposed by the use of implants and devices for dental and medical applications, novel techniques are needed to examine their mechanical reliability. Methods for ascertaining mechanical properties and adhesion have been stimulated by developments of bio-inspired coatings (i.e., HAp) on metallic substrates. Methods to quantitatively determine the mechanical properties of these thin coatings on substrates or free-standing films are required. There have been constant developments and improvements in equipment capable of extracting the mechanical properties of the structure or film and also the adhesion of the coating to an underlying substrate [61].

11.6.1 Adhesion and Wear Test Methods

Adhesion testing is essential to ensure the coating will adhere properly to the substrate to which it is applied. Many techniques have been used for the measurement of adhesion of coatings and films [61]. The most popular test methods for measuring the adhesive bonding to substrates include pull-off, cross-cut, indentation scratching (increasing load), and pin-on-disk.

11.6.2 Instrumented Nanoindentation

The starting point and method of choice by many practitioners in the dental and biomedical field for measuring the mechanical properties of coatings and implants is nanoindentation. The instrumentation and methodology have developed in leaps and bounds over the years and it is now considered a simple and effective means of obtaining meaningful measures of hardness, H, and Young's modulus, E, of coatings. The main requirements for obtaining the best possible results with such testing are in sample preparation, calibration of equipment and corrections for thermal drift, initial penetration, frame compliance, and indenter tip shape [62–65].

In the well-known conventional microindentation testing, a load is applied through a diamond tip of known geometry (typically Rockwell, Vickers, or Knoop) into the material surface and then removed and the area of the residual impression is measured by optical means to give the material hardness, an example of which is shown by Kealley et al. [66] for HAp ceramics (Figure 11.3). In nanoindentation testing, by contrast, the size of the residual impressions is often only a few microns and this makes it very difficult to obtain a direct measure using optical techniques. A set load in the millinewton (mN) range is applied to the indenter in contact with the specimen. As the load is applied, the penetration depth is measured (nm range).

At maximum load, the area of contact is determined by the depth of the impression and the known angle or radius of the indenter. The result is a load–displacement curve, which yields contact pressure or hardness and Young's modulus from the shape of the unloading curve using software based on the model and indenter type (pointed, i.e., Berkovich, Vickers, Knoop, or spherical) [63]. Similarly, different types of loading and unloading methods can be used to extract desired properties as a function of depth of penetration [64,65,67]. The application of nanoindentation can also be directed toward the measurement of residual stress and film adhesion from direct indentation or transverse scratching as described by Fischer-Cripps [64].

In the nanoindentation analysis, the elastic modulus (E) is obtained from the contact stiffness S, which is the slope of the unloading portion of an indentation load (P)–displacement (h) curve at maximum load (Figure 11.3):

$$S = \frac{dP}{dh} = \frac{2}{\sqrt{\pi}} E^* \sqrt{A} \qquad (11.1)$$

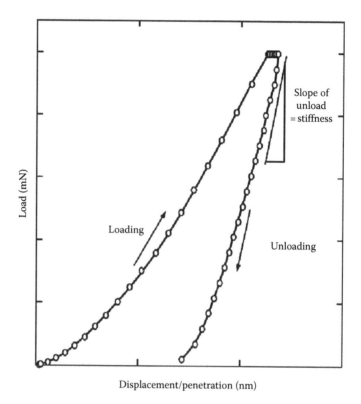

FIGURE 11.3
Indentation load–displacement curve for a pure hydroxyapatite using a Berkovich indenter.

with the contact area (A) at maximum load and the combined elastic modulus of the indenter and specimen (E*) expressed as

$$\frac{1}{E^*} = \frac{\left(1 - v_i^2\right)}{E_i} + \frac{\left(1 - v_s^2\right)}{E_s}$$ (11.2)

where E_i, v_i, and E_s, v_s are Young's modulus and Poisson's ratio of the indenter and specimen, respectively. For elastic contacts of axis-symmetric indenters (i.e., spherical, conical, and cylindrical punches), Equation 11.1 is valid. The hardness or mean contact pressure for indentation is obtained from the maximum load, P, over the projected contact area, A:

$$H = \frac{P}{A}$$ (11.3)

The contact area (A) is determined by the indenter geometry and the contact depth [64].

Nanoindentation provides a comprehensive assessment of hardness and Young's modulus of the coating as well as the elastic–plastic response from the loading and unloading curves based on the coating–substrate combination (e.g., soft/hard and rigid/compliant coating on soft/hard and rigid/compliant substrate). In addition, many tests can be performed in selected regions or in specific areas of interest that may show local variations in properties.

Nanoindenters have been employed to measure residual stress, coating adhesion from the load at which delamination occurs (taken from the pop-in that corresponds to a plateau or discontinuity in the load–displacement curve), and also to function as scratch testers depending on the equipment capability [64]. Likewise, viscoelastic and creep behavior can be examined [68,69] for softer materials and is particularly relevant in bone, dental composites/resins, and biological tissue studies.

11.6.3 Micro- and Nanoindentation in Medicine and Dentistry

The level of mineralization is shown to be in relationship with the mechanical properties of calcified tissue [70–74]. Diseases such as infectious dental caries and other developmental and genetic pathosis, which affect dental hard tissue, impose on the compositions and assemblages of tooth mineral. This in turn, alters the physical and mechanical behaviors of tooth. It has been shown that changes in the mechanical properties of the dental calcified tissues are associated with many of these pathologic conditions [75–77]. Enhancing the longevity of the restoration can be accomplished by matching the properties of restorative materials to the properties of teeth. As a result, data on the mechanical properties of teeth are needed. The knowledge gained is also vital to clinicians to help them understand how these tissues react under clinical loading conditions [78] and to help predict the behavior of the tooth/restoration interface [79].

To date, conventional compressive and tensile tests have been used to determine the mechanical data of teeth. Due to the relative quality and size of test specimens obtainable from a tooth, some limitations are associated with these tests. Preparation for these tests also becomes time consuming and difficult in order to achieve a number of ideal specimens. As a consequence of these problems, a wide variation in the results has been reported.

Since the early 1990s, nanoindentation has been used to examine dental hard tissues [80]. Nanoindentation allows the measurement of mechanical properties, hardness and elastic modulus, at the surface of a material. Compared with other conventional mechanical tests such as tensile, compressive shear strength bending and punch shear tests, the procedure for nanoindentation technique is much simpler, especially on small complex-shaped samples such as enamel, dentine, and cementum [81].

The measurements are relatively nondestructive when using indentation. It is also less time consuming for preparation as indentation test can be done on a bulk-polished specimen [81,82]. More importantly, the measurement of the mechanical properties of a very small selected region of the specimen is allowed, the dimension of which may be at a micrometer or even nanometer scale, which is important for the measurement of the local properties of nonhomogenous structures such as dental calcified tissues.

Choi et al. [83] investigated the variations of tensile and compressive stresses and the deformation in the nanocoating under microindentation simulated using three-dimensional FEA. The effects of the radius of indenter, thickness of the nanocoating and the Young's Modulus of the coating material were analyzed. The nanoindentation with spherical indenter was modeled as three-dimensional contact problem between two axisymmetric bodies, with an assumption that the dimension of the specimen to be indented is large compared with that of the deformed volume. For simplicity, the indenter was modeled as a rigid surface. Diamond was chosen as the indenter material as its Young's Modulus is much higher than that of the nanocoating materials and stainless steel substrate. Loading and unloading are the two subsequent steps used to simulate the whole nanoindentation process. During loading, the spherical indenter moves downward along the y-axis and penetrates the nanocoating. The depth of penetration is determined by the load applied. During unloading, the indenter returns to its initial position and the load applied is removed. The specimen was represented by a cylinder with such a large dimension that

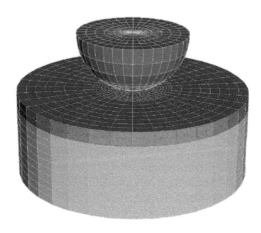

FIGURE 11.4
Three-dimensional finite element nanoindentation model. The load is applied at the top of the indenter. The indenter is resting on top of the nanocoating (dark grey). Restraints are usually applied at the bottom of the substrate (light grey) to prevent any movement of the model during nanoindentation testing.

any further increase in its dimension made no significant change in the stress distribution. In the vicinity of the contact region, a fine finite element mesh was employed to capture the localized stresses and deformation underneath the indenter. To reduce the computational stress and work for the finite element software, the fineness of the finite element mesh decreases away from the contact region. Both the coating and the substrate are considered to be homogeneous and isotropic. As strain hardening is not considered in this study, the coating and substrate are also considered to be linear elastic/perfectly plastic materials. It is also assumed that a perfect bonding condition occurs between the nanocoating and substrate. Nanoindentation simulations were carried out using a commercially available FEA package, STRAND7 (Figure 11.4).

Lucchini and coworkers [84] examined the damage mechanisms which accounts for the response of lamellar bone to nanoindentation tests, with particular regard to the decrease of indentation stiffness with increasing penetration depth and to the loss of contact stiffness during the unloading phase of the test. For this purpose, indentation experiments on bovine cortical bone samples along axial and transverse directions have been carried out at five penetration depths from 50 to 450 nm; furthermore, a continuum damage model has been implemented into finite element analyses, which are able to simulate indentation experiments. The commercial finite element code ABAQUS/Standard was used to run the numerical analyses. Given the transversely isotropic material behavior for the tissue, an axisymmetric mesh was used to simulate experiments along the axial direction; whereas, a three-dimensional finite element mesh was built to simulate indentation tests along the transverse direction. The models consist of an indenter, modeled as a rigid body, and a bone sample represented as a cylinder of 25 μm height and radius in the 2D mesh and a cylinder of 50 μm height and radius in the 3D one. A spheroconical, axisymmetric indenter with an internal angle of 70.3° and the same area-to-depth ratio as a three sided Berkovich pyramid was employed in the model. Displacement-controlled indentations are simulated along the axial as well as the transverse directions. Suitable mesh refinement was applied under the indenter tip. The 2D finite element mesh has 12,121 bi-linear elements of both types CAX3 and CAX4 with a characteristic element size in the refined region under the tip equal to 25.0 nm. Similarly, the 73,800 elements 3D mesh is denser in proximity of the contact area, where linear C3D8 and C3D6 8-node, hexahedral elements are used. Linear infinite elements (CINAX5 and CIN3D8) have been used to simulate the effect of an infinite solid domain in the region far from the indenter tip. They concluded that their present approach has shown the effectiveness of the coupling between indentation experiments and refined FEM simulations with a damage model aimed at obtaining a deeper knowledge of bone tissue mechanical behavior in the elastic and inelastic range and its relationship with the tissue constitution at multiple hierarchical levels.

To explore the effect of friction in simulations of bone nanoindentation, two-dimensional axisymmetric finite element simulations were performed by Adam and Swain [85] using a spheroconical indenter of tip radius 0.6 μm and angle 90°. The total FE domain was 60 μm × 60 μm (100 times the indenter tip radius). A graded mesh of reduced integration, linear 4-node axisymmetric elements (CAX4R) was used to discretize the domain. A preliminary mesh sensitivity analysis was performed to ensure that the simulation results were insensitive to mesh size in the indenter tip region. The model was loaded in two steps. The indenter was firstly subjected to a ramped 5 mN compressive load, followed by unloading to zero indenter force, in order to observe the indentation left in the bone upon removal of the load. During these steps, the lower edge of the bone was constrained vertically. An axisymmetric boundary condition was used along the symmetry axis beneath the indenter tip. In order to explore the effect of interface friction, a range of friction coefficients were simulated between indenter and bone. The models were solved using ABAQUS/Explicit version 6.7-1. They concluded that it is potentially important to include friction in nanoindentation simulations of bone if pile-up is used to compare simulation results with experiment.

Nanoindentation experimental tests and finite element simulations were employed by Carnelli and coworkers [86] to investigate the elastic–inelastic anisotropic mechanical properties of cortical bone. An axisymmetric, spheroconical indenter with an internal angle of 70.3°, a 300 nm end radius (estimated by the calibration curve on the fused silica reference sample) and the same area-to-depth ratio as a three sided Berkovich pyramid was employed in the model. The tip is modeled as a rigid body. Indentations along axial as well as transverse directions are simulated: an axisymmetric model was employed for indentation along the axial direction due to the transversely isotropic constitutive behavior of cortical bone; instead, a three-dimensional model was generated for indentations along the transverse direction. In the latter case, only a quarter of the bone tissue/indenter system was modeled by exploiting the symmetry of the material at 90° and the axisymmetry of the spheroconical indenter. The bone sample is represented as a cylinder of 25 μm high and radius in the 2D mesh and 50 μm high and radius in the 3D one. Suitable mesh refinement was applied under the indenter tip. The two-dimensional finite element mesh has 12,121 bi-linear elements of both types CAX3 and CAX4. The characteristic element size in the refined region under the tip is 25.0 nm. Similarly, the 73,800 elements three-dimensional mesh is denser in proximity of the contact area, where linear C3D8 and C3D6 8-node, hexahedral elements are used. The element characteristic length in the refined region under the indenter is 24.7 nm. In order to simulate the effect of an infinite solid domain in the region far from the indenter tip, linear infinite elements (CINAX5 and CIN3D8) have been used. The large deformations theory was used in the numerical models. Coulomb friction was assumed at the indenter–tissue interface. The commercial finite element code ABAQUS/Standard was used to conduct the analyses. They stated that their model provides a rigorous scientific linkage between the constitutive parameters representing the tissue's fundamental material properties in the context of the complex multiaxial stress field generated during an indentation test, which is increasingly being used for bone mechanics characterization and has potential for use as a diagnostic tool.

It is clearly shown that the potential of a jointed approach in which the coupling between indentation experiments and FEM simulations allows one to get a deeper knowledge of bone tissue mechanics.

Paietta et al. [87] investigated the contribution of structural hierarchy in PMMA-embedded lamellar bone to the variance in measured nanomechanical properties. Using Abaqus/CAE finite element code, a three-dimensional model for spherical nanoindentation of bone was developed. The contribution of bone's lamellar structure to the elastic indentation response was explored both through variation in indentation tip test location and tip size. Bone was modeled as containing alternating regions of lamellar bone (7 μm wide) and interlamellar bone (1 μm). Contact between the analytically rigid half spherical tips and an elastic quarter-space was modeled using hard, frictionless contact along with large deformation theory. The elastic quarter-space was $40 \times 40 \times 20\,\mu m^3$ in size with symmetry applied along the x–y plane along with a fixed bottom surface. A typical mesh consisted of approximately 30,000 8-noded hexahedral elements. For simulations where tip location was varied, a uniform mesh was utilized beneath the indenter tip to ensure consistent results. The model was validated by considering contact of the analytically rigid spherical tips with a homogeneous quarter-space, where the interlamellar and lamellar bone was assigned a single value for modulus. The mesh for simulations with variable tip size includes increased mesh density beneath the indentation tip. Simulations were performed to explore the contribution of tip location, where indentation tests were placed starting at the center of a lamella and progressively moved toward the center of the inter-lamellar region. Investigation of tip size employed indentation tests from 100 to 500 nm depths, with the indenter tip located at the center of a single lamella.

The FEM was applied by Toparlia and Koksalb [88] for studying the hardness and yield strength of dentin subjected to a nanoindentation process. The test with a spherical indenter was modeled as a contact problem between two axisymmetric bodies. In this study, three different samples of dentin were used and the specimen was modeled with 4950 4-node axisymmetric elements. The indenter was modeled as an undeformable surface and the radius of indenter was 10 μm. The simulations were performed using ABAQUS finite element code. The indentation region was small-modeled using edge-biased type. The contact constraint was enforced by the definition of the "master" and "slave" surface. A resultant load of 10 mN was applied as the surface pressure of the indenter. Increasing load from zero to the value of 10 mN and decreasing load from 10 mN to the zero simulated the indentation test. At each load increment, the program caries out a large number of iterations according to a specified convergence rate to reach an equilibrium and congruent configuration. The friction coefficient between the indenter and dentin surface is assumed to be zero and the dentin is assumed to be homogenous, isotropic and elasto-plastic in behavior. It was concluded that yield strength of dentin can be estimated using FEM to simulate indentation experiments using data of Young's modulus and hardness from the literature.

Dong and Darvell [89] examined the failure mode of ceramic structures under Hertzian indentation as well as the failure load and the tensile stress for crack initiation at the cementation surface as a function of the substrate modulus of elasticity and ceramic thickness. Discs of a glass–ceramic material were cemented to flat polymer substrates. The top surface of the ceramic–cement–substrate structure was loaded by a 20 mm radius spherical indenter until the initial failure of the ceramic occurred. The FEM was used to analyze the stress distribution under such Hertzian indentation as well as calculating the maximum tensile stress based on the experimentally observed failure load and contact radius. For the FEA study, an axisymmetric model was created based on the dimensions and the structure of the experimental specimens with the cement film thickness set at 50 μm and meshed with triangular elements. The validity of the FEM model was checked by comparison of the calculated and measured values of the contact radius. To simplify the FEM computation, some assumptions were made: that the interfaces were bonded perfectly (no delamination); that there was frictionless contact between the ceramic and the indenter; that the indenter was rigid enough; and that all materials were isotropic, linear, and homogeneous. With respect to the boundary conditions, points on the central axis (i.e., y-axis) of the spherical indenter, ceramic, and substrate were fixed horizontally (i.e., in the x-direction). The fineness of the meshing had been subjected to a convergence test to ensure sufficient accuracy.

The mechanical properties of thin films are commonly determined using nano- or ultra-microhardness indentation. Understanding the relationship of the measured data and the mechanical properties of the indented materials is of importance in order to obtain reliable mechanical properties, particularly of the thin films. Using FEA, the effects of the elastic modulus, yield strength, and strain hardening of the film on indentation data were analyzed by Gan and Ben-Nissan [67] and discussed for the indentation with 2, 8, 10, and 50 μm radius indenters. Elastic modulus of the films on a single ductile substrate shows relatively small influence whereas yield strength and strain hardening are found to have significant effect on the measured data.

11.6.4 Micro- and Nanocoatings and FEA: Past and Present

The FEM has been widely adopted to simulate the elastic and plastic deformations beneath a pointed indenter in nanoindentation test. Many advantages can be offered by FEA, for example, the experimental time can be reduced. Large-scale commercial codes for FEA software are currently available and they can provide simulation environment for different films and substrates, and physical models of indenter tips.

The effects of residual stress on the mechanical properties in DLC are investigated by Wei and Yang [90] using 2D FEA. In the model, the film thickness is 500 nm and the substrate thickness is 50 μm. The nanoindentation process is simulated by pressing a rigid conical indenter with a 70.3° half-included angle into the DLC/substrate system. The diamond tip was modeled with a Young's Modulus of 1140 GPa and Poisson's ratio of 0.07. The element in this model is CAX4R and the total number of elements was 20,000–30,000. The mesh near the indenter was finely remeshed to obtain sufficient accuracy. The analysis was implemented by the commercial software ABAQUS. The nonuniform stressing effect is investigated through different nanoindentation positions along the radial direction.

Rungsiyakull et al. [91] developed a new design framework to optimize the surface coating parameters in terms of the diameter of the titanium beads/particles and their volume fraction. As a key measure of the bone/implant interface stability and osseointegration, the average apparent bone density developed in the peri-implant region will be examined as a function of time. The FEA model used herein is based on the 3D solid model of a typical dental implantation setting. The model consists of a dental implant fixture, implant abutment, all-ceramic crown and a section of bone. The solid model of a human canine mandibular section was obtained from computerized tomography (CT) data and processed in Rhinoceros 3D. The complete 3D solid model was then converted into a 2D macroscale model by sectioning in the bucco-lingual direction. A mechanical load of 202.23 N was applied on the top of the crown at 2 mm offset horizontally from the center to the buccal side. To capture morphological features, 40 microscopic models were created to represent different coating scenarios for the localized cortical and cancellous regions with 20 models each. These microscopic models have 1 mm × 1 mm dimension and consist of the localized implant assemble, cortical or cancellous bones, and a layer of blood clot. The implant component of the model is comprised of a threaded section onto which are sintered (bonded) spherically shaped, atomized particles of Ti-6A1-4V alloy with diameters of 30, 50, 75, and 100 μm and volume fractions of 15%, 20%, 25%, 30%, and 35%, respectively. All the finite element analyses were performed using a commercial code ABAQUS with 3-node linear triangular plain strain elements (CPE3).

In order to study how the indenter tip radius affects the FEA of hard/soft coating on hard substrate, a two-dimensional (2D) axisymmetric model has been developed by Panich et al. [92] using the capacities of the ABAQUS finite element code. The hard/soft coating perfectly adhered to the substrate and was indented by a rigid conical indenter under condition of frictionless contact. The indenter has the same projected area–depth function as the standard Berkovich as it has a half angle of 70.3°. The rigid indenter was simulated by using few tip radii, including perfectly sharp tips and round tips of 0.2, 0.5, and 1.0 μm. The coating region and the adjacent substrate were finely meshed and continuously coarsened further away from the contact area. Both the coating and the substrate are considered to be isotropic which supposed to be linear elastic perfectly plastic material, as strain hardening is not considered in this case. The indenter was set to penetrate through the coated substrate boundary conditions that were applied to the centerline and bottom surface nodes, while the outermost side was assumed traction-free. Titanium was used as soft coating on the high-speed steel substrate, while titanium diboride (TiB_2) was used as hard coating.

The bone stress and strain distributions around thin HA-coated implants was observed by Aoki et al. [93] using three-dimensional FEA. A model of an implant in the mandible was developed using ANSYS FEA program. The thickness of the HA coating was reported to be 1 μm. The implants were inserted into a simplified mandible segment. The overall dimensions of this bone block were 14.0 mm (height), and 9.0 mm (width), and the bone was composed of a 2.5 mm thick cortical layer and cancellous bone. All materials used in the models were considered to be isotropic, homogeneous, and linearly elastic. The implant and bone were divided with 10-node tetrahedral structure solid (ANSYS solid 92). Forces of 100 N were applied axially and obliquely to the occlusal node at the center of the abutment. The oblique load was 45° to the vertical axis of the implant. The models were constrained in all directions at the node on the surface of the bone segment.

A novel approach that combines the indentation tests with nonlinear finite element modeling (FEM) was proposed by Zhang and coworkers [94] to estimate the elastic/plastic constitutive relation of plasma-sprayed HAp coatings on a Ti-6Al-4V substrate. Simulations for the Hertzian indentation on the Ti alloy control samples and HAp-coated Ti alloy implants were conducted by using the commercial finite element package, ABAQUS. Two-dimensional axisymmetric modeling was used for the indentation simulation under axially symmetric loading conditions, and smooth contact (no friction) was assumed. The 4-node bilinear axisymmetric quadrilateral (CAX4) elements were used in the analysis. In order to obtain the contact impression accurately, elements with fine size (10 × 10 μm) were used in the areas near the expected contact region. Linear elastic deformation was considered for the WC indenter. A post-test examination confirmed that no plastic deformation in the WC sphere occurred during indentation.

Using 3D FEA, Vlachos et al. [95] examined the behavior of coating–substrate system under ball indentation. In this analysis, different coating and substrate properties have been used for the models utilized in the simulations. FEA was performed using the implicit finite element code LS-NIKE3D. The model consists of three parts: the spherical indenter, the coating, and the substrate. The indenter is modeled using 1988 solid elements and isotropic elastic material properties. The radius of the sphere is 60 μm assuming an indenter diameter of 120 μm. The indenter was assumed as a nondeformable body. A nonfrictional contact was assumed in this analysis. The coating was modeled using 865 solid elements, assuming isotropic elastic-perfectly plastic material behavior. The coating model had the shape of a cylinder with a radius of 200 μm. The coating thickness was 10 μm for all models. The substrate was modeled using 4320 solid elements that exhibit isotropic elastic-perfectly plastic material behavior. The substrate had again the shape of a cylinder having a radius of 200 μm and a height of 190 μm. The imposed displacement was 560 nm and was applied in 56 equal increments. For each step, the force required to impose the corresponding displacement was calculated from the code and stored in an ASCII file.

Crack formation is investigated on both micro- and macroscale using spherical indenter tips by Thomsen et al. [96]. The coating used in this study is DLC deposited using a plasma-enhanced CVD technique (PECVD). Depth sensing indentation is used on the microscale and Rockwell indentation on the macroscale. A nonlinear elastic–plastic finite element model of the coating system which is loaded with a spherical indenter is used to simulated stress and displacement distributions in the material. In the numerical simulations the coating is assumed to be fully elastic, while the substrate is allowed to deform in an elastic–plastic manner. Frictionless contact between the indenter and the specimen is assumed. Since the prospective circle of contact, for an elastic–plastic contact, cannot be identified a priori, specialized contact elements are included in the finite element model. Nonlinear materials response and element contact are attained by applying the load to the indenter in increments. Due to the large strains involved, a geometric nonlinearity, or large displacement algorithm, is a required inclusion in the analysis. This "geometric" nonlinear analysis accounts for changes in stress distribution which result in a change in geometry of the model and becomes an important issue at large values of indentation strain where piling up may occur.

High stresses and complex deformation usually develop in thin films during indentation tests. Gan et al. [65] investigated the stresses and deformation in sol–gel derived zirconia films coated on stainless steel under spherical indentation using FEA. The stresses in the film under indentation and their variation as a function of the mechanical properties of the materials, the thickness of the film and radius of the indenter are studied. The indentation with a spherical indenter was modeled as an axisymmetric contact problem between two axisymmetric bodies. Simulations of the ultra-microhardness indentation were performed using the large strain, elastic–plastic features of the ABAQUS finite element code. To simulate the indentation process, the indenter was given a downward displacement which was specified as a series of steps. As the indenter moved downward into the specimen, the corresponding load was computed by summing the reaction force at the contact node points on the indenter. A perfect bonding condition between the film and substrate was assumed. Both the coating and substrate were assumed to be homogeneous and elastic/plastic.

Three-dimensional finite element simulation is applied by Wang and Bangert [97] to investigate the general behavior of coated samples subjected to an indentation test. MARC®, a commercially available finite element software package, was used to study the Vickers indentation process on the following two representative systems: copper-coated high-speed steel as an example for a soft layer on a hard substrate and titanium nitride-coated high-speed steel as the reverse situation. The calculation provides the deformation characteristics and material displacements under or after local load as well as the stress distribution inside the coating and across the interface into the substrate.

Mechanical properties of zirconia, hydroxyapatite, and alumina nanocoatings for biomedical applications were modeled by Gan and Ben-Nissan [67]. Understanding the relationship of the measured data and the mechanical properties of the indented materials is of importance in order to obtain reliable mechanical properties, particularly of the thin films. Using FEA, the effects of the elastic modulus, yield strength, and strain hardening of the film on indentation data are analyzed and discussed for the indentation with 2, 8, 10, and 50 μm radius indenters. Elastic modulus of the films on a single ductile substrate shows relatively small influence whereas yield strength and strain hardening are found to have significant effect on the measured data.

Bhattacharya and Nix [98] conducted a study on the elastic and plastic deformation associated with sub-micrometer indentation of thin films on substrates using the FEM. The effects of the elastic and plastic properties of both the film and substrate on the hardness of the film/substrate composite are studied by determining the average pressure under the indenter as a function of the indentation depth. Calculations have been made for film/substrate combinations for which the substrate is either harder or softer than the film and for combinations for which the substrate is either stiffer or more compliant than the film. Finite element simulation of the unloading portion of the load displacement curve permits the determination of the elastic compliance of the film/substrate composite as a function of indentation depth. The elastic properties of the film can be separated from those of the substrate using this information.

11.7 Conclusion

The dental and medical applications of materials and devices containing nanocoatings will increase in the next decade to be employed in implantable materials, slow drug-delivery systems, bone grafts, skin products, stem cell and biogenic material-containing scaffolds, and biologically active membranes.

Materials utilizing encapsulation or coating of therapeutic and nutritional products will increase. Slow drug delivery and targeted cancer treatment will use the nanocoatings and nanopowders. Coatings with unique electrical, magnetic, and optical properties will also be employed in diagnostic systems.

One of the major drawbacks of current synthetic implants is their failure to adapt to the local tissue environment. Until recently, surface modifications and tissue engineering has been directed toward taking advantage of the combined use of living cells and three-dimensional ceramic scaffolds to deliver vital cells to damaged sites in the body.

In the dental and biomedical fields, the surface modification of metallic materials, such as implants, aims to promote biocompatibility, inhibit wear, and reduce corrosion and ion release. Surface coatings offer the possibility of modifying the properties of a component, and therefore, improve both performance and reliability.

The mechanical properties of biomaterial micro- and nanocoatings and the substrate (i.e., implant) are strongly dependent on the film microstructure and deposition process as well as on the influence of interfacial constraint. Accurate measurement techniques are essential to determine the properties of the coatings as these properties can differ from the bulk. Above all, better techniques are required for quantitative measurements of adhesion strength, interfacial fracture toughness at the coating–substrate interface, hardness, and friction.

Measurement techniques such as instrumented nanoindentation is an essential tool for characterizing submicron coating properties, and further understanding of mechanical processes involved in these and other tests will help in obtaining valuable information, as well as in identifying the limitations of micro- and nanocoatings.

In addition, theoretical modeling approaches such as FEA are essential for the advancement in understanding thin film–substrate interfacial behavior, which may lead to better design and choice of thin film–substrate materials selection.

One of the major issues in biomedical materials research is the relationship between biological responses and surface properties of materials. Surface modification by thin film deposition has become an important tool for research aimed at understanding how structural and chemical surface properties influence material–biosystem interactions. As better understanding is achieved, one can expect that surface modifications for the purpose of controlling tissue response will open up avenues for developing new and superior implants and medical devices in a more systematic manner and at a faster rate than at present.

Bone–implant interactions, in addition to biological factors, are influenced by functionally applied multiaxial forces and biomechanics. The ultimate understanding of biological systems can only be accomplished with appropriate nanoscale mechanical properties of biogenic structures and the influence of the nanostructures and nanoloading on these biological systems.

Our drive to determine and measure mechanical properties of thin film at the nanoscale using FEA will inevitably open new avenues in understanding the influence of stresses and deformations in both the micro and nanoscale in the growth and repair mechanisms of biologic systems to allow us to use new materials, systems, and tools in tissue regeneration.

References

1. Ben-Nissan B, Choi AH. 2010. Nanoceramics for medical applications. In: *Advanced Nanomaterials*, Geckeler N (ed.). Weinheim, Germany: Wiley-VCH Verlag GmbH & Co. KgaA, Ch. 16, pp. 523–553.
2. LeGeros RZ. 1993. Biodegradation and bioresorption of calcium phosphate ceramics. *Clin Mater* 14:65–88.
3. Ducheyne P, Radin S, Heughebaert M, Heughebaert JC. 1990. Effect of calcium phosphate ceramic coatings on porous titanium: Effect of structure and composition on electrophoretic deposition, vacuum sintering and in vitro dissolution. *Biomaterials* 11:244–254.
4. Hench LL, West JK. 1990. The sol–gel process. *Chem Rev* 90:33–72.
5. Kokubo T, Kim HM, Kawashita M, Nakamura T. 2000. Novel ceramics for biomedical applications. *J Aust Ceram Soc* 36:37–46.
6. Ben-Nissan B, Chai CS. 1995. Sol–gel derived bioactive hydroxyapatite coatings. In: *Advances in Materials Science and Implant Orthopedic Surgery, NATO ASI Series, Series E: Applied Sciences*, Kossowsky R, Kossovsky N (eds.). Dordrecht, the Netherlands: Kluwer Academic Publishers, Vol. 294, pp. 265–275.
7. Chang MC, Ko CC, Douglas WH. 2003. Preparation of hydroxyapatite-gelatin nanocomposite. *Biomaterials* 24:2853–2862.
8. Zhang W, Liao SS, Cui FZ. 2003. Hierarchical self-assembly of nano-fibrils in mineralized collagen. *Chem Mater* 15:3221–3226.
9. Vinogradov SV, Bronich TK, Kabanov AV. 2002. Nanosized cationic hydrogels for drug delivery: Preparation, properties and interactions with cells. *Adv Drug Delivery Rev* 54:135–147.

10. Vinogradov SV, Batrakova EV, Kabanov AV. 2004. Nanogels for oligonucleotide delivery to the brain. *Bioconjug Chem* 15:50–60.

11. Park EJ, Brasuel M, Behrend C, Philbert MA, Kopelman R. 2003. Ratiometric optical PEBBLE nanosensors for real-time magnesium ion concentrations inside viable cells. *Anal Chem* 75:3784–3791.

12. Clark HA, Hoyer M, Philbert MA, Kopelman R. Optical nanosensors for chemical analysis inside single living cells. 1. Fabrication, characterization, and methods for intracellular delivery of PEBBLE sensors. *Anal Chem* 71:4831–4836.

13. Gavalas VG, Law SA, Ball JC, Andrews R, Bachas LG. 2004. Carbon nanotube aqueous sol-gel composites: Enzyme-friendly platforms for the development of stable biosensors. *Anal Biochem* 329:247–252.

14. Huang TT, Jones AS, He LH, Darendeliler MA, Swain MV. 2007. Characterisation of enamel white spot lesions using X-ray micro-tomography. *J Dent* 35:737–743.

15. Choi AH, Ben-Nissan B. 2007. Sol–gel production of bioactive nanocoatings for medical applications: Part II: Current research and development. *Nanomedicine* 2:51–61.

16. Lin CF. 2010. A computational protocol to the prediction of dental implant induced bone remodelling and its material design: A novel approach. PhD dissertation. University of Sydney, New South Wales, Australia.

17. Xing M, Zhong W, Xu X, Thomson D. 2010. Adhesion force studies of nanofibers and nanoparticles. *Langmuir* 26:11809–11814.

18. Elliott AM, Shetty AM, Wang J, Hazle JD, Jason Stafford R. 2010. Use of gold nanoshells to constrain and enhance laser thermal therapy of metastatic liver tumours. *Int J Hyperthermia* 26:434–440.

19. Lin D, Li Q, Li W, Duckmanton N, Swain M. 2010. Mandibular bone remodeling induced by dental implant. *J Biomech* 43:287–293.

20. Sassaroli E, Li KC, O'Neill BE. 2009. Numerical investigation of heating of a gold nanoparticle and the surrounding microenvironment by nanosecond laser pulses for nanomedicine applications. *Phys Med Biol* 54:5541–5560.

21. Verhulp E, Van Rietbergen B, Muller R, Huiskes R. 2008. Micro-finite element simulation of trabecular-bone post-yield behaviour—Effects of material model, element size and type. *Comput Methods Biomech Biomed Engin* 11:389–395.

22. Verhulp E, van Rietbergen B, Huiskes R. 2008. Load distribution in the healthy and osteoporotic human proximal femur during a fall to the side. *Bone* 42:30–35.

23. Chan KS, Lee YD, Nicolella DP, Furman BR, Wellinghoff S, Rawls HR. 2007. Improving fracture toughness of dental nanocomposites by interface engineering and micromechanics. *Eng Fract Mech* 74:1857–1871.

24. Isaksson H, Wilson W, van Donkelaar CC, Huiskes R, Ito K. 2006. Comparison of biophysical stimuli for mechano-regulation of tissue differentiation during fracture healing. *J Biomech* 39:1507–1516.

25. Taylor M, Verdonschot N, Huiskes R, Zioupos P. 1999. A combined finite element method and continuum damage mechanics approach to simulate the in vitro fatigue behaviour of human cortical bone. *J Mater Sci Mater Med* 10:841–846.

26. Vaillancourt H, Pilliar RM, McCammond D. 1995. Finite element analysis of crestal bone loss around porous-coated dental implants. *J Appl Biomater* 6:267–282.

27. Ben-Nissan B, Choi AH. 2006. Sol–gel production of bioactive nanocoatings for medical applications: Part I: An introduction. *Nanomedicine* 1:311–319.

28. Ben-Nissan B, Latella BA, Bendavid A. 2011. Biomedical thin films: Mechanical properties. In: *Comprehensive Biomaterials*, Ducheyne P, Healy KE, Hutmacher DW, Grainger DW, Kirkpatrick CJ (eds.). Oxford, U.K.: Elsevier, Vol. 3, pp. 63–73.

29. Choy KL. 2003. Chemical vapour deposition of coatings. *Prog Mater Sci* 48:57–170.

30. Kirk P, Pilliar R. 1999. The deformation response of sol–gel-derived thin films on 316L stainless steel using a substrate straining test. *J Mater Sci* 34:3967–3975.

31. Chen TS, Lacefield WR. 1994. Crystallisation of ion beam deposited calcium phosphate coatings. *J Mater Res* 9:1284–1296.
32. Anast M, Bell J, Bell T, Ben-Nissan B. 1992. Precision ultra-microhardness measurements of sol–gel derived zirconia thin films. *J Mater Sci Lett* 11:1483–1485.
33. Roest R, Eberhardt AW, Latella BA, Wuhrer R, Ben-Nissan B. 2004. Adhesion of sol–gel derived zirconia nano-coatings on surface treated titanium. In Bioceramics 16. *Key Eng Mater* 254–256:455–458.
34. Ebelmen J. 1846. Untersuchungen über die Verbindung der Borsaure und Kieselsaure mit Aether. *Ann Chim Phys Ser* 57:319–355.
35. Geffcken W, Berger E. 1939. Änderung des Reflexionsvermogens Optischer Gläser. German Patent 736411.
36. Roy DM, Roy R. 1954. An experimental study of the formation and properties of synthetic sepentines and related layer silicates. *Am Mineral* 39:957–975.
37. Schroeder H. 1965. Water-dispersed industrial and architectural coatings. *Paint Varnish Prod* 55:31–46.
38. Chai C, Ben-Nissan B, Pyke S, Evans L. 1995. Sol–gel derived hydroxyapatite coatings for biomedical applications. *Mater Manuf Process* 10:205–216.
39. Mazdiyasni KS. 1982. Powder synthesis form metal–organic precursors. *Ceram Int* 8:42–56.
40. Sakka S, Kamiya K, Makita K, Tamamoto Y. 1984. Formation of sheets and coating films from alkoxide solutions. *J Noncryst Solids* 63:223–235.
41. Yoldas BE. 1984. Wide-spectrum anti-reflective coatings for fused silica and other glasses. *Appl Opt* 23:1418.
42. Roy R. 1987. Ceramics by the solution–sol–gel route. *Science* 238:1664–1669.
43. Klein LC. (ed.). 1988. *Sol–Gel Technology for Thin Films, Fibers, Preforms, Electronics and Specialty Shapes*. Park Ridge, NJ: Noyes Publishing.
44. Scriven LE. 1988. Physics and application of dip coating and spin coating. *Mater Res Soc Symp Proc* 121:717–729.
45. Brinker CJ, Clark DE, Ulrich DR. (eds.). 1988. *Better Ceramics through Chemistry*, 3rd edn. Pittsburgh, PA: Materials Research Society.
46. Brinker CJ, Scherer GW. 1990. *Sol–Gel Science: The Physics and Chemistry of Sol–Gel Processing*. London, U.K.: Academic Press.
47. Floch HG, Belleville PF, Priotton JJ, Pegon PM, Dijonneau CS, Guerain J. 1995. Sol–gel optical coatings for lasers. *I Am Ceram Soc Bull* 74:60–63.
48. Percy MJ, Bartlett JR, Spiccia L, West BO, Woolfrey JL. 2000. The influence of b–diketones on hydrolysis and particle growth from zirconium (IV) N-propoxide in n-propanol. *J Sol–Gel Sci Technol* 19:315–319.
49. de Groot K, Geesink R, Klein CP, Serekian P. 1987. Plasma sprayed coatings of hydroxylapatite. *J Biomed Mater Res* 21:1375–1381.
50. Hench LL. 1991. Bioceramics: From concept to clinic. *J Am Ceram Soc* 74:1487–1510.
51. Ducheyne P, Hench LL, Kagan A II, Martens M, Bursens A, Mulier JC. 1980. Effect of hydroxyapatite impregnation on skeletal bonding of porous coated implants. *J Biomed Mater Res* 14:225–237.
52. Jarcho M. 1981. Calcium phosphate ceramics as hard tissue prosthetics. *Clin Orthop* 157:259–279.
53. Dorozhkin SV. 2009. Nanodimensional and nanocrystalline apatites and other calcium orthophosphates in biomedical engineering, biology and medicine. *Materials* 2:1975–2045.
54. Ben-Nissan B, Milev A, Vago R. 2004. Morphology of sol-gel derived nano-coated coralline hydroxyapatite. *Biomaterials* 25:4971–4976.
55. Bendavid A, Martin PJ, Comte C, Preston EW, Haq AJ, Ismail FSM, Singh RK, 2007. The mechanical and biocompatibility properties of DLC-Si films prepared by pulsed DC plasma activated chemical vapor deposition. *Diam Relat Mater* 16:1616–1622.
56. Dearnaley G, Arps JH. 2005. Biomedical applications of diamond-like carbon (DLC) coatings: a review. *Surf Coat Technol* 200:2518–2524.

57. Amin MS, Randeniya LK, Bendavid A, Martin PJ, Preston EW, 2009. Amorphous carbonated apatite formation on diamond-like carbon containing titanium oxide. *Diam Relat Mater* 18:1139–1144.

58. Bendavid A, Martin PJ, Randeniya L, Amin MS. The properties of fluorine containing diamond-like carbon films prepared by plasma-enhanced chemical vapour deposition. *Diam Relat Mater* 18:66–71.

59. Randeniya LK, Bendavid A, Martin PJ, Amin MS, Preston EW, Ismail FSM, Coe S. 2009. Incorporation of Si and SiO(x) into diamond-like carbon films: Impact on surface properties and osteoblast adhesion. *Acta Biomat* 5:1791–1797.

60. Uzumaki ET, Lambert CS, Belangero WD, Freire CMA, Zavaglia CAC. 2006. Evaluation of diamond-like carbon coatings produced by plasma immersion for orthopaedic applications. *Diam Relat Mater* 15:982–988.

61. Ben-Nissan B, Pezzotti G. 2002. Bioceramics: Processing routes and mechanical evaluation. *J Ceram Soc Japan* 110:601–608.

62. Field JS, Swain MV. 1993. A simple predictive model for spherical indentation. *J Mater Res* 8:297–306.

63. Field JS, Swain MV. 1995. Determining the mechanical properties of small volumes of material from submicrometer spherical indentations. *J Mater Res* 10:101–112.

64. Fischer-Cripps AC. 2002. *Introduction to Nanoindentation*. New York: Springer.

65. Gan L, Ben-Nissan B, Ben-David A. 1996. Modelling and finite element analysis of ultra-microhardness indentation of thin films. *Thin Solid Films* 290–291:362–366.

66. Kealley CS, Latella BA, van Riessen A, Elcombe MM, Ben-Nissan, B. 2008. Micro- and nano-indentation of a hydroxyapatite-carbon nanotube composite. *J Nanosci Nanotech* 8:3936–3941.

67. Gan L, Ben-Nissan B. 1997. The effect of mechanical properties of thin films on nano-indentation data: Finite element analysis. *Comput Mater Sci* 8:273–281.

68. Fischer-Cripps AC. 2004. A simple phenomenological approach to nanoindentation creep. *Mater Sci Eng A* 385:74–82.

69. Latella BA, Gan BK, Barbé CJ, Cassidy DJ. 2008. Nanoindentation hardness, Young's modulus, and creep behaviour of organic–inorganic silica-based sol–gel thin films on copper. *J Mater Res* 23:2357–2365.

70. Angker L, Nockolds C, Swain MV, Kilpatrick N. 2004. Correlating the mechanical properties to the mineral content of carious dentine— A comparative study using an ultra-micro indentation system (UMIS) and SEM-BSE signals. *Arch Oral Biol* 49:369–378.

71. Arends J, Ruben J, Jongebloed WL. 1989. Dentin caries *invivo*-combines scanning electron-microscopic and microradiographic investigation. *Caries Res* 23:36–41.

72. Featherstone JD, ten Cate JM, Shariati M, Arends J. 1983. Comparison of artificial caries-like lesions by quantitative microradiography and microhardness profiles. *Caries Res* 17:385–391.

73. Kodaka T, Debari K, Yamada M, Kuroiwa M. 1992. Correlation between microhardness and mineral content in sound human enamel. *Caries Res* 26:139–141.

74. Ten Bosch JJ, Angmar-Mansson B. 1991. A review of quantitative methods for studies of mineral content of intra-oral caries lesions. *J Dent Res* 70:2–14.

75. Angker L, Swain MW, Kilpatrick N. 2005. Characterising the micro-mechanical behaviour of the carious dentine of primary teeth using nano-indentation. *J Biomechanics* 38:1535–1542.

76. Mahoney E, Ismail FS, Kilpatrick N, Swain M. 2004. Mechanical properties across hypomineralized/hypoplastic enamel of first permanent molar teeth. *Eur J Oral Sci* 112:497–502.

77. Marshall GW Jr, Balooch M, Gallagher RR, Gansky SA, Marshall SJ. 2001. Mechanical properties of the dentinoenamel junction: AFM studies of nanohardness, elastic modulus, and fracture. *J Biomed Mater Res* 54:87–95.

78. Meredith N, Sherriff M, Setchell DJ, Swanson SA. 1996. Measurement of the microhardness and Young's modulus of human enamel and dentine using an indentation technique. *Arch Oral Biol* 41:539–545.

79. Kinney JH, Balooch M, Marshall SJ, Marshall GW Jr, Weihs TP. 1996. Hardness and Young's modulus of human peritubular and intertubular dentine. *Arch Oral Biol* 41:9–13.

80. Van Meerbeek B, Willems G, Celis JP, Roos JR, Braem M, Lambrechts P, Vanherle G. 1993. Assessment by nano-indentation of the hardness and elasticity of the resin-dentin bonding area. *J Dent Res* 72:1434–1442.

81. Waters NE. 1980. Some mechanical physical properties of teeth. In: *The Mechanical Properties of Biological Materials.* Vincent JFV, Currey JD, (eds.). Cambridge, U.K.: Cambridge University Press, pp. 99–134.

82. Mencick, J. 1992. *Strength and Fracture of Glass and Ceramics—Glass Science and Technology.* Amsterdam, the Netherlands: Elsevier.

83. Choi AH, Matinlinna JP, Ben-Nissan B. 2011. Three-dimensional finite element nanoindentation analysis of sol-gel derived nanocoatings. *Comp Meth Biomech Biomed Eng* (Submitted).

84. Lucchini R, Carnelli D, Ponzoni M, Bertarelli E, Gastaldi D, Vena P. 2011. Role of damage mechanics in nanoindentation of lamellar bone at multiple sizes: Experiments and numerical modelling. *J Mech Behav Biomed Mater* 4:1852–1863.

85. Adam CJ, Swain MV. 2011. The effect of friction on indenter force and pile-up in numerical simulations of bone nanoindentation. *J Mech Behav Biomed Mater* 4:1554–1558.

86. Carnelli D, Lucchini R, Ponzoni M, Contro R, Vena P. 2011. Nanoindentation testing and finite element simulations of cortical bone allowing for anisotropic elastic and inelastic mechanical response. *J Biomech* 44:1852–1858.

87. Paietta RC, Campbell SE, Ferguson VL. 2011. Influences of spherical tip radius, contact depth, and contact area on nanoindentation properties of bone. *J Biomech* 44:285–290.

88. Toparli M, Koksal NS. 2005. Hardness and yield strength of dentin from simulated nano-indentation tests. *Comput Methods Programs Biomed* 77:253–257.

89. Dong XD, Darvell BW. 2003. Stress distribution and failure mode of dental ceramic structures under Hertzian indentation. *Dent Mater* 19:542–551.

90. Wei C, Yang JF. 2011. A finite element analysis of the effects of residual stress, substrate roughness and non-uniform stress distribution on the mechanical properties of diamond-like carbon films. *Diam Relat Mater* 20:839–844.

91. Rungsiyakull C, Li Q, Sun G, Li W, Swain MV. 2010. Surface morphology optimization for osseointegration of coated implants. *Biomaterials* 31:7196–7204.

92. Panich N, Wangyao P, Surinphong S, Tan YK, Sun Y. 2007. Finite element analysis study on effect of indenter tip radius to nanoindentation behavior and coatings properties. *J Metals Mater Miner* 17:43–49.

93. Aoki H, Ozeki K, Ohtani Y, Fukui Y, Asaoka T. 2006. Effect of a thin HA coating on the stress/strain distribution in bone around dental implants using three-dimensional finite element analysis. *Biomed Mater Eng* 16:157–169.

94. Zhang C, Leng Y, Chen J. 2001. Elastic and plastic behaviour of plasma-sprayed hydroxyapatite coatings on a Ti-6Al-4V substrate. *Biomaterials* 22:1357–1363.

95. Vlachos DE, Markopoulos YP, Kostopoulos V. 2001. 3-D modelling of nanoindentation experiment on a coating-substrate system. *Comp Mech* 27:138–144

96. Thomsen NB, Fischer-Cripps AC, Swain MV. 1998. Crack formation mechanisms during micro and macro indentation of diamond-like carbon coatings on elastic-plastic substrates. *Thin Solid Films* 332:180–184.

97. Wang HF, Bangert H. 1993. Three-dimensional finite element simulation of Vickers indentation on coated systems. *Mater Sci Eng A* 163:43–50

98. Bhattacharya AK, Nix WD. 1988. Analysis of elastic and plastic deformation associated with indentation testing of thin films on substrates. *Int J Solids Struct* 24:1287–1298.

12

Application of Finite Element Analysis for Nanobiomedical Study

Viroj Wiwanitkit

Hainan Medical University, Haikou, People's Republic of China
and
Wiwanitkit House, Bangkok, Thailand

CONTENTS

12.1 Introduction 401
12.2 Overview of Finite Element Analysis and Its Application in Medicine...................402
12.3 Computational Nanomedicine with Finite Element Analysis 406
    12.3.1 Summary of Important Reports on the Application of Finite Element
        Analysis in Nanomedicine 407
12.4 Computational Finite Element Analysis Tools for
    Management of Biomedical Data 408
    12.4.1 Usefulness of Computational Finite Element Analysis in Nanomedicine......408
    12.4.2 Important Computational Finite Element Analysis Tool for
        Management of Biomedical Data 409
12.5 Common Applications of Computational Finite Element Analysis in General
    Nanomedicine Practice 410
    12.5.1 Examples of Nanomedicine Research Based on Computational Finite
        Element Analysis Application..412
        12.5.1.1 General Cases 412
12.6 Conclusion 418
References...418

12.1 Introduction

The integration of sciences leads to the advent of many discoveries and this is the trend of novel science in the present era. The link between physical and biological sciences has resulted in new approaches to solving many previously hard-to-explain problems. Combining two different sciences leads to a useful new science. A good example of such an integration is nanoscience, which is a multidisciplinary approach encompassing both physical and biological concerns to approaching to a situation. Nanoscience is accepted as "the science of the future." At present, there are many branches of nanoscience such as nanochemistry, nanopharmacology, nanoengineering, nanomedicine, etc., each of which is recognized for its usefulness.

Nanomedicine, the specific branch of nanoscience that covers medical aspects, is still a novelty in medicine. Based on advances in nanotechnology, diagnosis, treatment, and

TABLE 12.1

Usage of Computer-Based Approach for Problems in Nanomedicine

Applications	Details
1. Clarification	Computer-based approach can help clarify processes or phenomena in medicine. The examples are clarification of the nanostructure of a medical molecule (enzyme, hormone, antigen, antibody, etc.)
2. Prediction	Computer-based approach also helps predict or simulate processes or phenomenon in medicine. The predictions of the changes of structure after biomedical interaction are good examples (such as interaction between drug and receptor, interaction between antibody and pathogen, etc.)

Sources: Haddish-Berhane, N. et al., *Int. J. Nanomedicine*, 2(3), 315, 2007; Saliner, A.G. et al., *IDrugs*, 11(10), 728, October 2008.

prevention of disease can be easily done. Fast, highly accurate, and reliable medical activity can be expected. Today, nanomedicine is not "hope" but it is "fact." The applications of nanomedicine can be used in daily clinical practice (either in vivo, in vitro, or in silico [1]), and nanomedicine-based medical services are already in use in many medical institutes around the world.

Both in vivo and in vitro approaches are well-known forms in medicine. They are standard approaches in medical science. However, the so-called "in silico approach" might be a new concept in medical science. It is based on a computational approach. In silico technique might be described as a simulating or imaginary medical approach. This does not mean that in silico technique is not acceptable; in fact, it is very useful in medicine. The advantages of using this approach in medicine include shortening turnaround time, decreasing the cost of experiments, getting rid of confounding interference, and accessibility.

The in silico approach makes use of computational simulation techniques to help solve medical issues. A good example of this technique is "omics" science, which is well known in the biomedical society [2]. Application of computational technology can be performed for any kind of medicine including nanomedicine. As nanomedicine is a specific medicine that deals with very small objects, it might be an imaginary approach in another sense. Hence, the application of in silico techniques to help clarify and predict phenomena in nanomedicine is feasible (Table 12.1) [3,4].

At present, several computational techniques can be selected in nanomedicine research. However, an important technique that can be used is the finite element analysis. This can be used in studying medical structure and function at the nanolevel. This chapter discusses how finite element analysis can be useful in nanomedicine. In addition, a summary of important reports on the application of finite element analysis in nanomedicine is given.

12.2 Overview of Finite Element Analysis and Its Application in Medicine

Finite element analysis is a scientific term used to explain an analysis technique in engineering [5]. The main purpose of this engineering technique is to analyze structure or "element." It is one of the four main analyses for any phenomenon in engineering: finite different analysis, finite element analysis, finite volume analysis, and finite point analysis.

In brief, finite element analysis means analysis of an "element" by name, by separating the element into many exact numbers of small pieces of focused elements and using mathematical modeling of each separate piece for further matrix analysis. It can be useful for either solid or fluidic mechanical analyses. In fact, several kinds of elements can be analyzed such as rod elements, beam elements, composite elements, solid elements, spring elements, mass elements, rigid elements, viscous damping elements, etc.

Historically, finite element analysis has been acknowledged for more than 70 years beginning with the attempt to use specific numerical analysis methods accompanied with variational calculus for solving questions on vibration systems. It has since developed rapidly and become one of the main techniques for the analysis of structures in engineering. Today, it is accepted as a useful technique for the analysis of structural rigidity and defection. At first, it was widely used in mechanical engineering; however, due to the advantages of this analytical approach, it was further applied in the material sciences and then to other sciences. In the first period of development of finite element analysis, the technique seems to have been very difficult and complex process as it required manipulation of difficult mathematics, partial differential equations, and integral equations. However, the increasing use of computer science has eased the analysis. In the 40 years since the implementation of computers, more generalization of finite element analysis can be seen, and in the present day it has integrated into several engineering industries.

In the present computer era, computational modeling and designing of materials becomes the main issue of finite element analysis. Computational finite element analysis helps both prevention and correction activities. This can be helpful in product design, surveillance, and refinement. By focusing on prevention, computational finite element analysis can help solve the problem of unwanted manufacturing and construction. With prediction, the nonconformation piece will not be produced. On the other hand, computational finite element analysis also helps correct the problem of failed or erroneous pieces of work. It helps identify problems and design better pieces of work.

Finite element analysis can be a two- or three-dimensional model. Several parameters can be used in the simulation of a condition, and numerous algorithms or functions can be incorporated in analysis. The system can also be either linear or nonlinear. This means the high flexibility of computational finite element analysis in approaching a research question.

Finite element analysis can be very useful in engineering. It can be applied for engineering analysis. The widely used types of engineering analysis include the following:

1. *Structural analysis*: A very basic and direct usage of finite element analysis. The analysis can be based on either linear or nonlinear models as already discussed. Simple parameters are used, and it is usually primarily assumed for nonplastic deformity property of the analyzed elements. Stresses in the material can be simulated, and assessment of deformation is the main focused activity.

2. *Vibrational analysis*: A type of advanced analysis. The simulation of applying several vibrations to an element is performed. The assessment of responses of the material such as occurrence of resonance and subsequent failure is the main aim of this kind of analysis.

3. *Fatigue analysis*: Another kind of advanced analysis. The prediction of the durability of the element to effects of cyclic loading is primarily performed, and this is useful for prediction of crack propagation. This can give the tolerance value or lifespan of the studied materials.

TABLE 12.2

Application of Computational Finite Element Analysis in Medicine

Applications	Details
1. Structural approach	This application is used to deal with the question on structure. Since finite element analysis is a primary technique for material study, it can be applied to access any medical materials
2. Functional approach	This application is used to solve the problem of functional change due to several interferences (vibration, stress, or temperature). It is based on the basic principle of simulation in medicine

4. *Heat transfer analysis*: This advanced application is aimed at assessment of the conductivity or thermal fluid dynamics of the studied structure. Based on this analysis, a steady-state or transient transfer with constant thermoproperties in the material can be predicted.

5. *Electricity transfer analysis*: An advanced application aimed at assessment of the electrical dynamics of the studied structure. Based on this analysis, a steady state with constant electroproperties in the material can be predicted.

With the bridging between physical and biological science, the connection between engineering and medicine occurs. The hybrid science, medical engineering, is a complex science that needs both engineering and medical techniques for approaching the question.

Computational finite element analysis plays a role in several techniques. Although there are many computer-based analyses for medical research, at present a widely used analysis is computational finite element analysis. Structure is an important aspect of study in medical engineering, and the use of computational finite analysis to answer the questions raised in medical structure is not surprising. The advantages of computational finite element analysis can be applied in medicine (Table 12.2). In general, the use of computational finite element analysis in medicine can be either simple or advanced. The first level is simple usage. Simple usage can be employed for simple structural assessment and is based on the fundamental usefulness of finite element analysis in material science. The second level is advanced usage. This is the prediction of the structure after simulating change and is based on the in silico application of the interference, which can be vibration, stress, or temperature. This cannot be easily derived by other computational analytical programs. It should be noted that many medical phenomena are considered vibration, and this can be the cause of structural change. In addition, finite element analysis is also used as a basic concept for the development of many complex computational programs that can be applied in medicine.

At this point, the reader might still not be able to imagine the exact application of finite element analysis in medicine. To help better understand the topic, various common applications will be further discussed.

1. *Application in orthopedics*: There is little doubt that orthopedics is a main branch of medicine dealing with materials, bone, and bone replacement materials. Over the past 40 years, biomechanics studies have been widely performed and have become an important part of orthopedics research [6]. Huiskesand Chaosaid opines that "The method is now well established as a tool for basic research and for design analysis in orthopedic biomechanics, and the number of publications in which it is used is increasing rapidly [6]."

Following is a list of important new and interesting publications (in 2011) in this area:

a. Ma et al. [7] studied an important orthopedics problem, femoral head necrosis and femoral neck fractures via assessment; the relationship between the femoral head trabecular bone within the spatial structure and its biomechanics. In this work, a three-dimensional model of trabecular bone was obtained via computational finite element analysis technique [7].

b. Zhang et al. [8] used three-dimensional finite element models to study "the effect of vertebrae semidislocation on the stress distribution in facet joint and intervertebral disc of patients with cervical syndrome" and reported that "The vertebra semidislocation leads to the abnormal stress distribution of facet joint and intervertebral disc."

c. Fei et al. [9] used three-dimensional finite element models to explore "the biomechanical effects on adjacent vertebra of thoracolumbar osteoporotic vertebral compression fracture after percutaneous kyphoplasty with cement leakage."

d. Eichinger et al. [10] used three-dimensional finite element models to determine the effectiveness of screw hole inserts in empty locking screw holes for improving the strength and failure characteristics of locking plates.

e. Olson et al. [11] studied thermal effects of glenoid reaming during shoulder arthroplasty with the use of computational finite element analysis approach.

f. He et al. [12] studied the biomechanical mechanism of spinal three-column after interspinous process fusion by computational finite element analysis approach.

2. *Application in dentistry*: In addition to orthopedics, another widely used application of finite element analysis is in dentistry. Indeed, the knowledge of dental materials is important in filling of teeth. Dental material studies are widely performed and have become an important part of dental research.

Here are some important new and interesting publications (2011) in this area:

a. da Silva et al. [13] used three-dimensional finite element analysis to analyze the maxillary central incisor in two different situations of traumatic impact.

b. Campos et al. [14] used three-dimensional finite element analysis to analyze the pattern of ceramic crowns.

c. Naini et al. [15] used three-dimensional finite element analysis to analyze and assess tilted and parallel implant placement in the completely edentulous mandible.

d. Choi et al. [16] used three-dimensional finite element analysis to analyze Ti-6Al-4V and partially stabilized zirconia dental implants during clenching.

e. Panagiotopoulou et al. [17] used three-dimensional finite element analysis to analyze the mechanical significance of morphological variation in the macaque mandibular symphysis during mastication.

f. Chang et al. [18] used three-dimensional finite element analysis to analyze the effects of implant diameter and bone quality in short implants placed in the atrophic posterior maxilla.

g. Alikhasi et al. [19] used three-dimensional finite element analysis to analyze stress distribution around maxillary anterior implants.

h. Ghuneim [20] used three-dimensional finite element analysis to analyze tooth replica custom implants.

i. Dos Santos et al. [21] used three-dimensional finite element analysis to analyze the influence of different soft liners on stress distribution in peri-implant bone tissue during healing period.

j. Lin et al. [22] used three-dimensional finite element analysis to analyze and compare stress distribution to connector of lithia disilicate-reinforced glass-ceramic and zirconia-based fixed partial denture.

k. Rungsiyakull et al. [23] used three-dimensional finite element analysis to analyze the effects of occlusal inclination and loading on mandibular bone remodeling.

l. Benazzi et al. [24] used three-dimensional finite element analysis to analyze stress distributions in human molars in cases with occlusal wear information.

m. Han et al. [25] used three-dimensional finite element analysis to analyze biomechanical distribution of dental implants with immediate loading.

12.3 Computational Nanomedicine with Finite Element Analysis

Nanomedicine is the medical approach to things at a nanoscale. A common question asked is, "What is a nanoscale?" The major factor in determining nanoscale is the size or quantity. Viruses and hormones are good examples of medically related subjects at the nanoscale level. A virus is a very small object, and its size is usually measured in nanometers, which falls in the nanoscale. Viruses play an important role in medicine since they can cause infectious diseases. Hormones are very small secret biomaterial in human bodies. Their quantity is usually measured in nanogram per liter. This unit also falls into the nanoscale. Hence, nanoscale is not new but is well-known in several areas of medicine. However, due to the limitation of the sense organs of human beings, working within the nanoscale is usually difficult. Hence, several tools have to be developed to help solve the problem. There must be a technique to approach the nanolevel medical phenomenon. As noted earlier, the application of in silico technique is an acceptable solution [3,4].

In silico technique helps medical scientists solve many problems. In informatics, technology is already merged with nanoscience to form a newer science called "nanoinformatics" [26]. Nanoinformatics helps in approaching nanoscale phenomena in nanomedicine. Generally, a nanoinformatics approach can help solve existing problematic phenomenon as well as predict imaginary interactions. Focusing on nanomedicine, the computational approach is confirmed for its usefulness. Since processes in nanomedicine can be based on the same concepts as those in medicine, computational finite element analysis is probable and feasible. Both assessment of nanomaterial structure and simulation of interferences (vibration, stress, or temperature) on nanomaterial structure can be done. The simple and advanced usages can be similar to use in general medicine. The advanced use for clarification or prediction of kinetic things or processes in nanomedicine helps bring great progression to the field. More details on examples of important reports on this topic can be seen in the next section.

12.3.1　Summary of Important Reports on the Application of Finite Element Analysis in Nanomedicine

As earlier noted, computational finite element analysis can have great advantages in nano-medicine, and computational finite element analysis is presently widely used in it [27–30]. Here, the author provides a summary of important reports on the application of computational finite element analysis in nanomedicine (Table 12.3) [27,31–34].

1. *Application for clarification*: A good example of the application of computational finite element analysis is its usage in nanomaterials. Clarification of the structure of new nanomaterial in nanomedicine is possible. Many studies have been done on gold nanoparticles, a basic nanomaterial that is widely used in medicine, especially for diagnostic purposes. Lee et al. reported the assessment of their concept on using DNA-gold nanoparticle as markers for cell surface marker site [35]. In addition, clarification can also be applied to the nanostructure of biomaterials [36]. This is a basic nanometallology concept. The main application is on orthopedic and orthodontic materials as previously discussed.

2. *Application for prediction*: A good example of the application of computational finite element analysis is its usage in therapy. Prediction of the action of nanoparticles in therapeutic processes under different conditions is possible. Indeed, computational finite element analysis is a major technique used to assess inference-induced change of therapeutic nanoparticle during treatment. There are some interesting reports on this topic. Elliott et al. studied the use of gold nanoshells as adjuvant to laser thermal therapy in treatment of metastatic liver tumors [37]. Heat transfer simulation model was used in this work [37]. Elliott et al. state, "This indicates a potential to use nanoparticles to enhance both the safety and efficacy [37]." An interesting article describing the change of nanoshell during hyperthermia therapy was published by Liu et al [38]. Since nanoparticles have increased roles in novel treatments, especially for cancer treatments, the role of computational finite element analysis might also be enhanced.

In addition, since change is a commonly expected thing in any treatment, the application for prediction can also be useful in the assessment of the pharmacologic reaction of drug. It can be helpful in the nanosystem of drug delivery and targeting [32]. There are some

TABLE 12.3

Some Important Reports on Computational Finite Element Analysis in Nanomedicine

Authors	Details
Xing et al. [31]	Xing et al. used computational finite element analysis to analyze adhesion force of nanofibers and nanoparticles
Babincova and Babinec [32]	Babincova and Babinec explained the application of finite element analysis on magnetic drug delivery and targeting
Sassaroli et al. [33]	Sassaroli et al. used computational finite element analysis to analyze heating of a gold nanoparticle and the surrounding microenvironment by nanosecond laser pulses
Johnson et al. [34]	Johnson et al. used computational finite element analysis to analyze the effects of elastic anisotropy on strain distributions in decahedral gold nanoparticles
Yih et al. [27]	Yih et al. used computational finite element analysis in studying a nanoliter drug-delivery MEMS micropump with circular bossed membrane

available reports on this area. For example, Kim and Simon used computational finite element analysis to predict transport mechanisms in oral transmucosal drug delivery [39]. Wu et al. studied delivery and release of nitinolstent in carotid artery and their interactions via computational finite element analysis [40].

To summarize, finite element analysis is applicable for various sections of nanomedicine:

a. *In nanodiagnosis:* Such as modeling of the surface of nanomaterials used in development of nanobiosensors and study of the effect of magnetic fields or electricity on the nanofluidic tool

b. *In nanopharmacology:* Such as modeling of the surface of drugs and the study of the effect of direct stress, magnetic field, or electricity on drugs

12.4 Computational Finite Element Analysis Tools for Management of Biomedical Data

Based on computational technology, manipulation of data in medical science can be easier than it was in the past. Surface parameters are also an important group of data in biomedicine. Indeed, in physiology, characteristics and change in surface parameters of any biomaterial can significantly affect the biological process and reaction. Such process and reaction occur at a very small scale, usually nanoscale; hence, it is a main topic to be discussed in nanomedicine. The management of specific surface parameters data generated in nanomedical science is an important practice in nanomedicine, and advanced computational technology can help ease this practice.

Several nanomedical surface parameters can be seen in nanomedicine. Examples are the data on the nanostructure of the diagnostic particle, germ, drug, and cells. Understanding of the pathophysiology of an abnormality and designing of new diagnostic tools as well as therapeutic agents based on the nanostructure data can be more specific than previous classical practice. Manipulation of these biomedical data using computational tools is feasible. To get the most effective data management, selection of proper tools for specific work has to be done. Thus, it is necessary to understand the presently available tools. Here, we briefly summarize some important available tools that use the computational finite element analysis concept. These computational tools can be useful for work in nanomedicine.

12.4.1 Usefulness of Computational Finite Element Analysis in Nanomedicine

Finite element analysis uses an imaginary dividing of the element into small parts called nodes, which then make a framework grid called mesh. The generated mesh is the part to be further analyzed. In the past, this was based on very difficult mathematical calculations; however, with the use of computers, this has become much easier and application of interferences can be done. In nanomedicine, the assigned interferences can be heat, electricity, strain, pressure, etc. As discussed, computational finite element analysis is useful in manipulation of data in nanomedicine. Finite element analysis can be a fundamental concept for further development of programs for general usage. The reasons include the following: (a) finite element analysis is a basic concept in structure analysis and (b) structural problems are common in biomedicine. Using computational finite element analysis to solve a problem or to simulate a case is possible and can be an effective solution to biomedical research questions.

Focusing on the specific program for help with computational analysis, there are several available programs. A good one is MATLAB®, which is an effective measure with very few noises in processing [41]. It can be implemented for usage in finite element analysis of the model and widely used for structural clarification and prediction in reconstructive medicine at present [42]. Although MATLAB is an expensive program, it has several advantages and seems to be cost effective. The ways that MATLAB can be used to support the finite element analysis in nanomedicine problems are further discussed.

1. *Creating of graphical model*: MATLAB can help with the generation of graphical models for referencing in finite element analysis. Three-dimensional plots can be generated via MATLAB. The resulting graphical model can be assigned as either contour, mesh, or surface plots. For example, this is a case of using MATLAB for creation of a model of a small nano fragmented cell in blood. The MATLAB code in the MATLAB Command Window can be

```
>> [x, y, z] = peaks;
>> c = contour (x, y, z, 27);
>> clabel (c); title ('2 - D Contour plot of nano-fragmented cell
with clable');
>> figure (2)
>> c = contour3 (x, y, z 27);
>> clabel (c);
>> clabel (c); title ('3 - D Contour plot of nano-fragmented cell
with clabel');
>>
```

2. *Solving the problem of differential and integral equation*: As noted, the finite element analysis is usually based on differentiation and integration. MATLAB can help solve the calculation problems. This can also be useful in performing simulation. For example, this is a case of using MATLAB for solving the equation of desizing of viral particles after response to heat stress. An example of code is shown here.

```
>> p = - 8; delta = 0.08; y (1) = 6;
>> k = 0
>> for X = [delta: delta: delta: 0.5]
  k = k + 1;
  y (k + 1) = y (k) + p * y (k) * delta;
  end
>> x = [0: delta: 0.5];
>> y_true = 6 * exp (-8 * x);
>> plot (x, y '*', x, y_true);
>> legend (v'prediction', 'observation');
```

12.4.2 Important Computational Finite Element Analysis Tool for Management of Biomedical Data

There are several computational tools using computational finite element analysis for management of biomedical data. Some important tools that are specific to "nano" biomedical data are discussed.

1. *FEBio* [43]: FEBio is a finite element analysis based computation approach that is specifically designed for biomechanical applications. This tool was developed by Mass et al. in a Utah laboratory, and it was reported for its good properties compared with ABAQUS and NIKE3D [44]. It can help structural clarification of many structures such as musculoskeletal structures. It is based on three-dimensional finite element analysis. The software was first developed as an open source and is applicable for Windows. For more information, please visit http://mrl.sci.utah.edu/febio-overview

2. *PreView* [43]: PreView is a program that is designed as a preprocessor for FEBio. It is mainly used as geometry preparation for running of FEBio.

3. *PostView* [43]: PostView is a program that is designed as a postprocessor for FEBio. It is mainly used as a graphical user interface for finalized visualization.

4. *GAGSim3D:* GAGSim3D is a computer program developed for visualization of three-dimensional files. GAGSimSDa can be used for some statistical studies on vessel distribution (angles, lengths, etc.) Henningeret al. also proposed that GAGSim3D was useful in reduction of transmission electron microscopy artifacts [45].

5. *ANSYS:* ANSYS is an actual simulation tool. The simulation can be stress, thermal, etc. It is a computational finite element analysis tool that is useful for electromagnetic analysis. A good example of its usage in medicine is the transcranial magnetic stimulation analysis in cases with brain disease [46].

6. *COMSOL multiphysics:* COMSOL Multiphysics or FEMLAB is a specific informatics tool designed for computational finite element analysis. It also allows for interaction with MATLAB program. For example, it can be applied for the simultaneous solution of fiber excitation by computational models [47].

12.5 Common Applications of Computational Finite Element Analysis in General Nanomedicine Practice

1. *Situation 1:* Using computational finite element analysis to determine the flow pattern in flow cytometry

This is an example of using computational finite element analysis in nanomedicine. Basically, flow cytometry is a medical device used for counting small particles, especially blood cells. It is general practice in any hospital and the test is called complete blood count (CBC). In laboratory medicine practice, blood is collected from the patient by venipuncture, and the medical laboratory injects it into an automated cell counter for analysis. Structural analysis of blood cells is done automatically, and this is helpful in classifying the type of the blood cells and numerical counting as well as determining the abnormality (shape and content aberrations). In addition, sometimes analysis of more complicated cells (immature cells, pathogens, cancer cells) and particles (toxins, chemicals) can be done using flow cytometry. Computational finite element analysis plays an important role in the structural analysis of the tested blood cell. In brief, it is a mode of microfluidics, which can be assessed by the MEMS module. COMSOL Multiphysics can play its role at this point [48]. Flow field phenomena and dynamic sorting can

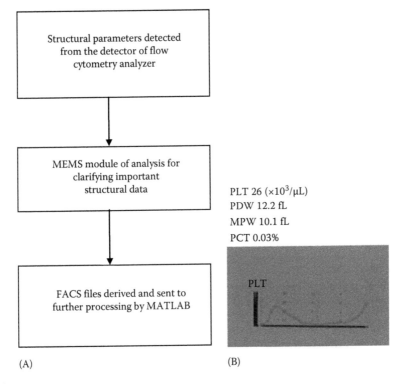

FIGURE 12.1
Roles of computational finite element analysis in flow cytometry analyzer: (A) brief steps and (B) example of data appearance. In this case, an example of flow cytometry analysis of platelets is given. The upper part is the file in finalized data on the platelet structure parameters and the lower part is the final graphical flow cytogram result.

be successfully done based on the tool [48]. With the use of the flow cytometer, the primary computational file is called FCS (flow cytometry standard) [49]. The FCS file can be further processed with the help of MATLAB to finalize the result into both graphical and numerical usage. The summary of the complex process is shown in Figure 12.1.

2. *Situation 2*: Using computational finite element analysis for dental filling material
    This is another example of using computational finite element analysis on medical structures. Basically, dental filling material is a widely used medical material for reconstruction of dental carries. This is widely used in many dental units. The main analysis is usually of the stress interference on dental materials [50,51].
    In primary classical practice, such a test has to be based on these steps:

a. Collection of tooth and dental materials

b. Stress application by mechanical instrument

c. Macroscopic and microscopic examination of the materials

However, with the use of computational finite element analysis approach, these steps can be modeled in silico.

12.5.1　Examples of Nanomedicine Research Based on Computational Finite Element Analysis Application

12.5.1.1　General Cases

Example 12.1　A study to determine the difference in electricity flux for planning nanoelectrial cardiac pacing wire

The heart is an important vital organ gland in human beings, and its main function is regulation of the circulation system. Normally, the heart beats regularly and continues from birth until death. Physiologically, the generation of the heartbeat is very complex. The start of the system is due to the in vivo physiological electrical circuit due to electrolytes. Generation of the first impulse beat occurs at the specific part of the heart namely the SA node, and then it passes to the atrium (upper heart) and to the ventricle (lower heart). Similar to many other organs in human beings, the problems of the heart can be seen and has become important in cardiology.

The problem of cardioelectrical systems can be seen, and this can result in the disease arrhythmia. In arrhythmia, the irregularity of the electrical circulation in heat can be seen. The treatment of arrhythmia is mainly to control the abnormality of the cardioelectrical system. Different means of treatment are available including several pharmacological substances. In the alternative approach, the correction of the abnormal cardioelectrical system is used. Indeed, this technique has been mentioned for a long time [52]. The electrical tool has been continuously developed, and the novel cardioelectrical tools at the nanolevel are already produced at present. The new nanomedical apparatus namely nonelectrical cardiac pacing wire is proposed for its usefulness in the implementation for correction of the arrhythmia [53]. It is also called artificial cardiac tissue [53]. However, implantation of such nanowire has to be based on the complete information of the cardioelectrical system in the abnormal cardiac tissue.

To get these data, the approach by means of finite element analysis can be applied. (Indeed, other techniques can also be used such as finite difference method and charge simulation method). We hereby present an example of clarification on the cardioelectrical system in an abnormal small part of heart tissue. In a small area, the model of the cardioelectrical system is shown in Figure 12.2. Assuming the cardioelectrical system is steady, the electrical potential is directly varied on distance, and the first equation is

$$\phi = \frac{50}{d},$$

where "ϕ" is equal to electrical potential and "d" is equal to distance

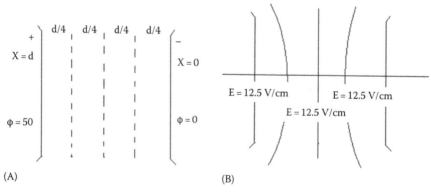

(A)　　　　　　　　　　　　　　　　　　　(B)

FIGURE 12.2
(A and B) Model of cardioelectrical circulation in an abnormal small part of heart tissue.

Considering the size parameter in Figure 12.2, we can see at each part, the electricity is equal to 12.5 V/cm. In this case, the use of electrical nanowire has to be set according to the finding. This concept can also be applied to assessment of other cardiac repair materials such as Dacron and expanded polytetrafluoroethylene (ePTFE) [53].

Example 12.2 A study to analyze the effect of fever on the nanomembrane of red blood cells

The effects of fever can be seen in various ways in the human body. Fever is the result of an inflammation process, which is a normal response in the immune system of the human body to any alien foreign body. While it is a useful process, it can seriously cause disadvantage and danger to our body. In cases of high fever, the pathology of the red blood cells can be seen, and this can be a cause of death [54]. The destruction of the red blood cell nanomembrane can be expected, and this leads to hemolytic episodes and rupture of the red blood cell.

Based on advanced nanotechnology, the modeling of the pathophysiology of fever-induced red blood cell abnormality is possible and is a case of applying a simple computational finite element analysis approach. Here we manipulate the dynamic function based on help from a simple computer program, Excel. Since red blood cells are very small and flat, the simple model of two-dimensional uniform thickness structure is primarily assumed. Starting with the boundary by the normal body temperature at 36.7°C, further manipulation by adding 1 more degree Celsius to see the effect on the membrane is done. In this work, a simple 3×3 grid is used to see the thermal changing from normal temperature to fever (which is medically defined at 38°C). The result can be seen in Figure 12.3. It can be seen that there is an irregular change in temperature on the nanomembrane of red blood cells, and this is concordant with the fact that there is no specific site on the membrane for hemolytic leakage of red blood cell in high fever. A recent study showed that the hyperthermia could result in generalized alteration of calcium influx, and this can be an explanation for the pathophysiology of fever-induced red blood cell abnormality [55].

Example 12.3 A study to analyze the effect of crush injury on small vascular injury

Injury can occur at any time, and this can be an important external insult to the human tissue. The crush-type injury is a direct pressure injury on the tissue, and this can cause

		36.7	36.7	36.7	
	36.7	36.7	36.7	36.7	36.7
	36.7	36.7	36.7	36.7	36.7
	36.7	36.7	36.7	36.7	36.7
		36.7	36.7	36.7	

(A)

		38	38	38	
	36.7	37.2581	37.38578	37.25762	36.7
	36.7	36.94471	37.02596	36.94423	36.7
	36.7	36.79334	36.82816	36.7931	36.7
		36.7	36.7	36.7	

(B)

FIGURE 12.3
Model of thermal change in red blood cell due to fever: (A) normal temperature and (B) with fever (38°C).

Circular cross-sections

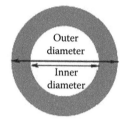

Outer diameter	Inner diameter	Moment	Moment of inertia (I)	Bending stress (Mc/l)
500	450	540	105507198	546555604

Clear/reset inputs

FIGURE 12.4

The reported online result of the exampled simulating case of crush injury to vascular structure. This is the appearance of the reported result in online interactive Online FEA Solver.

damage to the tissue. The degree of crush-type injury ranges from mild (such as simple touching) to severe (such as car accident). An important result of crush injury is the destruction of vascular structure and subsequent bleeding. To learn the pathophysiology of vascular response to injury, analysis of the vascular structure by computational finite element analysis is possible. Here, the author gives an example of assessment of the effect of crush injury on small vascular injury. A simple online finite element analysis interactive tool is used. The tool is Online FEA Solver, which is accessible at http://www.onlinefeasolver.com.

Here, the circular cross sectional parameters are given as outer and inner diameters equal to 500 and 450 nm, respectively. With the use of the Online FEA Solver, the online result can give the resulted value of the moment of inertia equal to 1055071985.91 nm⁴, and if the external applied moment due to injury is assigned to be 540 nm⁴, the further derived bending stress is equal to 0.000128 Mc/nm. The reported online result for this simulating case is presented in Figure 12.4. This technique can be very useful in studying of vascular dynamics in medicine. Finding the moment of inertia can be useful in prediction of flow in different vessels. This can be useful in the prediction of pathology in red cell abnormalities [56] and intravascular apparatus placement [57].

Example 12.4 A study on different concentrations of hydrogen on nanosensor diagnostic property

Basically, the nanosensor is the tool for nanodiagnosis. The development of the nanosensor has to be based on several novel nanomaterials. The change of ion composition is an important factor affecting the diagnostic property of the nanosensor. A good example is the effect of hydrogen, which is indirectly reflected by the pH on the nanosensor. Here we discuss the surface acoustic wave (SAW) nanosensor and its diagnostic properties at different hydrogen concentrations.

A case of hydrogen-sensitive electrode will be the focus. When hydrogen increases, the change in the crystal structure of nanomaterial composition at the

TABLE 12.4

Hydrogen Concentration (% per Nanomaterial) and Insertion Loss

Hydrogen Concentration (% per Nanomaterial)	Insertion Loss (dB)
0	45.7
10	44.2
20	43.12
30	42.99
40	42.71

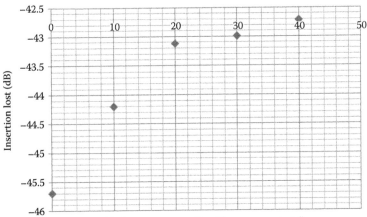

Hydrogen concentration (% per nanomaterial)

electrode can be expected, and this might result in a change in SAW nanosensor response. This can be expressed as an equation between hydrogen concentration and velocity of signal. The direct variation can be assigned as "concentration ∞ velocity." Then the velocity can be further interpreted in relation to frequency as "velocity = wavelength × frequency."

An example of a relationship between change in hydrogen concentration and finalized insertion loss is presented in Table 12.4. With the increase in the hydrogen concentration, the insertion loss increases, and this mean reduction of the wave is detected by SAW nanosensors. It can be seen that the effect of hydrogen concentration significantly affects the nanodiagnostic properties. This is a very big concern for several kinds of new nanosensors in medicine at present (examples of such nanosensors are hepatitis B antibody sensor [58], viral antigen nanosensor [59], etc.).

Example 12.5 A generation of graphical models for referencing in finite element analysis of nanofiber patch for nanotherapy of dead tissue

The use of synthesized nanomaterial as an artificial nanotissue to correct the pathological dead tissue is a novel approach in nanotherapy. A good example is the use of carbon nanofiber. The analysis of the material properties of developed nanofiber can be useful in planning reparative therapy. To start the analysis, an important step is to generate a graphical model for referencing in finite element analysis.

Here the author uses a simple computational modeling tool for creating a graphical model of carbon nanofiber. The NETGEN 4.4 is used. The simple assumption is (a) the

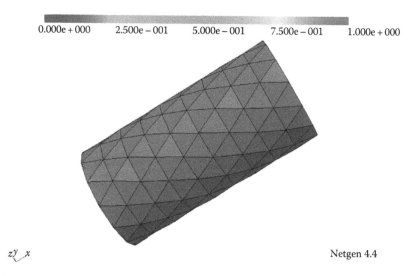

0.000e + 000   2.500e – 001   5.000e – 001   7.500e – 001   1.000e + 000

z/y x   Netgen 4.4

FIGURE 12.5
An example of derived graphical model of carbon nanofiber. In this case, the referenced parameters include 153 points, 369 elements, and 288 surf elements.

carbon nanofiber has a cylindrical shape and (b) a three-dimensional model is used for creation. An example of a result reported online is directly quoted and shown in Figure 12.5. Generally, the use of a carbon nanofiber patch can be effective in many nanotherapies such as use in repairing dead cardiac fiber [60]. With the use of further finite element analysis techniques, the prediction of response to stress such as heat can be possible (e.g., to predict response to heat, a model based on Joule heating and heat convection can be used [61].)

Example 12.6 A study on the quadrature surface of magnetic resonance imaging detection

Magnetic resonance imaging (MRI) is a widely used medical imaging technique. The main concern on the diagnostic property of the MRI is based on its detector. The quadrature surface nanostructure of the detector is an important focus. This structure is a circular loop and figure-of-eight or butterfly-shaped coils that are aimed at adjusting signal-to-noise-ratios [62]. Optimization of its structural condition is very useful [62].

In general, a quadrature rule is also based on an integration equation. The role of MATLAB for finite element analysis at this stage is acceptable. An example of a designed quadrature loop to integrate the function $f = x4$ on assigned triangular element is further discussed. Hence, we can manipulate the function based on MATLAB. Focusing on the code, it can be written as shown

```
[qPt,qWt] =quadrature(3,'TRIANGULAR',3);
for q=2:length(qWt)
xi=qPt(q);% quadrature point
% get the global coordinate x at the quadrature point xi
% and the Jacobian at the quadrature point, jac
...
f_int=f_int+x^4 * jac*qWt(q);
end
```

Example 12.7 A study of the electrostatic diffusion pattern of the new anticancer drug

At present, the advancements in nanomedicine have led to new nanotechnology-based pharmacology or nanopharmacology for management of many diseases. For various diseases, the use of nanopharmacology in nanooncology should be mentioned, especially since the design of new anticancer drugs in the present day aims at individualization and specificity. Focusing on the newly developed nanodrug, an important determinant for its effectiveness is the diffusion ability into the focused cells for therapeutic purpose. The electrostatic diffusion is an important pattern to be considered.

With basic graphical finite element analysis modeling, the first simple model of nanodrug can be derived as a spherical biomolecule (Figure 12.6). The primary assumptions include unit sphere with certain radius (r0) representing the nanodrug molecule and charge (q) is steady within the center of the sphere. The simple linearized Poisson-Boltzmann equation [63] might be used in the simulating case. The derived scattergram plots radius of sphere versus electrostatic potential can be presented as in Figure 12.6. The rapid change at the small radius level of the studied nanodrug with the steady state of electrostatic drug at a larger radius level can be observed, and this can help select the proper size of the nanodrug that will be most effective in cancericidal activity [64].

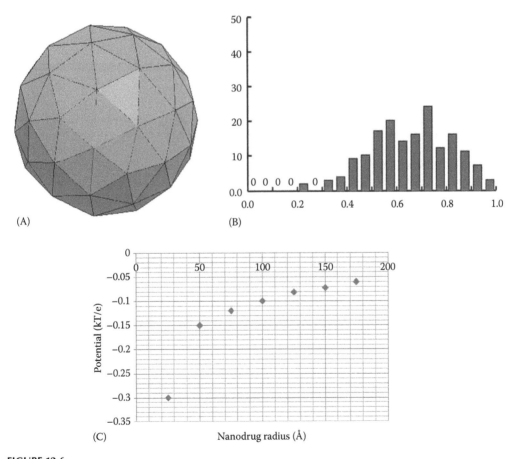

(A)

(B)

(C)   Nanodrug radius (Å)

FIGURE 12.6
An example of model of nanodrug for cancerous treatment: (A) simple molecule with generated mesh, (B) the mesh quality plot, and (C) scattergram plots radius of sphere versus electrostatic potential.

Example 12.8 A study on the nanobiopharmacological mechanism of chrisotherapeutic gold compounds

Several new nanopharmacological molecules have been developed in the last few years. The approval of their nanobiopharmacological properties before implementation on actual usage is required. Of several new nanomolecules, the new chrisotherapeutic gold compound is of interest. This is an application of gold nanoparticles in nanopharmacology. It is noted that this molecule can induce antiinflammatory effect, and prostaglandin E2 production can be observed in animal models [65]. However, its subcellular biomechanism, at the DNA level, is still unclear.

Here, the simple finite element analysis online bioinformatics tool for prediction of DNA binding is used. The tool is namely DDNA³ (accessed online at http://sparks.informatics.iupui.edu/yueyang/DFIRE/ddna3-DB-service.php) [66]. The process starts with the uploading of the referencing prostaglandin E2 molecule file in .PDB format, which can be derived via public databases, such as PubMed. After the simulating and prediction, it can be seen that the studied prostaglandin E2 has no specific DNA binding. Hence, it can be concluded that the exact nanobiopharmacological mechanism of chrisotherapeutic gold compounds should not involve the DNA level.

Indeed, Yamashitaet al. also performed a study on macrophage and observed no specific DNA binding in the pharmacological action of chrisotherapeutic gold compounds [65].

12.6 Conclusion

Computational technology is a very useful nanomedicine manipulation helping to clarify and answer many problems. Of several computational approaches, the use of computational finite element analysis is an important approach, which mainly helps assess the structure and further predicts functional changes due to various interferences. Computational finite element analysis can help medical scientists answer complex questions in nanomedicine via either structural or functional approaches. Clarification on the existing structures and prediction of the imaginary changes corresponding to applied simulations can be derived from using the computational finite element analysis technique. In addition, there are many available computational tools using finite element analysis for management of biomedical data at present. These new computational tools can be very useful for many studies in nanomedicine.

References

1. Gehlenborg N, O'Donoghue SI, Baliga NS, Goesmann A, Hibbs MA, Kitano H, Kohlbacher O, Neuweger H, Schneider R, Tenenbaum D, Gavin AC. Visualization of omics data for systems biology. *Nat Methods.* 2010 March; 7(3 Suppl):S56–S68.
2. Haarala R, Porkka K. The odd omes and omics. *Duodecim.* 2002; 118(11):1193–1195
3. Haddish-Berhane N, Rickus JL, Haghighi K. The role of multiscale computational approaches for rational design of conventional and nanoparticle oral drug delivery systems. *Int J Nanomedicine.* 2007; 2(3):315–331.
4. Saliner AG, Poater A, Worth AP. Toward in silico approaches for investigating the activity of nanoparticles in therapeutic development. *IDrugs.* 2008 October; 11(10):728–732.

5. Reddy JN. *Introduction to the Finite Element Method.* 2nd edn. Boston, MA: McGraw-Hill Science, 1993.
6. Huiskes R, Chao EY. A survey of finite element analysis in orthopedic biomechanics: The first decade. *J Biomech.* 1983; 16:385–409.
7. Ma X, Fu X, Ma J, Zhao Y, Wang T, Wang Z, Zhang Y, Dong B, Yang Y. A new method to reconstruct the spatial structure of human proximal femur and establishment of the finite element model. *Sheng Wu Yi Xue Gong Cheng Xue Za Zhi.* 2011; 28:71–75.
8. Zhang MC, Lü SZ, Cheng YW, Gu LX, Zhan HS, Shi YY, Wang X, Huang SR. Study on the effect of vertebrae semi-dislocation on the stress distribution in facet joint and intervertebral disc of patients with cervical syndrome based on the three dimensional finite element model. *Zhongguo Gu Shang.* 2011; 24:128–131.
9. Fei Q, Li QJ, Li D, Yang Y, Tang H, Li JJ, Wang BQ, Wang YP. Biomechanical effect on adjacent vertebra after percutaneous kyphoplasty with cement leakage into disc: a finite element analysis of thoracolumbar osteoporotic vertebral compression fracture. *Zhonghua Yi Xue Za Zhi.* 2011; 91:51–55.
10. Eichinger JK, Herzog JP, Arrington ED. Analysis of the mechanical properties of locking plates with and without screw hole inserts. *Orthopedics.* 2011; 34:19.
11. Olson S, Clinton JM, Working Z, Lynch JR, Warme WJ, Womack W, Matsen FA 3rd. Thermal effects of glenoid reaming during shoulder arthroplasty in vivo. *J Bone Joint Surg Am.* 2011; 93:11–19.
12. He D, Wu L, Chi Y, Zhong S. Facet joint plus interspinous process graft fusion to prevent postoperative late correction loss in thoracolumbar fractures with disc damage: finite element analysis and small clinical trials. *Clin Biomech* (Bristol, Avon). 2011; 26:229–237.
13. da Silva BR, Moreira Neto JJ, da Silva FI, de Aguiar AS. Three dimensional finite element analysis of the maxillary central incisor in two different situations of traumatic impact. *Comput Methods Biomech Biomed Eng.* 2011 September 7 [Epub ahead of print].
14. Campos RE, Soares CJ, Quagliatto PS, Soares PV, de Oliveira OB Junior, Santos-Filho PC, Salazar-Marocho SM. In vitro study of fracture load and fracture. Pattern of ceramic crowns: A finite element and fractography analysis. *J Prosthodont.* 2011; 20:447–455.
15. Naini RB, Nokar S, Borghei H, Alikhasi M. Tilted or parallel implant placement in the completely edentulous mandible? A three dimensional finite element analysis. *Int J Oral Maxillofac Implants.* 2011; 26:776–781.
16. Choi AH, Matinlinna JP, Ben-Nissan B. Finite element stress analysis of Ti-6Al-4V and partially stabilized zirconia dental implant during clenching. *Acta Odontol Scand.* 2011 August 5 [Epub ahead of print].
17. Panagiotopoulou O, Cobb SN. The mechanical significance of morphological variation in the macaque mandibular symphysis during mastication. *Am J Phys Anthropol.* 2011 August 8. doi: 10.1002/ajpa.21573 [Epub ahead of print].
18. Chang SH, Lin CL, Hsue SS, Lin YS, Huang SR. Biomechanical analysis of the effects of implant diameter and bone quality in short implants placed in the atrophic posterior maxilla. *Med Eng Phys.* 2012; 34(2):153–160.
19. Alikhasi M, Siadat H, Geramy A, Hassan-Ahangari A. Stress distribution around maxillary anterior implants as a factor of labial bone thickness and occlusal load angles: A 3D-finite-element analysis. *J Oral Implantol.* 2011 July 25 [Epub ahead of print].
20. Ghuneim WA. In situ tooth replica custom implant; a 3D finite element stress and strain analysis. *J Oral Implantol.* 2011 July 18 [Epub ahead of print].
21. Dos Santos MB, Consani RL, Mesquita MF. Influence of different soft liners on stress distribution in peri-implant bone tissue during healing period. A 3-D finite element analysis. *J Oral Implantol.* 2011 July 18 [Epub ahead of print].
22. Lin J, Shinya A, Gomi H, Shinya A. Finite element analysis to compare stress distribution of connector of lithia disilicate-reinforced glass-ceramic and zirconia-based fixed partial denture. *Odontology.* 2012; 100(1):96–99.

23. Rungsiyakull C, Rungsiyakull P, Li Q, Li W, Swain M. Effects of occlusal inclination and loading on mandibular bone remodeling: A finite element study. *Int J Oral Maxillofac Implants*. 2011; 26:527–537.

24. Benazzi S, Kullmer O, Grosse IR, Weber GW. Using occlusal wear information and finite element analysis to investigate stress distributions in human molars. *J Anat*. 2011; 219:259–272.

25. Han XL, Liu ZW, Li YT. Three dimensional finite element analysis of biomechanical distribution of dental implants with immediate loading. *Hua Xi Kou Qiang Yi Xue Za Zhi*. 2011; 29:121–124.

26. De La Iglesia D, Chiesa S, Kern J, Maojo V, Martin-Sanchez F, Potamias G, Moustakis V, Mitchell JA. Nanoinformatics: New challenges for biomedical informatics at the nanolevel. *Stud Health Technol Inform*. 2009; 150:987–991.

27. Yih TC, Wei C, Hammad B. Modeling and characterization of a nanoliter drug delivery MEMS micropump with circular based bossed membrane. *Nanomedicine*. 2005 June; 1(2):164–175.

28. Larsson G, Martinez G, Schleucher J, Wijmenga SS. Detection of nano-second internal motion and determination of overall tumbling times independent of the time scale of internal motion in proteins from NMR relaxation data. *J Biomol NMR*. 2003 December; 27(4):291–312.

29. Quintás G, Kuligowski J, Lendl B. On-line Fourier transform infrared spectrometric detection in gradient capillary liquid chromatography using nanoliter-flow cells. *Anal Chem*. 2009 May 15; 81(10):3746–3753.

30. Sartori A, Gatz R, Beck F, Rigort A, Baumeister W, Plitzko JM. Correlative microscopy: Bridging the gap between fluorescence light microscopy and cryo-electron tomography. *J Struct Biol*. 2007; 160:135–145.

31. Xing M, Zhong W, Xu X, Thomson D. Adhesion force studies of nanofibers and nanoparticles. *Langmuir*. 2010; 26:11809–11814.

32. Babincova M, Babinec P. Magnetic drug delivery and targeting: Principles and applications. *Biomed Pap Med Fac Univ Palacky Olomouc Czech Repub*. 2009; 153:243–250.

33. Sassaroli E, Li KC, O'Neill BE. Numerical investigation of heating of a goldnanoparticle and the surrounding microenvironment by nanosecond laser pulses for nanomedicine applications. *Phys Med Biol*. 2009; 54:5541–5560.

34. Johnson CL, Snoeck E, Ezcurdia M, Rodríguez-González B, Pastoriza-Santos I, Liz-Marzán LM, Hÿtch MJ. Effects of elastic anisotropy on strain distributions indecahedral gold nanoparticles. *Nat Mater*. 2008; 7:120–124.

35. Lee K, Drachev VP, Irudayaraj J. DNA-gold nanoparticle reversible networks grown on cell surface marker sites: Application in diagnostics. *ACS Nano*. 2011 March 22; 5(3):2109–2117.

36. Humphris AD, Zhao B, Catto D, Howard-Knight JP, Kohli P, Hobbs JK. High speed nanometrology. *Rev Sci Instrum*. 2011 April; 82(4):043710.

37. Elliott AM, Shetty AM, Wang J, Hazle JD, Jason Stafford R. Use of gold nanoshells to constrain and enhance laser thermal therapy of metastatic liver tumours. *Int J Hyperthermia*. 2010; 26(5):434–440.

38. Liu C, Mi CC, Li BQ. Energy absorption of gold nanoshells in hyperthermia therapy. *IEEE Trans Nanobiosci*. 2008 September; 7(3):206–214.

39. Kim KS, Simon L. Transport mechanisms in oral transmucosal drug delivery: Implications for pain management. *Math Biosci*. 2011 January; 229(1):93–100.

40. Wu W, Qi M, Liu XP, Yang DZ, Wang WQ. Delivery and release of nitinol stent in carotid artery and their interactions: A finite element analysis. *J Biomech*. 2007; 40(13):3034–3040.

41. Güçlü B. Low-cost computer-controlled current stimulator for the student laboratory. *Adv Physiol Edu*. 2007; 31:223–231.

42. Wong RC, Tideman H, Merkx MA, Jansen J, Goh SM, Liao K. Review of biomechanical models used in studying the biomechanics of reconstructed mandibles. *Int J Oral Maxillofac Surg*. 2011 April; 40(4):393–340.

43. Maas SA, Weiss JA. FEBIO User's Manual 2007 Available from: http://mrl.sci.utah.edu/software.php

44. Maas SA, Ellis BJ, Rawlins DS, Weiss JA. A comparison of FEBio, ABAQUS, and NIKE3D results for a suite of verification problems. Sci Institute Technical Report, University of Utah; Utah, UUSCI-2009–009.
45. Henninger HB, Maas SA, Shepherd JH, Joshi S, Weiss JA. Transversely isotropic distribution of sulfated glycosaminoglycans in human medial collateral ligament: A quantitative analysis. *J Struct Biol.* 2009 March; 165(3):176–183.
46. Cho YS, Suh HS, Lee WH, Kim TS. TMS modeling toolbox for realistic simulation. *Conf Proc IEEE Eng Med Biol Soc.* 2010; 2010:3113–3116.
47. Martinek J, Stickler Y, Reichel M, Mayr W, Rattay F. A novel approach to simulate Hodgkin-Huxley-like excitation with COMSOL Multiphysics. *Artif Organs.* 2008 August; 32(8):614–619.
48. Taylor JK, Ren CL, Stubley GD. Numerical and experimental evaluation of microfluidic sorting devices. *Biotechnol Prog.* 2008 July–August; 24(4):981–991.
49. Leif RC, Leif SB, Leif SH. CytometryML, an XML format based on DICOM and FCS for analytical cytology data. *Cytometry A.* 2003 July; 54(1):56–65.
50. Burgess JO, Summitt JB, Robbins JW. The resistance to tensile, compression, and torsional forces provided by four post systems. *J Prosthet Dent.* 1992; 68(6):899–903.
51. Drummond JL. In vitro evaluation of endodontic posts. *Am J Dent.* 2000; 13(Spec No): 5B–8B.
52. Roe BB, Bruns DL. The artificial cardiac pacemaker indications for implantation. *Calif Med.* 1964 December; 101:422–426.
53. Shah U, Bien H, Entcheva E. Microtopographical effects of natural scaffolding on cardiomyocyte function and arrhythmogenesis. *Acta Biomater.* 2010 August; 6(8):3029–3034.
54. Karle H. The pathogenesis of the anaemia of chronic disorders and the role of fever in erythrokinetics. *Scand J Haematol.* 1974; 13(2):81–86.
55. Föller M, Braun M, Qadri SM, Lang E, Mahmud H, Lang F. Temperature sensitivity of suicidal erythrocyte death. *Eur J Clin Invest.* 2010 June; 40(6):534–540.
56. Boryczko K, Dzwinel W, Yuen DA. Dynamical clustering of red blood cells in capillary vessels. *J Mol Model.* 2003 February; 9(1):16–33.
57. Cervera M, Dolz M, Herraez JV, Belda R. Evaluation of the elastic behaviour of central venous PVC, polyurethane and silicone catheters. *Phys Med Biol.* 1989 February; 34(2):177–183.
58. Lee HJ, Namkoong K, Cho EC, Ko C, Park JC, Lee SS. Surface acoustic wave immunosensor for real-time detection of hepatitis B surface antibodies in whole blood samples. *Biosens Bioelectron.* 2009 June 15; 24(10):3120–3125.
59. Bisoffi M, Hjelle B, Brown DC, Branch DW, Edwards TL, Brozik SM, Bondu-Hawkins VS, Larson RS. Detection of viral bioagents using a shear horizontal surface acoustic wave biosensor. *Biosens Bioelectron.* 2008 April 15; 23(9):1397–1403.
60. Genaidy A, Tolaymat T, Sequeira R, Rinder M, Dionysiou D. Health effects of exposure to carbonnanofibers: Systematic review, critical appraisal, metaanalysis and research to practice perspectives. *Sci Total Environ.* 2009 June 1; 407(12):3686–3701.
61. Park JG, Li S, Liang R, Fan X, Zhang C, Wang B. The high current-carrying capacity of various carbon nanotube-based buckypapers. *Nanotechnology.* 2008 May 7; 19(18):185710.
62. Kumar A, Bottomley PA. Optimized quadrature surface coil designs. *MAGMA.* 2008 March; 21(1–2):41–52.
63. Fogolari F, Brigo A, Molinari H. The Poisson–Boltzmann equation for biomolecular electrostatics: A tool for structural biology. *J Mol Recognit.* 2002; 15(6):377–392.
64. Lesyng B, McCammon JA. Molecular modeling methods. Basic techniques and challenging problems. *Pharmacol Ther.* 1993 November; 60(2):149–167.
65. Yamashita M, Ohuchi K, Takayanagi M. Effects of chrisotherapeutic gold compounds on prostaglandin E2 production. *Curr Drug Targets Inflamm Allergy.* 2003 September; 2(3):216–223.
66. Zhao H, Yang Y, Zhou Y. Structure-based prediction of DNA-binding proteins by structural alignment and a volume-fraction corrected DFIRE-based energy function. *Bioinformatics.* 2010 August 1; 26(15):1857–1863.

13

Finite Element Method for Micro- and Nano-Systems for Biotechnology

Jean Berthier

CEA Leti
and
University Joseph Fourier
Grenoble, France

CONTENTS

13.1 Introduction 424
   13.1.1 Biotechnology: A New Science Requiring a Multi-Physics Approach 424
   13.1.2 Role and Aims of Numerical Modeling in Biotechnology 425
13.2 Theoretical Basis 425
   13.2.1 Introduction 425
   13.2.2 Mass Conservation and Navier–Stokes Equations 426
   13.2.3 Stokes Equations 427
   13.2.4 Creeping Flow Reversibility 428
   13.2.5 Laminarity of Microfluidic Flows 429
   13.2.6 Pressure Drop 429
   13.2.7 Non-Newtonian Fluids 430
      13.2.7.1 Introduction 430
      13.2.7.2 Non-Newtonian Viscosity 432
      13.2.7.3 Non-Newtonian Pressure Drop 436
      13.2.7.4 Numerical Results for a Square Microchannel 437
      13.2.7.5 Non-Newtonian Networks 438
13.3 Shallow Channel or Hele–Shaw Model: A 2D Approach for a Pseudo
3D Microflow 440
   13.3.1 Introduction 440
   13.3.2 Model 440
   13.3.3 Application to a Microfluidic Resonator 442
13.4 Mixing in Microsystems 443
   13.4.1 Introduction 443
   13.4.2 Coflows 444
   13.4.3 Herringbone Structures 445
   13.4.4 Dean Flows 447
      13.4.4.1 Hydrodynamics of Dean Microflows 447
      13.4.4.2 Concentration in Dean Microflows 448

13.5 Cell Chips..448
    13.5.1 Single-Phase Flow Focusing...449
    13.5.2 Bifurcations and Pinched Channel Microflows 452
        13.5.2.1 Bifurcations..452
        13.5.2.2 Flow Focusing Combined with Bifurcations 454
        13.5.2.3 Pinched Flow Fractionation...455
        13.5.2.4 Example of PFF Coupled to a Flow Focusing Device 456
    13.5.3 Deterministic Lateral Displacement 457
        13.5.3.1 Introduction and Theory 457
        13.5.3.2 Numerical Model 461
    13.5.4 Trapping Chambers..462
    13.5.5 Microsystems for Cell Culture...464
13.6 Biochemical Reactions: DNA Recognition 465
    13.6.1 Introduction..465
    13.6.2 Heterogeneous Reactions 466
        13.6.2.1 Theoretical Background: Static Case...466
        13.6.2.2 Numerical Approach: Convection-Diffusion-Reaction Problem 470
13.7 Conclusions...471
References...473

13.1 Introduction

13.1.1 Biotechnology: A New Science Requiring a Multi-Physics Approach

Starting in the year 1980, biotechnology was as a new science at the boundary of physics and biology. The goal was to bring to biology, medical, and pharmaceutical research new automation tools to boost the development of new drugs, fabricate new body implants, perform DNA and protein analysis and recognition, and increase the potentialities of fundamental research. In reality, this vision, imagined by pioneers like Feynman [1], de Gennes [2], and Whitesides [3], has been extremely effective.

Historically, genomics and proteomics have been the first beneficiaries of the development of biotechnology, and it is the turn of cellomics now. These developments have spread beyond the domain of biotechnology and created a "cloud" of new applications in other domains such as bioinformatics, bioengineering, tissue engineering, etc. At the same time, the microfluidic techniques developed for biotechnology reached other domains, like materials science, optofluidics, microelectronics, and mechatronics.

The foreseen goals have required the downscaling of fluidic systems to the "convenient" size to work at the proper scale characteristic of a population of biologic targets. At the same time, it has been found that the downscaling was bringing economy in costly materials, fluids, and devices, that sensitivity was increased, and that operating times were greatly reduced by the integration of many functions on the same microchip. The increasingly large number of functions implemented on the same chip is associated to a multi-physics aspect. For example, a microsystem can require the use of microfluidics for actuating the flows through the chip, the use of magnetic beads to bind to the biological targets and concentrate them [4,5], or electric fields to induce dielectrophoretic or electrokinetic effects to separate the targets [6,7]; acoustics may also be used for the same purposes [8,9]. Polymerase chain reaction (PCR) aimed at the amplification of DNA strands require very localized thermal cycles [10,11], micromembranes or valves require micro-mechanical investigations [12,13], and the binding (capture) of targets to ligands is based on chemical recognition [5,14]. It is frequent that these phenomena are coupled, like microfluidics, thermics, and chemical reactions, for example.

13.1.2 Role and Aims of Numerical Modeling in Biotechnology

The historical trend of most new domains in physics is to start with experiments and a theoretical basis; modeling and simulation come later. Biotechnology is no exception. Presently, the development of biotechnology, and micro and nano-technologies in general, is sufficiently mature for numerical modeling and simulation to develop quickly.

In microsystems for biotechnology, due to flow laminarity, the velocity field can be extremely well calculated. On the other hand, phenomena like dielectrophoresis, electrokinetics, thermics, and chemical reactions are also well prone to modeling. However, all the numerical problems are far being solved: At the present stage, the complexity essentially stems from the coupling of the multi-physics involved with the microsystems, from the multi-phase flow aspects—tracking an interface is still complex and time consuming—and from the difficulty to model the transport of "large" and sometimes deformable objects like cells and vesicles.

In this chapter, we have tried to give the reader an overview of the modeling developments in biotechnology. First, we recall the theoretical grounds for microfluidics, with an introduction to non-Newtonian rheology—because the colloids used in biotechnology are often viscoelastic; then we take the example of cell chips in which cell concentration and separation is made by tailored microfluidics; finally, we present DNA hybridization in a microfluidic system.

13.2 Theoretical Basis

13.2.1 Introduction

From the most general point of view, pure fluid flows are determined by the knowledge of velocities $U = \{u_i, i = 1, 3\}$, pressure P, density ρ, viscosity μ (or η), specific heat C_p, and temperature T. For each fluid, density, viscosity, and specific heat are related to pressure and temperature (or enthalpy) via characteristic equations of state (EOS):

$$\rho = f(P, T)$$

$$\mu = g(P, T) \tag{13.1}$$

$$C_p = h(P, T)$$

Pressure and temperature characterize the number and the state of the molecules that are present in a given volume. Equations of state are generally complicated, but they can be approximated by analytical functions if the domain of variation of the parameters (P and T) is not too large. Thus, we are left with five unknowns: u_x, u_y, u_z, P, and T. These unknowns are related by a system of three equations: (1) a scalar equation for the mass conservation, (2) a vector equation for the conservation of momentum, and (3) a scalar equation for the conservation of energy. In biotechnology, fluid flows are most of the time isothermal or variation of temperature is negligible. Note that this is not the case for micro-chemistry where chemical reactions are seldom isothermal, or for specific biological protocols like polymerase chain reaction (PCR) to amplify DNA strands. If temperature is constant or nearly constant, we have to deal with four unknowns, u_x, u_y, u_z, and P, with the help of the mass conservation equation and the conservation of momentum equation, plus the EOS: $\rho = f(P)$, $\mu = g(P)$. Some authors give the name Navier–Stokes equations to the whole system, others restrict this name to the second equation (momentum).

Remark that if the fluid transports micro or nanoparticles, the preceding development must be adapted; the fluid characteristics depend on the transported species concentration c

$$\rho = f(P,T,c)$$

$$\mu = g(P,T,c) \tag{13.2}$$

$$C_p = h(P,T,c)$$

and an additional equation for the transport of concentration must be added.

13.2.2 Mass Conservation and Navier–Stokes Equations

The theoretical basis for hydrodynamics is the Navier–Stokes and mass conservation equations [15]. The conservation of mass (continuity) is based on the conservation of mass in any infinitesimal volume:

$$\frac{\partial \rho}{\partial t} + \frac{\partial (\rho u)}{\partial x} + \frac{\partial (\rho v)}{\partial y} + \frac{\partial (\rho w)}{\partial z} = 0 \tag{13.3}$$

This equation may be written as

$$\frac{\partial \rho}{\partial t} + u\frac{\partial \rho}{\partial x} + v\frac{\partial \rho}{\partial y} + w\frac{\partial \rho}{\partial z} + \rho\left[\frac{\partial u}{\partial x} + \frac{\partial v}{\partial y} + \frac{\partial w}{\partial z}\right] = 0 \tag{13.4}$$

and under a vector form

$$\frac{D\rho}{Dt} + \rho \nabla \cdot \vec{V} = 0 \tag{13.5}$$

where the operator D/Dt is

$$\frac{D}{Dt} = \frac{\partial}{\partial t} + u\frac{\partial}{\partial x} + v\frac{\partial}{\partial y} + w\frac{\partial}{\partial z} \tag{13.6}$$

in a three-dimensional (3D) Cartesian coordinate system. Liquids may generally be considered as incompressible, and the mass conservation equation is then reduced to

$$\nabla \cdot \vec{V} = 0 \tag{13.7}$$

In Cartesian coordinates, we have

$$\frac{\partial u}{\partial x} + \frac{\partial v}{\partial y} + \frac{\partial w}{\partial z} = 0 \tag{13.8}$$

The second equation is the momentum conservation equation (or Navier–Stokes equation). The change of momentum in a fluid element is equal to the balance between inlet momentum, outlet momentum, and exerted forces [2]:

$$\rho \frac{Du}{Dt} = -\frac{\partial \sigma_x}{\partial x} + \frac{\partial \tau_{xy}}{\partial y} + \frac{\partial \tau_{xz}}{\partial z} + F_x \qquad (13.9)$$

Normal stress and tangential stress are for most fluids (called Newtonian fluids, see discussion later) given by the constitutive relations

$$\sigma_x = P - 2\mu \frac{\partial u}{\partial x} + \frac{2}{3\mu}\left(\frac{\partial u}{\partial x} + \frac{\partial v}{\partial y} + \frac{\partial w}{\partial z}\right) \qquad (13.10)$$

$$\tau_{xy} = \mu\left(\frac{\partial u}{\partial y} + \frac{\partial v}{\partial x} + \frac{\partial w}{\partial z}\right)$$

where μ is the dynamic viscosity. Combining (13.9) and (13.10), and extending the formulation to the three-dimensional (3D) case, yields the Navier–Stokes equation:

$$\rho\left(\frac{\partial u}{\partial t} + u\frac{\partial u}{\partial x} + v\frac{\partial u}{\partial y} + w\frac{\partial u}{\partial z}\right) = -\frac{\partial P}{\partial x} + \mu\left(\frac{\partial^2 u}{\partial x^2} + \frac{\partial^2 u}{\partial y^2} + \frac{\partial^2 u}{\partial z^2}\right) + F_x$$

$$\rho\left(\frac{\partial v}{\partial t} + u\frac{\partial v}{\partial x} + v\frac{\partial v}{\partial y} + w\frac{\partial v}{\partial z}\right) = -\frac{\partial P}{\partial y} + \mu\left(\frac{\partial^2 v}{\partial x^2} + \frac{\partial^2 v}{\partial y^2} + \frac{\partial^2 v}{\partial z^2}\right) + F_y \qquad (13.11)$$

$$\rho\left(\frac{\partial w}{\partial t} + u\frac{\partial w}{\partial x} + v\frac{\partial w}{\partial y} + w\frac{\partial w}{\partial z}\right) = -\frac{\partial P}{\partial z} + \mu\left(\frac{\partial^2 w}{\partial x^2} + \frac{\partial^2 w}{\partial y^2} + \frac{\partial^2 w}{\partial z^2}\right) + F_z$$

The vectorial notation is

$$\rho \frac{D\vec{V}}{Dt} = -\nabla P + \mu \Delta \vec{V} + \vec{F} \qquad (13.12)$$

where $\vec{V}$ is the velocity vector (u,v,w) and F is the body force per unit volume.

13.2.3 Stokes Equations

At very low velocities, inertial forces become very small compared to the viscous forces. The Reynolds number

$$\text{Re} = \frac{V d}{v} \qquad (13.13)$$

is much smaller than 1—because the average velocity V and the characteristic dimension d are small—and the inertia terms on the left of Equation 13.11 may be neglected [5,16]. In this regime, the Navier–Stokes equation reduces to the Stokes equation:

$$\frac{\partial p}{\partial x} = v\left[\frac{\partial^2 u}{\partial x^2} + \frac{\partial^2 u}{\partial y^2} + \frac{\partial^2 u}{\partial z^2}\right] + F_x$$

$$\frac{\partial p}{\partial y} = v\left[\frac{\partial^2 v}{\partial x^2} + \frac{\partial^2 v}{\partial y^2} + \frac{\partial^2 v}{\partial z^2}\right] + F_y \qquad (13.14)$$

$$\frac{\partial p}{\partial z} = v\left[\frac{\partial^2 w}{\partial x^2} + \frac{\partial^2 w}{\partial y^2} + \frac{\partial^2 w}{\partial z^2}\right] + F_z$$

In the case where the external force is just the gravity force, the simplification is considerable because the system (13.14) is now linear

$$\nabla p = v\Delta\vec{V} + \rho g\nabla z \qquad (13.15)$$

where we have used the notation $\Delta = \nabla^2$ for the Laplacian operator. By taking the rotational of (13.15), and using the following mathematical relations

$$curl\,(grad\,P) = \nabla\times\nabla p = 0$$

$$\Delta\left(curl\,\vec{A}\right) = curl\left(\Delta\vec{A}\right)$$

we obtain

$$\Delta\left(\nabla\times\vec{V}\right) = \Delta\vec{\omega} = 0 \qquad (13.16)$$

where ω is the vorticity of the flow. Thus, in the Stokes formulation, vorticity is a harmonic function [17] and the problem can be solved in the vorticity-streamline formulation as soon as the values of the vorticity on the boundaries are known. Remark that harmonic functions are an interesting class of mathematical functions; in particular, they cannot have a minimum or maximum inside the computational domain: the extremums are always at the boundaries. In our case here, the vorticity is generated by the shear at the wall and is maximum or minimum (negative) at the walls. That would not be the case for turbulent flows where free vortices evolve in the flow domain.

13.2.4 Creeping Flow Reversibility

Stokes equations are linear. Stokes formulation for creeping flows is very attractive because an apparently complex problem can be simplified to a linear formulation. Besides linearity, Stokes equation is reversible [16]; that is, a change of the velocity u to its opposite—u on the boundaries of the domain—will result in a change of all the velocities to their opposite.

An example of this reversibility property can be done by considering a cylinder in a microflow, as sketched in Figure 13.1. The calculation of the flow has been performed with the numerical software COMSOL using the complete Navier–Stokes equations [18]. The flow lines are shown in Figure 13.1, for an inlet velocity of 1 mm/s from left to right. If the flow is reversed, the pattern of the flow lines is exactly the same. This is typically a case

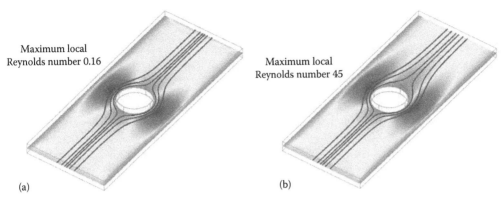

Maximum local
Reynolds number 0.16

Maximum local
Reynolds number 45

(a)

(b)

FIGURE 13.1
Contour plot of liquid velocities in a microchamber obstructed by a vertical cylinder; the continuous lines are streamlines originating at the left boundary: (a) at low Reynolds number, there is a complete symmetry; reversing the direction of the flow does not change the streamlines; (b) at medium Reynolds number, the symmetry breaks down and there is no more reversibility of the flow field.

where the Stokes equations are sufficient to describe the flow because the Reynolds number is much smaller than 1 anywhere in the computational domain.

On the other hand, if the inlet velocity is progressively increased, the inertial effects are increasing, and so is then Reynolds number. The flow field progressively loses its symmetry. The Stokes equations are then not sufficient to describe the flow.

The property of reversibility is very important because it shows that, at very low velocities, it is not possible to design a microfluidic "diode" where the pressure drop would be small in one direction and large in the opposite direction. A microfluidic "diode" requires that the fluid velocity must be sufficiently large to be outside the Stokes hypothesis, or, as will be shown later, the fluid must be non-Newtonian.

13.2.5 Laminarity of Microfluidic Flows

We have seen that microflows in microsystems are most of the time laminar—this is even more the case when the dimensions are reduced toward the nanoscale—and in such laminar flows, the flow streamtubes are very persistent as in Figure 13.2 which shows the merging of differently colored aqueous flows. The length of the domain where the different flows are keeping their identities depends on the diffusion coefficient and the flow velocity. We will analyze the merging of miscible microflows in detail in Section 13.4.

13.2.6 Pressure Drop

Flow channels in microsystems are often rectangular. This is due to the microfabrication process. Pressure drops in such channels have been largely documented [5,19–23]. There exist a few expressions of the laminar pressure drop in a rectangular channel. One of the most accurate is that of Bahrami et al. [22] that takes into account the aspect ratio $\varepsilon = \min (w/d, d/w)$, where d and w are, respectively, the depth and width of the channel. The pressure drop is related to the flow rate by

$$\Delta P = RQ, \tag{13.17}$$

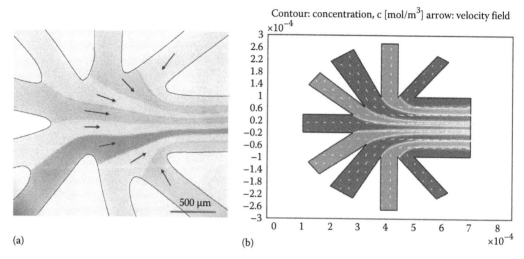

(a)   (b)

FIGURE 13.2
(a) High laminarity of microflows: the different fluid streams flow in parallel, without mixing. Remark the location of the stagnation points which depends on the flow rates. (b) Numerical simulation with COMSOL showing a progressive slow mixing.

where R is the flow resistance

$$R = \frac{4\mu L}{wd\,\min(w^2, d^2)q(\varepsilon)},$$ (13.18)

and the function q is the form factor given by

$$q(\varepsilon) = \frac{1}{3} - \left(\frac{64}{\pi^5}\right)\varepsilon \tanh\left(\frac{\pi}{2\varepsilon}\right).$$ (13.19)

A good agreement with the theoretical formula is obtained by a 3D calculation with COMSOL (Figures 13.3 and 13.4) [24]. In particular, Figure 13.4 shows a comparison between the closed form formula (13.18) and the numerical results.

13.2.7 Non-Newtonian Fluids

13.2.7.1 Introduction

The first microfluidic systems for biotechnology used conventional liquids, like water, or aqueous solutions and organic liquids, like mineral oil. Even with dilute species, these fluids are Newtonian, that is, their viscosity depends only on the concentration and temperature. Recently, the use of non-Newtonians or viscoelastic fluids has become increasingly widespread, for example, with the use of whole-blood (not diluted) for home test systems or liquid polymers like alginates and agarose for cell encapsulation devices [25–28]. Hence, it has become necessary to understand the viscoelastic behavior and be able to predict its effect on the flow field and pressure drop.

Slice: velocity field (m/s) arrow: velocity field

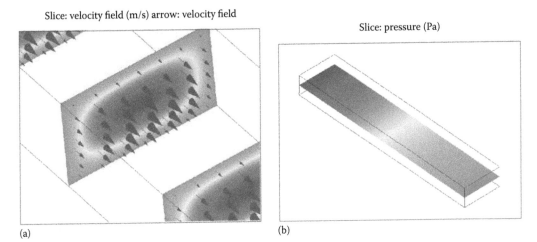

Slice: pressure (Pa)

(a)   (b)

FIGURE 13.3
(a) Contour plot of the velocities in a cross section of the channel, with velocity vectors; (b) pressure map in the channel mid-plane.

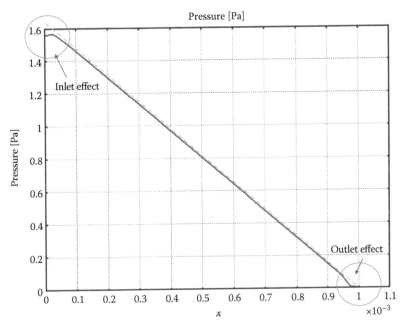

FIGURE 13.4
Comparison between COMSOL and analytical law of the pressure profile in a rectangular channel of aspect ratio ½ (continuous line COMSOL, dotted line analytical model).

Viscosity of usual (Newtonian) liquids depends on the temperature T. If micro or nanoparticles are transported by the fluid, their concentration has an important influence on the viscosity. The general expression of the viscosity is

$$\eta = f(T, c) \tag{13.20}$$

Note that the usual notation for viscosity is μ in hydrodynamics and η in rheology. This general category of fluids is called Newtonian. Non-Newtonian fluids—like polymeric solutions—have a more complicated viscosity law as will be shown in the next section.

13.2.7.2 Non-Newtonian Viscosity

Polymeric liquids used in microfluidic systems are more or less viscoelastic depending on their concentration. Basically, the viscosity of viscoelastic polymeric liquids depends on the concentration in polymers, temperature, and shear rate

$$\eta = \eta(c, T, \dot{\gamma}) \tag{13.21}$$

where $\dot{\gamma}$ is the shear rate (which will be defined in the next section).

13.2.7.2.1 Influence of Concentration

In rheology of polymers, the very general Martin's relation [29,30] usually applies for polymeric solutions

$$\eta_{sp} = \left(c[\eta]\right)e^{k'[\eta]c} \tag{13.22}$$

where η_{sp} is the specific viscosity defined by

$$\eta_{sp} = \frac{\eta - \eta_0}{\eta_0} \tag{13.23}$$

where η_0 is the viscosity of the carrier fluid alone. In (13.22), $[\eta]$ is the intrinsic viscosity and k' the Huggins coefficient. For dilute polymeric solutions, a Taylor expansion yields the Huggins law:

$$\eta_{sp} = c[\eta] + k'\left(c[\eta]\right)^2 \tag{13.24}$$

For semi-dilute solutions, more terms in the expansion of (13.24) should be kept. However, it has been shown that the specific viscosity can generally be approached by the power law:

$$\eta_{sp} = a\left(c[\eta]\right)^n \tag{13.25}$$

Taking alginate solutions as an example, it can be shown that relation (13.25) fits well the experimental results with $a = 0.1$, $[\eta]$ of the order of 300–800 mL/g, and n of the order of 3–4 depending on the type of alginate [28,31]. Finally, the viscosity of the polymeric solution depends on the concentration by the law

$$\eta = \eta_0\left[1 + a\left(c[\eta]\right)^n\right] \tag{13.26}$$

13.2.7.2.2 Influence of Temperature

As a general rule, the viscosity of a polymeric liquid decreases with temperature. The Vogel–Fulcher–Tamman (VFT) hyperbolic relation is often used to describe the thermal dependency of the viscosity [30] and writes

$$\log \eta = A + \frac{B}{T - T_0} \tag{13.27}$$

where A and B are experimentally determined coefficients (Figure 13.5). The change in viscosity with temperature of a polymeric liquid is such that it is important to always check the temperature before performing experiments with polymeric liquids.

13.2.7.2.3 Influence of Shear Rate

Besides its dependency on concentration and temperature, the viscosity of polymeric solutions decreases with the shear rate of the flow, because, in high shear regions, the long polymer chains align with the flow; such a behavior is called shear thinning. Many different laws have been proposed for the non-Newtonian viscosity depending on the carrier liquid and polymers. The most common law, often valid for small to medium shear rates is the Ostwald law or "power law" [29,30]. It is recalled that the viscosity of a "power law" fluid has the form

$$\eta = K \dot{\gamma}^{n-1} \tag{13.28}$$

where
  K and n are constants
  $\dot{\gamma}$ is the shear rate

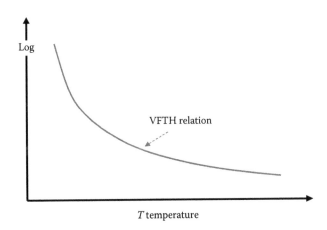

FIGURE 13.5
Thermal dependency of the viscosity of a polymeric liquid.

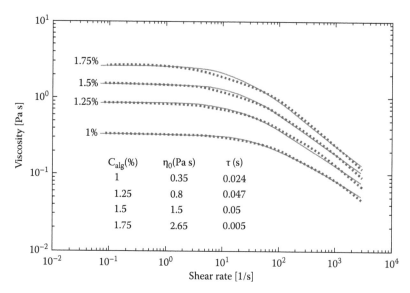

FIGURE 13.6

Viscosity of Keltone HV alginate solutions versus shear rate: the dots correspond to the experimental results, the continuous line to the Carreau–Yasuda model. The four curves correspond to four alginate concentrations: 1, 1.25, 1.5, 1.75 wt%. Alginate viscosity increases with the concentration and decreases with the shear rate. The relaxation times are deduced from a fit of (13.29) on the different experimental curves.

In the domain of medium shear rates, for shear thinning liquids, like xanthan or alginate, it is common to use the Carreau–Yasuda relation [29]

$$\eta = \eta_0 [(1 + (\tau \, \dot{\gamma})^\alpha]^{\frac{m-1}{\alpha}} \tag{13.29}$$

where
η_0 is the viscosity at zero shear rate
τ a relaxation time
α is a constant
m depends on the concentration of the solution

Again, taking alginates as an example, the Carreau–Yasuda relation fits well the experimental measurements as shown in Figure 13.6. The relaxation times are deduced from a fit of (13.29) on the different experimental curves.

Note that Carreau–Yasuda law and Ostwald law produce similar results for many polymeric solutions, like alginates for example (Figure 13.7).

In the general relation (13.21), concentration and temperature can often be considered as data, because concentration and temperature are often constant. In order to close the system, an expression of the shear rate $\dot{\gamma}$ has to be introduced. The shear rate can be derived from the expression of the deformation tensor. Liquid flows are characterized by their deformation tensor D (or rate-of-deformation or rate-of-strain tensor) [32]

$$D = \frac{1}{2}\left(\nabla V + \nabla V^T\right) \tag{13.30}$$

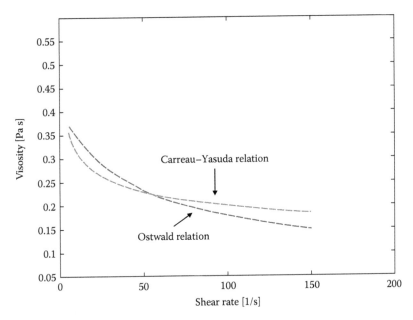

FIGURE 13.7
Comparison between the Carreau–Yasuda relation for the viscosity of liquid alginates and the power law (Ostwald relation).

where V is the velocity vector field. One defines the shear rate associated to a fluid deformation by

$$\dot{\gamma} = \sqrt{2D:D}. \tag{13.31}$$

In a 3D case, the general Cartesian expression for $\dot{\gamma}$ is

$$\dot{\gamma} = \sqrt{\frac{1}{2}\left[4\left(\frac{\partial u}{\partial x}\right)^2 + 4\left(\frac{\partial v}{\partial y}\right)^2 + 4\left(\frac{\partial w}{\partial z}\right)^2 + 2\left(\frac{\partial u}{\partial y} + \frac{\partial v}{\partial x}\right)^2 + 2\left(\frac{\partial v}{\partial z} + \frac{\partial w}{\partial y}\right)^2 + 2\left(\frac{\partial u}{\partial z} + \frac{\partial w}{\partial x}\right)^2\right]}. \tag{13.32}$$

The one-dimensional shear rate corresponding to a flow along a planar solid surface is

$$\dot{\gamma} = \frac{\partial u}{\partial y}. \tag{13.33}$$

In the case of a rectangular microchannel, $v = w = 0$, and u is only a function of y and z; then (13.32) simplifies to

$$\dot{\gamma} = \sqrt{\left(\frac{\partial u}{\partial y}\right)^2 + \left(\frac{\partial u}{\partial z}\right)^2}. \tag{13.34}$$

From a numerical standpoint, the dependency of the viscosity η with the shear rate requires an iterative solver corresponding to a nonlinear problem, just because η is a function of the unknowns u, v, and w.

13.2.7.3 Non-Newtonian Pressure Drop

The pressure drop determination of non-Newtonian fluid flows is complicated. Indeed, for a Newtonian fluid, the force balance on a control volume of a rectangular channel of width w, depth d, and wall surface S can be expressed as

$$\Delta P = \frac{1}{wd} \int_S \tau_w \, ds \tag{13.35}$$

where τ_w is the wall friction. For a 2D case, and a Poiseuille–Hagen flow, the wall friction is simply given by

$$\tau_w = \frac{6\mu \bar{U}}{d} \tag{13.36}$$

where
  μ is the viscosity
  $\bar{U}$ the average velocity

Substitution of (13.36) in (13.35) yields

$$\Delta P = \frac{12\mu L \bar{U}}{d^2} \tag{13.37}$$

where L is the length of the control volume. However, in the case of a non-Newtonian fluid, Equation 13.35 becomes a complicated integral

$$\Delta P = \frac{1}{wd} \int_S \eta(\dot{\gamma}_w) \, \dot{\gamma}_w \, dx \, dy \, dz \tag{13.38}$$

where $\dot{\gamma}_w$ is the wall shear rate. The only case for which a closed-form formulation exists is that of a cylindrical duct in which a "power law" fluid (Ostwald fluid) circulates. Using (13.28), the friction is expressed by

$$\tau = \eta \dot{\gamma} = K \dot{\gamma}^n \tag{13.39}$$

In such a case, the solution has been formally given by Rabinowitsch and Mooney [29,30]

$$\Delta P_{RM} = \frac{2^{(n+2)} L K}{w} \left(\frac{3n+1}{n} \right)^n \left(\frac{\bar{U}}{w} \right)^n \tag{13.40}$$

where K and n are the constants of the "power law" fluid. Hence, the hydraulic resistance is

$$R_{RM} = \frac{2^{(n+2)} L K}{w^3 d} \left(\frac{3n+1}{n} \right)^n \left(\frac{\bar{U}}{w} \right)^{n-1} \tag{13.41}$$

Relation (13.42) shows that the hydraulic resistance is not a geometrical constant and depends on the flow velocity. This is a drastic difference between Newtonian and non-Newtonian fluids that has important consequences on microfluidic networks [31]. Inspired by the cylindrical approach, approximated relations have been given for rectangular channels [33–36], leading to the expression

$$\Delta P = \frac{2^{3n+2} \, K L}{w^{n+1}} \left(\frac{c_1}{n} + c_2 \right)^n \bar{U}^n \tag{13.42}$$

where the geometric coefficients c_1 and c_2 are given by

$$c_1 = \frac{1}{2(1+\varepsilon)^2 \left[1 - \dfrac{32}{\pi^3 \cosh\left(\dfrac{\pi}{2\varepsilon}\right)} \right]}$$

$$c_2 = \frac{3}{2(1+\varepsilon)^2 \left[1 - \dfrac{192\varepsilon \tanh\left(\dfrac{\pi}{2\varepsilon}\right)}{\pi^5} \right]} - c_1 \tag{13.43}$$

and ε is the aspect ratio of the rectangular channel ($\varepsilon < 1$). The hydraulic resistance of a rectangular channel is then

$$R = \frac{2^{3n+2} \, K L}{w^3 d} \left(\frac{c_1}{n} + c_2 \right)^n \left(\frac{\bar{U}}{w} \right)^{n-1} \tag{13.44}$$

As expected, the hydraulic resistance depends on the flow conditions, namely, $\bar{U}$.

13.2.7.4 Numerical Results for a Square Microchannel

In a cylindrical tube, a laminar flow field of a Newtonian fluid is determined by the Poiseuille–Hagen relation, and the velocity profile is quadratic. This is not the case for a non-Newtonian viscoelastic fluid. Figure 13.8 shows the difference of velocity profile between the Newtonian and non-Newtonian case calculated with the COMSOL numerical program.

If we use the Ostwald expression $\tau = \eta \dot{\gamma}^n$, the Fanning friction factor can be expressed as $16/Re_{NN}$, where Re_{NN} is a non-Newtonian Reynolds number [29]. However, to our knowledge, there is no closed-form expression for a Carreau fluid for rectangular channels, and one must rely on numerical modeling. In the following section, we show some consequences of this change of frictional pressure drop.

Figure 13.9 shows a comparison between the literature results (Muzychka et al. [33], Kozicki et al. [34], Miller et al. [35]) and COMSOL 3D calculation for a $100\,\mu m \times 100\,\mu m$ channel.

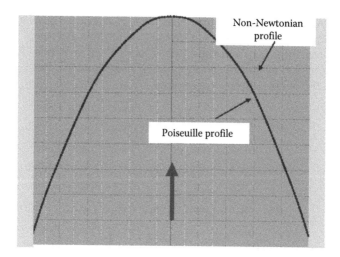

FIGURE 13.8
Newtonian and non-Newtonian velocity profiles in a cylindrical tube.

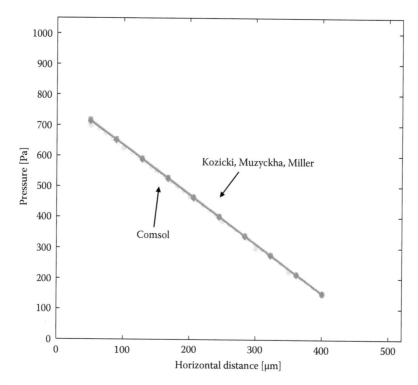

FIGURE 13.9
Non-Newtonian pressure profiles in a $100\,\mu m \times 100\,\mu m$ square channel of length $L = 500\,\mu m$.

13.2.7.5 Non-Newtonian Networks

An assembly of capillary tubes or microchannels is called a microfluidic network. Networks are now currently used in biotechnology, for example, to perform blood separation [37], concentration gradients [38], flow separation, and microporous needles [5,39]. Most of the time, the flow rate in each branch depends on the flow resistances of all the branches of the system. The system is then difficult to design. If the fluid is Newtonian, and only one fluid is used, it can be shown that the viscosity of the fluid does not affect the flow distribution [5]. However, if the liquid is non-Newtonian, the design of the system must be adjusted specifically [31]. Besides, the flow rate distribution will change with any change in the inlet conditions. In the Newtonian case, the hydraulic resistance R of a branch of rectangular cross section (a, b) is

$$R = R(\eta, L, a, b) = \frac{\Delta P}{Q} \tag{13.45}$$

where
  P is the pressure
  Q is the flow rate

In a non-Newtonian case, we can write symbolically the implicit expression

$$R = R(\eta(Q), L, a, b) = \frac{\Delta P}{Q} \tag{13.46}$$

A simple numerical calculation shows that after one bifurcation, the distribution of flow rates changes between a Newtonian and non-Newtonian fluid (Figure 13.10). This is even more the case after two bifurcations: The flow rate inside the smaller channels is dramatically reduced. The reason for this behavior is obvious: When the flow rate is small in a channel, the shear rate is small and the viscosity high. The flow redirects into the larger channels.

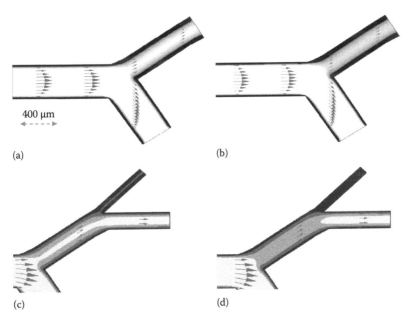

FIGURE 13.10
Flow in microfluidic networks: (a) and (c) Newtonian fluid, (b) and (d) non-Newtonian fluid (alginate 2%wt).

13.3 Shallow Channel or Hele–Shaw Model: A 2D Approach for a Pseudo 3D Microflow

13.3.1 Introduction

Flat or shallow channels—channels whose aspect ratio d/w is very small—are common in biotechnology. For example, they are currently used to force biologic targets to come to contact to the wall and to bind to immobilized ligands on the solid surface. It is recalled here that, due to the high laminarity of the microflows, only diffusion can trigger the contact of transported targets with a solid wall. A 3D approach to such problems is not practical. In order to describe precisely the vertical velocity profile, very small meshes are required, and a large horizontal domain cannot be covered with such small meshes. Clearly, the situation is close to a 2D situation, with a vertical quadratic velocity profile everywhere in the channel. An approach to this problem has been investigated by many researchers, Hele–Shaw first, Kirby, Schiliething and others [40] assuming that the pressure does not change with the vertical coordinate, an approximated solution is that of a potential flow in the 2D horizontal coordinate systems combined with a quadratic vertical profile:

$$\vec{u} = -\frac{1}{2\mu} z(d-z)\nabla p. \tag{13.47}$$

We shall see next that we can avoid the approximated two-dimensional (2D) potential flow by rigorously solving the 2D problem with a finite element formulation.

13.3.2 Model

Clearly, friction on the two horizontal walls cannot be disregarded; it is even the dominant friction force. In such a case, the 2D Navier–Stokes equations need to be modified [41]. For a steady state flow, let us consider the following equation:

$$\rho(\vec{u}.\nabla)\vec{u} = -\frac{\partial P}{\partial x} + \mu\left(\frac{\partial^2 u}{\partial x^2} + \frac{\partial^2 u}{\partial y^2}\right) - \frac{12\mu}{d^2}\vec{u} \tag{13.48}$$

The additional force in the right-hand side of (13.48) is the friction on the upper and bottom solid surfaces. Indeed, integrating this term on a $w\,d\,\Delta x$ parallelepiped domain leads to

$$\frac{1}{wd}\int_0^w \int_x^{x+\Delta x} \int_{-d/2}^{d/2} \frac{12\mu}{d^2} u\, dx\, dy\, dz = \frac{12\mu}{d^2}\frac{\Delta x}{d}\int_{-d/2}^{d/2} u\,dy = \frac{12\mu}{d^2}\frac{\Delta x}{}U \tag{13.49}$$

which is the friction force on two parallel plates on a length Δx (Figure 13.11). Using (13.49) in a 2D formulation produces identical flow rates and pressures than would a 3D calculation. Remark that the maximum velocities are not identical because the 2D calculation averages the velocity in the vertical z-direction. Hence, the maximum 2D velocity is only two-thirds of that obtained by the 3D calculation (Figures 13.12 and 13.13).

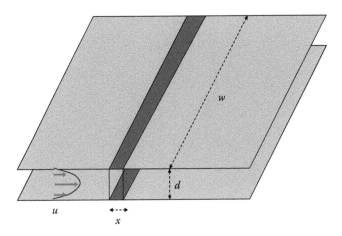

FIGURE 13.11
Sketch of a Hele–Shaw microchamber.

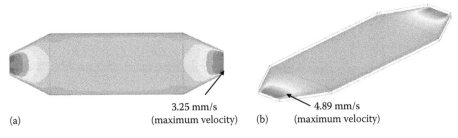

3.25 mm/s
(maximum velocity)

4.89 mm/s
(maximum velocity)

(a)

(b)

FIGURE 13.12
Flow velocity map obtained with (a) the 2D Hele–Shaw model, (b) a 3D model. In (b) the visualization plane is set at mid-height. Note that the ratio of the velocities between (a) and (b) is 2/3.

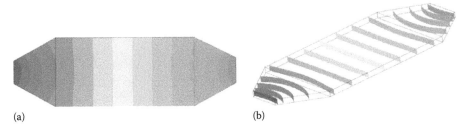

(a)

(b)

FIGURE 13.13
Pressure distribution in the microchamber: (a) 2D with Hele–Shaw model; (b) full 3D model.

A good agreement is also obtained by using a 2D Helle–Shaw calculation when the aspect ratio is less than 1/3. For larger aspect ratios (1/3 to 1), the agreement is a little less satisfactory, but this method still yields an approximation of the pressure drop (Figure 13.14). In a typical case of a channel of width $w = 100\,\mu m$, length $L = 400\,\mu m$, and flow rate $Q = 1\,\mu L/mn$, the analytical and 3D-COMSOL pressure profiles are nearly indiscernible for any aspect ratio, whereas the 2D-HS model is adequate for aspect ratios less

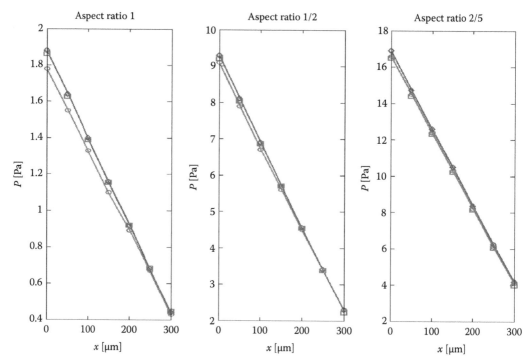

FIGURE 13.14

Comparison of the axial pressure profile in a rectangular channel between analytical expression (squares), 3D-COMSOL (continuous line with diamonds) and 2D-HS-COMSOL (continuous line with circles) formulations for three different aspect ratios 1, 0.5, and 0.4.

than 1/3 approximately: Relative errors of 2%, 3%, and 5% are, respectively, found for aspect ratios of 0.2, 0.5, and 1. Note that the pressure drop is much larger for a small aspect ratio channel under the same flow rate conditions.

13.3.3 Application to a Microfluidic Resonator

Microfluidic resonators have appeared recently under the impulse of the MIT and Caltech. They combine the high sensitivity of micro-cantilevers for detecting minute increase of mass and the liquid environment required in most biotechnological devices [42,43]. Let us recall that micro-cantilevers detect extremely small masses by the analysis of the change of frequency associated to the immobilization of a mass. However, if they are extremely efficient in an open (air) environment, the presence of liquids like water creates a damping that impedes the sensitivity of the device. A solution is to introduce a liquid channel in the interior of the cantilever blade. This solution is made possible by the improvements in microfabrication. Applications to DNA immobilization and to cell culture in microfluidic resonators have been recently demonstrated [44]. A typical microfluidic resonator is shown in Figure 13.15 [45].

The average flow velocity can be calculated using the shallow channel approach. Figure 13.16 shows that Helle–Shaw formulation is well adapted to this type of geometry.

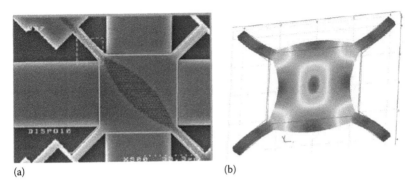

(a)   (b)

FIGURE 13.15
(a) View of a microfluidic resonator from CEA-Leti with the microfluidic channel embedded diagonally; (b) modeling of the Lamé mode of resonance of a microplate fixed in four angles, showing a maximum amplitude in the middle of the four edges. (From Agache, V., Blanco-Gomez, G., Baleras, F., and Caillat, P., An embedded microchannel in a MEMS plate resonator for ultrasensitive mass sensing in liquid, *Lab Chip*, 11, 2598–2603, 2011. Reproduced by permission of The Royal Society of Chemistry.)

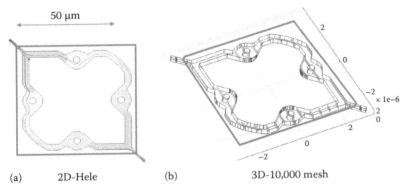

(a)   2D-Hele   (b)   3D-10,000 mesh

FIGURE 13.16
Microfluidic resonator with two microfluidic channels running alongside the edges; (a) shallow channel formulation; (b) full 3D calculation.

13.4 Mixing in Microsystems

13.4.1 Introduction

Mixing is difficult in microsystems because of the high laminarity of the flows. If only passive methods are used—in order to keep the system simple—only diffusion contributes to mixing. From a numerical standpoint, convection-diffusion equations for microflows are difficult to solve, mainly because of the large ratio between convection and diffusion. Consider the convection-diffusion problem described by the equation

$$\frac{\partial c}{\partial t} + \vec{V}.\nabla c = \nabla.(D\nabla c) + f \qquad (13.50)$$

where

  c is the concentration in the transported species
  D the diffusion coefficient
  f a source term

On one hand, the mesh Peclet number representing the ratio between the convection and the diffusion, defined by

$$Pe = \frac{\|V\|\Delta x}{2D} \tag{13.51}$$

where Δx is the size of the mesh and $\|V\|$ the norm of the velocity, is often larger than 1, principally because the diffusion coefficient is small, of the order of $10^{-10}\,\mathrm{m^2/s}$. The numerical scheme is then unstable; oscillations appear where steep gradients are present, mostly at the domain boundaries, and these oscillations may propagate to the whole domain; as a result, negative concentrations appear. Hence, extremely small meshes should be used—most of the time this is not possible—or stabilization is required to achieve the calculation of the concentration field. The easiest way to stabilize the calculation is to add an "artificial" isotropic diffusion

$$D_{art} = \lambda\|V\|\Delta x \tag{13.52}$$

where λ is a tuning parameter, the value of which is comprised between 0 and 0.5. Indeed, for $\lambda = 0.5$,

$$Pe = \frac{\|V\|\Delta x}{2(D + D_{art})} = \frac{\|V\|\Delta x}{2D + \|V\|\Delta x} < 1 \tag{13.53}$$

However, the use of isotropic artificial diffusion is disastrous in microfluidics due to the high laminarity of the flow and the small dimensions. Upwind schemes like the streamline upwind/Petrov–Galerkin model (SUPG) are more adapted to the problem. A review of upwind method has been performed by Volker and Knobloch [46]. In fact, the use of artificial diffusion should be reduced to a minimum, which necessitates a trial and error process. We recommend using some SUPG diffusion combined to the smallest possible isotropic artificial diffusion.

13.4.2 Coflows

In order to illustrate the difficulty of mixing at the microscale, we consider the merging of two miscible flows in the geometry of a T-junction (Figure 13.17). At the macroscale, turbulent structure would develop and enhance the mixing of the two flows. At the microscale, the flowstreams are laminar, and mixing is based on diffusion which is a slow process. The two flows run side by side with progressive mixing by cross diffusion.

  In the "diffusing zone," the streamlines are parallel and directed along the x-axis; the substance/solute progressively diffuses in the perpendicular y direction, and there is a growing distance $\delta(x)$ of concentration gradient. It can be shown that the concentration profile is given by the relation [5,16]

$$c(x, y) = \frac{1}{2}c_0\left(1 - erf\,\frac{y\sqrt{U}}{\sqrt{4Dx}}\right) \tag{13.54}$$

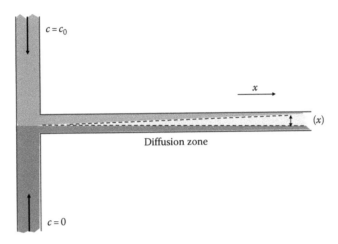

FIGURE 13.17
Sketch of the mixing of two miscible fluids in a T-junction.

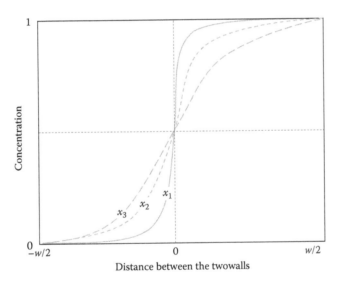

FIGURE 13.18
Concentration profile of diffusing species at three different locations in the channel.

where U is the mean flow velocity. Typical concentration profiles in different cross sections are shown in Figure 13.18.

From a numerical standpoint, it is essential to reduce as much as possible the isotropic artificial diffusion. Figure 13.19 compares the concentration field for two values of the artificial diffusion coefficient.

13.4.3 Herringbone Structures

Herringbone structures are ridges or troughs that are etched inside a microchannel, at the bottom or at the top of the channel (Figure 13.20). Herringbone structures are used to modify the flow field inside a microchannel [47,48]. They locally induce vorticity that facilitates mixing: Figure 13.21 shows the transverse component of the velocity showing

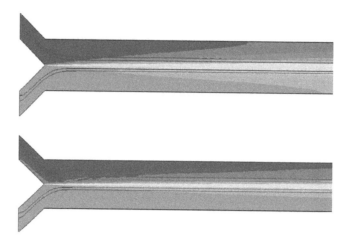

FIGURE 13.19
Concentration contours obtained with two different additional artificial diffusion ($D = 10^{-10}$ m²/s and $V_{inlet} = 2$ mm/s).

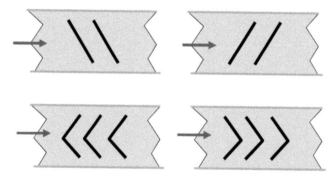

FIGURE 13.20
Typical shapes of herringbone structures.

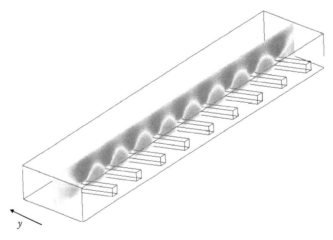

FIGURE 13.21
Vy-component of velocity in a one-sided herringbone channel.

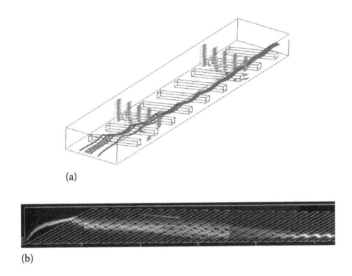

(a)

(b)

FIGURE 13.22
(a) One-sided herringbone structures transport 10 µm particles toward a corner of the channel (COMSOL); (b) trajectories of HL60 cells in passivated and P-selectin-coated channels. (From Choi, S., Karp, J.M., and Karnik, R., Cell sorting by deterministic cell rolling, *Lab Chip.*, 34, 12(8), 1427–1430, 2012. Reproduced by permission of The Royal Society of Chemistry.)

a change of direction in front of the ridges. Besides, herringbone structures induce a distortion of the streamlines, which result in a change of the particles trajectories in the channel. Figure 13.22 shows that relatively large particles traveling near the bottom of the channel are progressively pushed toward an angle of the channel. The results are in close agreement with the experimental results of Sungyoung Choi et al. [48].

13.4.4 Dean Flows

Fluid velocities are small in microsystems. The Reynolds number is almost always smaller than 50, and most of the time smaller than 1. Inertial forces are negligible at very low Reynolds number (Re < 1); however, at medium Reynolds number (Re ~ 5–50), some inertial effects can appear. Particularly, in the case of a strongly curved microchannel, inertial forces can induce a centrifugal effect. This is called the Dean effect, and this is the subject of this section.

13.4.4.1 Hydrodynamics of Dean Microflows

It has been observed long ago by W.R. Dean [49] that flows in curved channels present a centrifugal effect at medium or large Reynolds numbers. This effect is called the Dean effect. In this section, we investigate the Dean effect in microfluidics and show how it can be used in biotechnological applications.

Usually, Reynolds numbers are very small in microsystems for biotechnology—most of the time smaller than 1—and the flow is highly laminar. These types of flows are well adapted to the transport of very small molecules like DNA but not cells because of sedimentation. Recently, medium ranges of Reynolds numbers (Re ~ 5–50) have been investigated [50–53]. These flows are strong enough to carry cells, but still completely laminar. It has been shown that such flows in curved or bended tubes produce two opposite vortices—still in the laminar domain—with spiral streamlines in the curved regions due to

Slice y-velocity [m/s]

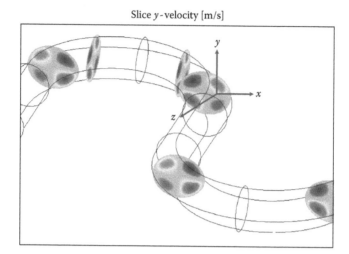

FIGURE 13.23
Vy-contour plot for a flow in a cylindrical curved tube (Re = 5) (COMSOL numerical software).

a centrifugal effect. These spiraling streamlines are used to guide and concentrate particles or cells and to enhance mixing [50].

Let us consider the example of Figure 13.23. The main component of the velocity is directed along the direction of the tube, but the transverse component of the velocity (here V_y), which is zero at low Reynolds number, is positive in two quadrangles and negative in the other two. Hence, there is a recirculating component of the velocity that induces a spiral flow.

Let us define the Dean number by

$$De = U \, R/v \, \sqrt{R/R_c} = \mathrm{Re} \, \sqrt{R/R_c} \tag{13.55}$$

where R is the radius of the cylindrical channel and R_c its curvature radius. For a square or rectangular channel, one must replace R by the hydraulic radius R_H. The inertia-induced spiral motion is noticeable when the Dean number is larger than 1.

Another way of pinpointing this spiraling motion is by plotting the vorticity contour plot, as shown in Figure 13.24.

13.4.4.2 Concentration in Dean Microflows

Dean flows have the property to induce mixing by centrifugal effect. Let us consider the case described in Figure 13.25, where two aqueous microflows merge before the curved part of the channel, with one of the two flows having a given concentration in a solute species while the other flow having a zero concentration in the same species. Just after merging, the concentration isolines are vertical. After one turn of the channel, the Dean effect has considerably modified the concentration distribution, a little like the Baker's transform [9,54].

13.5 Cell Chips

Cell chips are becoming an essential tool for biologists and biotechnological and pharmaceutical companies. These chips are very efficient to study the reactions of a cell or

Slice: verticity [1/s]

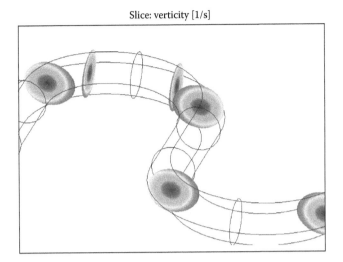

FIGURE 13.24
Vorticity contour plot for a flow in a cylindrical curved tube, Re = 5 showing the centrifugal effect (COMSOL numerical software).

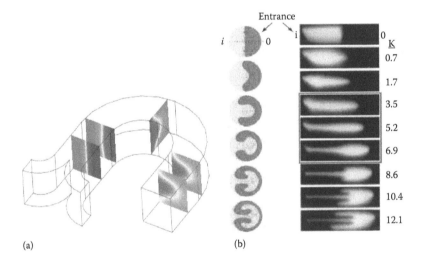

FIGURE 13.25
(a) Concentration in a Dean tube (k ~1) showing the effect of the centrifugal force; (b) Experimental observations. (Data from Sundarsan, A.P. and Ugaz, V.M., Multivortex micromixing, PNAS, 103(9), 7228–7233, Copyright 2006, National Academy of Sciences, USA, reproduced with permission.)

a group of cells to chemical or biochemical agent. Applications are numerous, ranging from the study of the cellular response to different concentrations of reagents and drugs—such as antibiotics—to cell differentiation—especially stem cells—and to cellular communication—the biochemical species produced by cells and migrating in the fluid domain.

13.5.1 Single-Phase Flow Focusing

Focusing cells or particles in a microfluidic channel is an essential step for all cell chips. Two types of flow focusing exist: the two-phase flow focusing that is used to encapsulate

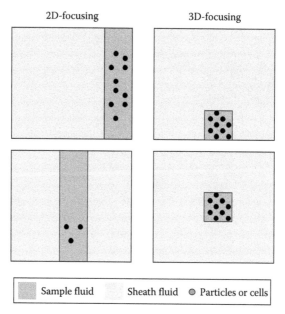

FIGURE 13.26
Principle of single-phase flow focusing.

biological objects and cells [55,56] and the single-phase flow focusing that is used to concentrate or focus a "beam" of liquid inside a sheath flow [57,58]. The biologic targets transported by the flow are focused or confined in a fraction of the cross section of the channel. Depending on the device, the focusing can be made along a wall of the microchannel or in a pinched streamflow (Figure 13.26). In the first case, the flow rate ratio is

$$\frac{Q_1}{Q} = \frac{1}{2}\frac{w_1}{w} \tag{13.56}$$

where Q_1 and Q are, respectively, the sample fluid flow rate and the total flow rate, and w_1 and w are, respectively, the width of the focused region and the total width of the channel. In the second case, the characteristic size of the pinched flow R is

$$\frac{Q_1}{Q} \approx \frac{9}{4}\frac{R^2}{wd} \tag{13.57}$$

Figure 13.27 shows how a side sheath flow can focus the fluid of interest along a wall of the channel. According to (13.57), the focusing of the flow by the sheath flow can be tuned by adjusting the sheath flow rate. Figure 13.28 shows a 2D calculation of the same type of focusing, demonstrating how particles (10 μm) transported by the carrier fluid and randomly dispersed in the carrier fluid are deflected toward the outside wall by the sheath flow. Figure 13.29 is a similar calculation with a focusing in the channel center by two symmetrical sheath flows. Note the narrow focusing of the randomly distributed particles.

Three-dimensional focusing can be achieved by a more elaborate device like that shown in Figure 13.30, initially proposed by Kennedy and colleagues [57].

Depending on the different sheath flow rates, the focusing may be adjusted inside the channel cross section, as shown in Figure 13.31.

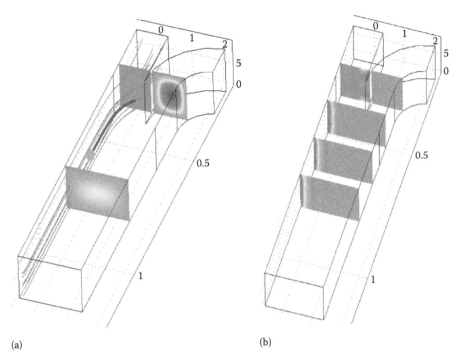

(a)   (b)

FIGURE 13.27
Single-phase flow focusing. (a) streamlines (continuous line) are concentrated along the wall under the action of the incoming sheath flow; a trajectory of a 10 μm particle is shown with the streamlines on the left of the picture; (b) concentration in slices across the channel showing the focusing.

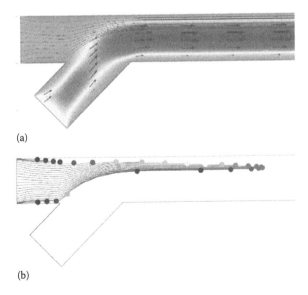

(a)

(b)

FIGURE 13.28
(a) Velocity field in a single phase focusing device; (b) 10 μm diameter particle trajectories showing the focusing (COMSOL).

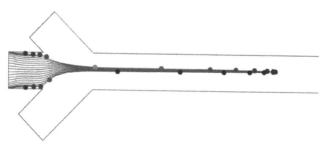

FIGURE 13.29
Focusing by pinching the flow using two sheath flows.

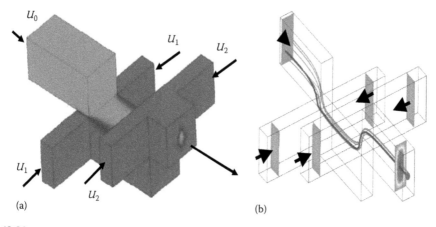

FIGURE 13.30
(a) Principle of 3-D single-phase flow focusing; (b) streamlines obtained by using COMSOL numerical program, showing the focusing.

13.5.2 Bifurcations and Pinched Channel Microflows

In cell chips, special geometrical designs are introduced to separate and isolate different categories of cells or micro and nanoparticles. In this section, we present two types of designs: the branched channel based on channel bifurcations and the pinched channel based on sudden enlargement of the main channel.

13.5.2.1 Bifurcations

In this paragraph, we consider a main channel with branches perpendicular to the main channel. Because the flow is laminar, the streamlines (and pathlines for a steady flow) are well defined (Figure 13.32). It is possible to determine the widths (δ_1, δ_2, δ_3 in Figure 13.32) that correspond to the stream bifurcating in each side branch.

At low flow rates, we make use of the assumption already used earlier that a cell/spherical particle follows the streamline passing by its centroid [59,60]. Let us consider spherical particles focused near a wall and approaching a bifurcation (Figure 13.33).

For simplicity, let us assume a 2D situation and neglect the effect of the channel depth. It can be shown that the complete 3D calculation produces the same result as the 2D calculation. The velocity field can be approximated by the Poiseuille–Hagen quadratic profile

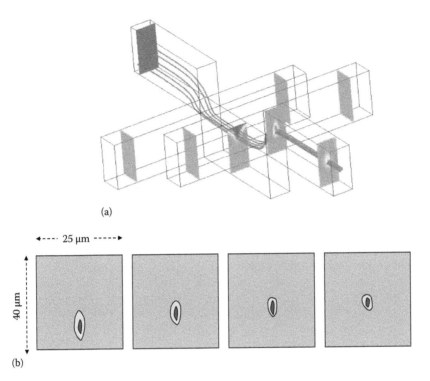

(a)

(b)

FIGURE 13.31
(a) Slices showing the concentration in the device and the focusing at the outlet; (b) different focusing obtained by tuning the different sheath flow rates.

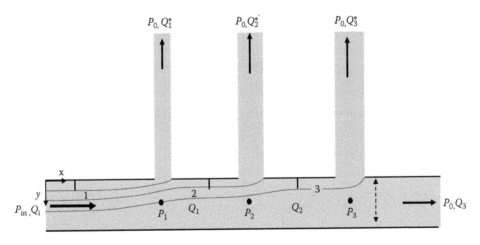

FIGURE 13.32
Flow in a branched network.

$$u(y) = 6U\,\frac{y\left(w-y\right)}{w^2} \tag{13.58}$$

where U is the mean axial velocity, related to the flow rate by $U = Q/(dw)$. At the bifurcation, the flow rate conservation equation requires that

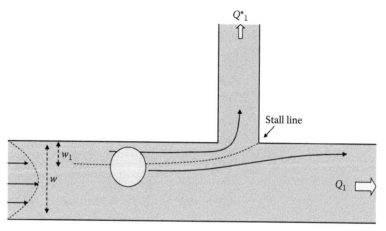

FIGURE 13.33
Trajectories of spherical particles at a bifurcation.

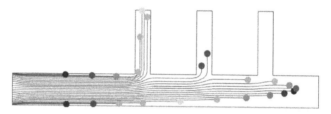

FIGURE 13.34
Computation of the 10 μm spherical particle trajectories in a branched network (COMSOL).

$$d \int_0^{w_1} u(y)\,dy = Q_1^* \tag{13.59}$$

where Q_1^* is the flow rate in the secondary channel. Using (13.58) and (13.59),

$$2\left(\frac{w_1}{w}\right)^3 - 3\left(\frac{w_1}{w}\right)^2 + \frac{Q_1^*}{Q} = 0 \tag{13.60}$$

Assuming that the flow rate distribution in the network is known, Equation 13.60 can be solved to produce the threshold width w_1. Wall-focused particles with a diameter D<2 w_1 will turn into the side channel. A 2D numerical example is presented in Figure 13.34, showing the bifurcation of the particles.

13.5.2.2 Flow Focusing Combined with Bifurcations

It has been found that the efficiency of "branched channel" is considerably increased by first focusing the particles or cells along the wall. A sheath flow first focuses the cells or particles, and then the separation is done at each bifurcation. Figure 13.35 shows a calculation of the trajectories of 10 μm diameter particles in a 300 μm/s carrier flow.

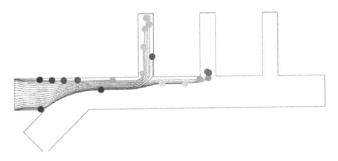

FIGURE 13.35
Particles trajectories in a device combining flow focusing and bifurcations.

13.5.2.3 *Pinched Flow Fractionation*

Separation of cells is fundamental for the study of a precise type of cell. Cells are often sorted out according to their size. A simple yet efficient method is that of "pinched channel," sometimes called pinched flow fractionation, or simply PFF (Figure 13.36).

Pinched channel geometry has been found to be an efficient way to separate particles and cells according to their size [61–63]. A first step is to concentrate all cells or particles alongside a wall. This step is called 2D focusing and has been presented in the preceding section. The targets are then transported toward a sudden enlargement. Small particles have their mass center closer to the wall than that of larger particles. In the enlargement, their trajectories will be different, the small and large targets not belonging to the same trajectory (Figure 13.37). Let us denote w_1 the half-width of the pinched channel, w_2 the half-width of the enlarged channel, and d the sphere (particle or cell) diameter; then the homothetic rule yields

$$\frac{y}{w_2} \approx \frac{w_1 - d/2}{w_1} \tag{13.61}$$

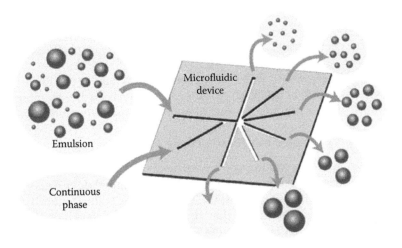

FIGURE 13.36
The principle of cell sorting by pinched channel method. (Data from Maenaka, H., Yamada, M., Yasuda, M., and Seki, M., Continuous and size-dependent sorting of emulsion droplets using hydrodynamics in pinched microchannels, *Langmuir*, 24(8), 4405–4410. Copyright 2008 American Chemical Society, reproduced with permission.)

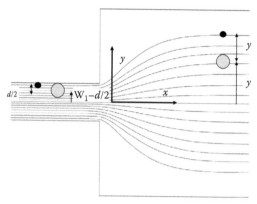

FIGURE 13.37
Principle of pinched channel.

which leads to

$$y \approx \left(w_1 - d/2\right)\frac{w_2}{w_1} \tag{13.62}$$

If we write relation (13.62) for two different types of cells, characterized by a difference of diameter Δd, the increase of vertical distance between the two trajectories is

$$\Delta y \approx \frac{\Delta d}{2}\frac{w_2}{w_1} \tag{13.63}$$

Such a behavior can be obtained using COMSOL with a simplified expression of the drag force

$$F_{drag} \approx C_D\left(V_f - V_p\right) \tag{13.64}$$

where C_D is the drag coefficient ($C_D \approx 6\pi\,\eta\,R_H$) and V_f and V_p are, respectively, the fluid and particle velocities.

The modeling with COMSOL in three dimensions is shown in Figure 13.38. Streamlines and trajectories for 5 µm radius spherical particles have been added to the graph, showing that the hypotheses of the simplified analytical model are approximately justified.

13.5.2.4 Example of PFF Coupled to a Flow Focusing Device

An interesting example of investigation of the efficiency of a PFF device has been made by Srivastav and colleagues [64] using for the first time a 30 channel device and a continuous distribution of spheres in the range 4–34 µm. The principle is illustrated in Figure 13.39 with only 12 outlet channels. The carrier flow rate is Q_1 and the sheath flow rate is Q_2. The drainage flow rate Q_3 is aimed to increase the separation efficiency in the outlet region.

Depending on the location of their centroid, particles follow different pathlines. These pathlines are shown in Figure 13.40. Particles circulating randomly in the inlet flow Q_1 are first focused alongside the wall by Q_2 and separated in the pinched chamber. Separation distance is maintained further downstream by the use of the drainage flow rate Q_3.

The separation efficiency of PFF devices for cell separation is shown in Figure 13.41.

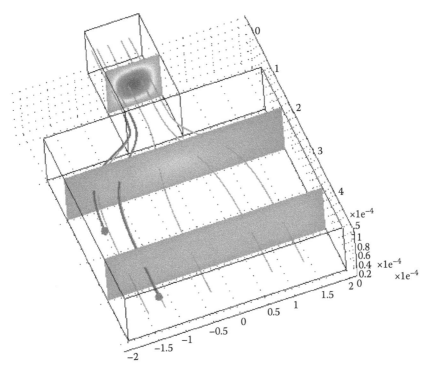

FIGURE 13.38
Pinched channel and sudden enlargement: streamlines (continuous grey lines) and trajectory separation of particles according to their focusing at the wall (continuous dark lines). Note that particle trajectories are not exactly identical to streamlines.

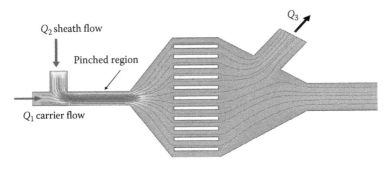

FIGURE 13.39
Velocity contour plot in a PFF device with streamlines.

13.5.3 Deterministic Lateral Displacement

13.5.3.1 Introduction and Theory

Deterministic arrays are a type of micro-devices for sorting out particles and cells [65–67]. The principle of deterministic lateral displacement (DLD) is related to the particle shifting of streamline caused by an obstacle. This principle is illustrated in Figure 13.42. A small particle—or cell—slides between the rows of pillars and, on average, follows a straight line. A large particle cannot slide between the pillars, because it is trapped by its streamline

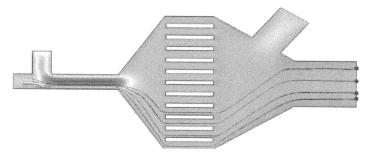

FIGURE 13.40
Trajectories of four spherical particles showing the separation efficiency of the PFF (continuous blue lines).

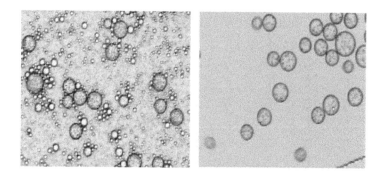

FIGURE 13.41
Separation of dispersed particulate suspension by pinched channel device. (Data from Maenaka, H., Yamada, M., Yasuda, M., and Seki, M., Continuous and size-dependent sorting of emulsion droplets using hydrodynamics in pinched microchannels, *Langmuir*, 24(8), 4405–4410. Copyright 2008 American Chemical Society, reproduced with permission.)

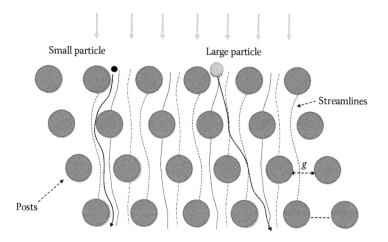

FIGURE 13.42
Sketch of a deterministic array: small particles stay in their own stream channel, whereas large particles are forced to shift to the next stream channel at the approach of a post.

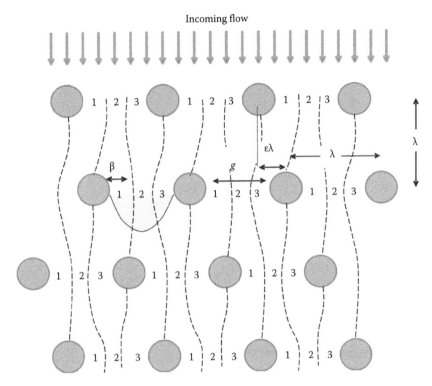

FIGURE 13.43
Schematic of a DLD device with $\varepsilon = 1/3$.

and laterally shifted when approaching the pillars. On average, it follows an oblique path. Hence, small and large particles do not behave the same. There is a size threshold that separates particles going straight and diagonally.

Let us analyze the situation in more detail. In the schematic Figure 13.43, rows of posts have been placed with a shift ε at each level (here the shift $\varepsilon = 1/3$). The flow can be decomposed in $n = 1/\varepsilon$ streamlines (here $n = 3$, and the streamlines are noted 1, 2, 3) having the same flow rate, equal to ε times the flow rate between two pillars. Between two neighboring pillars, the flow may be divided in n sub-flows occupying a width β of the gap m. Consider a small spherical particle in the stream channel 1. Always using the assumption that a particle or a cell follows the streamline passing by its centroid, because the diameter of the particle D is less than 2β, the particle is transported in its initial stream channel and globally has a straight trajectory. On the other hand, a particle of diameter $D > 2\beta$ cannot stay in channel 1 when passing through the gap between the two pillars; it is forced into the next channel, for example, channel 2. And this motion is repeated at each row of pillars. Globally, the particle follows a diagonal trajectory with an angle ε. In the following, we derive an expression for the critical particle diameter D_c, above which a particle is diagonally deviated.

The critical particle diameter is then

$$D_c = 2\beta \qquad (13.65)$$

Let us now calculate the value of β. Using the notations of Figure 13.43, we have

$$\int_0^{\beta} u(x)\,dx = \varepsilon \int_0^{m} u(x)\,dx \tag{13.66}$$

Assuming a parabolic velocity profile between two pillars, the velocity $u(x)$ can be expressed as

$$u(x) = u_{\max}\left[\frac{m^2}{4} - \left(x - \frac{m}{2}\right)^2\right] \tag{13.67}$$

Upon substitution of (13.67) in (13.66) and integration, the width β is solution of the cubic equation

$$\left(\frac{\beta}{m}\right)^3 - \frac{3}{2}\left(\frac{\beta}{m}\right)^2 + \frac{\varepsilon}{2} = 0 \tag{13.68}$$

And, using (13.65), the critical diameter is solution of

$$\left(\frac{D_c}{m}\right)^3 - 6\left(\frac{D_c}{m}\right)^2 + 4\varepsilon = 0 \tag{13.69}$$

A plot of D_c/m versus ε is shown in Figure 13.44.

A very interesting application of DLD is given in [66] (Figure 13.45), where a cell is progressively deviated into a lysis solution and is eventually lysed with chromosome and cell contents being separated. In such a device, the cell is initially larger than the critical radius and is deviated into the high concentration lysis solution. Once lysed, the chromosome is still larger than the critical radius and continues on an oblique trajectory whereas the cellular contents of small size follow a straight horizontal path. Chromosome is then

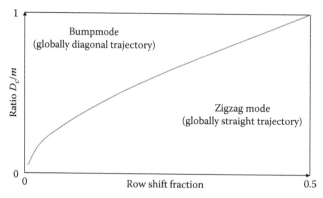

FIGURE 13.44
Plot of D_c/m as a function of ε.

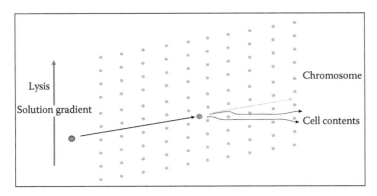

FIGURE 13.45
View of the DLD of a cell toward a lysis solution resulting in the lyse of the cell and the separation of the chromosome from other cell contents.

separated from the rest of the cellular contents. This is a very interesting way to purify the DNA content; besides this approach has a generic character.

13.5.3.2 Numerical Model

DLDs are complicated to model, especially in a 3D approach, for the mere reason that they necessitate a large number of meshes, due to their geometrical complexity. Most of the time, the approach will be 2D. Figure 13.46 shows the streamlines—which are identical to the pathlines for a steady flow—in a DLD microdevice. The three domains sketched in Figure 13.43 are delimited by the dark continuous lines in the figure. These

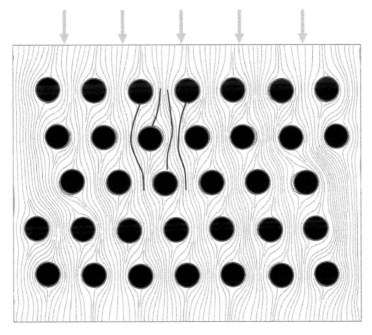

FIGURE 13.46
Flow in a micro DLD device with the streamlines.

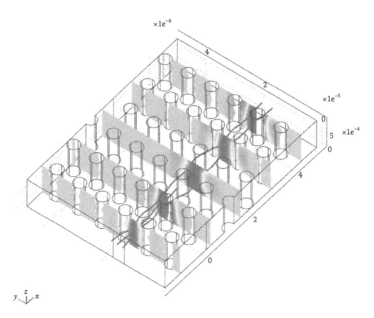

FIGURE 13.47
Concentration stream transported through a DLD (COMSOL).

lines are specific streamlines delimiting the three regions, confirming the approach of Section 13.5.3.1.

However, if the computational domain is small enough, a full 3D calculation can be done, as shown in Figure 13.47, where a local source of concentration has been introduced at the inlet.

13.5.4 Trapping Chambers

Recirculation microchambers are fluid chambers placed alongside the main channel. Depending on the conditions, the main flow may induce a recirculating vortex in the chamber that traps cells or particles transported by the carrier fluid. Figure 13.48 shows a design proposed by Shelby et al. [68], and Figure 13.49 another design by Manbachi et al. [69].

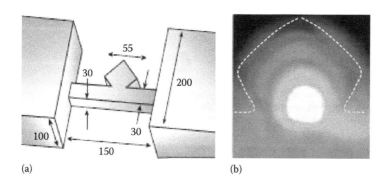

(a)   (b)

FIGURE 13.48
Schematic of a recirculation chamber (a), with the trapping of particles (b). (Reproduced by permission from Macmillan Publishers Ltd. *Nature*, Shelby, J.P., Lim, D.S.W., Kuo, J.S., and Chiu, D.T., High radial acceleration in microvortices, 425, 38, Copyright 2003.)

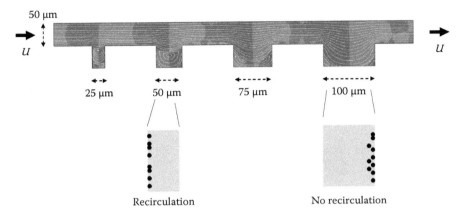

FIGURE 13.49
Trapping of particles in recirculation grooves. (From Manbachi, A., Shrivastava, S., Cioffi, M., Chung, B.G., Moretti, M., Demirci, U., Yliperttula, M., and Khademhossini, A., Microcirculation within grooved substrates regulates cell positioning and cell docking inside microfluidics channels, *Lab Chip*, 8, 747–754, 2008. Reproduced by permission of The Royal Society of Chemistry.)

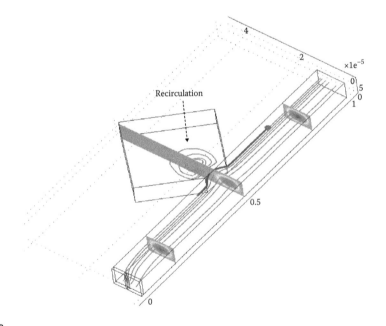

FIGURE 13.50
Recirculation in a diamond-shaped microchamber (Re = 1). The flow is focused along the recirculation chamber opening by the main flow. Streamlines are marked by the continuous lines and a 10 μm particle focused at the wall cannot enter the chamber (continuous dark lines).

Fluid flow recirculation is not easy to obtain in microsystems, principally because of the important friction on the walls. In the case illustrated in Figure 13.50, the friction of the fluid on the lower and upper horizontal walls reduces the possibilities of recirculation. In such a case, the Reynolds number of the main flow must be large enough and the "gate" or "opening" of the chamber must be small. Figure 13.50 shows a numerical calculation of the flow field performed with the finite element method (FEM) COMSOL. A compromise between

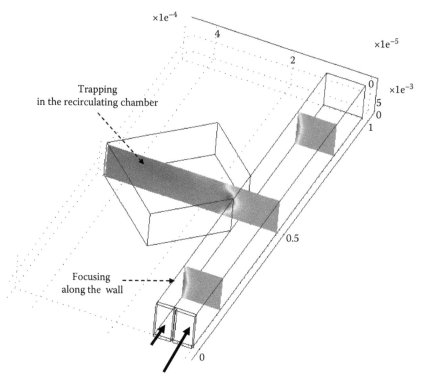

FIGURE 13.51
Slices of concentration showing an accumulation of small diffusivity particles ($D = 10^{-11}$ m^2/s) in the recirculating chamber.

the need for a narrow opening to promote recirculation in the chamber and the need for a sufficiently large opening to let the particles enter the chamber must be found. In parallel, a compromise between the need for sufficiently large main channel velocities to promote recirculation in the chamber and the need for sufficiently small particle velocity to allow for entering the chamber must be found. This complex situation requires numerical modeling.

Concentration calculation using small diffusion coefficients associated to large molecules shows the buildup of concentration in the recirculation chamber—when the conditions are such that there is a recirculation (Figure 13.51).

13.5.5 Microsystems for Cell Culture

The control of chemical delivery in microfluidic cell culture is a major research topic at the present time. The solution for the precise delivery of a chemical signal to a large group of cells without disturbing the cellular environment is a challenge. On one hand, the conventional use of passive diffusion gradients leads to overly slow and very approximate signal delivery. On the other hand, active methods using convective flow profoundly disturb the cellular environment: Cells can be removed by the flow, shear stress modifies the direction of cellular chemotaxis, and the natural cell signals are blurred; active methods can deliver the chemical signal, but their effect on the cell culture is too invasive [70,71].

A proposed solution is to combine the two approaches: A convective microchannel is used to quickly transport toward a microchamber the chemical signals, and a nanoporous

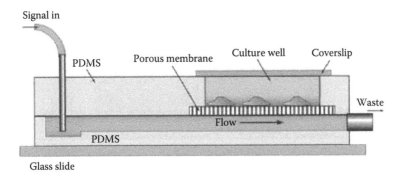

FIGURE 13.52
Schematic view of the cell culture system. (From VanDersal, J.J., Xu, A.M., and Melosh, N.A., Rapid spatial and temporal controlled signal delivery over large cell culture, *Lab Chip*, 11, 3057–3063, 2011. Reproduced by permission of The Royal Society of Chemistry.)

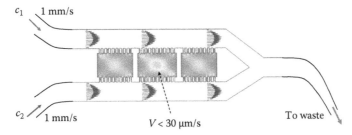

FIGURE 13.53
Velocity field inside the microsystem: in the cell culture microchambers, the fluid velocity is smaller than $30\,\mu m/s$.

membrane (porosity $\sim$ 10%) is used to protect the microchamber from flow convective motions (Figure 13.52) [70]. Chemical signals can then be delivered quickly to a large cell culture area without the drawbacks of the conventional active or passive methods.

Some other solutions are investigated, using micro-apertures instead of the nanoporous membrane, which makes the microfabrication easier. A calculation can easily be done with COMSOL for the system described in Figure 13.53.

Even if the velocities in the "feeding" channels are large (1 mm/s), the velocities inside the chambers stay small (less than $30\,\mu m/s$).

On the other hand, the diffusion of species to the cell culture chambers is quite fast, as shown in Figure 13.54. A bolus of concentration at the "top" inlet propagates very fast in the system: in less than 0.3 s, it affects the cell culture.

Moreover, a progressive gradient can be achieved across the system, as shown in Figure 13.55, where a stable (steady) gradient is obtained across the cell chambers.

13.6 Biochemical Reactions: DNA Recognition

13.6.1 Introduction

One of the principal aims of microsystems for biology is to perform biological reactions. DNA biochips are a typical example of such reactions. In this section, we focus on the

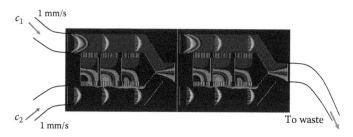

FIGURE 13.54
A bolus of concentration at the inlet reaches the cell culture areas in less than 0.3 s.

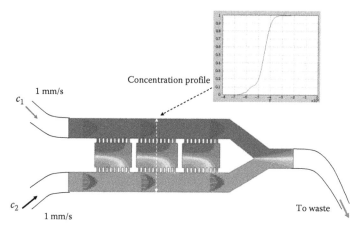

FIGURE 13.55
Concentration map in the system.

hybridization reaction and on its modeling and simulation in microsystems [72,73]. In microsystems, hybridization reactions are performed at the walls, on an active (functionalized) surface. This type of reaction is called heterogeneous reaction.

13.6.2 Heterogeneous Reactions

Heterogeneous reactions are reactions that occur at the contact of a solid wall, to the difference of homogeneous reactions that occur in the bulk. DNA hybridization follows a Langmuir law. We first present the theory of langmuirian reactions, and then we show examples of calculation of heterogeneous reactions.

13.6.2.1 Theoretical Background: Static Case

A very important class of reactions in biotechnology is the adsorption of molecules on a solid functionalized surface. In particular, it is the case of DNA hybridization. In such a reaction, there are three components: first, a "free" substrate in a buffer fluid sometimes called "target" or "analyte," in concentration $[S]$; second, a surface concentration $[\Gamma]_0$ of ligands—or capture sites—immobilized on a functionalized surface; third, a product which is the surface concentration of adsorbed targets, which we denote by $[\Gamma]$ (Figure 13.56). Remark that $[S]$ is a volume concentration (unit mol/m³) whereas $[\Gamma]$

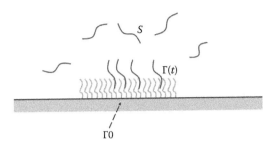

FIGURE 13.56
Adsorption of targets on a surface functionalized with immobilized ligands.

and $[\Gamma]_0$ are surface concentration (unit mol/m^2). Such a kinetics is called a Langmuir–Hinshelwood mechanism.

The reaction is weekly reversible because targets are constantly captured by ligands and they constantly dissociate (at a smaller rate). The reaction may be symbolized by

$$S \rightarrow \Gamma \tag{13.70}$$

$$\Gamma \rightarrow S$$

In the case of adsorption, the reaction rates are somewhat different to the definition of the usual chemical rates, mainly because the rate the immobilization of the substrate S depends not only on the volume concentration at the wall but also on the available sites for adsorption. Thus, we can write

$$v = \frac{d[S]}{dt} = k_{on}\left([\Gamma]_0 - [\Gamma]\right)[S]_w \tag{13.71}$$

$$v' = \frac{d[\Gamma]}{dt} = k_{off}[\Gamma]$$

where k_{on} and k_{off} are called, respectively, the adsorption and dissociation rates and $[S]_w$ is the concentration at the wall. For simplicity, we will note $\Gamma = [\Gamma]$, $c = [S]$, and $c_0 = [S]_w$. The net rate of adsorption is then

$$\frac{d\Gamma}{dt} = k_{on}c_0(\Gamma_0 - \Gamma) - k_{off}\Gamma \tag{13.72}$$

This last equation can be rewritten under the form

$$\frac{d\Gamma}{dt} = k_{on}c_0\,\Gamma_0 - \left(k_{on}c_0 + k_{off}\right)\Gamma \tag{13.73}$$

Equation 13.73 can be integrated and we obtain

$$\frac{\Gamma}{\Gamma_0} = \frac{k_{on}c_0}{k_{on}c_0 + k_{off}}\left[1 - e^{-\left(k_{on}c_0 + k_{off}\right)t}\right] \tag{13.74}$$

Using Equation 13.74, we obtain the surface concentration kinetics shown in Figure 13.57.

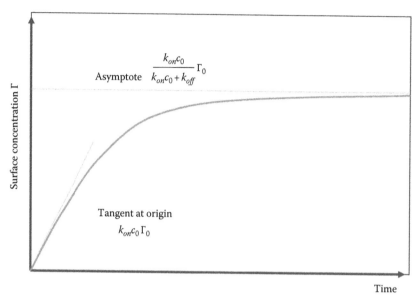

FIGURE 13.57
Kinetics of surface concentration from Equation 13.74.

At small times, the exponential term in Equation 13.74 can be developed in a Taylor expansion, and the surface concentration kinetics is the linear function of the time defined by

$$\Gamma = k_{on}c_0\Gamma_0\, t \qquad (13.75)$$

Equation 13.75 indicates that the kinetics described by the Langmuir equation is rapid if the term $k_{on}c_0$ is large, that is, when the adsorption constant on the surface and the concentration in molecules are large. For longer times, the surface concentration approaches an asymptotic value defined by

$$\frac{\Gamma_\infty}{\Gamma_0} = \frac{k_{on}c_0}{k_{on}c_0 + k_{off}} \qquad (13.76)$$

It can be verified in Equation 13.76 that in the case where k_{off} is zero, the asymptotic value is then Γ_0 and the surface becomes totally saturated. The larger the coefficient k_{off}, the smaller the value of Γ_∞/Γ_0.

Suppose that after the hybridization has reached its asymptotic value, the remaining targets or analytes in solution are suddenly washed out. Desorption is then the driving mechanism and the corresponding kinetics is schematized by Figure 13.58.

The starting time for desorption is the time t_a, and the surface concentration at this instant is Γ_a:

$$\Gamma_a = \frac{k_{on}\,c_0}{k_{on}\,c_0 + k_{off}}\Gamma_0 \qquad (13.77)$$

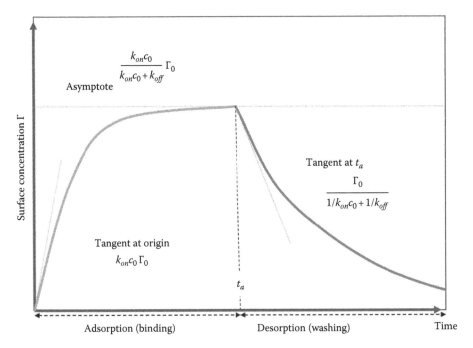

FIGURE 13.58
Kinetics of adsorption and desorption.

The Langmuir equation for desorption is

$$\frac{d\Gamma}{dt} = -k_{off}\,\Gamma \tag{13.78}$$

and the kinetics of desorption is

$$\frac{\Gamma}{\Gamma_a} = e^{-k_{off}\,(t-t_a)} \tag{13.79}$$

Desorption kinetics follows an inverse exponential law (Figure 13.58). The tangent to the desorption kinetic curve at $t = t_a$ is given by

$$\frac{\Gamma}{\Gamma_a} = 1 - k_{off}\left(t - t_a\right) \tag{13.80}$$

and the derivative at $t = t_a$ is

$$\frac{d\Gamma}{dt}\Big|_{t=t_a} = -k_{off}\,\Gamma_a = -\frac{k_{off}\,k_{on}\,c_0}{k_{on}\,c_0 + k_{off}}\,\Gamma_0 \tag{13.81}$$

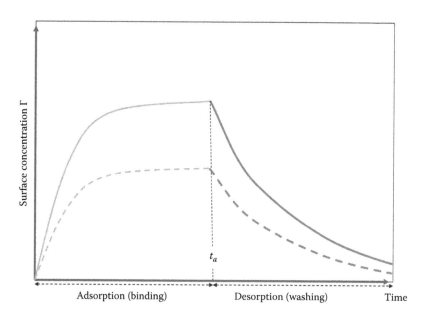

FIGURE 13.59
Different adsorption and desorption kinetics depending on the kinetic constants.

This last formula may be written under the following form:

$$\frac{d\Gamma}{dt}\bigg|_{t=t_a} = -\frac{\Gamma_0}{\dfrac{1}{k_{on}c_0} + \dfrac{1}{k_{off}}} \qquad (13.82)$$

When desorption follows adsorption, the kinetics of desorption depends not only on the desorption coefficient k_{off} but also on the values of Γ_0 and k_{on}. This property is shown in Figure 13.59 where different desorption kinetics are sketched, depending on the value of the saturation level.

13.6.2.2 Numerical Approach: Convection-Diffusion-Reaction Problem

From a numerical standpoint, solving a convection-diffusion-reaction problem requires to solve first for the velocity field $\vec{V}$, that is, solve the continuity and Navier–Stokes (or only Stokes) equations, store the results, and then solve the concentration equation using the previously calculated velocity field:

$$\frac{\partial c}{\partial t} + \vec{V}.\nabla c = \nabla.(D\nabla c) + f. \qquad (13.83)$$

In the case of a homogeneous (bulk) reaction, the source term f is the reaction rate. In the case of heterogeneous reactions, the source term f must be replaced by the wall condition

$$-D\frac{\partial c}{\partial n} = \frac{d\Gamma}{dt} \qquad (13.84)$$

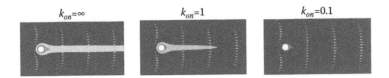

FIGURE 13.60
Comparison of the effect of a functionalized micropillar on the transported concentration of DNA for three different values of the adsorption rate $k_{on} = \infty$, 1 and 0.1 m³/mol/s.

where n is the unit vector normal to the wall. As was established in the preceding section, the kinetics for the immobilized wall concentration is

$$\frac{d\Gamma}{dt} = k_{on}c_0 \Gamma_0 - \left(k_{on}c_0 + k_{off}\right)\Gamma. \tag{13.85}$$

Combining (13.84) and (13.85) yields

$$\Gamma(t) = \frac{k_{on}c_0 \Gamma_0 + D\dfrac{\partial c}{\partial n}}{k_{on}c_0 + k_{off}} \tag{13.86}$$

Let us present here some examples. Consider the case of a cylindrical micro-pillar perpendicular to a laminar flow. The pillar surface is active, that is, it has been coated with ligands, and a hybridization reaction occurs between the DNA strands transported by the flow and the ligands immobilized at the wall.

Figure 13.60 shows a concentration map in the fluid channel after one hour, in three cases: first, an adherence condition (a DNA strand is immobilized as soon as it touches the wall); second, a Langmuirian reaction at the wall with $k_{on} = 1$ and $k_{off} = 0$; and, third, a Langmuirian reaction at the wall with $k_{on} = 0.1$ and $k_{off} = 0$. Concentration depletion in the channel can be clearly seen in the figure, and it is more pronounced when the immobilization rate is larger.

The preceding result can be generalized to a pillared microchannel. Pillared microchannels are similar to ordered microporous media, the porosity of which can be tuned by the size, distance, and arrangement of the pillars. Some pillared microsystems can be of very small dimensions, as shown in Figure 13.61, representing a microfluidic bypass channel in a microfluidic resonator [45].

Figure 13.62 shows the efficiency of capture of a pillared microchannel depending on the value of k_{on} calculated with COMSOL.

13.7 Conclusions

In this chapter, we have seen in a few examples the potentialities of FEM numerical approach for the design of microfluidic systems for biotechnology. Microflows, chemical and biochemical reactions, and concentration transport can be modeled by using FEM

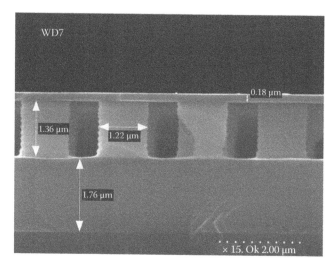

FIGURE 13.61
Pillared channel of very small dimension in a microfluidic resonator. (Courtesy V. Agache, CEA-Leti.)

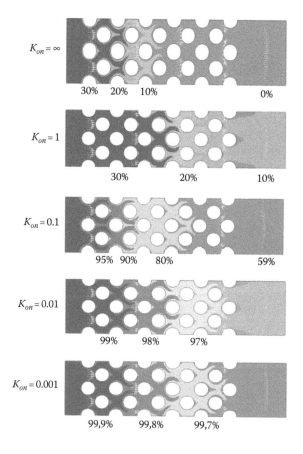

FIGURE 13.62
Comparison of the effect of a functionalized pillared microchannel on the transported concentration of DNA for five different values of the adsorption rate $k_{on} = \infty$, 1, 0.1, 0.01 and 0.001 m³/mol/s.

methods. However, as we pointed out in the introduction of this chapter, much remains to be done in the domain of multi-phase microflows where tracking interface motion and pinning is essential, and in the domain of the transport of large, deformable particles (vesicles, globules) where modeling the steric aspects and deformation associated to the local shear is a real challenge.

References

1. R.P. Feynman, Chapter 6, in *Building Biotechnology*, Y.E. Friedman (ed.), 3rd edn., Logos Press, Washington, DC, 2008.
2. P.-G. de Gennes, Les objets fragiles. Plon, Paris, 1994.
3. G.M. Whitesides, Chapter 9, in *Biotechnology and Materials Science—Chemistry for the Future*, L.M. Good (ed.), ACS Publications, Washington, DC, 1988.
4. D.W. Inglis, R. Riehn, J.C. Sturm, and R.H. Austin, Microfluidic high gradient magnetic cell separation, *J. Appl. Phys.*, 99(8), 2006.
5. J. Berthier and P. Silberan, *Microfluidics for Biotechnology*, 2nd edn., Artech House, Boston, MA, 2011.
6. J.C. Giddings, Field-flow fractionation: Analysis of macromolecular, colloidal, and particulate materials, *Science*, 260, 1456–1465, 1993.
7. G.H. Markx and R. Pethig, Dielectrophoretic separation of cells: Continuous separation, *Biotechnol. Bioeng.*, 45(4), 337–343, 1995.
8. H. Bruus, *Ultrasound Standing Wave Action on Suspensions and Bio-Suspensions in Micro- and Macrofluidic Devices*, Lecture notes for the advanced CISM school, Udine, Italy, 7–11 June 2010.
9. J. Berthier, *Microdrops and Digital Microfluidics*, William-Andrew Publishing, New York, 2008.
10. S.D. Senturia, Microsystem design. Chapter 22, *Microsystems for DNA Amplification*, Springer-Verlag, Berlin, Germany, 2002.
11. P.M. Holland, R.D. Abramson, R. Watson, and D.H. Gelfand, Detection of specific polymerase chain reaction product by utilizing the 5'–3'exonucleaseactivity of Thermusaquaticus DNA polymerase, *Proc. Natl. Acad. Sci. USA*, 88, 7276–7280, 1991.
12. C.J.M. Van Rijn, *Nano and Micro Engineered Membrane Technology*, Elsevier, Amsterdam, Boston, MA, 2004.
13. M. Allain, J. Berthier, S. Basrour, and P. Pouteau, Electrically actuated sacrificial membranes for valving in microsystems, *J. Micromech. Microeng.*, 20, 035006, 2010.
14. E. Harlow and D. Lane, *Using Antibodies: A Laboratory Manual*, Cold Spring Harbor Laboratory Press, New York, 1999.
15. A. Bejan, *Convection Heat Transfer*, John Wiley & Sons, New York, 1984.
16. P. Tabeling, *Introduction to Microfluidics*, Oxford Press, U.K., 2005.
17. W. Rudin, *Real and Complex Analysis*, 2nd edn., Mc Graw-Hill, New-York, 1966.
18. COMSOL numerical software: http://www.comsol.com.
19. E.W. Washburn, The dynamics of capillary flows, *Phys. Rev.*, 273–283, 1921.
20. R.K. Shah and A.L. London, *Laminar Flow Forced Convection in Ducts*, Academic Press, New York, 197, 1978.
21. S. Bendib and O. Français, Analytical study of microchannel and passive microvalve: Application to micropump simulation, *Proceeding, Design, Characterisation, and Packaging for MEMS and Microelectronics 2001*, Adélaide, Australia, pp. 283–291, 2001.
22. M. Bahrami, M.M. Yovanovich, and J.R. Culham, Pressure drop of fully-developed, laminar flow in microchannels of arbitrary cross section, *Proceedings of ICMM 2005, Third International Conference on Microchannels and Minichannels*, Toronto, Ontario, Canada, June 13–15, 2005.
23. I.E. Idel'cik, *Memento des pertes de charges*. Eyrolles, Paris, 1960.

24. J. Berthier, R. Renaudot, P. Dalle, G. Blanco-Gomez, F. Rivera, and V. Agache, COMSOL assistance for the determination of pressure drops in complex microfluidic channels, *Proceedings of the 2010 COMSOL European Conference*, Paris, France, November 2010.

25. L. Anna, N. Bontoux, and H.A. Stone, Formation of dispersions using flow-focusing microchannels, *Appl. Phys. Lett.*, 82, 364–366, 2003.

26. D. Luo, S.R. Pullela, M. Marquez, and Z. Cheng, Cell encapsules with tunable transport and mechanical properties, *Biomicrofluidics*, 1, 034102, 2007.

27. V.L. Workman, S.B. Dunnett, P. Kille, and D.D. Palmer, Microfluidic chip-based synthesis of alginate microspheres for encapsulation of immortalized human cells, *Biomicrofluidics*, 1, 014105, 2007.

28. J. Berthier, S. Le Vot, P. Tiquet, N. David, D. Lauro, P.Y. Benhamou, and F. Rivera, Highly viscous fluids in pressure actuated Flow Focusing Devices, *Sensor. Actuat. A*, 158, 140–148, 2010.

29. R.B. Bird, R.C. Armstrong, and O. Hassager, *Dynamics of Polymeric Liquids*, Wiley, New York, 1976.

30. J.F. Steffe, *Rheological Methods in Food Process Engineering*, 2nd edn., Freeman Press, East Lansing, MI, 1982.

31. J. Berthier, S. Le Vot, P. Tiquet, F. Rivera, and P. Caillat, On the influence of non-Newtonian fluids on microsystems for biotechnology, *Proceedings of the 2009 Nanotech Conference*, Santa Clara, CA, May 13–17, 2009.

32. L.E. Malvern, *Introduction to the Mechanics of a Continuus Medium*, Prentice-Hall, Englewood Cliffs, NJ, 1969.

33. Y.S. Muzychka and J.F. Edge, Laminar non-newtonian fluid flow in non-circular ducts and microchannels, *J. Fluid Eng.*, 130(11), 111–201, 2008.

34. W. Kozicki, C.H. Chou, and C. Tiu, Non-newtonian flow in ducts of arbitrary cross-sectional shape, *Chem. Eng. Sci.*, 21, 665–679, 1966.

35. C. Miller, Predicting non-newtonian flow behaviour in ducts of unusual cross section, *Ind. Eng. Chem. Fundam.*, 11, 534–628, 1972.

36. F. Delplace and J.C. Leuliet, Generalized Reynolds number for the flow of power law fluids in cylindrical ducts of arbitrary cross-section, *Chem. Eng. J.*, 56, 33–37, 1995.

37. M. Kersaudy-Kerhoas, R. Dhariwal, M.P.Y. Desmulliez, and L. Jouvet, Blood flow separation in microfluidic channels, *Proceedings of the First European Conference on Microfluidics—Microfluidics 2008*, Bologna, Italy, December 10–12, 2008.

38. S.K.W. Dertinger, D.T. Chiu, N.L. Jeon, and G.M. Whitesides, Generation of gradients having complex shapes using microfluidic networks, *Anal. Chem.*, 73, 1240–1246, 2001.

39. J. Berthier, F. Rivera, P. Caillat, and F. Berger, Dimensioning of a new micro-needle for the dispensing of drugs in tumors and cell clusters, *Proceedings of the 2005 Nanotech Conference*, Anaheim, CA, May 8–12, 2005.

40. H. Schlichting and K. Gersten, *Boundary Layer Theory*, 8th ed., Springer-Verlag, Secaucus, NJ, 2004.

41. H. Bruus, *Theoretical Microfluidics*, Oxford University Press, New York, 2008.

42. S. Son, W.H. Grover, T.P. Burg, and S.R. Manalis, Suspended microchannel resonators for ultralow volume universal detection, *Anal. Chem.*, 80, 4757–4760, 2008.

43. P.-S. Lee, J.H. Lee, N.Y. Shin, K.-H. Lee, D.K. Lee, S.M. Jeon, D.H. Choi, W. Hwang, and H.C. Park, Microcantilevers with nanochannels, *Adv. Mater.*, 20, 1732–1737, 2008.

44. M. Godin, F.F. Delgado, S. Son, W.H. Grover, A.K. Bryan, A. Tzur, P. Jorgensen, K. Payer, A.D. Grossman, M.W. Kirschner, and S.R. Manalis, Using buoyant mass to measure the growth of single cells, *Nat. Methods*, 7, 387–390, 2010.

45. V. Agache, G. Blanco-Gomez, F. Baleras, and P. Caillat, An embedded microchannel in a MEMS plate resonator for ultrasensitive mass sensing in liquid, *Lab Chip*, 11, 2598–2603, 2011.

46. J. Volker and P. Knobloch, On spurious oscillations at layers diminishing (SOLD) methods for convection-diffusion equations: Part 1- A review, *Comput. Methods Appl. Mech. Engrg.*, 196, 2197–2215, 2007.

47. W.B.J. Zimmerman, *Microfluidics: History, Theory and Applications*, Springer, New York, 2006.

48. S. Choi, J.M. Karp, and R. Karnik, Cell sorting by deterministic cell rolling, *Lab Chip*, 34, 12(8), 1427–1430, 2012.

49. W.R. Dean, Note on the motion of fluid in a curved pipe, *Phil. Mag.*, 20, 208–223, 1927.

50. D. Di Carlo, Review article, *Lab Chip*, 9, 3038–3046, 2009.

51. J. Berthier, Cellular biomicrofluidics, Chapter 20, in *Micro and Nanofluidics Handbook*, Taylor & Francis, Boca Raton, Florida, 2011.

52. A.P. Sudarsan and V.M. Ugaz, Multivortex micromixing, *PNAS*, 103(19), 7228–7233, 2006.

53. M. Lavine, Microfluidics: Streams swirled by Dean, *Science*, 312, 1281, 2006.

54. J.M. Ottino, *The Kinematics of Mixing: Stretching, Chaos, and Transport*, Cambridge University Press, Cambridge, England, 1989.

55. J. Atencia and D.J. Beebe, Controlled microfluidic interfaces, *Nature*, 437, 648–655, 2005.

56. S.L. Anna, N. Bontoux, and H.A. Stone, Formation of dispersions using flow focusing in microchannels, *Appl. Phys. Lett.*, 82(3), 364, 2003.

57. M.J. Kennedy, S.J. Stelick, S.L. Perkins, L. Cao, and C.A. Blatt, Hydrodynamic focusing with a microlithographic manifold: Controlling the vertical position of a focused sample, *Microfluid. Nanofluid.*, 7(4), 569–578, 2009.

58. G. Hairer and M.J. Vellekoop, An integrated flow-cell full sample stream control, *Microfluid. Nanofluid.*, 7(5), 647–658, 2009.

59. M. Yamada and M. Seki, Microfluidic particle sorter employing flow splitting and recombining, *Anal. Chem.*, 78, 1357–1362, 2006.

60. M.Yamada and M. Seki, Hydrodynamic filtration for on-chip particle concentration and classification utilizing microfluidics, *Lab Chip*, 5, 1233–1239, 2005.

61. H. Maenaka, M. Yamada, M. Yasuda, and M. Seki, Continuous and size-dependent sorting of emulsion droplets using hydrodynamics in pinched microchannels, *Langmuir*, 24(8), 4405–4410, 2008.

62. D. Di Carlo, D. Irimia, R.G. Tompkins, and M. Toner, Continuous inertial focusing, ordering, and separation of particles in microchannels, *PNAS*, 104(48), 18892–18897, 2007.

63. M. Yamada, H. Nakashima, and M. Seki, Pinched flow fractionation: continuous size separation of particles utilizing a laminar flow profile in a pinched microchannel, *Anal. Chem.*, 76, 5465–5471, 2004.

64. A. Srivastav, T. Podgorski, and G. Coupier, Efficiency of size-dependent separation by pinched flow fractionation, *Microfluid. Nanofluid.*, On-line publication, DOI 10.007/s10404-012-0985-8, April 2012.

65. J.A. Davis, D.W. Inglis, K.J. Morton, D.A. Lawrence, L.R. Huang, S.Y. Chou, J.C. Sturm, and R.H. Austin, Deterministic hydrodynamics: Taking blood apart, *PNAS*, 103, (40), 14779–14784, 2006.

66. K.J. Morton, K. Loutherback, D.W. Inglis, O.K. Tsui, J.C. Sturm, S.Y. Chou, and R.H. Austin, Crossing microfluidic streamlines to lyse, label and wash cells, *Lab Chip*, 8, 1448–1453, 2008.

67. D.W. Inglis, J.A. Davis, R.H. Austin, and J.C. Sturm, Critical particle size for fractionation by deterministic lateral displacement, *Lab Chip*, 6, 655–658, 2006.

68. J.P. Shelby, D.S.W. Lim, J.S. Kuo, and D.T. Chiu, High radial acceleration in microvortices, *Nature*, 425, 38, 2003.

69. A. Manbachi, S. Shrivastava, M. Cioffi, B.G. Chung, M. Moretti, U. Demirci, M. Yliperttula, and A. Khademhossini, Microcirculation within grooved substrates regulates cell positioning and cell docking inside microfluidics channels, *Lab Chip*, 8, 747–754, 2008.

70. J.J. VanDersarl, A.M. Xu, and N.A. Melosh, Rapid spatial and temporal controlled signal delivery over large cell culture, *Lab Chip*, 11, 3057–3063, 2011.

71. M.M. Sedeh, J.M. Khodadadi, K.A. Cramer, M. Hanson, and J.W. Wong, CFD-assisted design and optimization of pharmacokinetic microfluidic system, *Proceedings of the 2011 NSTI Nanotech Conference*, Boston, MA, June 13–16, 2011.

72. P. Gong and R. Levicky, DNA surface hybridization regimes, *PNAS*, 105(14), 5301–5306, 2008.

73. A. Halperin, A. Buhot, and E.B. Zhulina, On the hybridization isotherms of DNA microarrays: The Langmuir model and its extensions, *J. Phys.: Condens. Matter.*, 18(18), S463–S490, 2006.

14

Design of the Nanoinjection Detectors Using Finite Element Modeling

Omer G. Memis

Northwestern University, Evanston, Illinois

Hooman Mohseni

Northwestern University, Evanston, Illinois

CONTENTS

14.1 Introduction...478
14.2 Photodetection and Imaging in the Short-Wave Infrared Spectrum.........479
14.3 Single Photon Detectors: Importance and Applications.............................480
14.4 Nanoinjection Detector...481
    14.4.1 Layer and Band Structure...482
    14.4.2 Theory of Operation 483
14.5 Development of a Finite Element-Method Simulator 483
    14.5.1 Self-Consistent 3-D Core Physics..484
    14.5.2 Reduction to 2-D Cross Modeling with Cylindrical Symmetry 484
    14.5.3 Nonlinear Effects 485
14.6 Simulation Results for Electro-Optical Modeling.......................................488
    14.6.1 Optimization for Device Structure and Doping Levels...................488
    14.6.2 Dark Current versus Voltage..488
    14.6.3 Photocurrent and Amplification (Optical Gain).............................488
    14.6.4 Quantum Efficiency..489
    14.6.5 Potential Distribution and Electric Field 490
    14.6.6 Device Heating..491
14.7 Measured Performance of Nanoinjection Detectors....................................491
14.8 Measured Performance of Nanoinjection Imagers 494
    14.8.1 Current–Voltage and Gain–Voltage before and after Integration ...496
    14.8.2 Bandwidth of Hybridized Test Pixels 497
    14.8.3 Noise Performance of the Hybridized Test Pixels 497
    14.8.4 Spatial Sensitivity of Hybridized Test Pixels...................................498
    14.8.5 Temperature Behavior of Hybridized Test Pixels 499
    14.8.6 Comparison of the Signal-to-Noise Level of the Nanoinjection Imager with a Commercial Camera 499
    14.8.7 Signal-to-Noise Analysis of the Nanoinjection Imager...................500
14.9 Conclusion 503
References...503

14.1 Introduction

The nanoinjection photon detectors have been recently developed to address the main trade-off in using nanoscale features for light detection. Even though nanoscale sensors offer high sensitivity, their interaction with visible or infrared light is severely limited by their miniscule sizes. Nanoinjection detectors solve this problem by implementing a novel structure with highly sensitive nanometer sized pillars, "nanoinjectors," on large, thick absorption layers. The large, thick absorption layers ensure that the incoming light is captured with high efficiency, and that the nanoinjectors are responsible for sensing and amplifying these signals. With this nontraditional, nonplaner geometry, high sensitivity and high efficiency can be simultaneously attained, which can satisfy a growing need for identifying and counting photons in many modern applications.

The nanoinjection detectors were developed toward operation in short-wave infrared (SWIR) domain from 0.95 to 1.65 μm. They were designed to exploit the properties of the type-II band alignment in InP/GaAsSb/InGaAs structure to achieve higher internal amplification, lower noise levels, and high speed operation.

The layer structure with type-II band alignment and the nonplaner geometry of nanoinjection detectors required careful design and optimization through a custom built three-dimensional nonlinear finite element method (FEM) simulation, which provided the mathematical groundwork to implement the stationary, parametric, and transient simulation of nonlinear differential equations. We have based our simulation core on the drift-diffusion equations in two and three dimensions. For devices with cylindrical symmetry, we have also implemented a 2-D cross section simulator, which gives accurate 3-D modeling results in the timeframe needed for a simpler 2-D simulation. To improve accuracy under different conditions, several nonlinearities were implemented. Incomplete ionization of dopants, bimolecular recombination, Auger recombination, nonlinear mobility, impact ionization, thermionic emission, hot electron effects and surface recombination effects were included. Temperature effects, both in the form of variable operating temperature, and thermal heat generation and dissipation, were also added to the model.

Based on the results of the FEM simulations, the nanoinjection detectors were fabricated and measured. Amplification factors reaching 10,000 have been recorded, together with low dark current densities at room temperature. Noise suppression behavior is observed at amplification factors up to 4000+, which lower the detector noise to values below the theoretical shot noise limit. The devices, when properly surface-treated, show bandwidths exceeding 3 GHz with an impressive time-uncertainty (jitter) of 15 ps. These properties make the nanoinjection photon detectors extremely suitable for demanding imaging applications, which require high efficiency, high sensitivity, and high uniformity, in addition to many applications such as nanodestructive material inspection, high-speed quantum cryptography, or medical optical imaging.

Arrays of nanoinjection detectors were designed, processed, and hybridized to form focal plane array infrared cameras with 320 by 240 pixel arrays. The nanoinjection imagers show responsivity values in excess of 2500 A/W. Measured imager noise was 28 electrons at a frame rate of 1950 fps. Compared to commercial SWIR imagers, the imagers show two orders of magnitude improved signal-to-noise ratio (SNR) at thermoelectric cooling temperatures.

14.2 Photodetection and Imaging in the Short-Wave Infrared Spectrum

Infrared (IR) spectrum has always offered possibilities beyond the human visual capabilities. IR detectors have found their place in many applications, and have been indispensible tools for looking at objects from a different perspective to investigate less-evident properties. Thus, there has been a lot of ongoing research about IR detectors, exploring different approaches with a wide range of material systems.

The short-wave infrared (SWIR) detectors play in important role in many modern world applications spanning diverse fields. In the Internet backbone, they have been converting 1.3 and 1.55 µm optical bits to electrical bits, and recently their field of view extended into the "last mile" toward end-users with commercial fiber-optic links to individual residential sites. In military, arrays of SWIR detectors have been utilized into SWIR night-vision systems, which rely on the intense night-glow that can illuminate the scenery even when there is complete darkness in visible spectrum. In biophotonics, they help realize the noninvasive imaging methods, for example, optical coherence tomography systems utilizing SWIR exploit the low scattering properties of >1 µm light to see the previously unreachable, thick parts of the cornea. In addition to these existing applications, there exist newly emerging applications where SWIR detectors can be the enabling technologies. One particular example is quantum cryptography, which is in desperate need of an efficient and fast SWIR single photon detector (SPD) that can provide sufficient SNR even in the arrival of a single photon.

The steady advancements in technology constantly increased the requirement for better and more sensitive detectors for ever more demanding applications. This has propelled the research toward ultimate sensors, the SPDs. SPDs are ultra-low noise devices with increased sensitivity to be able to detect the minimum energy quantum of light, the photon. SPDs can sense and count individual photons and can be utilized in many emerging applications, where the only available signal is in the order of several photons.

For imaging purposes, the SWIR band provides a lot of benefits over the conventional visible imaging. SWIR imaging systems are sensitive to portions of the spectrum spanning from 1 to 3 µm.[1] This region has several important features: It is compatible with glass optics, as the glass transmission window extends almost to the end of SWIR spectrum. To see clearly in dark, it can make extensive use of the night glow, which can be orders of magnitude brighter than visible light, even in moonless nights.[2] SWIR night glow, or night sky radiance, is due to the chemical reactions involving hydroxyl groups in the upper atmosphere, mainly chemoluminescence, radiative recombination, and collisional de-excitation.[3] This spectrum also becomes a nice complement to visible and near-infrared (NIR) regions, and can provide information important for multispectral imaging.[4] The multispectral information can be extremely useful for military applications such as camouflage detection or friend-or-foe detection. Furthermore, SWIR spectrum is eye-safe,[5] as it cannot be focused on retina, so accidental exposure of SWIR light does not present any hazard to the operators or bystanders. This property becomes very valuable in military field applications and medical environments.[6]

Here in this work, we present a novel approach for a bio-inspired SWIR SPD, the "nanoinjection photon detector", conceptually based on the mechanism of light detection in the rod cells. To realize this detector, first a comprehensive three-dimensional nonlinear simulation environment was developed. Using this model, the layer structure, doping levels and geometry were optimized. The wafers for the optimized structure were grown by metal-organic chemical vapor deposition. In parallel, novel process flows were designed to fabricate the nanoinjection detectors.

14.3 Single Photon Detectors: Importance and Applications

Many modern applications such as medical instruments, imaging systems, and tele-communication devices have demanding requirements to achieve state-of-the-art performance. However, in many cases the performance of the system suffers from the limitations imposed by the photodetector. Improving the sensitivity of photodetectors is of the utmost importance in these systems, and the ultimate target for researchers is the detection of a single quantum of light, the photon.[7] It represents a very small amount of energy, less than 10^{-18} J in the infrared spectrum, so detectors specialized for individual photon detection need to have very low noise levels. These detectors are called "single photon detectors."

An ideal SPD can be used in almost every application that utilizes a regular photo-detector. The improved SNR that SPDs present would increase the performance of the system, resulting in deeper medical noninvasive probing, longer range fiber links, more accurate nondestructive parts inspection for aircraft safety, less error-prone satellite communications or clearer, higher resolution night vision images for military. All existing applications, whether in military, commercial, environmental or medical, will benefit tremendously.

In parallel to improving existing applications, SPDs are quickly becoming the enabling technology for many emerging applications which were previously thought impossible. One example of these technologies is quantum computing,[8] in which bits can assume many states, and perform calculations on these multiple states at once.[9] They therefore promise a great deal of improvement in calculation speed and efficiency and can potentially break down all existing barriers of traditional computation.

Quantum cryptography[10] relies on the quantum theory to provide secure communications. The current public–private key encryption schemes rely on the mathematical infeasibility of some inverse calculations, which will be significantly challenged by the paradigm shift to be introduced by quantum computing.[11] Quantum cryptography, on the other hand, will be immune from the effects of this vast change.[12]

Another example is quantum ghost imaging,[13] where the scenery in a remote location can be completely reconstructed from correlating the acquired signals, using reflected entangled photon pairs.

To achieve efficient single photon detection, the detectors need to possess some important properties, which can be very demanding to achieve simultaneously. The first of these properties is the quantum efficiency. As stated before, the quantum efficiency tells what percentage of the incoming photons gets converted to electron–hole pairs and extracted from the detector. It is logical to try to achieve high quantum efficiency, so as to maximize the signal.

Another important property is the spectral response. In this regard, a broadband detector that can cover from deep ultraviolet to SWIR (e.g., 100 nm–2 μm) is highly desirable. SPDs such as photomultiplier tubes or superconducting single photons come close, but others such as the avalanche photodetectors fall short in this claim.

A low dark count rate is also an essential property of a SPD, as it signifies the number of false counts per second. As each detector exhibits a certain amount of dark current and resulting noise, the noise sometimes exceeds the threshold of detection which gets registered as a valid signal. A high amount of false counts can become disastrous, as they can ruin an image in night vision imaging, trigger false alarms in homeland security or contaminate encrypted messages, possibly garbling the transmission.

Demanding applications have forced the speed of detection to become one of the important parameters of SPDs. In this regard, achieving high-speed transmission corresponds to many requirements. Two such parameters would be the bandwidth (BW) and rise-time of the detector. As in other systems, more bandwidth and faster rise-times are desirable.

In addition to BW and rise-times, fast recovery time is essential for high-speed detection. The time required to re-arm the device should be as short as possible, as this also puts a constraint on the maximum attainable transmission rate similar to bandwidth. In certain systems (e.g., APDs), the detector still registers some counts after the pulse, where this response slowly decays. However, during the decay, the SPD cannot be rearmed as it still registers a high number of counts. This effect is called afterpulsing, and it is a significant limitation on InGaAs/InP-based APDs.

At very high speeds, timing jitter can become the limiting factor. Timing jitter is defined as the unwanted time-uncertainty of the detector. Even when given an optical signal with an exact period of T, the electrical signal coming from the detector would have some time uncertainty that makes the period $T \pm \Delta T$. This is due to the subtle changes in the detection such as the point of absorption or the transit time.

One final property of an ideal SPD is the capability of photon number resolving. By definition, a detector can be called a SPD if it can detect and respond to one photon. This allows SPDs with very limited dynamic range, where the response is quantified simply as no photons or some photons (i.e., one or more). However, many applications require more dynamic range than that, and therefore a new SPD subclass was introduced: photon resolving SPDs, or single photon counters. In single photon counters, the amplitude of the response is varied based on the number of absorbed photons, offering at least some dynamic range.

14.4 Nanoinjection Detector

The design of the nanoinjection detector is inspired by the eye. The eye is extremely sensitive, and it owes this sensitivity to two main reasons:

1. Coupling a micron-scale absorbing volume with nanoscale sensing elements for improved sensitivity
2. Having a significant internal amplification to boost the miniscule photon energy to a detectable signal level above the system noise floor

We incorporated these principles in a novel semiconductor platform, called the nanoinjection photon detector. Structurally, we made the device analogous to the highly sensitive rod cells in the eye: The detector features large regions to absorb and channel the photoexcited carriers to nanoinjectors where the amplified flow of carriers is controlled.

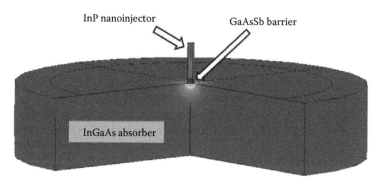

FIGURE 14.1
Device geometry of a nanoinjection photon detector.

We have also engineered the band structure such that it would provide an internal amplification. The amplification method (i.e., nanoinjection) is designed to have stability based on an internal negative feedback.

The structure of the nanoinjection photon detector is highlighted in Figure 14.1, including large absorbing volume and small nanoinjectors.

14.4.1 Layer and Band Structure

As highlighted in Figure 14.1, the detector features nanoinjectors on a large InGaAs absorption layer. The device is based on InP/GaAsSb/InGaAs material system and therefore has type-II band alignment (Figure 14.2), with the GaAsSb layer acting as a barrier for electrons and a trap for holes. As detailed in the reference,[14] the most tested active layer structure consists of 1000 nm $In_{0.53}Ga_{0.47}As$ (n-doped), 50 nm $GaAs_{0.51}Sb_{0.49}$ (p-doped), and 500 nm InP (n-doped) from bottom to up.

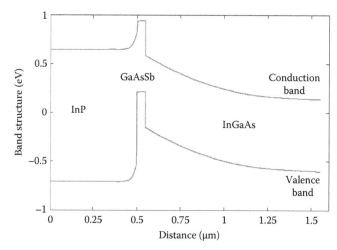

FIGURE 14.2
Band structure of a nanoinjection photon detector at the central axis.

14.4.2 Theory of Operation

The band structure across the central axis of the device experiences a type-II band alignment as depicted in Figure 14.2 when properly biased with 0.5–1 V. The conduction band incorporates a GaAsSb barrier to limit the flow of electrons from InP side to InGaAs side. The valence band, on the other hand, has a well structure, bound by the higher potential of InP and InGaAs layers.

Due to the doping level and work function of each layer, the device generates an internal electric field in the InGaAs region, which gets stronger when the device is biased correctly. Upon absorption, photons generate electron–hole pairs in the large absorption region. The electrons and holes are separated by the internal electric field of the device. Holes are attracted to the nanoinjector, which presents a potential trap for holes due to the type-II band alignment. A single photo-generated hole in the absorption region is equivalent to a charge density of 1.4×10^{-3} C/m³. However, when trapped inside the 50 nm high and 100 nm wide diameter nanoinjector, the same hole creates an effective charge density of more than 400 C/m³. Therefore, the impact of the hole increases by more than five orders of magnitude. Equivalently, the small volume of the trap represents an ultra-low capacitance, and hence the entrapment of a single hole leads to a large change of potential and produces an amplified electron injection, similar to a single electron transistor (SET). Our detailed simulations show that a single hole can alter the potential barrier by more than 52 mV. This is significantly higher than the thermal fluctuation energy of carriers at 300 K, and hence a high SNR is possible even at room temperature.

Device relaxation is achieved by thermal recombination of the trapped holes. The holes that are trapped in the GaAsSb well will eventually recombine, relaxing the bands and restoring the current to the low, dark-current values.

As the multiplication mechanism is purely applied to one carrier, the amplification noise can be very small in nanoinjection detectors. One possible explanation for this low-noise behavior of the device is the negative potential feedback mechanism in the device.[15] Even though the voltage of the barrier is mainly controlled by hole flux, the injected electrons also play an important role in the regulation of barrier layer voltage. Compared to holes, which are trapped in the GaAsSb barrier for a relatively long time, electrons have a very short but finite transit time through the barrier. During the transit time, they lower the local potential and increase the barrier height. The increase in barrier height opposes the flow of electrons and reduces the transmission probability.

14.5 Development of a Finite Element-Method Simulator

To quantify and optimize the device structure and performance, we have developed a custom FEM-based simulation model. The model was created in *Comsol Multiphysics*,[16] which provides mathematical groundwork to implement the stationary, parametric, and transient simulation of nonlinear differential equations.

We have based our design on the core drift-diffusion equations in two and three dimensions. For devices with cylindrical symmetry, we have also implemented a 2-D cross section simulator, which, given accurate 3-D modeling, results in the timeframe needed for a simpler 2-D simulation.

To improve accuracy under different conditions, several nonlinearities were implemented. Incomplete ionization of dopants, bimolecular recombination, Auger recombination,

nonlinear mobility, impact ionization, thermionic emission, hot electron effects, and surface recombination effects were included. Temperature effects, both in the form of variable
operating temperature, and thermal heat generation and dissipation, were also added to
the model.

14.5.1 Self-Consistent 3-D Core Physics

In the core of the simulation lie the drift-diffusion equations.[17] They are given by the following set of Poisson and continuity equations:

$$\nabla \cdot (\varepsilon_0 \varepsilon_r \nabla \phi) = -q(p - n + N) \tag{14.1}$$

$$\nabla \cdot (\mu_n \nabla E_{Fn}) = qR$$
$$\nabla \cdot (\mu_p \nabla E_{Fp}) = -qR \tag{14.2}$$

where
   ε_0 is the permittivity of space
   ε_r is the relative permittivity of the material
   Φ is the potential
   n and p are carrier concentrations for electrons and holes respectively
   N is the net doping density (donor density minus acceptor density)
   q is the electron charge
   μ_n and μ_p are the mobilities of electrons and holes
   E_{Fn} and E_{Fp} are the quasi-Fermi levels for electrons and holes
   R is the net recombination-generation rate

Here in these equations, the quasi-Fermi levels are used to combine the effect of both
electric field drift and carrier diffusion processes.

14.5.2 Reduction to 2-D Cross Modeling with Cylindrical Symmetry

In order to improve the time required to solve the problem while maintaining accuracy,
we needed to decrease the complexity of the problem. One common way to achieve this is
to exploit the symmetries in the device structure. In our case, the nanoinjection detector
incorporates cylindrical symmetry, and hence we can reduce the three-dimensional
equations into two-dimensional ones. However, additional terms need to be added to the
equations to model the radial changes in current and charge densities in 3-D.
   Starting from the original drift-diffusion equations in 3-D, we have expanded them into
radial and tangential. The tangential terms then become zero as the device has cylindrical
symmetry, which transforms the resulting equations into a simpler form. This simpler
equation is then converted to the form that Comsol Multiphysics uses.
   The gradient of any function f with cylindrical symmetry in cylindrical coordinates is
given by

$$\nabla f = \frac{\partial f}{\partial r} \vec{a}_r + \frac{\partial f}{\partial z} \vec{a}_z \tag{14.3}$$

where $\partial f / \partial z = 0$, $\vec{a}_r$ and $\vec{a}_z$ are unit vector in r and z directions, respectively. Then, the
divergence of a vector field $\vec{A} = A_r \vec{a}_r + A_z \vec{a}_z$ is given by

$$\nabla \cdot \vec{A} = \left(\frac{\partial}{\partial r} + \frac{1}{r} \right) A_r + \frac{\partial}{\partial z} A_z. \tag{14.4}$$

As the drift-diffusion equations are of the form $\nabla \cdot (k\nabla f) = c$, we can combine the preceding divergence equation with the gradient equation to get

$$\nabla \cdot (k\nabla f) = \frac{\partial}{\partial r} \left(k \frac{\partial f}{\partial r} \right) + \frac{k}{r} \frac{\partial f}{\partial r} + \frac{\partial}{\partial z} \left(k \frac{\partial f}{\partial z} \right) = c \tag{14.5}$$

which can be transformed into the Comsol form

$$\nabla \cdot (k\nabla f) = \nabla_* \cdot (k\nabla_* f) + \begin{pmatrix} \dfrac{k}{r} \\ 0 \end{pmatrix} \cdot \nabla_* f = c \tag{14.6}$$

where ∇_* is the Cartesian nabla operator.

14.5.3 Nonlinear Effects

Since the semiconductor materials exhibit many nonlinear effects, implementing some of the related nonlinear effects improves the accuracy of the simulation. Hence, the following nonlinearities were added:

Incomplete ionization of dopants: The ionization ratio of the doping in the semiconductor materials show a strong dependence on the degeneracy of the material, temperature, and the difference between the quasi Fermi level and the band. Therefore, equations[18]

$$N_D^+ = \frac{N_D}{1 + \dfrac{1}{2} \exp\left(\dfrac{E_{Fn} - E_D}{kT} \right)}$$

$$N_A^- = \frac{N_A}{1 + \dfrac{1}{4} \exp\left(\dfrac{E_A - E_{Fp}}{kT} \right)} \tag{14.7}$$

are used.

Bimolecular recombination: Our simulations have indicated that the bimolecular recombination yields better results than SRH recombination, which has a similar form. The equation is given by

$$R_{bimolecular} = B(np - n_0 p_0) \tag{14.8}$$

where
   B is the bimolecular recombination constant
   n and p are carrier densities
   $n_0 p_0$ is the equilibrium carrier density product

Auger recombination: Effective at higher carrier concentrations, the Auger recombination can be the dominant process in high current devices such as lasers. As such,

Auger recombination plays a much smaller role in our detector. However, specific parts in our device are still susceptible to strong Auger recombination, especially at corners where carriers get congested. We have used the following equation to formulate Auger recombination:[19]

$$R_{Auger} = C(n+p)(np - n_0 p_0) \tag{14.9}$$

where
    C is the Auger recombination constant
    n and p are carrier densities
    $n_0 p_0$ is the equilibrium carrier density product

Nonlinear mobility: The simple equation for electron mobility is

$$v_n = \mu_n E \tag{14.10}$$

where the carrier velocity v_n does not get saturated at any electric field E, and hence, mobility μ_n stays constant. A more realistic approach takes into consideration the velocity overshoot and saturation,[20]

$$\mu_n = \frac{\mu_{n0} + \dfrac{v_{sn}}{E_{0n}} \cdot \left(\dfrac{E_t}{E_{0n}}\right)^3}{1 + \left(\dfrac{E_t}{E_{0n}}\right)^4} \tag{14.11}$$

where
    μ_{n0} is the low field mobility
    v_{sn} is the saturation velocity
    E_{0n} is the overshoot field strength
    E_t is the total local electric field

Here, the parameters v_{sn} and E_{0n} are determined empirically. The equation for holes is the analogous of this equation.

Impact ionization: The impact ionization is very much a field-dependent process, whose strength grows super-exponentially with increasing field. Therefore it requires high voltages to achieve intense electric fields and it is not a pronounced effect in our nanoinjection detectors. Still, we have implemented impact ionization to improve accuracy in our model, using[21]

$$G_{imp} = (a_n n \mu_n + a_p p \mu_p) \cdot E \tag{14.12}$$

$$a_n = a_{n\infty} \cdot \exp\left(\frac{-E_{cn}}{E}\right)$$

$$a_p = a_{p\infty} \cdot \exp\left(\frac{-E_{cp}}{E}\right) \tag{14.13}$$

where
    G_{imp} is the per-volume generation rate due to impact ionization
    n and p are carrier densities
    μ_n and μ_p are the carrier mobilities

a_n and a_p are the impact ionization strengths
$a_{n\infty}$ and $a_{p\infty}$ are the impact ionization coefficients
E_{cn} and E_{cp} are the field coefficients
E is the local electric field

The impact ionization coefficients and field coefficients are extracted from material-specific experimental data.

Thermionic emission: Drift-diffusion equations are perfectly applicable to bulk material and homojunctions. However, their application to heterojunctions reduces accuracy, as the conduction and valence band discontinuities are not taken into consideration. The primary mechanism of transport over hetero interfaces is thermionic emission, which explicitly depends on the discontinuities in band structure when calculating the rate of transport across the interface. Assuming a heterojunction where the conduction band of material 2 is higher than the conduction band of material 1 by an amount E_{cbarr}, this effect can be modeled using the internal boundary equation[22]

$$J_{TE,n} = A_r T^2 \left[\exp\left(E_{Fn1} - E_{C1} - E_{Cbarr}\right) - \exp\left(E_{Fn2} - E_{C2}\right)\right] \tag{14.14}$$

where
A_r is the Richardson coefficient
T is the temperature
E_{Fn}s are the quasi Fermi levels at either side
E_Cs are the conduction bands at either side
E_{Cbarr} is the conduction barrier due to the heterojunction discontinuity

Note that this equation takes into consideration the net thermionic electron flow across the interface, and the net flow becomes zero under equilibrium conditions. To account for the thermionic injection of holes, a similar equation is also needed for the net hole flow across the interface.

Hot electron effects: The carriers, which are more energetic to thermally generated counterparts, are called hot particles. They are usually formed by energizing a particle by letting it speed up an electric field, or by ballistically emitting a carrier in a material with lower energy band where the carrier becomes an energetic, or hot, particle. Similar to heat transfer, hot particles can distribute their energy. Hence, an energy-transfer approach is more appropriate instead of the default particle flow model. However, the energy flow can be converted into a particle flow with correction terms to account for the differences. In our model, we have re-derived the cylindrical to Cartesian conversion equation, and implemented the following equation:

$$\nabla \cdot (k' \nabla(Tf)) = \nabla_* \cdot \left(k' T \nabla_* f + k' f \begin{pmatrix} \partial T/\partial r \\ \partial T/\partial z \end{pmatrix} \right) + \begin{pmatrix} k' T/r \\ 0 \end{pmatrix} \cdot \nabla_* f + \frac{k'}{r} \frac{\partial T}{\partial r} f \tag{14.15}$$

where ∇_* is the Cartesian nabla operator.

Surface recombination effects: Similar to bulk recombination effects, the surfaces provide additional sites where recombination can occur. These sites are created by surface states, which are extremely dependent on the state of the surfaces. Quantification of this effect is not easy, but an approximation for surface recombination in n-doped layers can be made by using[20]

$$-qD_n p_n \vec{n} \cdot \nabla E_{Fp}\Big|_{surface} = qS_p \frac{(np - n_0 p_0)}{(n + p + 2\sqrt{n_0 p_0})} \tag{14.16}$$

where

　　q is the electron charge
　　D_n is the diffusion constant
　　p_n is the hole charge density in n-doped layer
　　S_p is the surface recombination strength
　　n and p are charge densities
　　$n_0 p_0$ is the equilibrium charge density product

A similar approximation can be written for p-doped layers by changing "p" with "n" in the equation.

Temperature effects: The temperature effects in our simulation model are of two categories: temperature-dependent modeling where the operation temperature can be varied to evaluate the effects of cooling or heating, and heat generation and dissipation mechanisms to alter the local temperatures. The former is implemented by converting every equation or variable to temperature-dependent forms. The latter is implemented by Joule heating and steady-state heat transfer equations.

14.6 Simulation Results for Electro-Optical Modeling

Using the three-dimensional nonlinear simulation model developed, we have started investigating our design and optimizing it.

14.6.1 Optimization for Device Structure and Doping Levels

The device geometry and layer structure were optimized with our simulation model multiple times. The layer thicknesses, layer compositions (AlGaAsSb vs. GaAsSb) and individual layer doping levels were optimized to yield the highest optical gain with the lowest dark current at room temperature (Figure 14.3).

14.6.2 Dark Current versus Voltage

The dark current behavior of the devices were simulated and later compared to experimental measurements.

　　The simulation results predicted that devices with 5 μm diameter would exhibit around 0.5 μA of dark current at a bias of 1 V (Figure 14.4). The model showed that sub-micron devices would have much lower dark current values. For example, 100 nm diameter devices were expected to have less than 20 nA dark current at the same bias (Figure 14.5).

14.6.3 Photocurrent and Amplification (Optical Gain)

In our simulation model, one of the first functions implemented was the capability to illuminate the device. Based on this capability, we have simulated cases where the device was illuminated from the top or the bottom side.

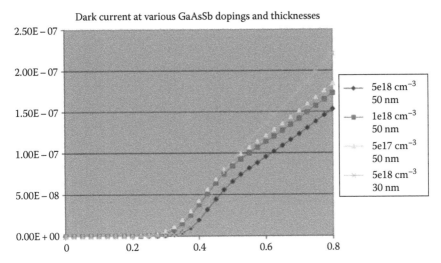

FIGURE 14.3
Simulation result for dark current at different base doping and thicknesses.

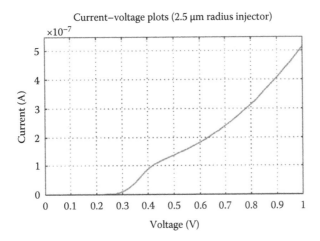

FIGURE 14.4
Current–voltage plot for a 5 µm diameter nanoinjection detector.

Using the results of CW illumination in the model, we have calculated the expected optical gain the devices would exhibit (Figure 14.6). Here, the optical gain is defined as the ratio of the number of injected electrons to the number of the absorbed photons. In our simulations, we have seen optical gain values of around 1000 for unpassivated first generation devices, 10 for passivated first generation devices, and more than 100,000 for passivated second generation devices.

14.6.4 Quantum Efficiency

Since nanoinjection detector features an internal amplification method that is tightly coupled with the detection method, we have expected to have challenges figuring out the actual electrical gain and quantum efficiency of the device. Because of this reason, we have

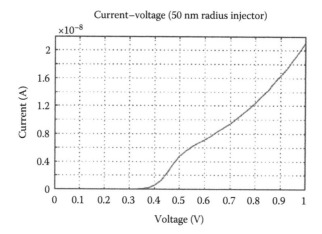

FIGURE 14.5
Current–voltage plot for a 100 nm diameter nanoinjection detector.

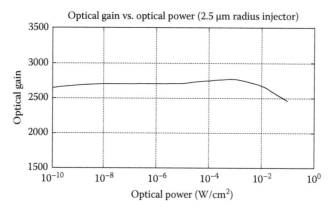

FIGURE 14.6
Simulation result for optical gain at different optical powers for a 5 μm diameter nanoinjection detector.

defined and used optical gain instead of the electrical gain. However, simulations could shed light on this challenge, helping us to isolate the quantum efficiency of our detector.

To find the simulated internal quantum efficiency, we have performed two simulations, one in dark and another with a specific optical illumination. In these simulations, we have performed boundary integration on the hole current that is thermionically emitted from InGaAs to GaAsSb due to photo generation. In addition to this current, we have volumetrically integrated the holes that are optically generated in this volume. Finally, we have taken the ratio of these values to find the internal quantum efficiency. These calculations resulted in quantum efficiency values around 70%.

14.6.5 Potential Distribution and Electric Field

To understand the electrical transport properties inside the device, the potential and electric field distribution was studied extensively (Figure 14.7).

The simulations revealed the favored paths inside the device for carrier transport. For example, in devices with unetched GaAsSb layers, this layer presented a low impedance

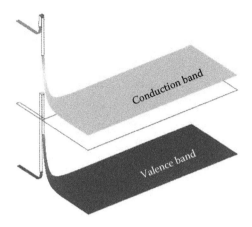

FIGURE 14.7
Simulated band structure in the 2-D cross-section of a nanoinjection detector.

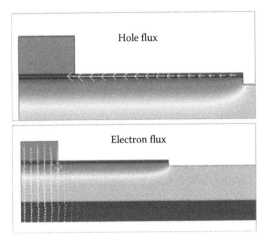

FIGURE 14.8
Different pathways for electron and hole flow reduce the recombination probability in a nanoinjection detector.

path for holes to reach the nanoinjectors, leading to spatial separation of electron and holes inside the device and reducing the recombination probability (Figure 14.8).

14.6.6 Device Heating

The joule heating inside the device was also simulated. However, the results did not show any significant heating at room temperature, even at locations with high current density (i.e., corners) (Figure 14.9).

14.7 Measured Performance of Nanoinjection Detectors

We measured the dark current, photo-response and spectral noise power of the devices at room temperature. At each data step, the dark current and spectral noise power

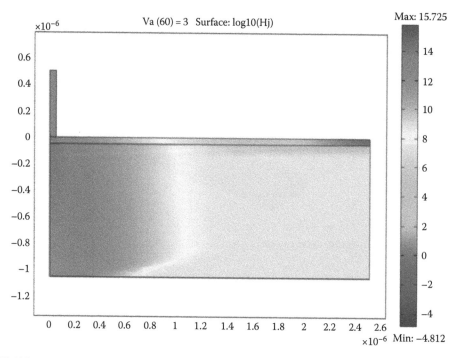

FIGURE 14.9
Simulation showing the heat generation inside the detector.

measurements were taken simultaneously, quickly followed by photocurrent measurements. For accurate laser power calibration, a commercial PIN detector was placed inside the setup as a separate experiment and its response was measured to accurately quantify the laser power reaching the sample.

Devices with 30 μm active diameter and 10 μm nanoinjector showed dark current values around 1 μA and internal amplification values exceeding 3000 at 0.7 V[14] (Figure 14.10). The DC current measurements, when coupled with the optical gain measurements, yielded a unity gain dark current density of less than 900 nA/cm² at 1 V.

Similar to reference,[23] the spectral noise power after amplification in unpassivated devices was measured with a spectrum analyzer around 1.5 kHz, which is beyond the 1/f noise knee but lower than the measured bandwidth of the device of about 4 kHz. The measured spectral power was compared to predicted spectral noise density due to Poissonian shot noise with amplification ($2qM^2 I_{int}\Delta f$), where we have observed noise suppression similar to the Fano effect.[24] This phenomenon is shown to result from temporal correlation mechanisms influencing particle flow, such as Coulomb blockade[25] or Pauli exclusion principle.[26] The strength of shot noise suppression (or enhancement) is quantified by the Fano factor[27] γ as

$$\gamma = \frac{I_n^2}{I_{shot}^2} = \frac{I_n^2}{2q I_{DC}\Delta f}$$

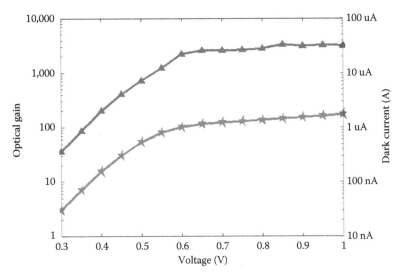

FIGURE 14.10
Current–voltage and optical gain–voltage plots for a nanoinjection detector at room temperature.

where
  I_n is the measured standard deviation of current, or current noise
  I_{shot} is the Poissonian shot noise
  q is the electron charge
  I_{DC} is the average value of current
  Δf is the bandwidth

For devices with internal amplification, the denominator of the right-hand side needs to be modified into $2qM^2I_{DC}\Delta f$. This is because of the amplification that applies to both the signal and the noise, and hence the noise power needs to be scaled by M^2 to conserve SNR.

The devices showed high internal amplification values around 3000 at 0.7 V and 5000 at around 1 V. The measured Fano factor was $F \sim 0.55$ (Figure 14.11).

The noise equivalent power (NEP) of the devices was measured as 4.5 fW/Hz$^{0.5}$ at room temperature without any gating, using the relation

$$P_{NEP} = \frac{I_{n,meas}}{\Gamma}$$

where Γ is the responsivity.

The spatial response of the unpassivated devices was measured using a surface scanning beam with ~1.5 μm diameter, and 10 nm step resolution. Despite such a high gain, the device showed a very uniform spatial response, primarily due to the low internal electric field in our devices. The measured response decreases rapidly beyond a radius of about 8 μm. This property meant that two-dimensional arrays of nanoinjection detectors would not need pixel isolation methods, and that became one of the bases for the nanoinjector imagers.

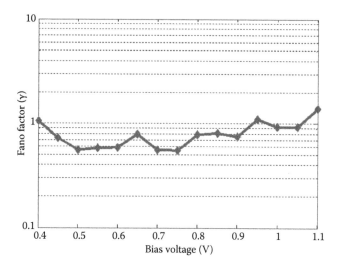

FIGURE 14.11
Variation of noise versus voltage for a 10 μm diameter detector.

The measured bandwidth of the unpassivated devices was around 3–4 kHz, much different compared to the lifetime in the GaAsSb trap layer (~1 ns with our doping levels), which we believe will be the ultimate constraint on bandwidth. We have attributed this difference to the existence and influence of surface traps.

When the devices were passivated, we observed a drastically different behavior. The gain decreased significantly to values around 10 and the spatial response extended to beyond 100 μm when the devices were not confined by hard-etching the trap layer. However, the bandwidth of these devices exceeded 3 GHz. The rise-time values of 200 ps were measured.

Passivated nanoinjection detectors with bandwidths exceeding several gigahertz exhibited ultra-low jitter values of ~14 ps at room temperature,[28] which is a record breaking performance (Figure 14.12). The transient response of the detector was studied by exploring the relation between lateral charge transfer and jitter, where we found out that the jitter is primarily transit time limited.

14.8 Measured Performance of Nanoinjection Imagers

After individual nanoinjection detectors were designed, fabricated, and evaluated, we worked on arrays of nanoinjection detectors to form focal plane array infrared cameras. The fabrication included building an array of detectors, putting indium bumps on both the detector and the read-out array, and integrating them using indium bump bonding (Figures 14.13 and 14.14).

Regarding design parameters, we chose to build 320 by 240 pixel arrays with 30 μm pixel pitch. We chose an off-the-shelf read-out-integrated circuit (ROIC), ISC9705 from Indigo, for hybridization. It is a 320 × 256 pixel ROIC with 30 μm pixel pitch. The full well capacity is 18×10^6 electrons and the ROIC can process up to 346 fps at full resolution, and up to 15,600 frames with reduced region-of-interest with cooling.

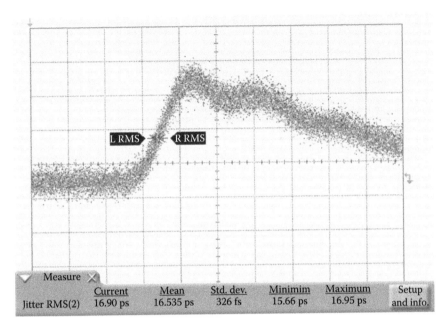

FIGURE 14.12
High-speed pulse response of passivated 10 μm nanoinjection detector, showing a jitter of 16.9 ps.

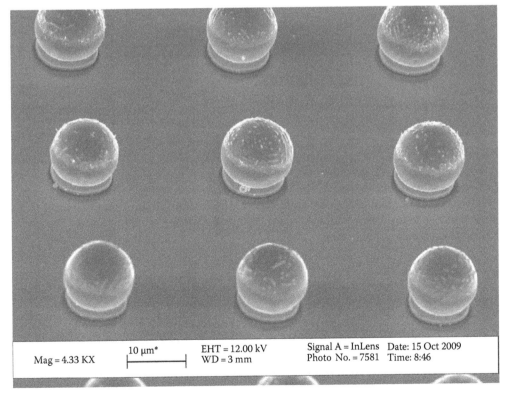

FIGURE 14.13
SEM image of the detectors with indium bumps on top.

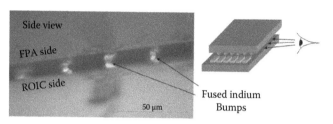

FIGURE 14.14
Microscope image of the gap between two dummy wafers integrated using indium bump bonding.

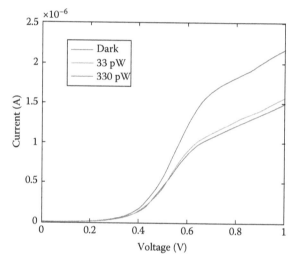

FIGURE 14.15
Current voltage plot of a hybridized, unpassivated 10 μm nanoinjection detector in the focal plane array.

Several measurements were performed on the hybridized devices.[29] The text pixels were directly tested for current, gain, and noise versus voltage (Figure 14.15). The spatial responsivity was measured and mapped. The imager performance and the SNR were compared to a high-end commercial PIN-detector based infrared camera. The signal-to-noise of the nanoinjection detector was evaluated under different photon fluxes to identify the excess noise and overall electronic noise.

14.8.1 Current–Voltage and Gain–Voltage before and after Integration

The sample was placed in a custom-built visible/infrared microscope setup, which doubled as a beam collimator for the laser beam. A tunable laser source at $\lambda = 1.55\,\mu m$ was used as modulated optical source with adjustable attenuation. The actual power reaching the sample was measured with a calibrated PIN detector. Before hybridization, the detectors were tested with a 1 μm radius Ni-W probe and a low noise current preamplifier (Stanford Research Systems SR-570). After hybridization, the test detectors with external wirebonding pads were soldered to a coaxial cable, and tested using

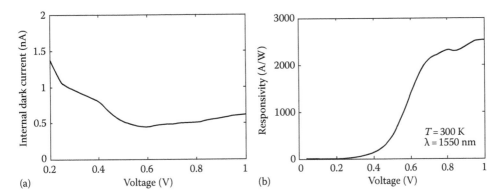

FIGURE 14.16
(a) Internal dark current (before amplification) versus bias voltage, and (b) the responsivity versus bias voltage.

SR-570, the output of which was monitored by an Agilent multimeter for dark and photo DC measurements (Figure 14.16).

14.8.2 Bandwidth of Hybridized Test Pixels

The test detectors with external wirebonding pads were soldered to a coaxial cable, and tested using SR-570, the output of which was monitored by an Agilent multimeter for dark and photo DC measurements and Stanford Research Systems SR-770 FFT Spectrum Analyzer. A calibrated laser was focused on the devices. The laser output was modulated, and the frequency was varied. The spectrum of the electrical signal was analyzed, and the bandwidth was marked as the frequency where amplitude of the fundamental frequency dropped by 3 dB. The device bandwidth was 3.6 kHz at 0.7 V.

14.8.3 Noise Performance of the Hybridized Test Pixels

After the test detectors with external wirebonding pads were soldered to a coaxial cable, the signal was amplified by SR-570 trans-impedance amplifier, and recorded by an Agilent multimeter for dark and photo DC measurements and Stanford Research Systems SR-770 FFT Spectrum Analyzer for noise. The power spectral noise was measured at 2.5 kHz, below the bandwidth of the devices, 3.6 kHz.

The excess noise factor F was calculated using the relation (Figure 14.17)

$$F = \frac{I_{n,measured}^2}{I_{n,expected}^2} = \frac{(I_{n,meas}^2/\Delta f)}{2qI_{int}G^2} \tag{14.17}$$

where
  $(I_{n,measured}/\Delta f)$ is the measured spectral noise power
  q is the electron charge
  I_{int} is the internal dark current (before amplification)
  G is the gain or internal amplification

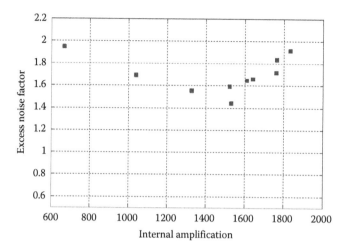

FIGURE 14.17
Excess noise factor F of a hybridized nanoinjection detector at different internal amplification values.

14.8.4 Spatial Sensitivity of Hybridized Test Pixels

The sample was placed in the custom-built visible/infrared microscope setup with an average optical power of 5 nW incident on the sample. The laser spot focused onto the sample with a spot size of about 2 μm. Focused laser spot was scanned over the sample using motorized drivers with nanometer resolution (Figure 14.18). A computerized setup was used to control instruments through Labview.

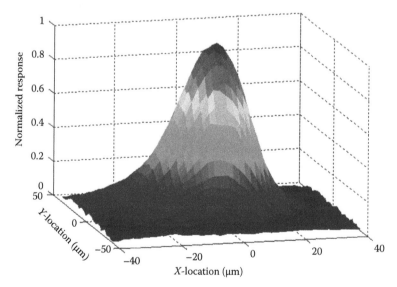

FIGURE 14.18
Spatial sensitivity map of a hybridized, unpassivated nanoinjector detector in the focal plane array.

14.8.5 Temperature Behavior of Hybridized Test Pixels

After hybridized FPA was placed on the LCC and wirebonded, the imager was placed in a cryostat facing a multispectral ZiSe window. The wirebonds connected to test pixels were soldered to the cryostat feedthroughs. The cryostat was pumped down, and a large area over the sample was illuminated using a laser source at 1.55 μm. The response of the device was amplified by SR-570 and recorded under darkness and with illumination. The power incident on the sample was found using room temperature photoresponse. The sample was than cooled down, and dark and photocurrent measurements were taken at every temperature.

The dark current and internal amplification values of hybridized test pixels were measured at different temperatures ranging from 300 to 78 K. The dark current decreased monotonically until 175 K, where a change in device behavior was observed either due to surface leakage or tunneling through the barrier. The Arrhenius plot indicates that activation energy of the dark current is 0.26 eV, which is in perfect agreement with the heterojunction injection barrier height of ~0.25 eV (InP–GaAsSb interface)[30] (Figures 14.19 through 14.21).

14.8.6 Comparison of the Signal-to-Noise Level of the Nanoinjection Imager with a Commercial Camera

The imager was placed in the camera (ROIC Evaluation Kit, from Indigo), and the camera dewar was pumped down. The imager was cooled down to −75°C. The devices were biased to −500 mV and the detector common was set to 6–6.5 V. The integration time was set to 0.5 ms. For the evaluation kit with nanoinjection imager, a Canon EF 50 mm f/1.8 lens was used. The same integration time setting was used in the commercial infrared camera (AlphaNIR). The lens used with AlphaNIR camera is IR-compatible 50 mm f/1.8. For both cameras, the lenses were used in maximum aperture, f/1.8. The cameras were calibrated using the same uniform light box. Before taking the actual images, the dynamic range of both cameras were equalized and fixed, by disabling the auto gain and contrast in both cameras while they were imaging the same scene. The acquired images (Figure 14.22) were

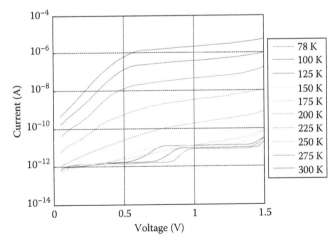

FIGURE 14.19
Dark current of the hybridized unpassivated nanoinjection detector at temperatures from 300 to 78 K.

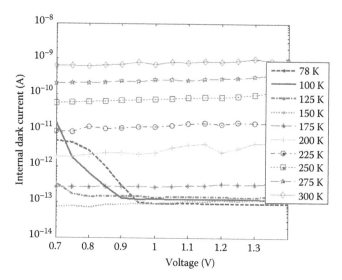

FIGURE 14.20
Internal dark current (before amplification) of the hybridized unpassivated nanoinjection detector at temperatures from 300 to 78 K.

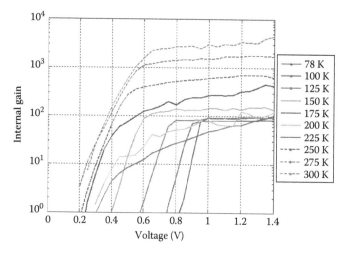

FIGURE 14.21
Internal gain of the hybridized unpassivated nanoinjection detector at temperatures from 300 to 78 K.

opened in MATLAB®, and the signal was calculated by subtracting dark scene from the illuminated scene, and standard deviation of each pixel was calculated from its time evolution.

The measured sensitivity of the nanoinjection detector was two orders of magnitude higher than the commercial PIN camera (1656 vs. 17), as shown in the comparison images and SNR histograms[31] (Figure 14.23).

14.8.7 Signal-to-Noise Analysis of the Nanoinjection Imager

The nanoinjection IR camera setup was similar to SNR measurements. A collimator was used together with an optical fiber with calibrated light output and the collimated beam

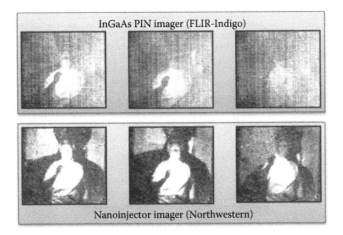

FIGURE 14.22
Short-wave infrared images, taken simultaneously by a commercial InGaAs PIN imager and the Northwestern nanoinjection imager.

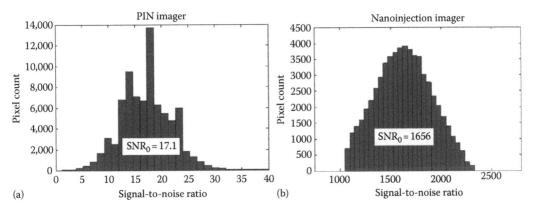

FIGURE 14.23
Signal-to-noise comparison of the images taken simultaneously by a commercial InGaAs PIN imager (a) and the Northwestern nanoinjection imager (b).

was focused onto the focal plane array. Different levels of laser output at 1.55 μm were mapped from the fiber and images were acquired at a reduced region of interest for highest frame rates. The images were imported into MATLAB, and at every illumination level, the signal was calculated by subtracting the dark scene from the illuminated scene. The standard deviation of each pixel was calculated from the time evolution. After the results were plotted, the curve was fitted to the relation[32] between SNR and photon count (Figure 14.24):

$$SNR = \frac{S_{photo}}{P_{noise}} = \frac{\phi t_{int} G^2}{G^2 F + \dfrac{\sigma}{\phi t_{int}}} = \frac{N^2}{FN + \dfrac{\sigma}{G^2}} = \frac{N^2}{FN + \sigma'_{overall}},\tag{14.18}$$

$$\sigma'_{overall} = \sigma_{nano-injection} + \frac{\sigma_{ROIC}}{G^2},\tag{14.19}$$

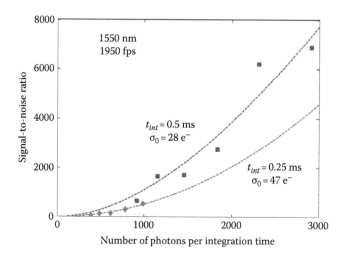

FIGURE 14.24

SNR of the nanoinjection imager versus number of photons, gathered using images acquired at different photon fluxes at a frame rate of 1950 fps. The analysis of the data reveals that the read-out noise is 28 e⁻ at 0.5 ms integration time and 47 e⁻ at 0.25 ms.

$$N = \phi t_{int} = \frac{P_{opt} t_{int}}{E_{ph}}, \tag{14.20}$$

where

S_{photo} is the signal power
P_{noise} is the noise power
Φ is the photon flux per second
t_{int} is the integration time
G is the gain (multiplication factor) of the detector
F is the excess noise factor
σ is the electronic noise
N is the number of photons
$\sigma_{overall}$ is the overall input noise of the system
$\sigma_{nanoinjection}$ is detector noise
σ_{ROIC} is ROIC noise
P_{opt} is the calibrated optical power
E_{ph} is the photon energy

In contrast to PIN-based detectors, the nanoinjection imagers amplify the signal before it reached the read-out electronics. Due to this, the imager performance was not limited by the electronic noise of the read-out integrated circuits. At an integration time of 0.5 ms and a frame rate of 1950 fps with reduced region-of-interest, an analysis of SNR versus number of photons[29] revealed that the overall electronic noise of the imager (28 e⁻) was much less than the read-out noise of (575 ~ 870 e⁻), and that the overall imager excess noise F was less than 1. When the integration time is halved, the electronic noise was almost twice (47 e⁻). This is in agreement with the fact that expected electronic noise scales down inversely with integration time.[33] The integration capacitor of our off-the-shelf ROIC prevents longer integration times. However, we expect the imager overall noise to be less than 2 e⁻ RMS at $t_{int} = 10$ ms and frame rate of 100 fps (Figure 14.24).

14.9 Conclusion

With a nontraditional geometry composed of nanoscale sensing and amplification nodes on thick absorption layers, the nanoinjection detectors offer high-sensitivity photon detection and amplification. However, due to their nonplanar design, type-II band alignment and the coupled detection/amplification mechanisms, the design and development of the nanoinjection devices required detailed nonlinear three-dimensional FEM simulations.

The FEM simulation provided a multi-physics environment for stationary, parametric, and transient simulations, all based on the drift-diffusion equations in two and three dimensions. The model incorporated several nonlinearities such as the incomplete ionization of dopants, bimolecular recombination, Auger recombination, nonlinear mobility, impact ionization, thermionic emission, hot electron effects, surface recombination effects, and temperature effects.

Once the devices were optimized through simulations, the nanoinjection devices were fabricated through micro-and nanofabrication. The measurements showed amplification factors reaching 10,000 with low dark current densities and Fano noise suppression. The passivated nanoinjection devices had bandwidths exceeding 3 GHz with a jitter of 15 ps. Nanoinjection detector arrays of 320-by-256 pixels were hybridized to form focal plane array infrared cameras. With a pixel level responsivity exceeding 2500 A/W, the nanoinjection focal plane arrays showed a noise level of 28 electrons at a frame rate of 1950 fps. These imagers showed two orders of magnitude improved SNR compared to commercial SWIR imagers, at thermoelectric cooling temperatures. These demonstrated capabilities of the nanoinjection detectors and imagers make them excellent candidates, extremely suitable for demanding applications such as optical tomography, satellite imaging, nanodestructive material inspection, high-speed quantum computing and cryptography, night vision imaging, and machine vision for process control.

References

1. P. Colarusso, L. H. Kidder, I. W. Levin, J. C. Fraser, J. F. Arens, and E. N. Lewis, *Applied Spectroscopy* **52**, 104A (1998).
2. J. Battaglia, R. Brubaker, M. Ettenberg, and D. Malchow, *Proceedings of the SPIE* **6541**, 654106 (2007).
3. M. Vatsia, Atmospheric optical environment, *Research and Development Technical Report* ECOM-7023 (1972).
4. D. A. Fay, A. M. Waxman, M. Aguilar, D. B. Ireland, J. P. Racamato, W. D. Ross, W. W. Streilein, and M. I. Braunl, *Proceedings of the Third International Conference on Information Fusion*, TuD3–3 (2000).
5. M. H. Ettenberg, M. Blessinger, M. O'Grady, S. -C. Huang, R. M. Brubaker, and M. J. Cohen, *Proceedings of the SPIE* **5406**, 46 (2004).
6. Y. Du, X. H. Hu, M. Cariveau, X. Ma, G. W. Kalmus, and J. Q. Lu, *Physics in Medicine and Biology* **46**, 167 (2001).
7. A. Migdall and J. Dowling, *Journal of Modern Optics*, **51**, 1265 (2004).
8. D. Deutsch, *Proceedings of the Royal Society of London A* **400**, 97 (1985).
9. T. P. Spiller, *Proceedings of the IEEE* **84**, 1719 (1996).

10. R. J. Hughes, D. M. Alde, P. Dyer, G. G. Luther, G. L. Morgan, and M. Schauer, *Contemporary Physics* **36**, 149 (1995).
11. P. W. Shor, *SIAM Journal on Scientific and Statistical Computing* **26**, 1484 (1997).
12. C. Elliott, *IEEE Security & Privacy* **2**(4), 57–61 (2004).
13. T. B. Pittman, Y. H. Shih, D. V. Strekalov, and A. V. Sergienko, *Physical Review A* **52**, R3429 (1995).
14. O. G. Memis, A. Katsnelson, S. C. Kong, H. Mohseni, M. Yan, S. Zhang, T. Hossain, N. Jin, and I. Adesida, *Applied Physics Letters* **91**, 171112 (2007).
15. O. G. Memis, A. Katsnelson, S. C. Kong, H. Mohseni, M. Yan, S. Zhang, T. Hossain, N. Jin, and I. Adesida, *Optics Express* **16**, 12701 (2008).
16. Comsol Multiphysics, Comsol, Inc. http://www.comsol.com
17. R. E. Owens and H. Y. Zhang, *InP and Related Materials, Fourth International Conference,* 397 (1992).
18. D. A. Neamen, *Semiconductor Physics and Devices: Basic Principles,* (McGraw-Hill, New York, 2003).
19. J. Piprek, *Semiconductor Optoelectronic Devices: Introduction to Physics and Simulation,* (Academic Press, San Diego, CA, 2003).
20. M. Sze, *Semiconductor Devices: Physics and Technology* (John Wiley & Sons, New York, 1980).
21. A.G. Chynoweth, *Physical Review* **109**, 1537 (1958).
22. K. Horio and H. Yanai, *IEEE Transactions on Electron Devices* **37**, 1093 (1990).
23. F. Z. Xie, D. Kuhl, E. H. Bottcher, S. Y. Ren, and D. Bimberg, *Journal of Applied Physics* **73**, 8641 (1993).
24. P. J. Edwards, *Australian Journal of Physics* **53**, 179–192 (2000).
25. G. Kiesslich, H. Sprekeler, A. Wacker, and E. Scholl, *Semiconductor Science and Technology* **19**, S37-S39 (2004).
26. I. A. Maione, M. Macucci, G. Iannaccone, G. Basso, B. Pellegrini, M. Lazzarino, L. Sorba, and F. Beltram, *Physical Review B* **75**, 125327 (2007).
27. A. Wacker, E. Schöll, A. Nauen, F. Hohls, R. J. Haug, and G. Kiesslich, *Physica Status Solidi* C **0**, 1293–1296 (2003).
28. O. G. Memis, A. Katsnelson, H. Mohseni, M. Yan, S. Zhang, T. Hossain, N. Jin, and I. Adesida, *IEEE Electron Device Letters* **29**(8), 867, (2008).
29. O. G. Memis, J. Kohoutek, W. Wu, R. M. Gelfand, and H. Mohseni, *IEEE Photonics Journal* **2**, 858 (2010).
30. C. R. Bolognesi, M. W. Dvorak, N. Matine, O. J. Pitts, and S. P. Watkins, *Japanese Journal of Applied Physics* **41**, 1131 (2002).
31. O. G. Memis, J. Kohoutek, W. Wu, R. M. Gelfand, and H. Mohseni, *Optics Letters* **35**(16), 2699 (2010).
32. B. E. A. Saleh, M. M. Hayat, and M. C. Teich, *IEEE Transactions on Electron Devices* **37**(9), 1976–1984 (1990).
33. M. M. Hayat, W. L. Sargeant, and B. E. A. Saleh, *IEEE Journal of Quantum Electronics* **28**(5), 1360–1365 (1992).

15

Finite Element Method (FEM) for Nanotechnology
Application in Engineering: Integrated Use
of Macro-, Micro-, and Nano-Systems

Radostina Petrova

Technical University of Sofia, Sofia, Bulgaria

P. Genova

Bulgarian Academy of Science, Sofia, Bulgaria

M. Tzoneva

Technical University of Sofia, Sofia, Bulgaria

CONTENTS

15.1 Introduction..506
15.2 Basic Idea and Applied Methods...506
15.3 Examples 508
    15.3.1 Clamps for Operating with Micro or Nano Objects 510
        15.3.1.1 Background of the Problem...510
        15.3.1.2 Aim of the Example 510
        15.3.1.3 Description of the Finite Element Model............................511
        15.3.1.4 Solution for a Set of Materials 512
        15.3.1.5 Numerical Experiments and Results 512
        15.3.1.6 Conclusions on the Cited Numerical Example.................517
    15.3.2 Analysis and Design of Parallel Micro-Robot Structures.............519
        15.3.2.1 Why Tripod Structure?..520
        15.3.2.2 Aim of the Provided Example...520
        15.3.2.3 Description of the Established Mathematical
                Model and of the Developed Solution Strategy..................522
        15.3.2.4 Numerical Examples for Outlining the Accessible
                Zone of the Vertex of the Controlled Link 526
        15.3.2.5 FEM of a Spatial Three-Pod Mechanism with Three
                Built-in Drives..530
        15.3.2.6 Conclusions on the Preceding Numerical Example 533
    15.3.3 Kinematics of a Macro-Mechatronic System 538
        15.3.3.1 Analytical Solution 538
        15.3.3.2 Finite Element Solution 544
        15.3.3.3 Conclusions on the Preceding Example 546

15.3.4 Analysis and Synthesis of Mechanical Bioreactors
   for Bio-Nanotechnologies 547
   15.3.4.1 Short Review of Existing Devices for Bioreactor Systems 547
   15.3.4.2 Kinematics of Spatial Mechanisms in Biotechnologies....................551
   15.3.4.3 Conclusions on the Preceding Example 555
References...556

15.1 Introduction

"The step from micro-technology to nano-technology requires more than a reduction of the size by a factor of a thousand. If you want to move precisely in the nano-world, you don't succeed by perfecting proven techniques." This statement of Handelsblat is cited by Klocke and Gesang (2002) and describes briefly the innovative approach in nanoscience.

Therefore, to provide successful design of mechanisms for operating with micro and nano objects, we need to understand the basic ideas and goals set in service of macro-space, to refine and complete the final technological operation, using various micro-/nano-structures.

In addition, it must not be forgotten that operating with micro-/nano robots is directly related to the type of driving devices (actuators). As there is no precise definition and distinction between different conceptions of actuators, micro-actuators, driving devices, micro-machines, and nano-machines are in the known to the authors of literature sources, related to micro-electro-mechanical systems (MEMS), the authors make the following assumptions:

- The term "micro" defines the size itself as a property of the object—micro, small, large, etc.
- The actuator is a set of physical bodies (elements) that when activated by an external energy source implements a single or continuous mechanical movement (or action) as a result of the changes in their body size and form. An actuator can operate as an energy converter also.
- Micro-motor is analogous to the electric drive, but based on the principle of action of the electroactive polymer actuators.
- Nano-machines are devices which are the smallest in size of all MEMS devices, even smaller than those assembled of a molecule.

Micro- and nano-robots are assembled of these micro-devices and other micro-manipulators. In some cases the micro- and nano-machines themselves act as robots. The idea of the size of a micro-robot is illustrated by the photos in Figures 15.1 and 15.2, where a nano-motor and a nano-manipulator of a match size are given.

15.2 Basic Idea and Applied Methods

The idea of the chapter is to examine several examples of cooperation between macro- and micro-robots, as shown in Figure 15.3. The authors focus their attention to the accession of micro-robot downstream to the macro-robot, to its driving and control.

FIGURE 15.1
Photo of a nano-motor.

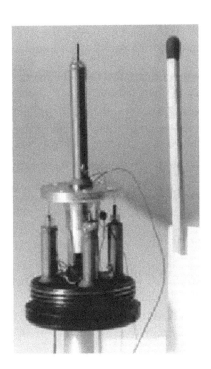

FIGURE 15.2
Photo of a nano-manipulator.

More substantial benefits of the cooperation between micro- and macro-robots are

- The two robots are autonomous. They can be produced independently by different companies. Various combinations of different types and different sizes depending on the specific defined purposes can be done.
- The macro-robot, without cooperation with a micro-robot, can be used for other industrial operations requiring adequate accuracy.

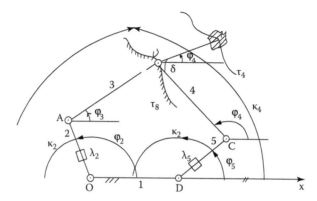

FIGURE 15.3
Scheme of integrated micro- and macro-robots.

- Multiple stationary operations, such as injection, pipette, installing of micro-units, etc., can be manipulated like this.

Among the main disadvantages of the preceding cooperation are

- The macro-robot must be positioned and then the micro-robot must be orientated inside its small workspace.
- Besides specific technological process sensors, the use of cameras is often required to guarantee precise identification of the position of the manipulated object and the micro-robot itself.
- The operations of the two robots are necessary to be consistent.

The methods used to illustrate the idea of cooperation between macro-, micro-, or nano-robots can be classified as follows:

- Some of the cited examples are used as source data construction drawings of existing mechanisms.
- Classical analytical approach for determining the kinematic parameters of the macro-, micro-, and integrated mechanisms.
- Finite element method (FEM) for static analysis, modal, harmonic, and transient analysis of the objects and structures, time-history analysis of mechanisms.
- Multiparametric optimization.
- Neural networks (NNs).

15.3 Examples

The basic ideas of the chapter are illustrated through some examples.

The first one examines a three-finger clamp for manipulating with tools, micro and nano objects. The operation is performed by piezoceramic actuators. Optional types of methodology for their design, both through a displacement analysis by FEM software package and through a selection of eligible pseudo-optimal solutions using NNs, are proposed. The authors examine the clamp structure, varying many different materials and selecting those which guarantee a predefined displacement of each finger of the clamp, provoked by similar forceful impact on the clamping frame. They try to implement proposed algorithm of design of similar clamping devices to the provided example.

The second example shows an algorithm for establishment of the accessible space by a monitored vertex in the face of the proboscis of a tripod mechanism with built-in rotary/linear drives. The aim is to define criteria for structural design of UPU couples and to develop a decision strategy of control and management of the motion of the monitored sensor. The established algorithm is realized through writing a computer code in the MATLAB® environment. It is applicable for solving the inverse problem of kinematics for space mechanisms of that type. The idea is to use the rotary/liner drives for establishment of the macro-configuration of the mechanism and later to move the sensor more precisely (micro-motion) to the exact micro-disposition through piezo-actuators. In addition a CAD/CAE model of a mechanism of that type is developed in SolidWorks environment. All its dimensions are numerically introduced. This model is used in the context to show how a trace path of the monitored vertex can be obtained, regarding the true boundary and contact conditions and loading. This model is user-friendly and easily adaptable, regarding dimensions, material, loads, fixtures, and drives implemented on the mechanism. All dynamic phenomena in the structured system can also be examined through this finite element model.

The third example shows a detailed kinematic analysis of a hybrid macro-micro robot with a five-link two-crank closed structure and the established solution of direct problem of kinematics (DPK) through basic geometry and kinematic laws. The geometrical conditions of full rotation of the two input links are set forth, the accessible spaces for such a MS are outlined, and the transfer functions (TFs) of the actuators incorporated into the links are derived assuming that the system linearization is legitimate. The aim of the present example is to solve the inverse problem of kinematics for a macro mechatronic system without linearization and the one for a micro mechatronic system, which is redundant, with two extra degrees of freedom (DoFs). The basic algorithms for designing a control strategy are presented. The problem is considered in two stages: solving the inverse problem of kinematics considering the macro-mechatronic system and solving the inverse problem of the kinematics of the micro-mechatronic system. A numerical example is considered on the basis of derived formulas. In addition a finite element spatial model of the aforementioned mechanism is presented. The CAD model is created in SolidWorks environment. The piezo-actuators are situated in the holes of the links. They increase the length of each link up to 1/100. Three different cases of driving this mechanism are compared. The dead weight of the links as well as a loading force of 1 N, modeling the action on the mechanism of the operated object, are regarded in calculating the reaction in the hinges through the given finite element solution. By that time the friction in the hinges is neglected as it is too small and this the first iteration stage of the promoted numerical simulation.

The last example is focused research on bioreactors. Most of the bioreactors are designed with only one axis of rotation. They put the cell growth under the influence of one force vector only, due to which they provide a physical signal only in the direction of this one force vector. Hence, cells are not inclined to growth of three-dimensional (3D) cell cultures. Bioreactor systems have controllable motors and monitoring sensors to control the processes in the bioreactor chamber, i.e., they are mechatronic systems. The object of the example is a bioreactor device with spatial mechanisms. The authors analyze the possibilities of implementation of spatial mechanisms with one or two DoFs, i.e., an extensive analysis of coupler's spatial motion is required. Furthermore, the possibility of using the coupler's rotation around its own axis when both of the kinematic couples to which it is linked are spherical ones is studied.

15.3.1 Clamps for Operating with Micro or Nano Objects

15.3.1.1 Background of the Problem

In the late twentieth century began the use of robotics for micro/nano manipulation processing. A new type of robots with appropriate manipulating systems (MSs) and clamping devices, with new controlling configurations and sensors (Ionescu et al., 2004, Kostadinov et al., 2004, 2005a,b, Tiankov, 2006a,b) have been developed for implementing different operations with micro/nano objects in medicine and engineering. For successful accomplishment of manipulations, the need for smaller displacements and higher precision of disposition and impact has been arisen. Thus a new type of actuators, electromechanical or electromagnetic piezo-actuators (Kostadinov et al., 2004, 2005a,b, Tiankov, 2006a) has come into use. Here, a three-finger clamp, called later "clamping device," for micro and nano objects and tools, realized by three identical modules, arranged in 120° (Figure 15.4) is studied through FEM. Each of the modules implements a radial moving of its finger, not greater than tenths of a millimeter through the deformation of its elements. Piezoceramic actuators, causing element deformation, are integrated in these modules. More details on the clamp configuration and its functions are provided in Petrova et al. (2011).

15.3.1.2 Aim of the Example

The aim of the example is to demonstrate how to choose the most appropriate material from a set of predefined ones through stress and deformation analysis of the frame of the described

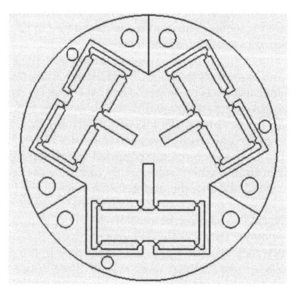

FIGURE 15.4
Scheme of the clamping device.

clamp. The gripping of micro objects is achieved by the motion of the "proboscis" to the clamp center, arisen by built-in piezoceramic actuators. Therefore, a given displacement at a given load/force impact on the frame of the clamping device is guaranteed.

15.3.1.3 Description of the Finite Element Model

First a 3D model of the module has been created using the existing design documentation. The 3D model has been created through the software SolidWorks (SolidWorks—Users Manual, 2009). Figure 15.4 shows a scheme of the clamping device.

Second, the 3D model has been transformed to a model of spatial finite elements, through software package SW-Simulation, a toolbox of SolidWorks (SolidWorks Simulation—Training Manual, 2010). Spatial finite elements of type tetrahedron of first order have been used. Through the FEM (Zienkiewicz, 1972), a more precise computer modeling of stress, strain, and displacement distribution in structure and identification of endangered areas is performed.

The defined boundary conditions have been set as follows: attachment at all screw openings—type "fixed" (Figure 15.5). The loading is implemented through uniformly distributed forces and simulates the pressure on the inner radial sides of the modules' frames (see Figure 15.6). The magnitudes of these surface loads are equal to the magnitude of the impact of the piezo-crystal on frame inner surfaces. These in-built piezo-crystals are not shown on the figures.

The input data are organized into two sets:

1. A set of different types of suitable materials, introduced by their modules of elasticity and coefficients of Poisson
2. A set of loads, equal to the known values of the forces of the actuators

The combinations between the two data sets are infinite.

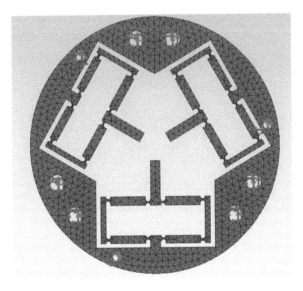

FIGURE 15.5
Finite element 3D model of the clamping module.

The number of eligible solutions and extraction of pseudo-optimal options from the set of eligible solutions has been limited by following the criteria:

- Radial displacement of the "proboscis" within a predefined range
- Minimum stresses in the vulnerable zones

15.3.1.4 Solution for a Set of Materials

The authors illustrate the proposed algorithm in the following text (Figure 15.6).

The modulus of elasticity, E, and Poisson's ratio, μ on one side and the force value on the other are set as input data. The software calculates the shear modulus, according to the known relation $G = \dfrac{E*0.5}{1+\mu}$.

The distribution of von Misses stress, shown in Figure 15.7, enables easy determination of stress endangered areas.

Figure 15.8 shows the vertex nodes of the surface of the "proboscis," in which radial displacements (along axis Y, shown in the figure) are monitored. Values of these movements are given in Tables 15.1 and 15.2.

15.3.1.5 Numerical Experiments and Results

Figures 15.9 and 15.10 show graphical and numerical results of FEM analysis. Figure 15.9a shows a general picture of the distribution of the full nodal displacements of the model. Figure 15.9b shows a similar motion at the end of the "proboscis." "Shade" or "loop" shows the nondeformed shape.

Figure 15.10 shows nodal displacements along radial axis (radial displacements toward the center). For the detailed figure, the radial displacement is parallel to axis Y. Depending on the chosen material, the radial displacement can be either positive (to the center) or negative.

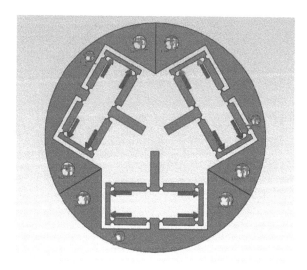

FIGURE 15.6
3D model of the loaded clamping module.

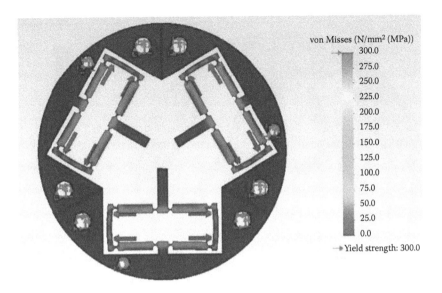

FIGURE 15.7
Distribution of the von Misses stress in the studied module.

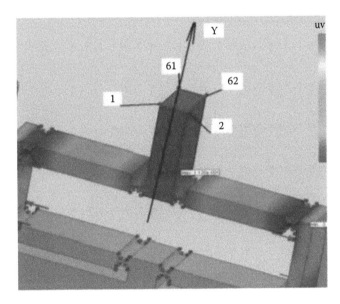

FIGURE 15.8
Numbers of the nodes in the vertexes of the surface of the "proboscis," whose movement is monitored.

The finite element numerical experiments are conducted in the following three stages:
Stage I: Modulus of elasticity E is varied
The obtained results are summarized in Table 15.1.

Stage II: Poisson's ratio is varied
Poisson's ratio is additionally varied in order to refine the study. In case of pure aluminum, it is equal to 0.22 and for aluminum alloys it is around 0.3. The results are shown in Table 15.2.

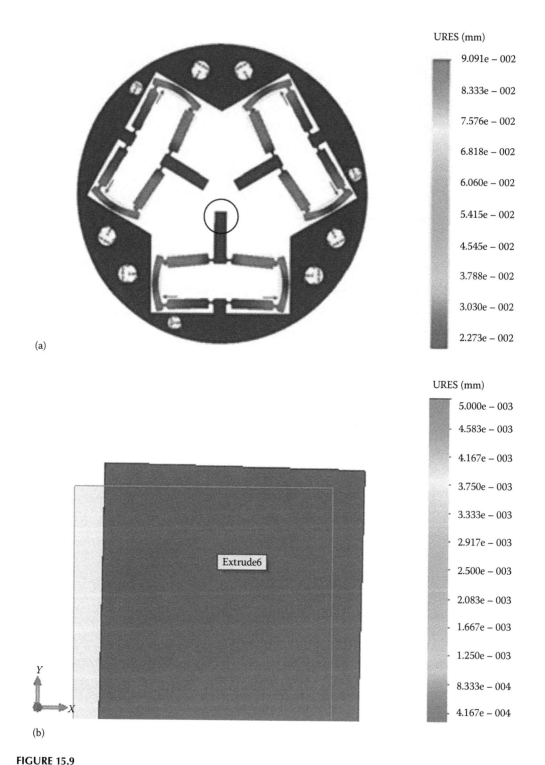

URES (mm)

9.091e − 002

8.333e − 002

7.576e − 002

6.818e − 002

6.060e − 002

5.415e − 002

4.545e − 002

3.788e − 002

3.030e − 002

2.273e − 002

(a)

URES (mm)

5.000e − 003

4.583e − 003

4.167e − 003

3.750e − 003

3.333e − 003

2.917e − 003

2.500e − 003

2.083e − 003

1.667e − 003

1.250e − 003

8.333e − 004

4.167e − 004

Extrude6

Y

X

(b)

FIGURE 15.9
Distribution of the nodal displacement in finite element model: (a) Scheme of the deformed shape of the clamp and nondeformed shadow shape and (b) displacement at the end of the "proboscis" (in detail).

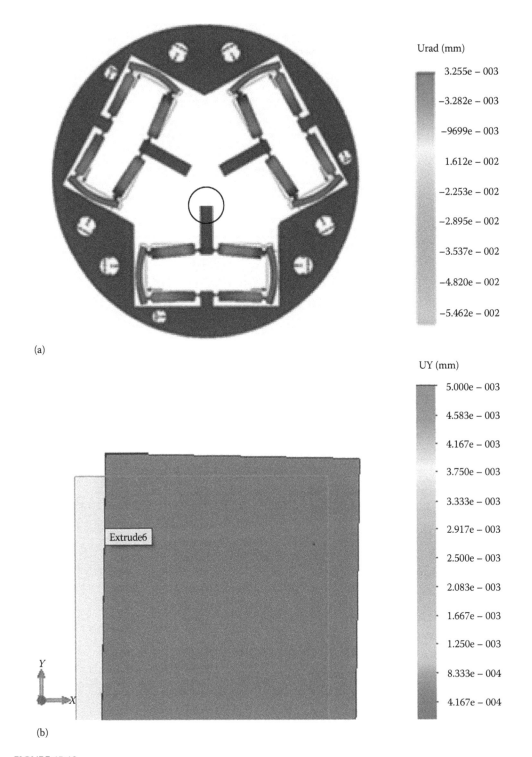

FIGURE 15.10
Distribution of the radial nodal displacement in finite element model: (a) Scheme of the deformed shape of the clamp and nondeformed shadow shape and (b) Radial displacement of the surface at the end of the "proboscis."

TABLE 15.1

Numerical Data of Averaged Radial Displacement (along Axis Y) of the Surface of the "Proboscis" Shown in Figure 15.8 While Modulus of Elasticity E Is Varied

Poisson's Ratio	0.2	0.22	0.25
Modulus of Elasticity, GPa	**Averaged Radial Displacement, mm**		
200	0.039069	0.035004	0.027429
150	0.051551	0.046354	0.036575
100	0.078051	0.069699	0.054869
90	0.086819	0.077708	0.060962
85	0.091824	0.082281	0.064548
80	0.097420	0.087293	0.068582
75	0.104070	0.091175	0.073154
70	0.111610	0.100010	0.078379
65	0.120210	0.106210	0.084611
60	0.130230	0.116390	0.093173
55	0.141700	0.126970	0.099739
50	0.156280	0.140010	0.109710

TABLE 15.2

Numerical Data of Averaged Radial Displacement (along Axis Y) of the Surface of the "Proboscis," Shown in Figure 15.8 While Poisson Ratio μ Is Varied

Poisson's Ratio	Radial Displacement (along Axis Y), mm				
	Node 1	Node 2	Node 61	Node 62	Average Value
Modulus of elasticity $E = 70$ GPa					
0.20	0.1084	0.1063	0.1080	0.1073	0.1067
0.22	0.0953	0.0954	0.0950	0.0946	0.0943
0.25	0.0772	0.0745	0.0769	0.0759	0.0748
Modulus of elasticity $E = 69$ GPa					
0.20	0.1099	0.1078	0.1095	0.1089	0.1082
0.22	0.1024	0.1011	0.1019	0.1017	0.1015
0.25	0.0778	0.0751	0.0775	0.0764	0.0075
Modulus of elasticity $E = 68$ GPa					
0.20	0.1118	0.1096	0.1114	0.1107	0.1101
0.22	0.9871	0.9741	0.9835	0.0981	0.9777
0.25	0.0801	0.0773	0.0797	0.0787	0.0777

Stage III: Additional studies

In numerical simulation using material brass, type free-cutting brass with material characteristics: $E = 97$ GPa, $G = 37$ GPa, and $\mu = 0.31$ the monitored displacements are in node $1 \rightarrow 0.11490$ mm; in node $2 \rightarrow 0.11590$ mm; in node $61 \rightarrow 0.11460$ mm; and in node $62 \rightarrow 0.11570$ mm. The average displacement of the surface is 0.11767 mm.

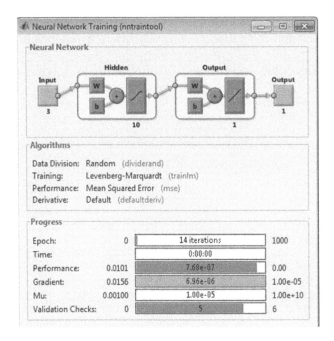

FIGURE 15.11
Created neural network—basic view, used algorithms, progress.

Stage IV: Model of NN

After testing a set of materials through FEM, a one-direction relation is established between the magnitude of the force, the modulus of elasticity and Poisson's ratio of the material on one side and the radial displacement of the "proboscis" on the other.

We will show how in combination with the other numerical methods such as NNs (Haykin, 1998), the FEM can be transformed into a powerful mathematical tool for design. Because of the great number of possible solutions of this design problem, it is appropriate to use NNs to narrow the range of possible solutions and later to choose the most applicable (the optimum) one.

Therefore, an NN of type "Fitting Function through software product MATLAB, package Neural Networks" (Hagan et al., 2002, Beale et al., 2010) is set up. It has 10 hidden layers (Figure 15.11).

In the cited case, the input data include the magnitude of the force, modulus of elasticity, and Poisson's ratio of real materials. The target data include the radial displacements of the "proboscis" obtained through FEM. The neural net automatically removes all outcomes that do not meet the predefined radial displacement criterion and ensures compliance with the first optimization criterion.

The results of NN calculation are shown in Figure 15.12.

If all these calculations were made using traditional solving methods, for example "pure" FEM, they would have taken much more computer time and human resources.

15.3.1.6 Conclusions on the Cited Numerical Example

Based on the performed numerical experiments, conclusions about the relation between physical characteristics of the material/material constants and monitored displacement as well as about applicable solutions are made.

The authors hope to prove that a comprehensive solution of structural design problem can be obtained by combining software running different versions of FEM with those based on NNs.

The developed algorithm is unique and applicable to a wide range of design problems.

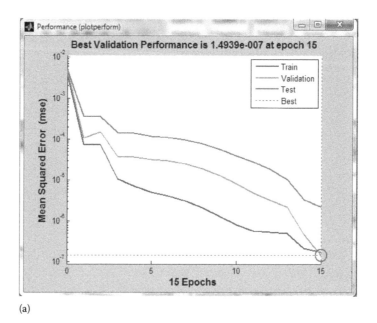

(a)

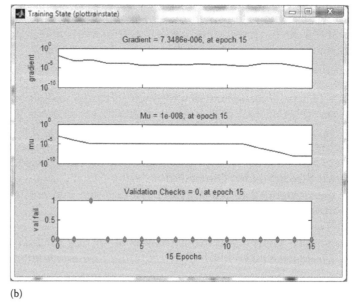

(b)

FIGURE 15.12

Graphs of the results obtained by trained, tested, and validated network: (a) best validation performance, (b) training characteristics of the designed network,

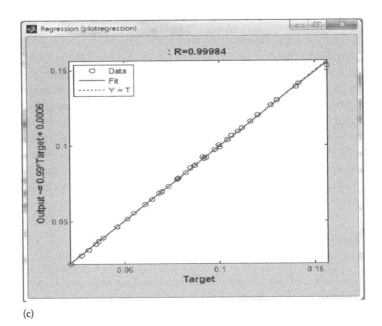

(c)

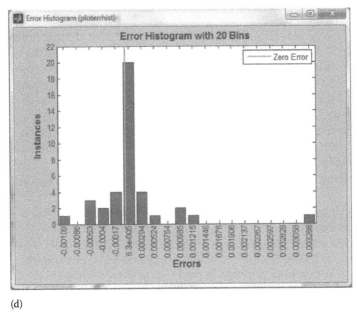

(d)

FIGURE 15.12 (continued)
Graphs of the results obtained by trained, tested, and validated network: (c) regression function, and (d) error histogram.

15.3.2 Analysis and Design of Parallel Micro-Robot Structures

An analysis of the accessible space of the monitored vertex in the face of the proboscis of a tripod mechanism with built-in rotary/linear drives is presented in this example. The aim is to define criteria for structural design of UPU couples of fourth grade and to develop a strategy for decisions on the control and management of the motion of the monitored sensor. The established algorithm is applicable for solving the direct problem of kinematics for space mechanisms of that type. The idea is to use the rotary/liner drives for establishment of the macro-configuration of the mechanism and later to move the sensor more precisely (micro motion) to a predefined exact disposition, regarding its spatial orientation toward fixed coordinate system, through piezo-actuators.

The algorithm presented in the following regards the micro-motion of the tripod structure only. It is a brief geometrical solution realized through the MATLAB software (first part of the provided algorithm). Based on the numerical data, calculated through the first stage, a CAD/CAE model for studying the structure motions is established. This model can be used for prediction and control of the vertex motion and is a foundation of a developed strategy for motion management. The CAD/CAE model is implemented through SolidWorks software tools (second part of the algorithm).

15.3.2.1 Why Tripod Structure?

Starting in the last two decades of the twentieth century up to the present-day manipulating systems (manipulators, MS) with parallel structures are particularly relevant (Chanhee Han et al., 2002, Han Sung Kim et al., 2005, Sadjadian et al., 2005, Conconi et al., 2009, Kanaan et al., 2009, Hexapod & Tripod, 2010).

Their main advantages are compactness, loading capacity, technological profile, and possibility to mount their drives on a relatively rigid unit. Solving the inverse problem of kinematics for this type of manipulators is elementary, i.e., the algorithms for their control are relatively simple. The main drawbacks of these manipulators are the narrowly accessible space as well as the numerous singular configurations of the structure. Manipulators with parallel structures can be implemented as tripods and more frequently as sixpods, having three or six independent degrees of freedom. Later several types of parallel manipulators are illustrated. Figure 15.13 (Chanhee Han et al., 2002) shows a tripod structure, and Figure 15.14 a sixpod one (Hexapod & Tripod, 2010). Authors who studied the structure shown in Figure 15.13 have investigated ways of avoiding singular situations of the parallel MS through technological gaps in the joints.

Another interesting structure of three-arm delta robot is shown in Figure 15.15. Conconi et al. (2009) prove that use of parallelogram "pods" instead of linear ribs reduces the number of hazardous singular configurations, increases the level of precision of operation, and expands the accessible space.

There have been a lot of publications in the field of MSs with parallel structures. Without claiming that they are aware of all, the authors can define some not entirely explored issues and some other issues that have not been in the focus of the researchers yet. Among them are

- Hybridization between parallel MS (PMS), servicing the macro-space, and PMS for micro-operation processing.
- Solving the direct kinematic problem, which is significantly more difficult than finding a solution of the inverse one for this type of manipulators. Outlining the accessible space enables solving a lot of optimization problems related to strategies of monitoring and control of structure behavior of these mechanisms.

15.3.2.2 Aim of the Provided Example

In the following text, the authors present an analysis of the macro-accessible space by a vertex sensor disposed in the head of the proboscis of a tripod structure (Figure 15.16):

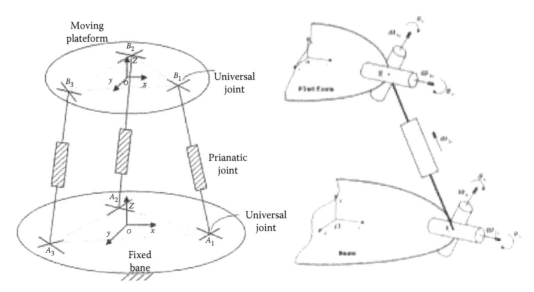

FIGURE 15.13
Tripod structure of the type UPU Joint. (From Chanhee Han et al., *Mech. Mach. Theory,* 37, 787, 2002.)

FIGURE 15.14
Sixpod structures. (From http://www.physikinstrumente.com/en/products/primages.php?sortnr=700881&pic view=2#gallery)

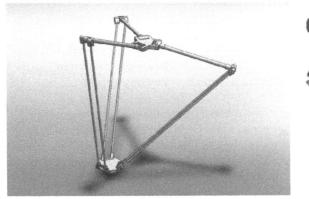

FIGURE 15.15
Photos of three-arm delta robots.

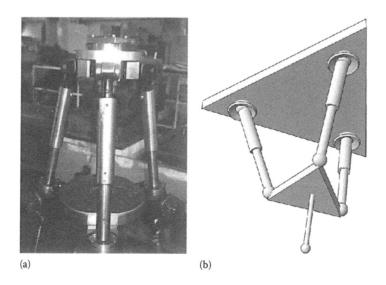

(a)   (b)

FIGURE 15.16
A general 3D-view of the studied tripod structure: (a) a photo of an MS, similar to the studied one (From Han, C. et al., *Mech. Mach. Theory*, 37, 787, 2002.) and (b) 3D CAD model of the studied MS structure.

- With built-in rotary drives, operating the rotation angles of the pods
- With built-in linear drives, operating the lengths of the pods

The stretches of the proboscis along the three coordinate axes and the rotation angles of the pods are calculated. They depend on the characteristics of the built-in drives. Similarly, the accessible space of the sensor vertex disposed in the proboscis head is outlined. The lengths of the pods are varied within a range of 10% of their maximum length.

This information is of extreme importance during the process of structural design of kinematic couples, which in this case are of the type of UPU couples (see Figure 15.13). All decisions on management strategies are carried out in an "off-line" regime.

15.3.2.3 Description of the Established Mathematical Model and of the Developed Solution Strategy

15.3.2.3.1 Geometric Solution of Problem

To determine the accessible space of the sensor vertex for this designed MS, a spatial geometric problem is initially solved. The scheme of the mechanism is shown in Figure 15.17. All denoted at the given scheme lengths are assumed to be constant during one calculating cycle.

For a mechanism with rotary drives, the angles between the pods and the base vary and for a mechanism with linear drives lengths of the pods do. The base denoted (Ab1, Ab2, Ab3) is fixed. It is assumed that the nodes (points Abi, $i = 1 \div 3$) coincide with the vertexes of an equilateral triangle. Fixed Cartesian coordinate system (CS) XYZ, whose origin is at the point Ab1 and the X-axis coincides with the segment Ab1–Ab2 and is introduced for more convenient derivation of equations. It is assumed that the mobile platform is an equilateral triangle, whose vertexes Api, $i = 1 \div 3$ are joined to the pods through UPU-joints. The executive link AC-C (Figure 15.17) is fixed perpendicularly to the mobile platform and point

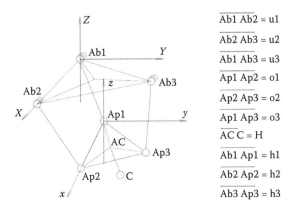

$$\overline{Ab1\ Ab2} = u1$$
$$\overline{Ab2\ Ab3} = u2$$
$$\overline{Ab1\ Ab3} = u3$$
$$\overline{Ap1\ Ap2} = o1$$
$$\overline{Ap2\ Ap3} = o2$$
$$\overline{Ap1\ Ap3} = o3$$
$$\overline{AC\ C} = H$$
$$\overline{Ab1\ Ap1} = h1$$
$$\overline{Ab2\ Ap2} = h2$$
$$\overline{Ab3\ Ap3} = h3$$

FIGURE 15.17
Geometrical solution of the solved spatial mechanism.

C coincides with its geometric (mass) center. All links' lengths are introduced as constants in the written MATLAB code. All used parameters are shown in Figures 15.17 and 15.18. The mathematical solution of the problem of identification of all applicable configurations of the mechanism structure while varying the angles between the pods and the base is implemented in two stages by successively solving two nonlinear systems of equations.

- *First stage*: A pseudo-plane closed four bar mechanism is geometrically described. The problem is reduced to the standard problem of finding geometrical relations between the coordinates of its joins, its generalized coordinates, and the lengths of the links. The point is that instead of the actual lengths of the units Ap1–Ap2 and Ab2–Ap2, the lengths of their projections Ap1–Ap2′ and Ab2–Ap2′ (Figure 15.18) are used. At that stage, the angles denoted di, $i = 1 \div 3$ vary and the program code itself retains only physically assessable solutions according to specified criteria. Then the algorithm continues with solving the system equations of the second stage, but the decision is sought only among initially examined and marked as applicable during the first-stage solutions.

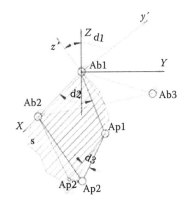

Ab1 Ab2 Ap2
determine the plans s, which is rotated at an angle equal to b1, around axis X.

Coordinate axes X and z', are situated in the plane s, while axis y' is perpendicular to s.

The angle between the axis X and the axis $\overline{Ab1\ Ap1}$ is equal to d2.

Ap2′ is the projection of Ap2 on s.

$$Ap2'\ Ap2 = h = O1^* \sin d3$$
$$Ap2'\ Ab2 = h2'$$
$$Ap2'\ Ap1 = O1' = O1^* \cos d3$$

FIGURE 15.18
Geometrical scheme of the solved pseudo-2D four bar mechanism.

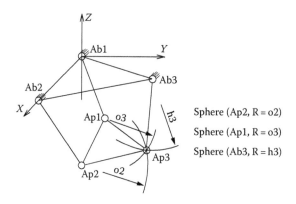

FIGURE 15.19
Geometrical scheme of the second stage in solving the spatial geometry problem.

- *Second stage*: A nonlinear spatial mathematical system for finding the intersecting point of three spheres, each centered in the geometrical center of the couples Ap2, Ap1, and Ab3 and with radii equal, respectively, to O2, O3, and h3 (Figure 15.19) is numerically solved. Based on the input data from the first stage of the calculation algorithm, i.e., the coordinates of the couples Ap2, Ap1, and Ab3, all mathematically possible solutions are defined. As it is not possible to find a physically applicable solution for all input data sets, the software displays an appropriate message and drops the data set out of the group of applicable configurations of the mechanism. The problem is defined and solved like an optimization problem of three different parameters, which are the searched spatial coordinates of point Ap3. The target function is a system of three nonlinear functions whose minima are sought.

The result of the solution is a set of all assessable physical configurations of the MS at predefined lengths of the pods.

The geometric solution is implemented through MATLAB environment, while some of the library codes of the optimization package are implemented in the written new software code.

15.3.2.3.2 *Implementing the Design Solution, Regarding Some Constructive Requirements*

The obtained solution through the written computer code is a database, containing the entire information on the spatial configuration of the structure, including coordinates of points Ap1, Ap2, Ap3, AC, C in the fixed Cartesian coordinate system, all elements of transferring matrixes between the fixed and the used floating coordinate systems or vice versa, depending on the mission. All data are processed and exported to MS Access or to MS Excel.

Some additional restrictions, due to the construction of base UPU couples, Ab1, Ab2, and Ab3, are imposed. Along the directions ai, $i = 1 \div 3$ (Figures 15.20 and 15.21), rotary drives, which enable pods' rotation around the axes xi, $i = 1 \div 3$, are placed. Measuring and defining of the rotation angle a3 is shown in detail in Figure 15.21. The same is the case for the other two angles a1 and a2. The following restriction on the angles ai, $i = 1 \div 3$, is imposed: $30° < ai < 150°$. The rotation around axes yi, $i = 1 \div 3$ is denoted with bi

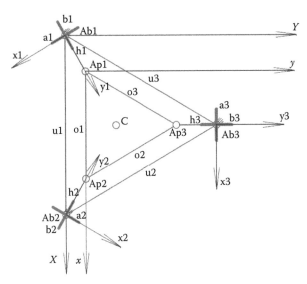

FIGURE 15.20
Scheme of the studied mechanism with given UPU joints.

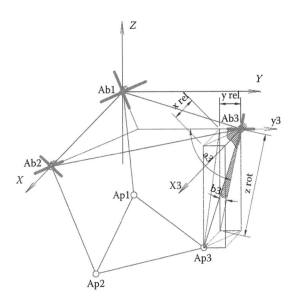

FIGURE 15.21
Scheme of calculating the construction range of angles ai and bi, $i = 1–3$ in UPU-joints.

(Figures 15.20 and 15.21). This rotation depends only on the geometrical configuration of the structure, but the constructive design of the joints does not allow the angle to be greater than 45°, i.e., $-45° < bi < 45°$, $i = 1 \div 3$.

Further the decisions that do not meet the preceding conditions are eliminated in MS Excel (MS Access) environment. This additionally reduces the number of applicable solutions.

15.3.2.4 Numerical Examples for Outlining the Accessible Zone of the Vertex of the Controlled Link

15.3.2.4.1 Mechanism with Built-in Rotary Drives

Some numerical examples for the mechanism with the following lengths of the units: $ui = 60$, $oi = 40$, $hi = 50$, $i = 1 \div 3$, and $H = 30$ in mm are given in the following text. The accessible space of vertex C is outlined.

Let us assume that one pod (in this case pod Ab1-Ap1) is fixed (angle a1 = −89.91°) and the other one (denoted Ab2-Ap2) rotates about an axis $x2$ within a range of −83.59° to −52.10°. The sign "−" indicates the relative displacement of the plate couple Ap2 toward the corresponding base kinematic couple in the local coordinate system Abi, xi, yi (Figures 15.20 and 15.21). Figure 15.22 shows two possible configurations of the mechanism and the trace path of vertex C under the preceding conditions.

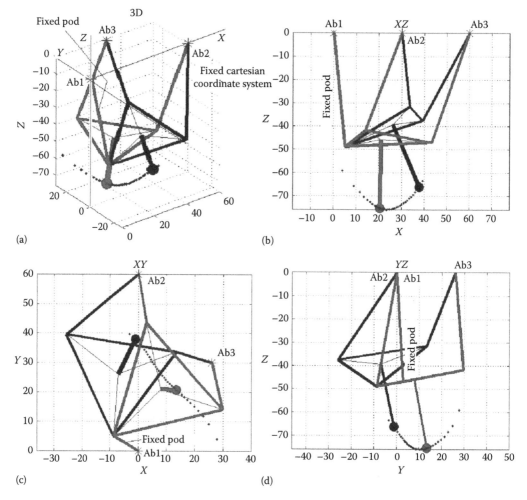

FIGURE 15.22

Two different spatial structure configurations and the trace path of vertex C for a mechanism with one fixed pod and one operating rotary drive: (a) 3D scheme, (b) projection of the mechanism in XZ plane, (c) projection of the mechanism in XY plane, and (d) projection of the mechanism in YZ plane.

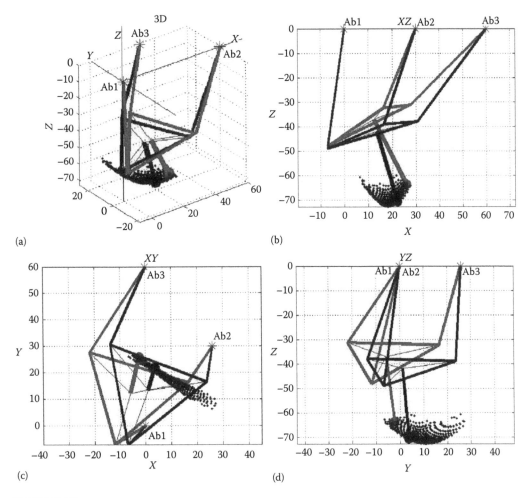

FIGURE 15.23
Two different structure spatial configurations and trace path of vertex C for a mechanism with two operating rotary drives: (a) 3D scheme, (b) projection of the mechanism in *XZ* plane, (c) projection of the mechanism in *XY* plane, and (d) projection of the mechanism in *YZ* plane.

A case of two simultaneously operating rotary drives is shown in Figure 15.23. The first operating rotary drive is built-in pod 1 and angle a1 varies in between 71.98° and 80°. The second rotary drive operates angle a2 in between −59.93° and −47.45°. The accessible by vertex C zone is a part of the spatial surface, which is shown in the Figure 15.23a–d.

If an operation of the three rotary drives, located at the base couples Abi, is assumed and the angles of rotation of the three pods vary under the condition $30° < a i < 150°$, $i = 1 \div 3$, the outlined accessible zone of vertex C of the implementing unit is given in Figure 15.24. It is a spatial domain, whose approximate size along the axes is as follows: along *X*—63.1 mm; along *Y*—56.8 mm; and along *Z*—62.5 mm. The domain is roughly symmetric about a plane, parallel to *YZ*, displaced at a coordinate *X* = 30 mm. The volume of the accessible domain is about one third of the volume of the space occupied by the structure itself.

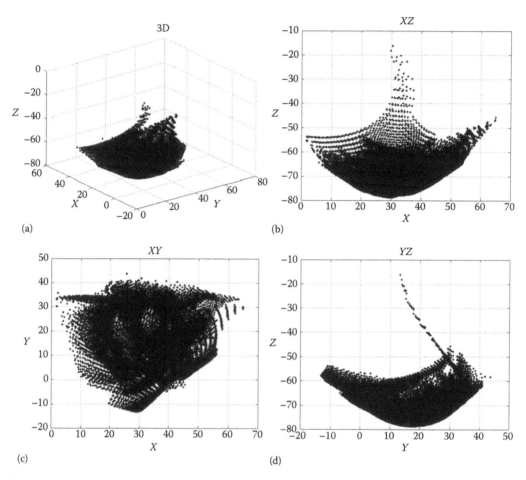

FIGURE 15.24
Accessible by vertex C zone while there are three operating rotary drives: (a) 3D picture of the accessible by vertex C zone, (b) projection of the accessible zone in XZ plane, (c) projection of the mechanism in XY plane, and (d) projection of the mechanism in YZ plane.

15.3.2.4.2 Mechanism with Built-in Linear Drives

The following numerical results are about a mechanism with the following dimensions in millimeters: $ui = 60$; $oi = 40$; $hi = 45 \div 50$; $i = 1 \div 3$; and $H = 30$. The accessible zone of vertex C is outlined.

If it is assumed that

- The two pods (pod Ab1-Ap1 and pod Ab2-Ap2) are of a constant length equal to 90% of the maximal pods' length, i.e. equal to 45 mm.
- The length of the third pod (pod Ab3-Ap3) varies in-between 90% and 100%.
- The angle between the fixed pod Ab1-Ap1 and the base is equal to 30°.
- The angle between the pod Ab2-Ap2 is 70°.
- Some of the applicable configurations of the structure and the accessible zone of vertex C are shown in Figure 15.25.

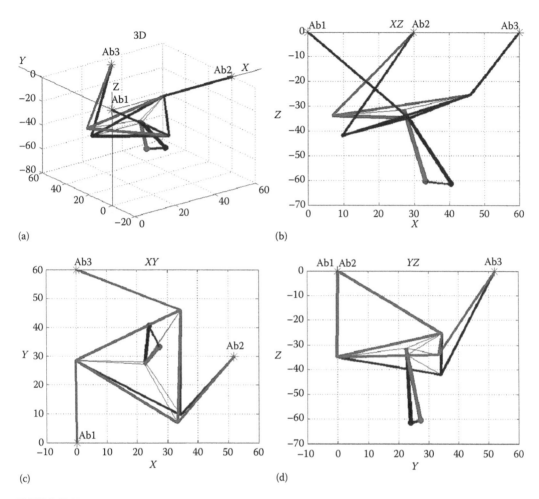

FIGURE 15.25
Trace path of vertex C of the implementing unit while only the one linear drive is in operation: (a) 3D picture of the accessible by vertex C zone, (b) projection of the accessible zone in *XZ* plane, (c) projection of the mechanism in *XY* plane, and (d) projection of the mechanism in *YZ* plane.

In Figure 15.26 is shown the case when there are two simultaneously operating linear drives—the one built-in pod 2 and the one built-in pod 3. The lengths of these pods vary in the range of 45–50 mm, and it is assumed that the length of pod 1 is constant during the simulation. It is equal to the minimum of 45 mm. It is assumed that the angles between the pods 1 and 2 and the base are constant and equal to 30°, while the angle between the pod 3 and the base varies in between 0° and 70°. These values determine the limits of applicable configurations of this spatial mechanism. By now the constructive requirements on the joint between the pod 3 and the base are neglected. The accessible zone of vertex C is a spatial surface.

If the three built-in linear drives are in operation, the lengths of each pod vary in the range of 90%–100% of its maximal length, i.e., their limits are 45 ÷ 50 mm. In addition, it is accepted that the angles between the pods and the base are fixed during the simulation. Then the accessible by vertex C zone is given in Figure 15.27. The charts are obtained for a predefined discrete growth of the pods' lengths and that causes the

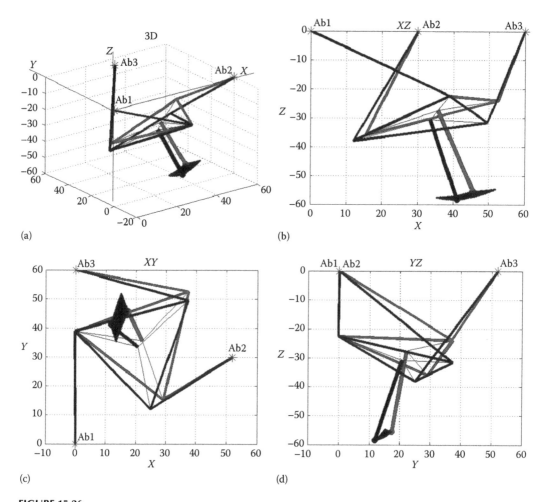

FIGURE 15.26
Two different spatial structure configurations and accessible zone of vertex C for a mechanism with two operating linear drives: (a) 3D picture of the accessible by vertex C zone, (b) projection of the accessible zone in XZ plane, (c) projection of the mechanism in XY plane, and (d) projection of the mechanism in YZ plane.

gaps in the final chart of the accessible zone. In the cited example, it is situated in a space of the following limits: along axis x—32 to 52 mm; along axis y—2 to 20 mm; along axis z—−56 to −62 mm.

15.3.2.5 FEM of a Spatial Three-Pod Mechanism with Three Built-in Drives

After outlining the accessible zone of the vertex C of the implementing unit, some numerical examples of a similar spatial three-pod mechanisms with built-in rotary or built-in linear drives are provided.

The following examples are of mechanism with 6 UPU kinematic couples situated in both ends of the three pods. The dimensions of the mechanism are similar to the earlier presented ones:

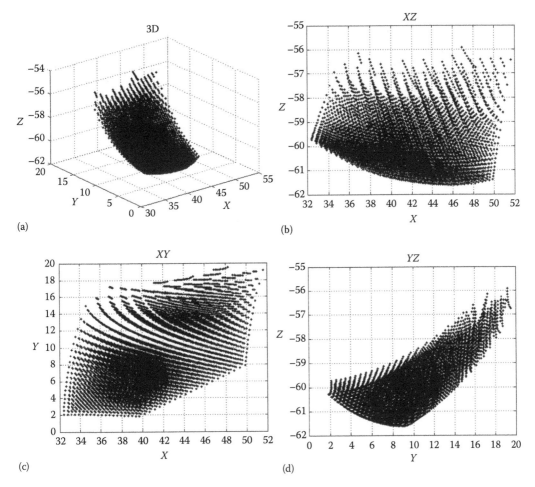

(a)

(b)

(c)

(d)

FIGURE 15.27

Accessible by vertex C zone while there are three operating linear drives: (a) 3D picture of the accessible by vertex C zone, (b) projection of the accessible zone in *XZ* plane, (c) projection of the mechanism in *XY* plane, and (d) projection of the mechanism in *YZ* plane.

- The fixed vertexes of the base are disposed in the vertexes of an equilateral triangle, which side is equal to 60 mm.
- The longitudinal length between the hinges in both ends of each pod varies in between 40 and 50 mm.
- True constructive dimensions of all units are implemented in the provided CAD model.
- The moving plate is an equilateral triangle with a side of 60 mm.
- The length of the implementing unit is 30 mm.
- The trajectory of its vertex is traced.

It is assumed that the mechanism is made of aluminum alloy.

The provided CAD model is developed in SolidWorks environment (SolidWorks—Users Manual, 2009). The established FEM is made through SW Simulation environment (SolidWorks Simulation—Training Manual, 2010) and the simulation of the motion of the mechanism is performed through SW Motion (SolidWorks Motion Studies, 2011).

In addition, the mechanism is loaded by gravity forces in +Z direction and a vertical (in +Z direction) force of 1 N, applied in the vertex and modeling the action of the dead weight of the operated object.

Finite element model is created by tetrahedrons of order 1.

The aim of the cited examples in the later sections is just to show the functions of the trace path and of the displacement of the vertex as well as the rotation of the plate at predefined laws of operation of the drives. The type of software used prevents the impact of body deformations on the kinematic parameters of the motion.

Simulating and studying the velocity, acceleration, contact forces between the units, stress, and deformed shapes of all elements is the next step in this study. The authors intend to create a methodology which will enable a detailed investigation on the dynamic characteristics of the motion and a performance of some constructive (related to shapes and dimensions and with an impact on stress distribution) and some driving (related to the characteristics of drives and with an impact on acceleration, dynamic reactions, displacement versus time, etc.) optimization criteria on the structure.

15.3.2.5.1 FEM of a Spatial Three-Pod Mechanism with Three Built-in Rotary Drives

The established FEM is shown in Figure 15.28. The three arcs at the pods' joint show the disposition of the three rotary drives. Next is shown a 3D view of the traced path of the vertex of the implementing unit. Of course it strongly varies on the predefined laws of the drives, disposition of UPU couples, and a lot of other structural parameters. It is assumed that the displacement laws in hinges with built-in rotary drives are of sine oscillating functions. The amplitudes of these functions are introduced in degrees and are equal to: of drive 1°–5°; of drive 2°–3°; and of drive 3°–4°. The frequency varies as follows: for drive 1: 0.2 Hz; for drive 2: 0.5 Hz; and for drive 3: 0.75 Hz (see Figure 15.29).

Figure 15.30 shows the function of the displacement of the traced vertex—its magnitude as well as its projections along coordinate axis. In Figure 15.31 is given the magnitude of angular displacement of the implementing unit during the motion.

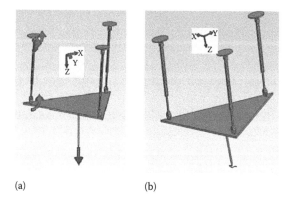

(a)	(b)

FIGURE 15.28
A 3D view of the developed CAD-CAE model of a spatial three-pod mechanism with three built-in rotary drives: (a) disposition of rotary drives and (b) sketch of the vertex trace path during the operation of the three drives.

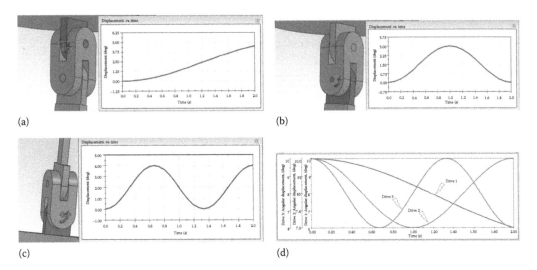

(a)

(b)

(c)

(d)

FIGURE 15.29
Characteristics of all built-in rotary drives and corresponding changes in hinges' angles: (a) characteristic of drive 1, (b) characteristic of drive 2, (c) characteristic of drive 3, and (d) angular displacement in the hinges with built-in drives.

15.3.2.5.2 FEM of a Spatial Three-Pod Mechanism with Three Built-in Linear Drives

The established FEM with built-in linear drives is shown in Figure 15.32. The three arrows at pods mark the disposition of the three linear drives. Next to that figure is shown the 3D sketch of the traced path of the vertex of the implementing unit. The blue arrow marks the dead weight of the orated object.

It is assumed that the variations of the lengths of the pods are sine oscillating functions. The amplitude of these functions is introduced in millimeters and is as follows: for drive 1–5 mm; for drive 2–3 mm, and for drive 3–4 mm. The frequency of the driving functions varies as follows: for drive 1–0.2 Hz; for drive 2–0.5 Hz; and for drive 3–0.75 Hz (see Figure 15.33). The variations of pods' lengths due to the operation of each drive are given in the same figure.

Figure 15.34 shows the function of the displacement of the traced vertex—its magnitude as well as its projections along coordinate axis while the three drives are operating simultaneously. In Figure 15.35, the magnitude of angular displacement of the implementing unit during the motion is given.

15.3.2.6 Conclusions on the Preceding Numerical Example

The presented methodology uses basic geometric spatial relationships between units of the studied mechanism while implementing the first part based on computer code developed in MATLAB environment. Thus, the direct problem of kinematics of MS is solved. The presented algorithm is suitable and easily adaptable for all spatial MSs of that type. Through this methodology, the accessible zone of the sensor vertex of the implementing unit can be found for some given initial conditions such as its geometric configuration, the type of the drives (rotary or linear), the number of simultaneously operating drives, the range of variation of the angles (for a mechanism with built-in rotary drives), or the lengths of the pods (for a mechanism with built-in linear drives). Consequently, it is easy the transferring functions of the first and the second order for each generalized coordinate to be derived.

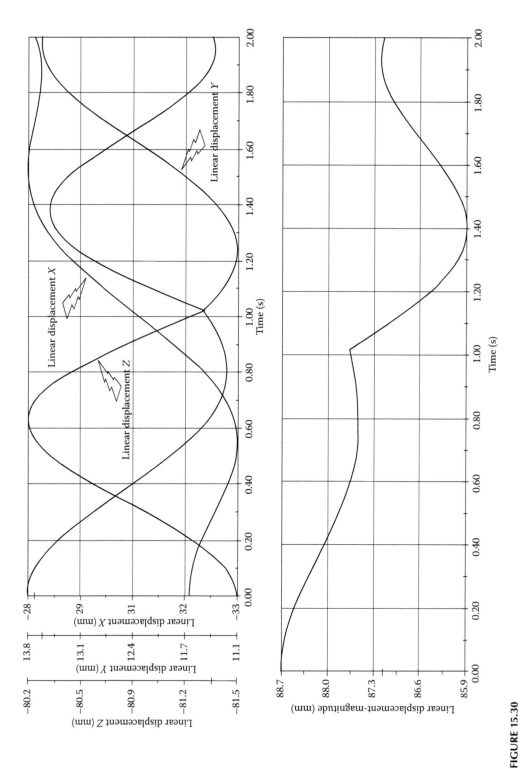

FIGURE 15.30
Linear displacement of vertex C, compared to the fixed base (in relative coordinates).

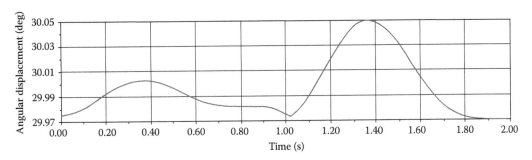

FIGURE 15.31
Angular rotation of implementing unit with vertex C, compared to the fixed base.

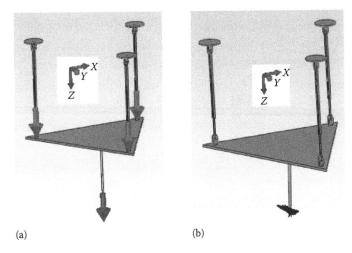

(a)   (b)

FIGURE 15.32
A 3D view of the developed CAD-CAE model of a spatial three-pod mechanism with built-in three linear drives:
(a) disposition of linear drives and (b) sketch of the vertex trajectory during the operation of the three drives.

The development of this mathematical code enables the authors to solve several spatial kinematic tasks, illustrated by the aforementioned examples for MS with three rotary or three linear drives. Depending on the number of operating drives and additionally imposed restrictions on the angles of pods' rotation and on varying their lengths, the accessible zone is either a spatial path (for one operating drive) or a spatial surface (for two simultaneously operating drives) or a spatial domain with changing characteristics (for three operating drives).

The post-developed finite element models are applicable for solving precise numerical problems and are applicable in the second part of the investigating methodology. They are easily adaptable and user-friendly—the dimensions of the units can be varied and even the material and the laws of operating drives can be easily changed. The initial configuration can also be easily reconstructed. The advantage of this approach is that the gravity and the additional loading are included in FE solution, which can be used for precise calculation of dynamic reaction of the mechanism as a whole and for performing some optimizing changes for better operation of the system itself.

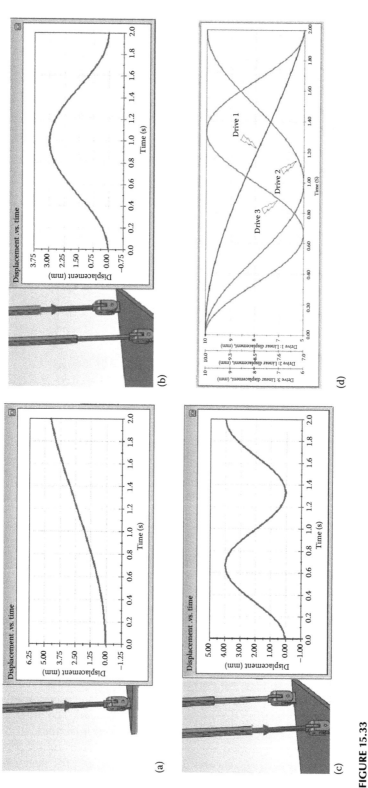

FIGURE 15.33
Characteristics of all built-in linear drives and corresponding changes in pods' lengths: (a) characteristic of drive 1, (b) characteristic of drive 2, (c) characteristic of drive 3, and (d) linear displacement and corresponding change of pods' lengths.

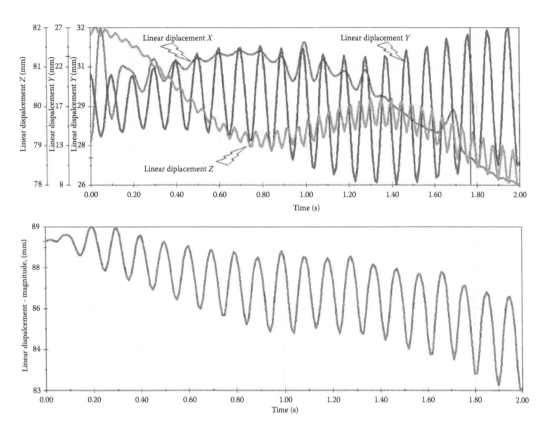

FIGURE 15.34
Linear displacement of vertex C, compared to the fixed base (in relative coordinates).

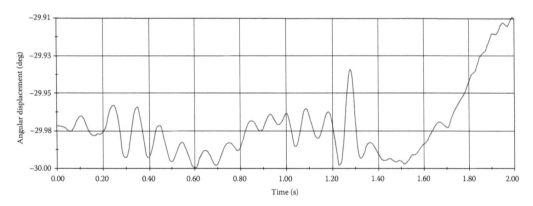

FIGURE 15.35
Angular rotation of implementing unit with vertex C, compared to the fixed base.

The authors intend to focus their further research in this area on

- Hybridization between parallel manipulating structures, servicing the macro-space and driven by conventional rotary/linear drives, on one side and parallel MSs, servicing the micro-space and driven by piezo-actuators

- Optimization problems related to management strategy, such as finding a trace path with minimum deviations from a predefined one, with a maximal level of uniformity of the motion of all units of the MS, with minimum number of picks in velocity and acceleration functions during a given integral time for executing the operation

Based on the huge number of applicable solutions of the earlier-defined problems and the necessity of handling large databases, aiming deriving of optimal decision according to some predefined kinematic, dynamic, constructive, stress, and deformation criteria, the use of NNs could be appropriate. As it has been proved in the previous example, this opportunity for developing an interdisciplinary methodology combines the advantages of the classical methods of mechanics, theory of mechanisms and machines, and strength of materials with newer methods such as FEM and NNs. The combination works very well in the field of numerical modeling, simulation, and processing control of mechanical systems, saving human and computer resources.

15.3.3 Kinematics of a Macro-Mechatronic System

15.3.3.1 Analytical Solution

The following example shows a detailed kinematic analysis of a hybrid macro-micro robot with a five-link two-crank closed structure and the established solution of DPK through basic geometry and kinematic laws. The geometrical conditions of full rotation of the two input links are set forth, the accessible space for such an MS are outlined, and the TFs of the actuators incorporated into the links are derived assuming that the system linearization is legitimate. Also, the conditions for unidirectionality of macro motions at the end of the cycle and micro motions at the start of the cycle are found. These problems are specific for the DPK.

The aim of the presented example is to solve the inverse problem of kinematics (IPK) for a macro-mechantronic system without linearization and the one for a micro-mechatronic system, which is redundant, with two extra DoFs. Conditions for unidirectionality of the basic links of a micro MS are taken as optimization conditions. The basic algorithms for designing a control strategy are presented. Calculated through the first iteration results can be used to choose the configuration of a macro MS at the start of a micro operation.

A numerical example is considered on the basis of formulas derived to solve the inverse problem of kinematics. The validation of the theoretical setup is realized by designing a virtual model of the robotized system and performing an analysis of the simulation results obtained.

15.3.3.1.1 Inverse Problem of Kinematics Considering the Macro-Mechatronic System

A macro-mechantronic system is based on a five-link mechanism shown schematically in Figure 15.36. On a macro level, the chain consists of five mobile links (1–5) with dimensions 11–15 and angular coordinates φ_2–φ_5. Those whose angular displacements are specified by angles φ_2 and φ_5 are chosen as active pairs. Note that the angular displacement characterizes a mechanism with 2 DoFs. Chain operational space is specified by arcs k_2–k_5, s_2, and s_5. To perform finishing operations in micro and nano ranges, a piezo-actuator is integrated in each mobile body (λ_1–λ_4).

Coordinates x and y of the point B are given for an IPK (note that velocities can also be specified), and a corresponding configuration of KC is sought. More specifically, we look

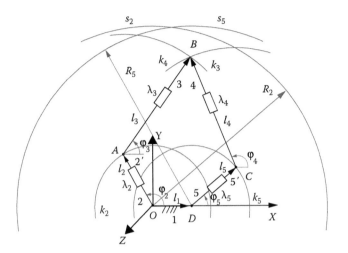

FIGURE 15.36
Two-crank closed five-link mechanism.

for the parameters of the actuating devices (AD)—angle φ_2 and the respective velocities. Point B is the cross point of the circles:

$$(x - x_A)^2 + (y - y_A)^2 = l_3^2,$$
$$(x - x_C)^2 + (y - y_C)^2 = l_4^2, \tag{15.1}$$

where

$$x_A = l_2 c\varphi_2, \quad y_A = l_2 s\varphi_2, \tag{15.2}$$

$$x_C = l_1 + l_5 c\varphi_5, \quad y_C = l_5 s\varphi_5.$$

Here one can consider either the nominal values of the dimensions of links l_i or dimension increase after actuator activation. After elementary transformations, we get the following via Equations 15.1 and 15.2

$$x^2 + y^2 - 2(xx_A - yy_A) = l_3^2 - l_2^2, \tag{15.3}$$

$$2(xx_C - yy_C) + 2(yy_A - xx_A) = C^2, \tag{15.4}$$

where $C^2 = l_3^2 - l_2^2 - l_4^2 + l_5^2$.

Equation 15.3 contains parameter φ_2, only, while Equation 15.4 contains parameter φ_5. The system is nonlinear but it can be reduced to quadratic equations

$$s\phi_3 = \frac{y - l_2 s\phi_2}{l_3}, \tag{15.5}$$

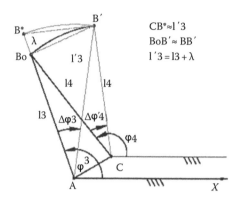

FIGURE 15.37
Operational area of actuators 3, 4, and 5.

$$s\phi_4 = \frac{y - l_5 s \phi_5}{l_4}, \tag{15.6}$$

which are used to calculate angles ϕ_3 and ϕ_4. The positional IPK is solved via those relations.

We consider a numerical example using equations for solving the IPK of a macro-mechantronic system. The following parameters are used as initial data for the structure shown in Figure 15.37:

$$l_1 = 20 \text{ mm}; \quad l_2 \equiv l_5 = 30 \text{ mm}; \quad l_3 \equiv l_4 = 50 \text{ mm}$$

$$\phi_2 = 140°; \quad \phi_3 = 50°; \quad \phi_4 = 130°; \quad \phi_5 = 40°. \tag{15.7}$$

Considering an absolute coordinate system XYZ (Figure 15.2), we find the following angles corresponding to the specified coordinates (20 and 65) of point B along axes X and Y:

$$\phi_2 = 115°, \quad \phi_3 = 49°, \quad \phi_4 = 116°, \quad \phi_5 = 42°. \tag{15.8}$$

We also find the angular displacements for a given initial state in the joints:

$$\Delta\phi_2 = 25°, \quad \Delta\phi_3 = 1°, \quad \Delta\phi_4 = 14°, \quad \phi_5 = 2°. \tag{15.9}$$

15.3.3.1.2 Inverse Problem of the Kinematics of the Micro-Mechatronic System

Using the actuators, we should attain a position with coordinates $x + \Delta x$ and $y + \Delta y$, i.e., we should realize additional delta-displacement. Hence, $\Delta x = a$ and $\Delta y = b$ are known. We look for values of λ_i, $i = 2, 3, 4,$ and 5, which would realize the position specified. However, we can solve the problem using one, two, etc., operating actuators. For small values of λ in the DPK, we derive the following equations:

$$a = \sum_{2}^{5} \lambda_j i_{jx}, \tag{15.10}$$

$$b = \sum_{2}^{5} \lambda_j i_{jy}, \tag{15.11}$$

where

$$i_{2x} = c\phi_2 - \frac{s\phi_3 c(\phi_4 - \phi_2)}{s(\phi_3 - \phi_4)}$$

$$i_{3x} = c\phi_3 - \frac{s\phi_3}{tg(\phi_3 - \phi_4)}$$

$$i_{4x} = \frac{s\phi_3}{s(\phi_3 - \phi_4)}$$

$$i_{5x} = \frac{s\phi_3 c(\phi_4 - \phi_5)}{s(\phi_3 - \phi_4)} \tag{15.12}$$

$$i_{2y} = s\phi_2 + \frac{c\phi_3 c(\phi_4 - \phi_2)}{s(\phi_3 - \phi_4)}$$

$$i_{3y} = s\phi_3 + \frac{c\phi_3}{tg(\phi_3 - \phi_4)}$$

$$i_{4y} = -\frac{c\phi_3}{s(\phi_3 - \phi_4)}$$

$$i_{5y} = -\frac{c\phi_3 c(\phi_4 - \phi_5)}{s(\phi_3 - \phi_4)} \tag{15.13}$$

Obviously, the system is redundant—there are two extra DoFs. Moreover, due to the closed kinematical chain, an area where point B can be positioned is bounded by each of the actuators. Areas of autonomous operation of actuators 3, 4, and 5 are outlined in Figure 15.37.

The strategy of control of micro motions is based on the analysis of the capabilities of each actuator and on the capabilities of each couple and triad of actuators, namely, 2 and 3, 2 and 4, 2 and 5, 3 and 4, 3 and 5, 4 and 5, 3, 4, and 5, etc.

Examine the operation of actuators 3, 4, and 5. Angles φ_2 and φ_5 are constants. The change of link length affects the change of angles φ_3 and φ_4, and it is found via the following equalities:

$$\Delta\phi_3 = \frac{c_x c\phi_4 + c_y s\phi_4}{l_3 s(\phi_4 - \phi_3)}, \tag{15.14}$$

$$\Delta\phi_4 = \frac{c_x c\phi_3 + c_y s\phi_3}{l_4 s(\phi_4 - \phi_3)}, \tag{15.15}$$

where

$$c_x = -\lambda_2 c\varphi_2 - \lambda_3 c\varphi_3 + \lambda_5 c\varphi_5 + \lambda_4 c\varphi_4$$

$$c_y = -\lambda_2 s\varphi_2 - \lambda_3 s\varphi_3 + \lambda_5 s\varphi_5 + \lambda_4 s\varphi_4. \tag{15.16}$$

If actuator 3 is activated, only, it follows from (15.14) that

$$\Delta\phi_3 = -\frac{\lambda_3}{l_3 tg(\phi_4 - \phi_3)}, \quad \lambda_3 \neq 0. \tag{15.17}$$

The change of φ_3 yields a change of $\varphi_4 - \Delta\,\varphi_4'$, which is found via the condition of existence of a kinematical chain (Figure 15.3), namely

$$c\Delta\phi_4' = \frac{l_3'^2 + l_4^2 - 4l_3'^2 s^2\,\dfrac{\Delta\phi_3}{2}}{2l_3'l_4} \tag{15.18}$$

The results are similar if actuator λ_4 is activated, only. Then, the following relations are found:

$$\Delta\phi_4 = -\frac{\lambda_3}{l_3 tg(\phi_4 - \phi_3)} \tag{15.19}$$

$$c\Delta\phi_3' = \frac{l_3^2 + l_4'^2 - 4l_4'^2 s^2\,\dfrac{\Delta\phi_4}{2}}{2l_3 l_4'}, \quad \lambda_4 \neq 0. \tag{15.20}$$

It is seen from (15.17) and (15.19) that the variations of angles φ_3 and φ_4 are opposite to one another, i.e., if one angle increases the other one decreases. Thus, a total unidirectional effect is not to be sought. When activating the actuator of link 5, the variation of the two angles takes the form

$$\Delta\phi_3 = \frac{\lambda_5 c(\phi_4 - \phi_5)}{l_3 s(\phi_4 - \phi_3)}, \quad \lambda_5 \neq 0 \tag{15.21}$$

$$\Delta\phi_4 = \frac{\lambda_5 c(\phi_5 - \phi_3)}{l_4 s(\phi_4 - \phi_3)}, \quad \lambda_5 \neq 0. \tag{15.22}$$

Hence, we can look here for total effects being unidirectional with the effects of the other two actuators.

Note that when a single actuator is operating, the solutions are sought along boundary circle arcs or along concentric arcs plotted for different values of λ_j. If there are two operating actuators, all points in the outlined areas are possible solutions. This fact is illustrated in Figure 15.38.

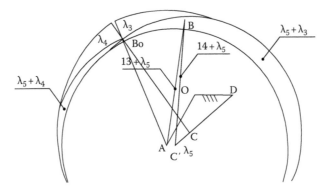

FIGURE 15.38
Working area of a five-link mechanism with two operating actuators.

A most general solution of IPK can be found via the system of Equations 15.7 and 15.8 completed by equalities

$$\operatorname{sgn} \dot{\phi}_3 = \operatorname{sgn} \Delta\varphi_3, \tag{15.23}$$

$$\operatorname{sgn} \dot{\phi}_4 = \operatorname{sgn} \Delta\varphi_4. \tag{15.24}$$

which guarantee elimination of the unidirectional looseness of the mechanical system. Velocities of links 2 and 5 of a macro MS are specified or found using the velocity of point B, while those of links 3 and 4—by differentiation of the equations of the vector contour of the kinematical chain. The following relations are found in this case:

$$\dot{\phi}_2 = \frac{l_2(\dot{x}c\phi_2 + \dot{y}s\phi_2) - (x\dot{x} + y\dot{y})}{l_2(xs\phi_2 - yc\phi_2)}, \tag{15.25}$$

$$\dot{\phi}_5 = \left[l_2(yc\phi_2 + xs\phi_2)\dot{\phi}_2 + l_2(\dot{y}s\phi_2 - \dot{x}c\phi_2) + \frac{l_5(\dot{x}c\phi_2 - \dot{y}s\phi_5)}{l_5(xs\phi_5 + yc\phi_5)} \right], \tag{15.26}$$

where

$$\dot{\phi}_3 = \frac{D_3}{D},$$

$$\dot{\phi}_4 = \frac{D_4}{D}. \tag{15.27}$$

$$D = l_3 l_4 s(\varphi_3 - \varphi_4),$$

$$D_3 = l_4(l_5\dot{\phi}_5 s(\varphi_5 - \varphi_4) + l_2\dot{\phi}_2 s(\varphi_4 - \varphi_2)),$$

$$D_4 = l_3(l_5\dot{\phi}_5 s(\varphi_5 - \varphi_3) + l_2\dot{\phi}_2 s(\varphi_3 - \varphi_2)). \tag{15.28}$$

15.3.3.2 Finite Element Solution

A finite element spatial model of the earlier-described mechanism is presented in the following text. The CAD model is created in SolidWorks environment. All linear dimensions are assumed to be equal to: $l_1 = 20\,mm$; $l_2 \equiv l_5 = 30\,mm$; $l_3 \equiv l_4 = 50\,mm$ (Figure 15.39). The links are made of aluminum alloy. The traced vertex is situated in hinge B. As this is a mechanism with two redundant DoFs, there are two rotary drives in hinges O and D. The piezo-actuators are situated in the holes of the links. They can increase the length of each link up to 1/100.

Three different cases of driving the mechanism are compared through SolidWorks Motion toolbox (SolidWorks Motion Studies, 2011). The trace paths of the vertex B, where an operating clamp is situated, and the driving laws for these three cases are given in Figure 15.40. In order to enable the entire circle motion of link 5 (CD) in case 3, the left part of the immovable link 1 (in point O) is flipped, compared to the cases 1 and 2. The rotary drives in cases 1 and 2 operate in oscillating laws, while the ones in case 3 are 5-power polynomials fitted through predefined points. The law of angular velocity and the law of angular acceleration of the two rotary drives in case 3 are also shown. The dead weight of the links as well as a loading force of 1 N, modeling the action on the mechanism of the object operated by the clamp, are regarded in calculating the reaction in the hinges. By that time the friction in the hinges is neglected as it is too small and this is the first iteration step of the promoted numerical simulation.

If it is accepted that due to the action of piezo-actuators the lengths of the links have increased with 1/100 up to $l_1 = 20\,mm$; $l_2 \equiv l_5 = 30.03\,mm$; $l_3 \equiv l_4 = 50.05\,mm$, the trace path of the vertex B changes a little. Mathematically this difference can be described through the following functions:

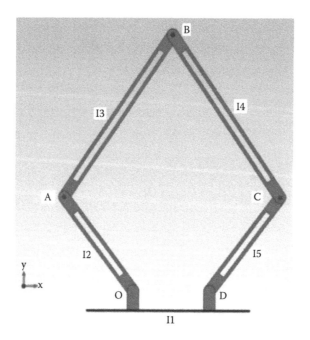

FIGURE 15.39
CAD/CAE model of the mechanism described in Section 15.3.3.

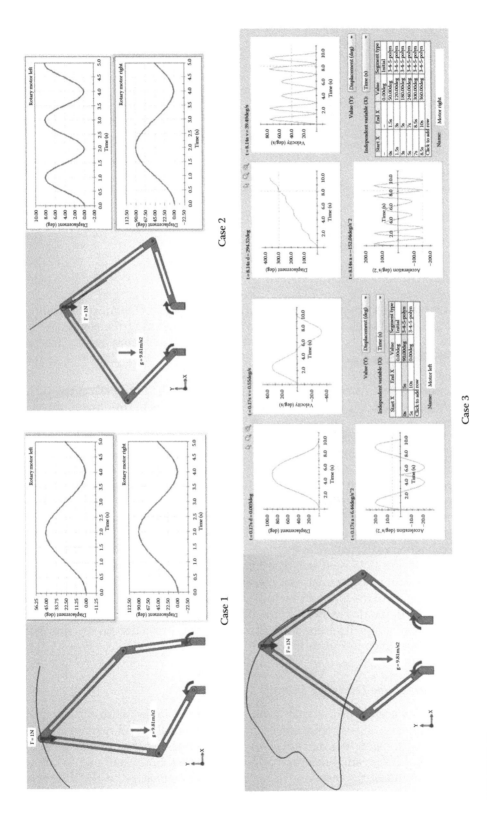

FIGURE 15.40
Trace paths of vertex D and laws of rotary drives for the three studied cases.

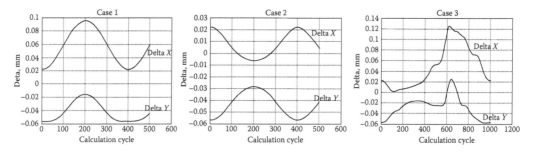

FIGURE 15.41
Difference functions of X and Y coordinates of trace paths of vertex D, due to action of piezo-actuators.

$$\Delta X = X_{WithoutPiezo} - X_{WithPiezo};$$

$$\Delta Y = Y_{WithoutPiezo} - Y_{WithPiezo};$$

$$(15.29)$$

Their graphs for the three discussed cases are given in Figure 15.41. It must be admitted that the projection axes X and Y, used in Figure 15.41 for comparing the results are not parallel to the axes of the global coordinate system shown in Figure 15.40. They are mutually perpendicular axes, whose plane orientation depends on the initial configuration of the mechanism.

The graphs of functions (15.29) prove that the coordinates of the trace path of the vertex can be precisely controlled, i.e., in the range of a few hundredths of millimeter, through piezo-actuators.

The linear kinematic parameters of the motion of vertex B (linear displacement, linear velocity, and linear acceleration) for the last two of the studied three cases are shown in Figures 15.42 and 15.43a. The differences between the values of these functions due to the action of piezo-actuators and the elongation of the links with 1/100th of their length are shown in the right side of corresponding figures.

As the spatial orientation of the operated object is important for the researchers, the angular motion of the clamp in vertex B is also examined. The graphs of rotation, angular velocity, and angular acceleration of the fixture of the clamp for the two cases presented in detail are given in Figures 15.44 and 15.45—left side. The comparison between the values of the function without or with acting piezo-actuators is presented at the right side of the same figures.

Due to the dead weight of the links of the studied mechanism and the dead weight of the operated object some dynamic reactions in the hinges of the mechanism are raised. The graphs of the reaction in hinge B joining links 13 and 14 and its projection on two mutually perpendicular axes are given in Figure 15.46.

15.3.3.3 Conclusions on the Preceding Example

Studying all provided graphs proves that through the use of linear piezo-actuators the kinematic parameters of a chosen point can be precisely controlled and managed. These examples do not aim to show the most even or the smoothest graphs of all kinematic parameters, including trace path, linear and angular velocities, linear and angular acceleration of the clamp and of the reactions in hinges, but to emphasize on the difference provoked by the operation of the piezo-actuators.

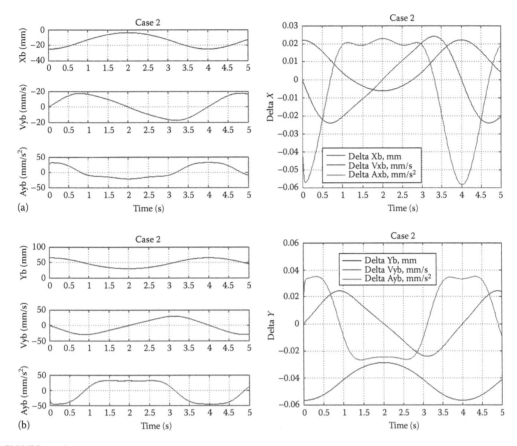

FIGURE 15.42
Kinematic parameters of motion of vertex D—case 2: (a) kinematic parameters along axis X and (b) kinematic parameters along axis Y.

15.3.4 Analysis and Synthesis of Mechanical Bioreactors for Bio-Nanotechnologies

15.3.4.1 Short Review of Existing Devices for Bioreactor Systems

Human body consists of over 200 types of cells which assemble organs such as skin, bones and muscles. In the middle of the last century the molecular biologist Edmund B. Wilson wrote in his book *The Cell in Development and Heredity* that "the key to every biological problem must finally be solved in the cell." Cells are about five times smaller than the smallest visible particle and they contain all the molecules necessary for an organism to live and reproduce. This fact prevents the scientists from seeing their structure, disclosing their molecular composition and understanding how their various components are functioning. Therefore, the "in vivo" methods cannot give an answer to these problems. Growth of human cells "in vitro," outside living organisms, allows for the investigation of basic biological and physiological phenomena such as controlling the normal life cycle and many of its mechanisms. The design of 3D cell cultures is necessary which are eligible for medical implants. Numerous experiments have shown that this is impossible using stationary bioreactor systems. It is also difficult to achieve with the known rotating bioreactors under the conditions on earth.

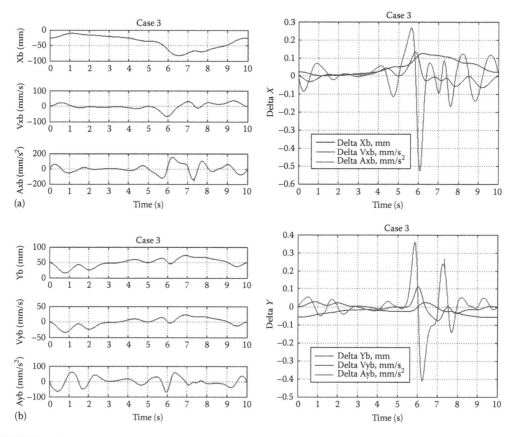

FIGURE 15.43
Kinematic parameters of motion of vertex D—case 3: (a) kinematic parameters along axis X and (b) kinematic parameters along axis Y.

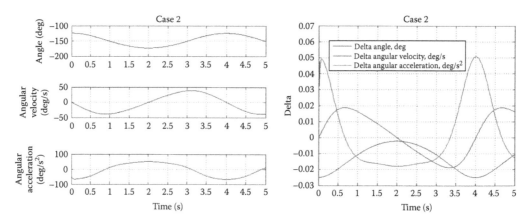

FIGURE 15.44
Angular kinematic parameters of motion of vertex D—case 2.

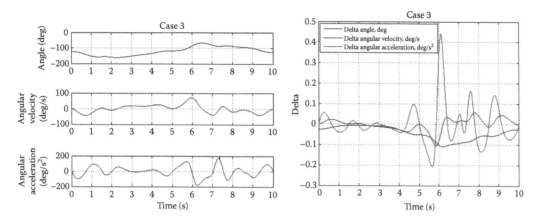

FIGURE 15.45

Angular kinematic parameters of motion of vertex D—case 3.

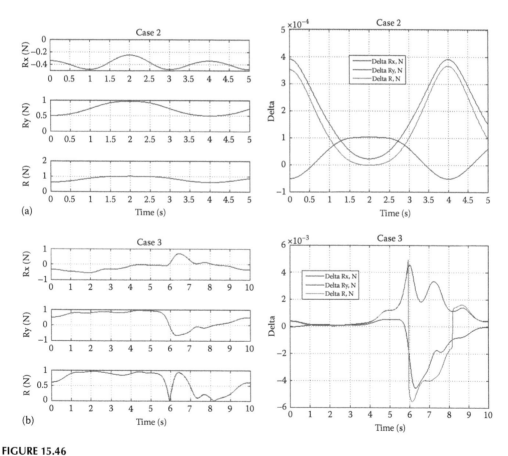

FIGURE 15.46

Reactions in hinge B, due to dead weight of the links and of the operated object: (a) reaction in hinge B—case 2 and (b) reaction in hinge B—case 3.

(a)

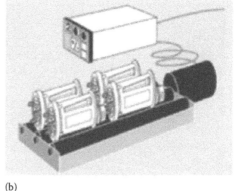
(b)

FIGURE 15.47
Rotating bioreactor with natural circulation of fluid. (a) A photo of NASA'S RWV (rotating wall vessel) bioreactor and (b) miniPERM bioreactor.

FIGURE 15.48
Photo of a rotating bioreactor by force fluid circulation.

The experiments done in simulated weightlessness achieved by free falling of an airplane are successful to some extent [Wolf A.—the study was funded by NASA], as well as those using microcarriers which play the role of skeleton constructions to which the cells are attached. Manley P. and Lelkes P. have developed an experimental device to study the motion of cell formations in rotating bioreactors, also funded by NASA. The more popular, commercially available rotating bioreactors with natural fluid circulation are shown in Figure 15.47a,b, and Figure 15.48 shows forced-circulation-loop bioreactors.

The main disadvantage of the existing rotating bioreactors is the permanent orientation of bioreactor's axis of rotation to the acceleration of gravity. This fact causes the precipitation of cell cultures, their clinging to the bioreactor walls that have fatal ending, and the inadequate exchange of substances between cells and media.

The dramatic advance in the fields of biochemistry, cell and molecular biology, genetics, medicine, biomedical engineering, and material science gave rise to the development of interdisciplinary scientific areas such as tissue engineering and the solution of organ problems by medical implants. To achieve satisfactory results in cell and tissue cultures, bioreactors have to operate under conditions as close as possible to "in vivo" conditions. The difficulties occurring with the known bioreactors are that they cannot provide a constant and regulated feeding and metabolic bioproducts removal. The growth of 3D cell formations requires physical and chemical bonds. The chemical bonds are accomplished mainly through the culture medium components. The physical bonds for the growth of cell and tissue cultures require the use of a bioreactor.

The known bioreactors are designed with only one axis of rotation. They put the cell growth under the influence of one force vector only, due to which they provide a physical signal only in the direction of this one force vector. Hence, cells are not inclined to growth of 3D cell cultures.

The bioreactor systems have controllable motors and monitoring sensors to control the processes in the bioreactor chamber, i.e., they are mechatronic systems. The aim of this study is mechanical devices, of bioreactor devices with spatial mechanisms. Spatial mechanisms with CKC are most often used for transmission. The output link (OL) performs a rotational motion (rocker or crank). This is the reason why in the traditional literature on the theory of mechanisms and machines, the kinematic and dynamic analysis of spatial motions of coupler links is poorly covered or not mentioned at all. We analyze the possibilities of implementation of spatial mechanisms with one or two DoFs with CKC and OL, i.e., a coupler in bioreactor devices, which required the extensive analysis of coupler's spatial motion. Furthermore, the possibility is studied of using the coupler's rotation around its own axis when both of the kinematic pairs (KP) to which it is linked are spherical ones of the third class.

15.3.4.2 Kinematics of Spatial Mechanisms in Biotechnologies

In Figure 15.49, the kinematic scheme is shown of a spatial four-bar linkage (SFL), wherein the planes of motion of the rotating links 2 and 4 make an angle α. This mechanism can function adequately if axes A and D are connected to the frame through bearings—KP of the fifth class, the joints B and C are spherical of the third and the fourth class (the latter is spherical with a pawl). If both KP are of the third class, then the rotation around the coupler's own axis is possible. In the traditional application of these mechanisms, this additional DoF is unwanted. The coupler 3 performs the spatial motion. The bioreactor chamber is mounted to it.

Figure 15.50 shows the kinematic scheme of an SFL, where the angle α is 90°, i.e., the trajectories of joints B and C are onto the planes Oxy and Oyz.

The scales (values) and angles necessary for the kinematic analysis are given in Figure 15.51.

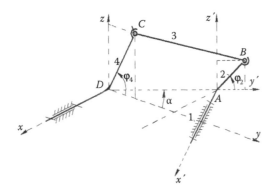

FIGURE 15.49
The spatial four-bar linkage wherein the planes of motion of the rotating links 2 and 4 make an angle α.

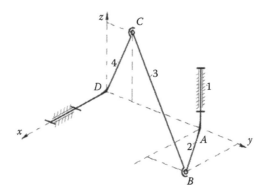

FIGURE 15.50
Kinematic scheme of a spatial four-linkage mechanism, $\alpha = 90°$.

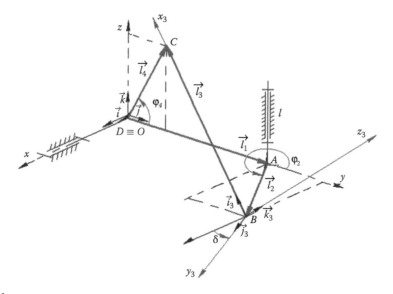

FIGURE 15.51
Kinematic scheme of the spatial four-bar linkage, wherein the planes of motion of the rotating links 2 and 4 make an angle $\alpha = 90°$.

The purpose of the kinematic analysis is to determine the position and velocity characteristics of the coupler's motion. The method of closed-loop vectors is applied. The position relations are

$$\vec{l_1} + \vec{l_2} + \vec{l_3} = \vec{l_4} \tag{15.30}$$

$$l_2 \sin \varphi_2 + l_3 \cos \alpha_3 = 0$$

$$l_1 + l_2 \cos \varphi_2 + l_3 \cos \beta_3 = l_4 \cos \varphi_4$$

$$l_3 \cos \gamma_3 = l_4 \sin \varphi_4, \tag{15.31}$$

out of which the unknown angle parameters of the coupler are determined:

$$\cos \varphi_4 = \frac{c}{a} \tag{15.32}$$

where

$$c = l_1^2 + l_2^2 - l_3^2 + l_4^2 + 2l_1 l_2 \cos \varphi_2$$

$$a = 2l_1 l_4 + 2l_2 l_4 \cos \varphi_2 \tag{15.33}$$

The direction cosines of the vector BC are

$$a_{11} = \cos \alpha_3 = -\frac{l_2}{l_3} \sin \phi_2$$

$$a_{21} = \cos \beta_3 = -\frac{l_1 + l_2 \cos \phi_2 - l_4 \cos \phi_4}{l_3}$$

$$a_{31} = \cos \gamma_3 = \frac{l_4 \sin \phi_4}{l_3} \tag{15.34}$$

The components of the coupler's angular velocity in its own coordinate system $Bx_3 y_3 z_3$ and in the absolute coordinate system $Oxyz$ are

$$\omega_{x_3} = -\frac{\cos \gamma_3}{\sin^3 \gamma_3} \left[\cos \alpha_3 \left(\dot{\beta}_3 \sin \beta_3 \sin \gamma_3 + \dot{\gamma}_3 \cos \beta_3 \cos \gamma_3 \right) \right.$$

$$\left. - \cos \beta_3 \left(\dot{\alpha}_3 \sin \alpha_3 \sin \gamma_3 + \dot{\gamma}_3 \cos \alpha_3 \cos \gamma_3 \right) \right]$$

$$\omega_{y_3} = -\frac{\cos \gamma_3}{\sin \gamma_3} \left(\dot{\alpha}_3 \cos \alpha_3 \sin \alpha_3 + \dot{\beta}_3 \cos \beta_3 \sin \beta_3 \right) + \dot{\gamma}_3 \left(1 - \sin^2 \gamma_3 \right)$$

$$\omega_{z_3} = \frac{1}{\sin \gamma_3} \left(\dot{\alpha}_3 \sin \alpha_3 \cos \beta_3 - \dot{\beta}_3 \cos \alpha_3 \sin \beta_3 \right) \tag{15.35}$$

$$\omega_x = \frac{\omega_z \cos \alpha_3 + \dot{\beta}_3 \sin \beta_3}{\cos \gamma_3}$$

$$\omega_y = \frac{\left(\dot{\alpha}_3 \sin \alpha_3 - \omega_z \cos \beta_3 \right) \sin \gamma_3}{\cos \alpha_3 \cos \gamma_3}$$

$$\omega_z = \frac{\dot{\alpha}_3 \sin \alpha_3 \sin \gamma_3 + \dot{\gamma}_3 \cos \alpha_3 \cos \gamma_3}{\sin \gamma_3 \cos \beta_3} \tag{15.36}$$

From relations (15.35) and (15.36), it follows that

- All the components of the coupler's velocity are periodic functions of the angle of rotation of the input shaft.
- A full rotation of the coupler around its own axis is not possible if the mechanism has one DoF.
- If both KP are from the third class (the mechanism has two DoFs), another actuator may be used, with the help of which the full rotation of the coupler, or the bioreactor chamber, respectively, will be performed around its own axis.

Several numerical experiments with varying input data are performed in MATLAB environment. The dimensions of the units provide full revolution of the driving unit. The graphs in Figure 15.52 show the results of simulations of the motion of the mechanism with the following input data: $l_1 = 0{,}3\,\text{m}$; $l_3 = 0{,}3\,\text{m}$; $l_4 = 0{,}15\,\text{m}$; $\omega_2 = 1\text{s}^{-1}$; and varying l_2. The functions of the climate projections of the angular velocity vector of the connecting rods in fixed coordinate system $Dxyz$ are shown in Figure 15.53.

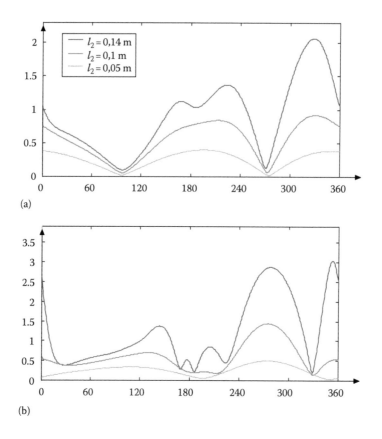

(a)

(b)

FIGURE 15.52
Graphs of angular velocity and angular acceleration of rod 13 versus the rotation of the driving unit. (a) Angular velocity of rod 13 during the full revolution of the driving unit for different values of 12. (b) Angular acceleration of rod 13 during the full revolution of the driving unit for different values of 12.

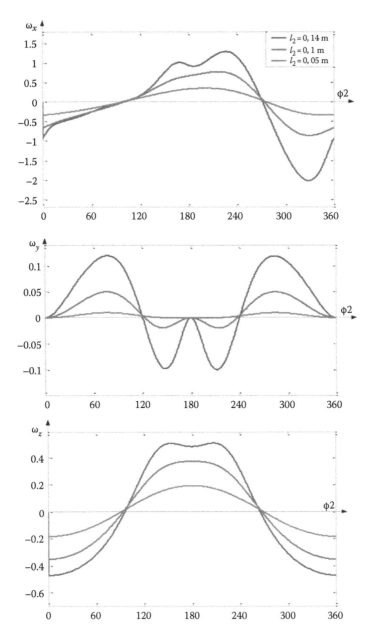

FIGURE 15.53
Projection of $\vec{\omega}_3$ about $Dxyz$.

15.3.4.3 Conclusions on the Preceding Example

The innovative solutions in this study are

- Use of the spatial motion of the coupler from the traditional SFL as an OL and in particular, as a carrier of the bioreactor chamber.
- Use of SFL with two spherical joints and two DoFs, wherein the full rotation of the coupler or the bioreactor chamber, respectively, around its own axis is performed

by a second actuator. The module of the bioreactor chamber with the actuator is connected to the coupler of the mechanism through a fixed connection.

- With reference to these solutions, detailed kinematic analyses are of the spatial motion of the SFL coupler, which are not available in the traditional literature on the theory of mechanisms and machines because of the use of these mechanisms for transmission between the two links with elementary rotating motions.

References

Beale M., M. Hagan, and H. Demuth. 2012. *Neural Network Toolbox™ , User's Guide for* Version 7.0.3 (Release 2012a), online only, http://www.mathworks.com/help/pdf_doc/nnet/nnet_ug.pdf

Chanhee Han et al. 2002. Kinematic sensitivity analysis of the 3-UPU parallel mechanism, *Mechanism and Machine Theory* 37, 787–798.

Conconi M. et al. 2009. A new assessment of singularities of parallel kinematic chains, *IEEE Transactions on Robotics* 25(4), 757–770, August.

Hagan M., H. Demuth, and M. Beale. 2002. *Neural Network Design*. Distributed by Campus Publishing Service, Colorado University Bookstore, Boulder, CO.

Han Sung Kim. 2005. Kinematic calibration of a Cartesian parallel manipulator, *International Journal of Control, Automation and Systems* 3(3), 453–460, September.

Haykin S. 1998. *Neural Networks: A Comprehensive Foundation*, 2nd edn., Prentice Hall, Englewood Cliffs, NJ.

Ionescu Fl., K. Kostadinov, and R. L. Hradynarski. 2004. Operation and control of a 6,5 DoF micro and nano robot for cell manipulations. In *Proceedings of the Mechatronics and Robotics Conference*, Aachen, Germany.

Kanaan D. et al. 2009. Singularity analysis of lower mobility parallel manipulators using Grassmann-Cayley algebra, *IEEE Transactions on Robotics* 25(5), 995–1004.

Klocke V. and T. Gesang. 2002. Nanorobotics for micro production technology, *Proceedings the International Society for Optical Engineering*, 4943, 132–141.

Kostadinov K., R. Iankov, Fl. Ionescu, and P. Malinov. 2004. Analysis of 2 DoF Piezo Actuated joint utilizing in the robot for micro and nano manipulations, *Romanian Journal of Applied Mechanics* 49, 413–416.

Kostadinov K., Fl. Ionescu, R. L. Hradynarski, and T. Tiankov, 2005a. Robot based assembly and processing micro/nano operations. In *Proceedings of the 4M2005, 1st International conference on Multi- Material Micro Manufacture*, Karlsruhe, Germany.

Kostadinov K., R. Kasper, and M. Al-Wahab. 2005b. Control approach for structured piezo actuated micro/nano manipulator. In *IUTAM Symposium*, Munich, Germany.

MATLAB Documentation for R 2012, online source, http://www.mathworks.com/help/techdoc/

Petrova R., P. Genova, and K. Kostadinov. 2011. Design of special clamp for manipulating with tools, micro and nano objects, optimization of the robots and manipulators, In *International Proceedings of Computer Science and Information Technology (IPCSIT)*, 8, 49–54, online source: http://www.ipcsit.com/vol8/8-S2.3.pdf

Sadjadian H. et al. 2005. Kinematic analysis of the hydraulic shoulder: A 3-dof redundant parallel manipulator. *Proceedings of the IEEE, International Conference on Mechatronics and Automation*, Niagara Falls, Ontario, Canada, July 2005.

SolidWorks—Users Manual, 2012. Dassault Systemes SolidWorksCorporation, online source: http://help.solidworks.com/

SolidWorks Motion Studies, 2012. Dassault Systemes SolidWorksCorporation, on-line reference guide to software tool SolidWorks Motion

SolidWorks Simulation—Training Manual, 2012. Dassault Systemes SolidWorksCorporation, Concord, MA; on-line reference guide to software tool SolidWorks Simulation and online source: http://www.solidworks.com/sw/support/808_ENU_HTML.htm

Tiankov T., A. Shulev, N. Elian, D. Chakarov, I. Roussev, M. Al-Wahab, and K. Kostadinov. 2006a. Experimental motion study of the mehchatronic system for manipulation and processing micro/nano operations. In *Nanoscience and Nanotehcnology*, Vol. 6, Heron Press Science Series, Sofia, Bulgaria.

Tiankov T., A. Shulev, D. Gotzeva, I. Roussev, and K. Kostadinov. 2006b. Investigation of piezo actuated micro/nano manipulator by digital speckle photography and interferometry. In *Proceedings of the International Conference on Automatics and Informatics' 06*, Sofia, Bulgaria.

Zienkiewicz O. C. 1972. *Finite Element Method in Engineering Science*, 2nd edn., McGraw-Hill Inc., New York.

16

Modeling at the Nano Level: Application to Physical Processes

Serge Lefeuvre

Eurl Creawave, Toulouse, France

Olga Gomonova

Siberian State Aerospace University, Krasnoyarsk, Russia

CONTENTS

16.1 Introduction...559
16.2 Filtration: An Ideal and a Fictitious Soil..560
    16.2.1 Porosity of Fictitious Soil...561
16.3 Meshing of Geometry Objects: Examples in 2D....................................567
    16.3.1 Examples of Objects..567
16.4 3D Modeling: Examples...575
16.5 3D and Capillaries..580
16.6 Fluid Flow in the Capillaries...580
16.7 Microwave Heating..581
16.8 Conclusion...582
References..583

16.1 Introduction

The processing of heterogeneous materials was, from the beginning of human activity, the fruit of experiments transmitted as hand-turns or empiric expressions. Nowadays, experiments remain compulsory but are integrated into more accurate descriptions such as finite elements description. The partial differential equations, solved in orthogonal spaces, are now solved in more complicated geometries thanks to finite element method (FEM), but the constants characteristic of heterogeneous materials keep usually a touch of empiricism. For instance, the use of polynomial approximations is still largely spread.

The modeling at the nano level is an attempt to achieve a more precise description of the blend, even if it remains impossible to describe the exact geometry of each nano grain. Apart from preserving the grain proportion of each component, the grain to grain description opens the way to the description of the surface activities. This point is crucial since more the volume is small, more its surface is active.

For a long time, the nano-level approach was understood as a pileup of spheres as shown in Figure 16.1.

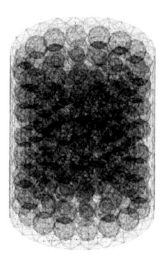

FIGURE 16.1
Model of a granular material.

One of the best ways of modeling heterogeneous materials respecting granulometry is to work at the level of nano grains. Mostly, during the investigation of materials and their properties, researchers are interested in calculating the main characteristics of these materials, such as porosity, permeability, and permittivity [1,7].

Working at the nano level implies working on large numbers of objects. One way to draw these objects is to start from meshes already used in FEM techniques. Meshes are very attractive because they are automatically produced and because they completely fill the domain. At last, because the mesher reduces the dispersion of the meshes to get a fair description, the use of nano objects deduced from these meshes leads to a monodisperse compound of nanoparticles.

In this chapter, it is shown how to transform meshes into objects and how to modify them just using matrix analysis. The size and the number of the objects taken into consideration are restricted by the memory number and the computational time of the computer and also by the need for readable figures. In most examples, 200–300 objects are taken into account depending on the number of PDEs to be solved after a general re-meshing.

The chapter is divided into two main parts: 2D and 3D modeling. Both of them begin with raw objects, straight transposition of meshes, and present modifications, namely, homothetic reduction to get capillaries, grain joins, and so on.

In dealing with large numbers of objects, a pioneer was Leibenzon who worked on the porosity of soils to understand, among others, the process of filtration. Some principal well-known facts on the theory of granulated materials rest on his famous works, e.g.; on Ref. [6].

16.2 Filtration: An Ideal and a Fictitious Soil

Particles of natural fluids in natural soil [6] (such as water, gas, oil) move through the pores of the soil; that is, these particles are transferred through the finest channels which are formed by the not closely contacted grains of the soil. Such kind of fluid motion in the soil is called *filtration* [2,3].

Viscosity of fluid is very important because of utterly small cross section of pores and because of slow velocities of fluids inside the pores. As a rule, the motion inside the pores is accepted to be laminar, but it can be like vortical transfer because of curving canals and modification of their cross section. Seeing that soil particles are the granules of awkward shapes and different sizes, it is possible to find solution of equations which describe the motion of the viscous fluid in such kind of medium. That is why some simplified models were constructed.

There are two kinds of soils: an ideal and a fictitious. In case of ideal soil, all pores are considered as cylindrical, and axes of these cylinders are parallel to each other. In case of fictitious soil, all its granules are supposed to be spheres of the same diameter.

16.2.1 Porosity of Fictitious Soil

Assume that we have some natural soil of volume V_1. All the granules of soil occupy volume V_2 of the volume V_1. Hence, a volume of pores in the V_1 equals $V_3 = V_1 - V_2$.

A value

$$m = \frac{V_3}{V_1} = \frac{V_1 - V_2}{V_1} = 1 - \frac{V_2}{V_1}$$

is called *porosity*. It is obvious that $0 < m < 1$.

Slichter determined a value of porosity of fictitious soil by means of simple geometrical way [6,8,9]. The value of porosity obviously depends on configurations of the spheres (which represent grains of the soil). As all the spheres have got the same size, the distance between the centers of two of the closest spheres equals to diameter of the sphere. Therefore, centers of each eight contacting spheres are situated in summits of a parallelepiped, all planes of which are rhombuses (Figure 16.2).

This rhombohedron is a basic model of the fictitious soil in Slichter's method. Studying of the geometrical properties of this model can give a possibility to calculate the value of porosity m. Different dispositions of the soils spheres have two limiting states. One of these states corresponds to the closest contact of spheres, another one implies not so close contact. But in both the situations, the spheres are contiguous.

It is evident that angle θ of the rhombohedron planes is included into interval [60°; 90°] (Figure 16.3). For every angle of the rhombohedron, there is another angle which is added up to 180°. That is why eight pieces of full spheres which are cut from eight concerned spheres form one whole sphere.

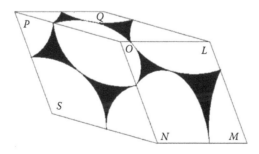

FIGURE 16.2
Model of a fictitious soil: rhombohedron. (After Leibenzon, L.S., *Motion of Natural Fluids in Porous Medium,* Technical and Theoretical Literature Publishing, Moscow, Leningrad, 1947; Slichter, C.S., Theoretical investigation of the motion of ground water, U.S. Geological Survey 19th Annual Report, Part II, 1899; Slichter, C.S., The motions of underground waters, U.S. Geological Survey Water Supply Paper 67, 1902.)

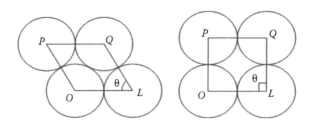

FIGURE 16.3
Limited dispositions of spheres.

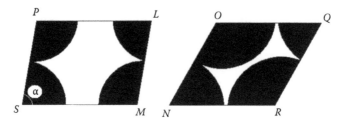

FIGURE 16.4
Diagonal cross sections of the rhombohedron.

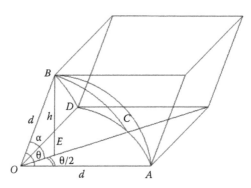

FIGURE 16.5
Intersections of the full sphere.

In Figure 16.4, there are diagonal sections *SPLM* and *NOQR* of the rhombohedron.

Let us obtain a value of the angle α of the parallelogram *SPLM*, for example. For this, we will circumscribe a full sphere from the vertex *O* of the rhombohedron. The radius of the sphere equals *d* (Figure 16.5).

The diagonal section, jointly with the faces *OAD* and *OAB*, crosses this full sphere in arcs which form spherical right-angled triangle *ABC* with right angle *BCA*. Perpendicular *BE* dropped on diagonal *OC* is the altitude *h* of the rhombohedron. From the triangle *ABC* follows

$$\cos AB = \cos BC \cdot \cos AC.$$

But $\overset{\frown}{AB} = \theta$, $\overset{\frown}{AC} = \dfrac{\theta}{2}$, $\overset{\frown}{BC} = \alpha$,

Hence,

$$\cos\alpha = \frac{\cos\theta}{\cos\dfrac{\theta}{2}}. \tag{16.1}$$

From (16.1), we get

$$\sin\alpha = \sqrt{1-\cos^2\alpha} = \sqrt{\frac{\cos^2\dfrac{\theta}{2}-\cos^2\theta}{\cos^2\dfrac{\theta}{2}}} = \sqrt{\frac{\sin^2\theta-\sin^2\dfrac{\theta}{2}}{\cos^2\dfrac{\theta}{2}}} = \sqrt{\frac{4\sin^2\dfrac{\theta}{2}\cos^2\dfrac{\theta}{2}-\sin^2\dfrac{\theta}{2}}{\cos^2\dfrac{\theta}{2}}} =$$

$$= \frac{\sin\dfrac{\theta}{2}}{\cos\dfrac{\theta}{2}}\sqrt{4\cos^2\dfrac{\theta}{2}-1}.$$

As $4\cos^2\dfrac{\theta}{2}-1 = 2(1+\cos\theta)-1 = 1+2\cos\theta$, that

$$\sin\alpha = \tan\frac{\theta}{2}\sqrt{2\cos\theta+1}.$$

From this expression, we obtain finally

$$\sin\alpha = \frac{2\sin\dfrac{\theta}{2}\cos\dfrac{\theta}{2}}{2\cos^2\dfrac{\theta}{2}}\sqrt{2\cos\theta+1} = \frac{\sin\theta}{1+\cos\theta}\sqrt{2\cos\theta+1}. \tag{16.2}$$

Further, from the right-angled triangle *BEO*, we can find

$$h = d\sin\theta. \tag{16.3}$$

As area of the base of the rhombohedron is $d^2\sin\alpha$, then volume of the rhombohedron equals

$$V_1 = hd^2\sin\theta.$$

With respect to formulas (16.2) and (16.3), the last expression becomes

$$V_1 = \frac{d^3\sin^2\theta\sqrt{1+\cos\theta}}{1+\cos\theta}. \tag{16.4}$$

V_2 is a sum of all eight pieces of full spheres located inside the rhombohedron; and as it was stated earlier, V_2 equals to the volume of one whole sphere:

$$V_2 = \frac{\pi d^3}{6}. \tag{16.5}$$

The porosity m, with respect to formulas (16.4) and (16.5), becomes

$$m = 1 - \frac{V_2}{V_1} = 1 - \frac{\frac{\pi d^3}{6}(1 + \cos\theta)}{d^3 \sin^2\theta\sqrt{1 + \cos\theta}}.$$

Substitute $\sin^2\theta = (1 - \cos\theta)(1 + \cos\theta)$ into the last expression and obtain the fundamental Slichter's formula:

$$m = 1 - \frac{\pi}{6(1 - \cos\theta)\sqrt{1 + 2\cos\theta}}. \tag{16.6}$$

It follows from the Slichter's formula that porosity of fictitious soil, which consists of spherical particles, does not depend on diameters of these particles; it depends only on their disposition and value of the angle θ.

The limit values of the angle θ are $60°$ and $90°$; therefore, an interval of theoretical porosity, taking into account formula (16.6), is

$$0.259 \le m \le 0.476.$$

As one can see from Figure 16.2, the area of a free space among the full spheres in a plane which contains centers of these spheres, equals S:

$$S = S_1 - S_2,$$

where
S_1 is an area of a rhombus
S_2 is a sum of areas of the circles' parts inside this rhombus

It is easy to see that all four parts of the circles inside the rhombus form one whole circle with area

$$S_2 = \frac{\pi d^2}{4}.$$

The area of rhombus S_1

$$S_1 = d^2 \sin\theta.$$

Hence,

$$S = \left(\sin\theta - \frac{\pi}{4}\right)d^2.$$

Slichter introduced the following relation

$$n = \frac{S}{S_1} = 1 - \frac{S_2}{S_1}$$

and called it *free space*. This value describes the area of a fluid which goes through the narrowest place of a pore channel. Taking into consideration expressions for S_1 and S_2, the value of n equals

$$n = 1 - \frac{\pi}{4\sin\theta}.$$

(16.7)

and it includes into the interval

$$0.0931 \le n \le 0.2146.$$

From the formula (16.7), one can see that the fictitious soil value of n does not depend on diameters of the granules.

These results were obtained by Slichter. We decided to improve his results, because we want to take into consideration not only the closest spheres' contact but also the different sizes of spheres. That is why we can inscribe spheres of a smaller diameter into the free space among the granules. We consider the following model.

If we take into consideration not only the closest disposition of soil granules but also different sizes of these granules, it will be able to inscribe new spheres of a smaller radius into free space among the granules. So, we can get two variants of dislocation of the grains (Figure 16.6a and b).

Let us obtain values of *free space* (n) and porosity (m) for the first variant of dispositions of the grains (Figure 16.6a). Let R be the radius of the grain of bigger size and r be the radius of a smaller grain. Based on uncomplicated mathematical calculations, one can obtain that

$$r = (\sqrt{2} - 1)R,$$

area of the square $OPQL$ is

$$S = 4R^2,$$

and value of the free space among the grains is

$$S_1 = 4R^2 - \pi R^2 - \pi r^2.$$

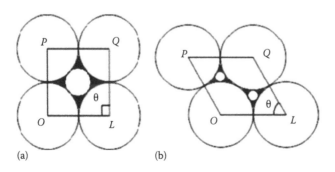

(a)   (b)

FIGURE 16.6
Improved model.

Further, volume of a cube, vertexes of which are located in the centers *P, Q, L, O* of whole spheres, equals

$$V = 8R^3,$$

and value of a free space among them is

$$V_1 = 8R^3 - \frac{4}{3}\pi R^3 - \frac{4}{3}\pi r^3.$$

Finally, one can obtain values of *free space* and porosity, respectively:

$$n = \frac{S_1}{S} = \frac{4R^2 - \pi R^2 - \pi R^2(\sqrt{2}-1)^2}{4R^2} = 1 - \frac{2-\sqrt{2}}{2}\pi,$$

$$m = \frac{V_1}{V} = \frac{8R^3 - \frac{4}{3}\pi R^3 - \frac{4}{3}\pi R^3(\sqrt{2}-1)^3}{8R^3} = 1 - \frac{5\sqrt{2}-6}{6}\pi.$$

Applying the same way of reasoning to the second case of dislocation of the grains (Figure 16.6b), one can obtain the following meanings of correspondent values:

$$r = \left(\frac{2\sqrt{3}}{3} - 1\right)R;$$

$$S = 2\sqrt{3}R^2;$$

$$S_1 = S - \pi R^2 - 2\pi r^2;$$

$$V = 4\sqrt{3}R^3,$$

$$V_1 = V - \frac{4}{3}\pi R^3 - \frac{8}{3}\pi r^3.$$

And values of *free space* and porosity in this case are as follows:

$$n = \frac{S_1}{S} = \frac{2\sqrt{3}R^2 - \pi R^2 - 2\pi R^2\left(\frac{2\sqrt{3}}{3}-1\right)^2}{2\sqrt{3}R^2} = 1 - \frac{17-8\sqrt{3}}{6\sqrt{3}}\pi,$$

$$m = \frac{V_1}{V} = \frac{4\sqrt{3}R^3 - \frac{4}{3}\pi R^3 - \frac{8}{3}\pi R^3\left(\frac{2\sqrt{3}}{3}-1\right)^3}{4\sqrt{3}R^3} = 1 - \frac{52\sqrt{3}-81}{27\sqrt{3}}\pi.$$

16.3 Meshing of Geometry Objects: Examples in 2D

The aforementioned analytical approach is very powerful but remains limited because it is unable to describe any size of pores, up to no pores at all.

Working on the model represented in Figure 16.1, one can confront some difficulties. One of them arises during changing the value of porosity: It is possible but complicated to get the needed value using only the spheres and circles, even varying radii. And in this case, the real value of "opened" porosity seems to be very approximate. Another problem follows from the previous one—time of calculating and simulation. Even working with 2D model (which is in fact an intersection of the sample), it takes too much time to resolve the given task. Therefore, taking into consideration all these conditions, authors decided to construct and work with another model of grain material.

To construct the corresponding model, the authors dealt with any automatic meshing, used in FEM processes, which provides the user with all the necessary mathematical tools.

Among these tools are matrix of coordinates of nodes and matrix of meshes. Using these matrices, it is possible to process all the needed geometrical transformations: reconstruction of the grains elements, homothetic transformations, displacement, etc., to match all the elements contained in a heterogeneous material which are already known by using electronic microscopy [5].

The meshes used in FEM open the way to a new type of description, since they let absolutely no vacuum. The counterpart is that the meshes have an angular geometry as, for instance, triangles or tetrahedrons. To counterbalance this inconvenience, it is possible to include, before meshing, objects of given shape such as circles or spheres or any handmade form built using Bezier method. Bezier method is very useful in shaping because it traces smooth shapes with a minimum number of variables and so facilitates later computations.

The geometric objects are produced by meshing: Each mesh is transformed into an object and then adjusted as needed using geometric transformation. This is done using Comsol-MATLAB® which exchanges their data quite easily: Comsol provides a *mesher* and a *solver* and *exports* data matrices to MATLAB which shapes and draws objects which, in turn, will be imported by Comsol.

16.3.1 Examples of Objects

This part of the paragraph gives different examples in order to show the flexibility of the method.

As it has already been said, the information about mesh and geometric objects is imported by Comsol Multiphysics into the main matrices: Pt, the matrix of nodes (i.e., the coordinates of each node), and Tr, matrix of meshes (which contains the list of all the nodes belonging to each mesh). For convenience, these matrices are contracted into one matrix TrP which contains, instead of the number of the nodes, the set of the coordinates, classified as x, y, and z. This matrix also has the interest to ignore the doubles, triples, etc., sometimes introduced by the mesher for its own convenience.

The matrix TrP contains all the necessary information to reshape and draw the objects.

The objects can be modified as needed. For instance, it is possible to reduce them by homothetic transformation in order to let appear capillaries or to join two neighbors or to divide them. The order of objects, the lines of *TrP*, is given by the meshing. It is possible to modify it as needed, for instance, to begin by the objects closed to a side or closed to a point.

All these operations are made in MATLAB and then imported by Comsol; all the nano objects are remeshed again for computing.

Example 16.1 Triangles in a square

Comsol draws a square width: 0.001 m, corner: (0,0). The chosen meshes give the size of the triangles. For instance, the ultrafine meshing produces 25,127 meshes, that is, 158^2. This amount of meshes would give as many triangles with a mean side of some nanometers if the initial square would have a side around 0.1 μm. The final meshing of this arrangement will be at least 300,000 meshes which can be solved on a Station. Here, in order to obtain simple figures, the coarse method is preferred. Then, it is necessary to run an application, for instance, electricity. Comsol exports data in a txt file to get a matrix of coordinates:

```
Pt = [...
0.0                    2.0E-4
0.0                    1.0E-4
6.4200256E-5           1.6731408E-4
0.0                    0.0
6.4200256E-5           6.731408E-5

....................................

7.461472E-4            9.097E-4
8.342704E-4            9.3145535E-4
6.922944E-4            8.1939995E-4
7.804176E-4            8.4115536E-4
8.685408E-4            8.629107E-4] ;
```

and a matrix of the numbers of elements (triangular):

```
Tr = [...
2        3        5
1        3        2
2        5        4
3        6        5
8        9        11

....................................

399      402      401
404      405      407
403      405      404
404      407      406
405      408      407] ;
```

Comsol Multiphysics also exports the values of the variable at each node, but this information is not needed at this point. Then, the following short file produces a matrix *TrP*

```
[line,col] = size (Tr) ;
TrP = zeros (line,6) ;
for tt = 1:line
pt1 = Tr (tt,1) ;
pt2 = Tr (tt,2) ;
pt3 = Tr (tt,3) ;
TrP (tt, :) = [Pt (pt1,1) Pt (pt2,1) Pt (pt3,1) Pt (pt1,2) Pt (pt2,2) Pt (pt3,2)] ;
end
TrP ;
```

which is

```
TrP = [...
0    6.4e-005    6.4e-005    1.0e-004    1.7e-004    6.7e-005
0    6.4e-005    0           2.0e-004    1.7e-004    1.0e-004
0    6.4e-005    0           1.0e-004    6.7e-005    0
..............................
6.4e-005         1.3e-004    6.4e-005    1.7e-004    1.3e-004 6.7e-005
1.0e-004         6.4e-005    1.6e-004    0           6.7e-005 6.7e-005];
```

Each line of the matrix *TrP* gives the summits of triangle; the coordinates are ordered: the *x*'s first and then the *y*'s.

Then, it is just to add, in a loop [1:line], the expression

```
fprintf(1,strcat('g_',num2str(tt),'=line2([',x1s,',',x2s,',',x3s,'],
[',y1s,',',y2s,',',y3s,']);n'))
```

where *x1s* and others are the string form of the abscissa's of the first node to get on the screen

```
g_1=line2([0,6.42e-005,6.42e-005],[0.0001,0.00016731,6.7314e-005]);
```

which will be imported by Comsol as a triangle to give Figure 16.7 (consists of 272 triangles).

This geometry may support a lot of physical expressions. For instance, it could be a capacitance with a given repartition of permittivity on each triangle. Since the size of the sample is very small compared to the wavelength, only the following electrical equation is considered, that is, electric equation for a capacitance:

$$\partial(\sigma + i\omega\varepsilon_0\varepsilon_r)\partial V = 0,$$

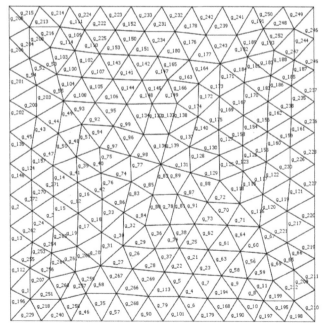

FIGURE 16.7
Triangles reconstructed from the mesh information.

The boundary conditions are the following: ground, electric potential, electric insulation, and distributed impedance

$$\vec{n} \cdot \vec{J} = (\sigma + i\omega\varepsilon_0\varepsilon_r)V.$$

which gives the possibility of surface properties different from volume ones.

The mean permittivity is obtained after integration of the normal current I_t over a boundary. For instance, if the upper side is at the potential V, the downer at the ground, and the left and right are isolated, I_t is

$$I_t = jC\omega V = j\varepsilon_0\varepsilon_m V$$

with $j^2 = -1$. In 2D, I_t and C have the meaning of a density in the z direction, and the exact length of the side disappears in the case of a square.

If the triangles have a pure permittivity and if there is a continuity between them, the value of I_t will be a pure imaginary number (e.g., 0.005596j [A/m]). But if the inside boundaries between the triangles are supposed to be lossy for any reason, as Comsol gives the possibility under the label "*Distributed impedance*," then the behavior of the square is changed.

Figure 16.8 shows the current lines (plain lines) and the electric field (arrows). The electric field points the influence of the conductive layer. The frequency measurement gives an insight on the influence of the conductive boundaries, that is,

$$f = 1e9\ [Hz] \rightarrow I_{(y=0.001)} = 49.868574 + 0.094599i\ [A/m].$$

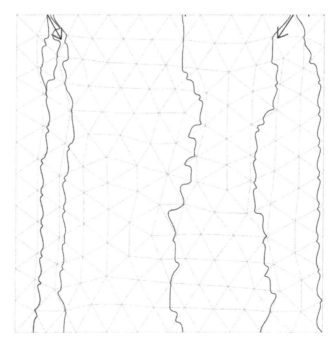

FIGURE 16.8
The lines of current are plain; the arrows stand for the electric field.

Note that the following conditions were chosen in this example:
for the triangles, $\varepsilon_r = 10$;
ground for $y = 0$;

$$V = 1 \text{ for } y = 0.001;$$

$$\sigma = 1e5 \text{ [S/m]}, \varepsilon_r = 10 \text{ for } x = 0, \text{ and } x = 0.001;$$

for internal boundaries, $\sigma = 1e5$ [S/m], $\varepsilon_r = 10$.

The disturbances are due to the conductivity of the side of the triangles compared to the permittivity $\omega\varepsilon$ of their surface.

Example 16.2 Capillaries in a square

Capillaries are introduced by reducing the size of the triangles. There are many possible ways to get this result. Among them, the homothetic reduction with the mass center as base point is the most simple because it is fully automated. Starting from *TrP* and giving *rh* as the reduction factor, the new nodes are immediately obtained. But this result is to be corrected so that the bases of the triangles on the side of the square form a straight line. Moreover, it is necessary to add an outside square to close the domain and get the following result.

To reconstruct triangles according to the meshing and to transform them into independent objects, each number of nodes correlates with its coordinate. After that, it is possible to apply any transformation, for instance, to change sizes of this triangles-objects applying homothety relative to the center of gravity of each triangle.

For this purpose, the mass center of each triangle (point *PtG*) using the well-known formula was calculated, and every point *PtG* was considered as a center of homothety. By changing a value of the homothety coefficient *rh*, we obtain needed sizes of the triangles:

```
% ptG - center of gravity of triangle
ptG=([Pt(pt1,1)+Pt(pt2,1)+Pt(pt3,1),
Pt(pt1,2)+Pt(pt2,2)+Pt(pt3,2)])/3;
% coefficient of homothety
rh=0.85;
xx=[Pt(pt1,1) Pt(pt2,1) Pt(pt3,1)];
yy=[Pt(pt1,2) Pt(pt2,2) Pt(pt3,2)];
% homothety
xx=ptG(1,1)+rh*(xx-ptG(1,1));
yy=ptG(1,2)+rh*(yy-ptG(1,2));
geomplot(poly1([xx], [yy]));
hold('on')
```

As a result, the following model was constructed (Figure 16.9). In this model, the triangles represent the nanoparticles separated by pores. Obviously, the mathematical treatment can provide any desired grading of the particles' size.

As noted, one can adjust a value of a free space among the triangles by changing the coefficient of homothety *rh*. This way, the needed value of porosity of the material can be obtained. Here, reduction ration *rh* = 0.8 (that corresponds to an approximate value of porosity 30%).

It is interesting to operate this drawing with Navier–Stokes equation because the triangles are not active in the flow transfer or just active through the friction on their sides (no slip) or leaking wall. With the first physical conditions this drawing produces the velocity field shown on Figure 16.10. The arrangement of the Figure 16.9 produces the velocity field shown on the Figure 16.10 on which the lighter areas represent the higher velocities.

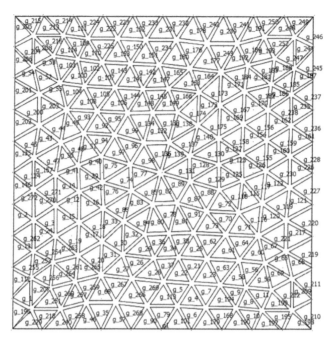

FIGURE 16.9
Capilleries in triangles.

Figure 16.10 shows the amplitude of the fluid velocity and the arrows its direction. Comsol Multiphysics provides the velocity in each output; this result immediately gives the mean value of the permeability function, among others, of the size of the capillaries. The conditions which were chosen for this example are as follows:
Incompressible Navier–Stokes equation:

$$\begin{cases} \rho(\vec{u} \cdot \partial)\vec{u} = \partial(\eta(\partial\vec{u} + (\partial\vec{u})^T)), \\ \partial \cdot \vec{u} = 0; \end{cases}$$

The triangles are inactive.
Capillaries: fluid density $= 1000$ [kg/m³], dynamic viscosity $= 0.01$ [Pa·s];
Boundaries: $x = 0.001$, $P = 0.715$ [Pa]; $x = 0$, $P = 0$; $y = 0$ and $y = 0.001$, wall;
The computed value of the integral velocity field is 1.223114e-10 [m²/s].

Example 16.3   Microwave sintering

The same geometry can be used to simulate the sintering of a mixture of large grains (triangles) coated with smaller ones supposed to fill the precedent capillaries [4,5]. The modeling uses two equations: one is for the electric capacitance and the other is for the heating. The source term of the heat equation is only in the capillaries (the very small nano grains have a good ability to catch the electrical energy and transform it into thermal energy while the larger grains do not have this ability). The large grains heat by conduction, and since the source is in the very heart, the heating is very fast. The surface of the square is not well heated; usually an infrared heating has to be added to get a fair homogeneity.
Figure 16.11 shows the temperature field with a lower temperature inside than on their boundaries. This is a typical requirement for sintering applications.

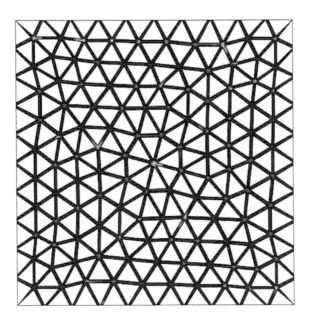

FIGURE 16.10
Velocity field.

FIGURE 16.11
Temperature field.

Conditions in this example:
Heat transfer equation

$$\rho C_p \partial T / \partial t - \partial(k \partial T) = \begin{cases} 0, \\ microwave\ pover; \end{cases}$$

Boundary conditions: thermal insulation, temperature;
For triangles $\varepsilon_r = 10$;
For capillaries $\varepsilon_r = 10 - i$;
Thermal conductivity $= 4$ [W/m·K].

Example 16.4 Introduction of a bean into the geometry

It may be necessary to add grains which are not produced by an initial meshing, that is, which are not triangles. They will be produced, for instance, by copying a microphotography. Nevertheless, the triangles will remain necessary to fulfill all the space. Figure 16.12 gives an example in which a large grain looks like a bean. The line is made only with ellipses (Bezier curves) and straight lines to facilitate the meshing before computing.

This drawing was obtained in three steps.

The first step was to draw a square with the bean, and then to extract the bean from the square so that the domain had a hole inside it. This is to separate the meshes (triangles) from the bean.

Then, the remaining space was meshed to give triangles and capillaries, exactly as previously. A rectangular envelop was added to close the domain.

In the third step, the bean was reintroduced after a slight reduction.

Navier–Stokes equation is a good tool to look at the behavior of this geometry. All the objects are supposed non-active and the boundary conditions are the same as the first

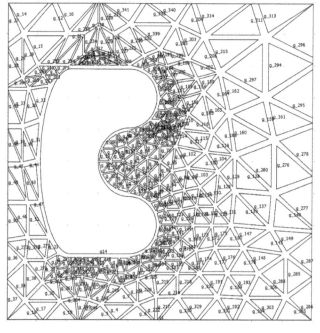

FIGURE 16.12
Bean inscribed into geometry.

FIGURE 16.13
Velocity field, the lighter areas represent higher velocities.

model. Figure 16.13 shows the fluid velocity, amplitude (shadow), and direction arrows. As expected, the bean is an obstacle with a shadow effect in the back and a badly irrigated area in the front.

Figure 16.14 is a lens effect on the upper left corner of the previous sample.

Obtained value of integral velocity field is 6.821822e-11 [m^2/s]. As compared with the preceding case, the fluid flow is normally decreased by the bean.

Example 16.5   Introduction of corks in the geometry

Starting from the capillary geometry, it is possible to add, by hand, new objects in their junctions. These objects could simulate corks, for instance in wet clay, which stop the flow. When they are heated, they dry and shrink, opening the way to the flow in the capillaries (Figure 16.15). For instance, they catch energy and their gas permeability increases with their temperature. At the beginning, the capillaries are full of gas which escapes when the permeability of the corks increases.

Gas is confined in capillaries closed by corks. Electrical energy opens the doors. Figure 16.16 shows the gas pressure at a given time.

16.4 3D Modeling: Examples

The principle of 3D modeling is very similar to 2D:

- To start with a cube, 1 mm side with a lower left summit at (0; 0; 0)
- To choose a coarse meshing and run any application
- To export the data in txt file

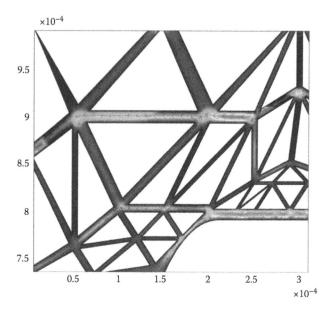

FIGURE 16.14
Enlargement of the upper left corner of Figure 16.13.

FIGURE 16.15
Distribution of gas pressure, the higher pressures are in the centre of the square and the lower on the sides.

The exported matrices look like

```
% Matrix of coordinates
Pt = [
0.0    0.0       0.0
0.0    2.5E-4    2.5E-4
```

FIGURE 16.16
Corks in nodes of capillary net.

```
0.0      0.0        5.0E-4
0.0      5.0E-4     5.0E-4
0.0      2.5E-4     7.5E-4
```

....

```
0.0       5.0E-4    5.0E-4
7.5E-4    5.0E-4    2.5E-4
7.5E-4    7.5E-4    5.0E-4
5.0E-4    5.0E-4    5.0E-4
0.0010    5.0E-4    5.0E-4];
```

and

```
% Matrix of elements (tetrahedrons)
Tr = [...
2      5     8    3
3      2     7    8
3      7     9    8
8      3     5    9
7      3     1    2
```

....

```
238    233   235   239
237    233   231   232
238    235   232   234
239    236   233   235
240    239   237   238];
```

As for the 2D case, it is convenient to form the matrix *TrP* which gathers, in each line, all the coordinates classified as *x*'s, *y*'s, and *z*'s. This matrix has 12 columns. Then, it is enough to add, in a loop on *TrP* lines, the following sentence

```
fprintf(1,strcat('g_',num2str(tt),'=tetrahedron3([',xxstr,';',yystr,';',
zzstr,']);n'))
```

to get on the screen

```
g_1=tetrahedron3([0 0 0.00025 0;0.00025 0.00025 0.00025 0;0.00025
0.00075 0.0005 0.0005]);
```

xxstr is the string corresponding to the four *xx*'s and similarly for *yy*'s and *zz*'s.

The set of all the g_i imported into Comsol produces a cube full of tetrahedrons (Figure 16.17).

Figures 16.18 and 16.19 give the result of the experimentation of this drawing seen as a capacitance: The ground is at the lower side ($z=0$), the potential at the higher ($z=0.001$), and the lateral sides are said to be isolated. The tetrahedrons have the same permittivity and all the internal boundaries have an electrical conductivity σ. The behavior of the capacitance is highly dependent on the ratio between σ and $\omega\varepsilon$. In the first figure, $\omega\varepsilon \gg \sigma$; it is the opposite in the second one. The current lines are those of a pure capacitance in Figure 16.18, that is perfectly straight when they are highly disturbed in Figure 16.19.

The equipotential surface, the grey voile on the Figure 16.19, also show the influence of the conductive boundaries. This kind of result helps to understand the influence of surface defects, for instance, in a sintering: The measurement of the output current versus frequency gives an insight into the internal boundaries.

To obtain these results, the following conditions were chosen:
For tetrahedrons $\varepsilon_r = 10$;
On the boundaries: $V=1$ for $z=0.001$; ground for $z=0$;

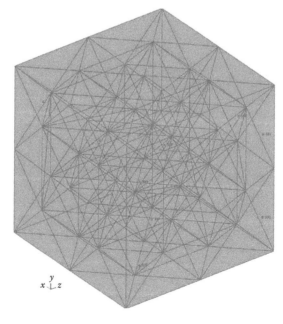

FIGURE 16.17
Cube full of tetrahedrons.

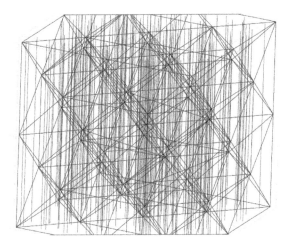

FIGURE 16.18
Current flows straight lines in the capacitance if the tetrahedrons have no surface conductivity.

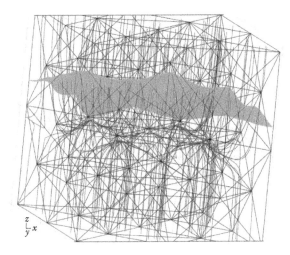

FIGURE 16.19
Deformed current lines and equipotential surface (grey voile) in the capacitance if the tetrahedrons have a high surface conductivity.

For lateral faces of the cube: isolation
Inside boundaries: $\sigma = 1000$, $\varepsilon_r = 10$.

The only difference between the two results is the frequency: 1e-6 Hz for the first, 1e9 Hz for the second. The use of non-continuous boundaries is lighter to be solved since it avoids the use of capillaries to simulate the grain joins.

At last, if the tetrahedrons have different physical values and if the repartition is known, this type of model gives a mean value of the cube behavior. When the repartition is not known, it is necessary to look to statistics, that is, to make different tries and choose.

16.5 3D and Capillaries

The capillaries are obtained by homothetic reduction of the tetrahedrons centered on the mass center. Since the tetrahedrons are not identical, this reduction produces irregularities which must be corrected for the tetrahedrons on the face. Elsewhere, it would be very difficult to close the domain.

The correction has to be introduced before the loop on *TrP* transforming the meshes in objects. In fact, the preliminary treatment was the following:

- The matrix of the reduced tetrahedrons is called *TrPr*.
- Two loops on the lines and the columns of *TrP* find the points located on the faces of the cube, that is, $x=0$ or $x=0.001$ and the same for y and z. Let (line_i,col_j) be one of these results. It indicates that the value of *TrPr* (line_i,col_j) which is surely not null (neither equal to 0.001) must be kept to zero (or 0.001) to reintroduce the given points exactly on this surface.
- In this way, the tetrahedrons on the face will exactly fit to initial cube as shown in Figure 16.20. This figure shows grains and capillaries, and in shadow the input into capillaries from one face. Here, reduction factor is $rh=0.8$, which corresponds to the value of porosity ≈ 7%.

16.6 Fluid Flow in the Capillaries

This cube is experimented by solving Navier–Stokes equations. The tetrahedrons are not active, input and output are on two opposite faces of the cube ($x=0$ and 0.001), and the remaining lateral faces are said to be a wall. Figure 16.20 shows also the input face, and Figure 16.21 gives an example of the velocity field.

The tetrahedrons are active but only the capillaries are supposed to catch electrical energy. The cube is seen as a capacitance. The heat conductivity is supposed to be low, which is the case in many applications, and the face of the cube is chemically isolated.

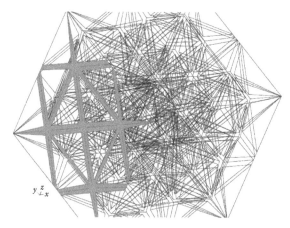

FIGURE 16.20
Cube full of reduced tetrahedrons with input capillaries (in grey).

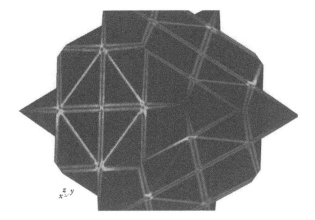

FIGURE 16.21
Velocity field.

16.7 Microwave Heating

Figures 16.22 and 16.23 show the result of the microwave heating of the capillaries full of liquid water, nanoparticles, or any kind of susceptor. The electrical energy transformed into heat energy is, by the same time, diffused to the tetrahedrons. The result is deeply

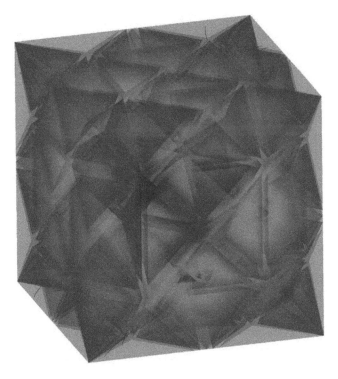

FIGURE 16.22
Temperature of the tetrahedrons surface.

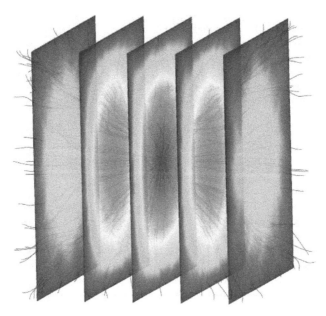

FIGURE 16.23
Net heating in light and lines of thermal currents.

dependant on the ratio between the input and the output power into and from the cap-illaries. Moreover, the density of the capillary network is also an important factor of homogeneity.

16.8 Conclusion

Automatic meshing and its associated nano objects is a convenient way to produce, in 2D and 3D as well, full spaces without any forgotten vacuum. As a consequence, it lets the boundaries play their specific behavior. At last, it is compatible with all the geometrical shapes, circles, spheres, and so on but also shapes extracted from experimentation as shown with the bean.

Finally, automatic meshing can be seen as a way to fill with nanoparticles spaces let empty between objects previously implanted.

Two different applications were described throughout the chapter:

1. Computation of mean values of physical constants of heterogeneous material, for instance, thermal and electric conductivity, permittivity, etc., with or without the grain joins influence

2. Evaluation of capillary flows leading to an inside knowledge of permeability, let-ting out of the computation the nanoparticles which act only by their surfaces, but keeping them

On these geometries, it is possible to use all the classical FEM used for drying, sintering, microwave heating [4], and so on.

As a consequence, when a problem is solved in a small domain, it is possible to enlarge the solution to larger domains, just by matrices' association. Comsol Multiphysics and MATLAB softwares were used (run on an HP Z800, 16 proc, 64 Gbits).

References

1. Akulich P.V. and Grinchik N.N., Modeling of heat and mass transfer in capillary-porous materials, *Journal of Engineering Physics and Thermophysics*, 71, 2, 1998, 225–233.
2. Barenblatt G.I., Entov V.M., and Ryzhik V.M., *Fluid Transfer in Natural Seams*, Nedra, Moscow, Russia, 1984 [in Russian].
3. Gubkin I.M., *Study of Oil*, Nauka, Moscow, Russia, 1975 [in Russian].
4. Lefeuvre S. and Gomonova O., Microwave heating at the grain level, *Proceedings of COMSOL Conference 2010*, Paris, France, November 17–19, 2010.
5. Lefeuvre S., Federova E., Gomonova O., and Tao J., Microwave sintering of micro- and nano-sized alumina powder, *Advances in Modeling of Microwave Sintering: 12th Seminar Computer Modeling in Microwave Engineering and Applications*, Grenoble, France, March 8–9, 2010, pp. 46–50.
6. Leibenzon L.S., *Motion of Natural Fluids in Porous Medium*, Technical and Theoretical Literature Publishing, Moscow, Leningrad, 1947 [in Russian].
7. Selyakov V.I. and Kadet V.V., *Percolation Models of Transfer Processes in Porous Media*, Nedra, Moscow, Russia, 1995.
8. Slichter C.S., Theoretical investigation of the motion of ground water, U.S. Geological Survey 19th Annual Report, Part II, 1899.
9. Slichter C.S., The motions of underground waters, U.S. Geological Survey Water Supply Paper 67, 1902.

Appendix A: Material and Physical Constants

A.1 Common Material Constants

TABLE A.1

Approximate Conductivity at 20°C

Material	Conductivity (S/m)
1. Conductors	
Silver	6.3×10^7
Copper (standard annealed)	5.8×10^7
Gold	4.5×10^7
Aluminum	3.5×10^7
Tungsten	1.8×10^7
Zinc	1.7×10^7
Brass	1.1×10^7
Iron (pure)	10^7
Lead	5×10^7
Mercury	10^6
Carbon	3×10^7
Water (sea)	4.8
2. Semiconductors	
Germanium (pure)	2.2
Silicon (pure)	4.4×10^{-4}
3. Insulators	
Water (distilled)	10^{-4}
Earth (dry)	10^{-5}
Bakelite	10^{-10}
Paper	10^{-11}
Glass	10^{-12}
Porcelain	10^{-12}
Mica	10^{-15}
Paraffin	10^{-15}
Rubber (hard)	10^{-15}
Quartz (fused)	10^{-17}
Wax	10^{-17}

TABLE A.2

Approximate Dielectric Constant and Dielectric Strength

Material	Dielectric Constant (or Relative), ε_r (Dimensionless)	Dielectric Strength, E (V/m)
Barium titanate	1200	7.5×10^6
Water (sea)	80	—
Water (distilled)	8.1	—
Nylon	8	—
Paper	7	12×10^6
Glass	5–10	35×10^6
Mica	6	70×10^6
Porcelain	6	—
Bakelite	5	20×10^6
Quartz (fused)	5	30×10^6
Rubber (hard)	3.1	25×10^6
Wood	2.5–8.0	—
Polystyrene	2.55	—
Polypropylene	2.25	—
Paraffin	2.2	30×10^6
Petroleum oil	2.1	12×10^6
Air (1 atm)	1	3×10^6

TABLE A.3

Relative Permeability

Material	Relative Permeability, μ_r
1. Diamagnetic	
Bismuth	0.999833
Mercury	0.999968
Silver	0.9999736
Lead	0.9999831
Copper	0.9999906
Water	0.9999912
Hydrogen (STP)	≈ 1.0
2. Paramagnetic	
Oxygen (STP)	0.999998
Air	1.00000037
Aluminum	1.000021
Tungsten	1.00008
Platinum	1.0003
Manganese	1.001
3. Ferromagnetic	
Cobalt	250
Nickel	600
Soft-iron	5000
Silicon-iron	7000

A.2 Physical Constants

Quantity	Best Experimental Value	Approximate Value for Problem Work
Avogadro's number (/kg mol)	6.0228×10^{26}	6×10^{26}
Boltzmann constant (J/k)	1.38047×10^{-23}	1.38×10^{-23}
Electron charge (C)	-1.6022×10^{-19}	-1.6×10^{-19}
Electron mass (kg)	9.1066×10^{-31}	9.1×10^{-31}
Permittivity of free space (F/m)	8.854×10^{-12}	$\dfrac{10^{-9}}{36\pi}$
Permeability of free space (H/m)	$4\pi \times 10^{-7}$	12.6×10^{-7}
Intrinsic impedance of free space (Ω)	376.6	120π
Speed of light in vacuum (m/s)	2.9979×10^{8}	3×10^{8}
Proton mass (kg)	1.67248×10^{-27}	1.67×10^{-27}
Neutron mass (kg)	1.6749×10^{-27}	1.67×10^{-27}
Planck's constant (J s)	6.6261×10^{-34}	6.62×10^{-34}
Acceleration due to gravity (m/s^2)	9.8066	9.8
Universal constant of gravitation (m^2/kg s^2)	6.658×10^{-11}	6.66×10^{-11}
Electron volt (J)	1.6030×10^{-19}	1.6×10^{-19}
Gas constant (J/mol K)	8.3145	8.3

Appendix B: Symbols and Formulas

B.1 Greek Alphabet

Uppercase	Lowercase	Name
A	α	Alpha
B	β	Beta
Γ	γ	Gamma
Δ	δ	Delta
E	ε	Epsilon
Z	ζ	Zeta
H	η	Eta
Θ	θ, ϑ	Theta
I	ι	Iota
K	κ	Kappa
Λ	λ	Lambda
M	μ	Mu
N	ν	Nu
Ξ	ξ	Xi
O	o	Omicron
Π	π	Pi
P	ρ	Rho
Σ	σ	Sigma
T	τ	Tau
Y	υ	Upsilon
Φ	ϕ, φ	Phi
X	χ	Chi
Ψ	ψ	Psi
Ω	ω	Omega

B.2 International System of Units (SI) Prefixes

Power	Prefix	Symbol	Power	Prefix	Symbol
10^{-35}	stringo	—	10^0	—	—
10^{-24}	yocto	y	10^1	deka	da
10^{-21}	zepto	z	10^2	hecto	h
10^{-18}	atto	A	10^3	kilo	k
10^{-15}	femto	f	10^6	mega	M
10^{-12}	pico	p	10^9	giga	G
10^{-9}	nano	n	10^{12}	tera	T
10^{-6}	micro	μ	10^{15}	peta	P
10^{-3}	milli	m	10^{18}	exa	E
10^{-2}	centi	c	10^{21}	zetta	Z
10^{-1}	deci	d	10^{24}	yotta	Y

B.3 Trigonometric Identities

$$\cot\theta = \frac{1}{\tan\theta}, \quad \sec\theta = \frac{1}{\cos\theta}, \quad \csc\theta = \frac{1}{\sin\theta}$$

$$\tan\theta = \frac{\sin\theta}{\cos\theta}, \quad \cot\theta = \frac{\cos\theta}{\sin\theta}$$

$$\sin^2\theta + \cos^2\theta = 1, \ \tan^2\theta + 1 = \sec^2\theta, \ \cot^2\theta + 1 = \csc^2\theta$$

$$\sin(-\theta) = -\sin\theta, \ = \cos(-\theta) = \cos\theta, \ \tan(-\theta) = -\tan\theta$$

$$\csc(-\theta) = -\csc\theta, \ \sec(-\theta) = \sec\theta, \ \cot(-\theta) = -\cot\theta$$

$$\cos(\theta_1 \pm \theta_2) = \cos\theta_1 \cos\theta_2 \mp \sin\theta_1 \sin\theta_2$$

$$\sin(\theta_1 \pm \theta_2) = \sin\theta_1 \cos\theta_2 \pm \cos\theta_1 \sin\theta_2$$

$$\tan(\theta_1 \pm \theta_2) = \frac{\tan\theta_1 \pm \tan\theta_2}{1 \mp \tan\theta_1 \tan\theta_2}$$

$$\cos\theta_1 \cos\theta_2 = \frac{1}{2}\left[\cos(\theta_1 + \theta_2) + \cos(\theta_1 - \theta_2)\right]$$

$$\sin\theta_1 \sin\theta_2 = \frac{1}{2}\left[\cos(\theta_1 - \theta_2) - \cos(\theta_1 + \theta_2)\right]$$

$$\sin\theta_1 \cos\theta_2 = \frac{1}{2}\left[\sin(\theta_1 + \theta_2) + \sin(\theta_1 - \theta_2)\right]$$

$$\cos\theta_1 \sin\theta_2 = \frac{1}{2}\left[\sin(\theta_1 + \theta_2) - \sin(\theta_1 - \theta_2)\right]$$

$$\sin\theta_1 + \sin\theta_2 = 2\sin\left(\frac{\theta_1 + \theta_2}{2}\right)\cos\left(\frac{\theta_1 - \theta_2}{2}\right)$$

$$\sin\theta_1 - \sin\theta_2 = 2\cos\left(\frac{\theta_1 + \theta_2}{2}\right)\sin\left(\frac{\theta_1 - \theta_2}{2}\right)$$

$$\cos\theta_1 + \cos\theta_2 = 2\cos\left(\frac{\theta_1 + \theta_2}{2}\right)\cos\left(\frac{\theta_1 - \theta_2}{2}\right)$$

$$\cos\theta_1 - \cos\theta_2 = -2\sin\left(\frac{\theta_1 + \theta_2}{2}\right)\sin\left(\frac{\theta_1 - \theta_2}{2}\right)$$

$$a\cos\theta - b\sin\theta = \sqrt{a^2 + b^2}\cos(\theta + \phi), \quad \text{where } \phi = \tan^{-1}\left(\frac{b}{a}\right)$$

$$a\sin\theta - b\cos\theta = \sqrt{a^2 + b^2}\sin(\theta + \phi), \quad \text{where } \phi = \tan^{-1}\left(\frac{b}{a}\right)$$

$$\cos(90° - \theta) = \sin\theta, \ \sin(90° - \theta) = \cos\theta, \ \tan(90° - \theta) = \cot\theta$$

$$\cot(90° - \theta) = \tan\theta, \ \sec(90° - \theta) = \csc\theta, \ \csc(90° - \theta) = \sec\theta$$

$$\cos(\theta \pm 90°) = \mp\sin\theta, \ \sin(\theta \pm 90°) = \pm\sin\theta, \ \tan(\theta \pm 90°) = -\cot\theta$$

$$\cos(\theta \pm 180°) = -\cos\theta, \ \sin(\theta \pm 180°) = -\sin\theta, \ \tan(\theta \pm 180°) = \tan\theta$$

$$\cos 2\theta = \cos^2\theta - \sin^2\theta, \ \cos2\theta = 1 - 2\sin^2\theta, \ \cos2\theta = 2\cos^2\theta - 1$$

$$\sin 2\theta = 2\sin\theta\cos\theta, \ \tan2\theta = \frac{2\tan\theta}{1 - \tan^2\theta}$$

$$\cos 3\theta = 4\cos^3\theta - 3\sin\theta$$

$$\sin 3\theta = 3\sin\theta - 4\sin^3\theta$$

$$\sin\frac{\theta}{2} = \pm\sqrt{\frac{1 - \cos\theta}{2}}, \quad \cos\frac{\theta}{2} = \pm\sqrt{\frac{1 + \cos\theta}{2}},$$

$$\tan\frac{\theta}{2} = \pm\sqrt{\frac{1 - \cos\theta}{1 + \cos\theta}}, \quad \tan\frac{\theta}{2} = \frac{\sin\theta}{1 + \cos\theta}, \quad \tan\frac{\theta}{2} = \frac{1 - \cos\theta}{\sin\theta}$$

$$\sin\theta = \frac{e^{j\theta} - e^{-j\theta}}{2j}, \quad \cos\theta = \frac{e^{j\theta} + e^{-j\theta}}{2}\left(j = \sqrt{-1}\right), \quad \tan\theta = \frac{e^{j\theta} + e^{-j\theta}}{j(e^{j\theta} + e^{-j\theta})}$$

$$e^{\pm j\theta} = \cos\theta \pm j\sin\theta \quad \text{(Euler's identity)}$$

$$1 \text{ rad} = 57.296°$$

$$\pi = 3.1416$$

B.4 Hyperbolic Functions

$$\cosh x = \frac{e^x + e^{-x}}{2}, \quad \sinh x = \frac{e^x - e^{-x}}{2}, \quad \tanh x = \frac{\sinh x}{\cosh x}$$

$$\cosh x = \frac{1}{\tanh x}, \quad \operatorname{sech} x = \frac{1}{\cosh x}, \quad \operatorname{csch} x = \frac{1}{\sinh x}$$

$$\sin jx = j\sinh x, \quad \cos jx = \cosh x$$

$$\sinh jx = j\sin x, \quad \cosh jx = \cos x$$

$$\sin(x \pm jy) = \sin x\cosh y \pm j\cos x\sinh y$$

$$\cos(x \pm jy) = \cos x\cosh y \mp j\sin x\sinh y$$

$$\sinh(x \pm y) = \sinh x\cosh y \pm \cosh x\sinh y$$

$$\cosh(x \pm y) = \cosh x\cosh y \pm \sinh x\sinh y$$

$$\sinh(x \pm jy) = \sinh x\cos y \pm j\cosh x\sin y$$

$$\cosh(x \pm jy) = \cosh x\cos y \pm j\sinh x\sin y$$

$$\tanh(x \pm jy) = \frac{\sinh 2x}{\cosh 2x + \cos 2y} \pm j \frac{\sin 2y}{\cosh 2x + \cos 2y}$$

$$\cosh^2 - \sinh^2 x = 1$$

$$\operatorname{sech}^2 + \tanh^2 x = 1$$

B.5 Complex Variables

A complex number can be written as

$$z = x = jy = r\angle\theta = re^{j\theta} = r(\cos\theta + j\sin\theta),$$

where

$$x = \operatorname{Re} z = r\cos\theta, \; y = \operatorname{Im} z = r\sin\theta$$

$$r = |z| = \sqrt{x^2 + y^2}, \quad \theta = \tan^{-1}\left(\frac{y}{x}\right)$$

$$j = \sqrt{-1}, \; \frac{1}{j} = -j, \; j^2 = -1$$

The complex conjugate of $z = z^* = x - jy = r\angle -\theta = re^{-j\theta} = r(\cos\theta - j\sin\theta)$
$(e^{j\theta})^n = e^{jn\theta} = \cos n\theta + j\sin n\theta$ (de Moivre's theorem)
If $z_1 = x_1 + jy_1$ and $z_2 = x_2 + jy_2$, then only if $x_1 = x_2$ and $y_1 = y_2$.
$z_1 \pm z_2 = (x_1 + x_2) \pm j(y_1 + y_2)$

$$z_1 z_2 = (x_1 x_2 - y_1 y_2) + j(x_1 y_2 + x_2 y_1) = r_1 r_2 e^{j(\theta_1 + \theta_2)} = r_1 r_2 \angle\theta_1 + \theta_2$$

$$\frac{z_1}{z_2} = \frac{(x_1 + jy_1)}{(x_2 + jy_2)} \cdot \frac{(x_2 - jy_2)}{(x_2 - jy_2)} = \frac{x_1 x_2 + y_1 y_2}{x_2^2 + y_2^2} + j\frac{x_2 y_1 - x_1 y_2}{x_2^2 + y_2^2} = \frac{r_1}{r_2} e^{j(\theta_1 - \theta_2)} = \frac{r_1}{r_2} \angle\theta_1 - \theta_2$$

$$\ln(re^{j\theta}) = \ln r + \ln e^{j\theta} = \ln r + j\theta + j2m\pi \; (m = \text{integer})$$

$$\sqrt{z} = \sqrt{x + jy} = \sqrt{r}(e^{j\theta/2}) = \sqrt{r}\angle\theta/2$$

$$z^n = (x + jy)^n = r^n e^{jn\theta} = r^n \angle n\theta \quad (n = \text{integer})$$

$$z^{\frac{1}{n}} = (x + jy)^{\frac{1}{n}} = r^{\frac{1}{n}} e^{\frac{j\theta}{n}} = r^{\frac{1}{n}} \angle\theta/2 + 2\pi m / n, \quad (m = 0, 1, 2, \ldots, n-1)$$

B.6 Table of Derivatives

$y =$	$\dfrac{dy}{dx} =$
c (constant)	0
cx^n (n any constant)	cnx^{n-1}
e^{ax}	ae^{ax}
$a^x (a > 0)$	$a^x \ln a$

$y =$	$\dfrac{dy}{dx} =$
$\ln x (x > 0)$	$\dfrac{1}{x}$
$\dfrac{c}{x^a}$	$\dfrac{-ca}{x^{a+1}}$
$\log_a x$	$\dfrac{\log_a e}{x}$
$\sin ax$	$a \cos ax$
$\cos ax$	$-a \sin ax$
$\tan ax$	$a \sec^2 ax = \dfrac{a}{\cos^2 ax}$
$\cot ax$	$-a \csc^2 ax = \dfrac{-a}{\sin^2 ax}$
$\sec ax$	$\dfrac{a \sin ax}{\cos^2 ax}$
$\csc ax$	$\dfrac{-a \cos ax}{\sin^2 ax}$
$\arcsin ax = \sin^{-1} ax$	$\dfrac{a}{\sqrt{1 - a^2 x^2}}$
$\arccos ax = \cos^{-1} ax$	$\dfrac{-a}{\sqrt{1 - a^2 x^2}}$
$\arctan ax = \tan^{-1} ax$	$\dfrac{a}{1 + a^2 x^2}$
$\text{arccot } ax = \cot^{-1} ax$	$\dfrac{-a}{1 + a^2 x^2}$
$\sinh ax$	$a \cosh ax$
$\cosh ax$	$a \sinh ax$
$\tanh ax$	$\dfrac{a}{\cosh^2 ax}$
$\sinh^{-1} ax$	$\dfrac{a}{\sqrt{1 + a^2 x^2}}$
$\cosh^{-1} ax$	$\dfrac{a}{\sqrt{a^2 x^2 - 1}}$
$\tanh^{-1} ax$	$\dfrac{a}{1 - a^2 x^2}$
$u(x) + \upsilon(x)$	$\dfrac{du}{dx} + \dfrac{d\upsilon}{dx}$
$u(x)\upsilon(x)$	$u \dfrac{d\upsilon}{dx} + \upsilon \dfrac{du}{dx}$

(continued)

$y =$	$\dfrac{dy}{dx} =$
$\dfrac{u(x)}{v(x)}$	$\dfrac{1}{v^2}\left(v\dfrac{du}{dx} - u\dfrac{dv}{dx}\right)$
$\dfrac{1}{v(x)}$	$\dfrac{-1}{v^2}\dfrac{dv}{dx}$
$y(v(x))$	$\dfrac{dy}{dv}\dfrac{dv}{dx}$
$y(v(u(x)))$	$\dfrac{dy}{dv}\dfrac{dv}{du}\dfrac{du}{dx}$

B.7 Table of Integrals

$$\int a\,dx = ax + c \quad \text{(c is an arbitrary constant)}$$

$$\int x\,dy = xy - \int y\,dx$$

$$\int x^n dx = \frac{x^{n+1}}{n+1} + c, \quad (n \neq -1)$$

$$\int \frac{1}{x}\,dx = \ln|x| + c$$

$$\int e^{ax} dx = \frac{e^{ax}}{a} + c$$

$$\int a^x dx = \frac{a^x}{\ln a} + c \quad \text{for } (a > 0)$$

$$\int \ln x\,dx = x\ln x - x + c \quad \text{for } (x > 0)$$

$$\int \sin ax\,dx = \frac{-\cos ax}{a} + c$$

$$\int \cos ax\,dx = \frac{\sin ax}{a} + c$$

$$\int \tan ax\,dx = \frac{-\ln|\cos ax|}{a} + c$$

$$\int \cot ax\,dx = \frac{\ln|\sin ax|}{a} + c$$

$$\int \sec ax \, dx = \frac{-\ln\left(\frac{1-\sin ax}{1+\sin ax}\right)}{2a} + c$$

$$\int \csc ax \, dx = \frac{\ln\left(\frac{1-\cos ax}{1+\cos ax}\right)}{2a} + c$$

$$\int \frac{1}{x^2+a^2} \, dx = \frac{\tan^{-1}\left(\frac{x}{a}\right)}{a} + c$$

$$\int \frac{1}{x^2-a^2} \, dx = \frac{\ln\left(\frac{x-a}{x+a}\right)}{2a} + c \quad \text{or} \quad \frac{\tanh^{-1}\left(\frac{x}{a}\right)}{a} + c$$

$$\int \frac{1}{a^2-x^2} \, dx = \frac{\ln\left(\frac{x+a}{x-a}\right)}{2a} + c$$

$$\int \frac{1}{\sqrt{a^2-x^2}} \, dx = \sin^{-1}\left(\frac{x}{a}\right) + c$$

$$\int \frac{1}{\sqrt{a^2-x^2}} \, dx = \frac{\sinh^{-1}\left(\frac{x}{a}\right)}{a} + c \quad \text{or} \quad \ln(x+\sqrt{x^2+a^2}) + c$$

$$\int \frac{1}{\sqrt{x^2-a^2}} \, dx = \ln(x+\sqrt{x^2+a^2}) + c$$

$$\int \frac{1}{x\sqrt{x^2-a^2}} \, dx = \frac{\sec^{-1}\left(\frac{x}{a}\right)}{a} + c$$

$$\int xe^{ax} dx = \frac{(ax-1)e^{ax}}{a^2} + c$$

$$\int x\cos ax \, dx = \frac{\cos ax + ax\sin ax}{a^2} + c$$

$$\int x\sin ax \, dx = \frac{\sin ax + ax\cos ax}{a^2} + c$$

$$\int x\ln x \, dx = \frac{x^2}{2}\ln x - \frac{x^2}{4} + c$$

$$\int xe^{ax} dx = \frac{e^{ax}(ax-1)}{a^2} + c$$

$$\int e^{ax}\cos bx \, dx = \frac{e^{ax}(a\cos bx + b\sin bx)}{a^2+b^2} + c$$

$$\int e^{ax} \sin bx \; dx = \frac{e^{ax}(-b\cos bx + a\sin bx)}{a^2 + b^2} + c$$

$$\int \sin^2 x \; dx = \frac{x}{2} - \frac{\sin 2x}{4} + c$$

$$\int \cos^2 x \; dx = \frac{x}{2} - \frac{\sin 2x}{4} + c$$

$$\int \tan^2 x \; dx = \tan x - x + c$$

$$\int \cot^2 x \; dx = -\cot x - x + c$$

$$\int \sec^2 x \; dx = \tan x + c$$

$$\int \csc^2 x \; dx = -\cot x + c$$

$$\int \sec x \; \tan x \; dx = \sec x + c$$

$$\int \csc x \cot x \; dx = -\csc x + c$$

B.8 Table of Probability Distributions

1. Discrete Distribution	Probability $P(X{=}x)$	Expectation (Mean) μ	Variance σ^2
Binomial $B(n, p)$	$\binom{n}{r} p^r (1-p)^{n-r} = \frac{n! \, p^r q^{n-r}}{r!(n-r)!}$ $r = 0, 1, \ldots, n$	np	$np(1-p)$
Geometric $G(p)$	$(1-p)^{r-1} p$	$\dfrac{1}{p}$	$\dfrac{1-p}{p^2}$
Poisson $p(\lambda)$	$\dfrac{\lambda^n e^{-\lambda}}{n!}$	λ	λ
Pascal (negative binomial) $NB(r, p)$	$\binom{x-1}{r-1} p^r (1-p)^{x-r},$ $x = r, r+1, \ldots$	$\dfrac{r}{p}$	$\dfrac{r(1-p)}{p^2}$
Hypergeometric $H(N, n, p)$	$\dfrac{\binom{Np}{r}\binom{N-Np}{n-r}}{\binom{N}{n}}$	np	$np(1-p)\dfrac{N-n}{N-1}$

2. Continuous Distribution	Density $f(x)$	Expectation (Mean) μ	Variance σ^2
Exponential $E(\lambda)$	$\begin{cases} \lambda e^{-\lambda x}, & x \geq 0 \\ 0, & x < 0 \end{cases}$	$\dfrac{1}{\lambda}$	$\dfrac{1}{\lambda^2}$
Uniform $U(a, b)$	$\begin{cases} \dfrac{1}{b-a}, & a < x < b \\ 0, & \text{elsewhere} \end{cases}$	$\dfrac{a+b}{2}$	$\dfrac{(b-a)^2}{12}$
Standardized normal $N(0, 1)$	$\varphi(x) = \dfrac{e^{\frac{-x^2}{2}}}{\sqrt{2\pi}}$	0	1
General normal	$\dfrac{1}{\sigma}\varphi\left(\dfrac{x-\mu}{\sigma}\right)$	μ	σ^2
Gamma $\Gamma(n, \lambda)$	$\dfrac{\lambda^n}{\Gamma(n)} x^{n-1} e^{-\lambda x}$	$\dfrac{n}{\lambda}$	$\dfrac{n}{\lambda^2}$
Beta $\beta(p, q)$	$a_{p,q} x^{p-1}(1-x)^{q-1}, \quad 0 \leq x \leq 1$ $a_{p,q} = \dfrac{\Gamma(p+q)}{\Gamma(p)\Gamma(q)}, \quad p > 0, q > 0$	$\dfrac{p}{p+q}$	$\dfrac{pq}{(p+q)^2(p+q+1)}$
Weibull $W(\lambda, \beta)$	$\lambda^\beta \beta x^{\beta-1} e^{-(\lambda x)\beta}, \quad x \geq 0$ $F(x) = 1 - e^{-(\lambda x)\beta}$	$\dfrac{1}{\lambda}\Gamma\left(1+\dfrac{1}{\beta}\right)$	$\dfrac{1}{\lambda^2}(A-B)$ $A = \Gamma\left(1+\dfrac{2}{\beta}\right)$ $B = \Gamma^2\left(1+\dfrac{1}{\beta}\right)$
Rayleigh $R(\sigma)$	$\dfrac{x}{\sigma^2} e^{\frac{-x^2}{2\sigma^2}}, \quad x \geq 0$	$\sigma\sqrt{\dfrac{\pi}{2}}$	$2\sigma^2\left(1-\dfrac{\pi}{4}\right)$

B.9 Summations (Series)

1. Finite element of terms

$$\sum_{n=0}^{N} a^n = \frac{1-a^{N+1}}{1-a}; \quad \sum_{n=0}^{N} na^n = a\left(\frac{1-(N+1)a^N + Na^{N+1}}{(1-a)^2}\right)$$

$$\sum_{n=0}^{N} n = \frac{N(N+1)}{2}; \quad \sum_{n=0}^{N} n^2 = \frac{N(N+1)(2N+1)}{6}$$

$$\sum_{n=0}^{N} n(n+1) = \frac{N(N+1)(N+2)}{3};$$

$$(a+b)^N = \sum_{n=0}^{N} NC_n a^{N-n} b^n, \quad \text{where } NC_n = NC_{N-n} = \frac{NP_n}{n!} = \frac{N!}{(N-n)!n!}$$

2. Infinite element of terms

$$\sum_{n=0}^{\infty} x^n = \frac{1}{1-x}, (\,|x|<1); \quad \sum_{n=0}^{\infty} nx^n = \frac{1}{(1-x)^2}, (\,|x|<1)$$

$$\sum_{n=0}^{\infty} n^k x^n = \lim_{a \to 0} (-1)^k \frac{\partial^k}{\partial a^k} \left(\frac{x}{x-e^{-a}} \right), (\,|x|<1); \quad \sum_{n=0}^{\infty} \frac{(-1)^n}{2n+1} = 1 - \frac{1}{3} + \frac{1}{5} - \frac{1}{7} + \cdots = \frac{1}{4}\pi$$

$$\sum_{n=0}^{\infty} \frac{1}{n^2} = 1 + \frac{1}{2^2} + \frac{1}{3^2} + \frac{1}{4^2} + \cdots = \frac{1}{6}\pi^2$$

$$e^x = \sum_{n=0}^{\infty} \frac{x^n}{n!} = 1 + \frac{1}{1!}x + \frac{1}{2!}x^2 + \frac{1}{3!}x^3 + \cdots$$

$$a^x = \sum_{n=0}^{\infty} \frac{(\ln a)^n x^n}{n!} = 1 + \frac{(\ln a)x}{1!} + \frac{(\ln a)^2 x^2}{2!} + \frac{(\ln a)^3 x^3}{3!} + \cdots$$

$$\ln(1 \pm x) = -\sum_{n=1}^{\infty} \frac{(\pm 1)^n x^x}{n} = \pm x - \frac{x^2}{2} \pm \frac{x^3}{3} - \cdots, \quad (\,|x|<1)$$

$$\sin x = \sum_{n=0}^{\infty} \frac{(-1)^n x^{2n+1}}{(2n+1)!} = x - \frac{x^3}{3!} + \frac{x^5}{5!} - \frac{x^7}{7!} + \cdots$$

$$\cos x = \sum_{n=0}^{\infty} \frac{(-1)^n x^{2n}}{(2n)!} = 1 - \frac{x^2}{2!} + \frac{x^4}{4!} - \frac{x^6}{6!} + \cdots$$

$$\tan x = x + \frac{x^3}{3} + \frac{2x^5}{15} + \cdots, \quad (\,|x|<1)$$

$$\tan^{-1} x = \sum_{n=0}^{\infty} \frac{(-1)^n x^{2n+1}}{2n+1} = x - \frac{x^3}{3} + \frac{x^5}{5} - \frac{x^7}{7} + \cdots, \quad (\,|x|<1)$$

B.10 Logarithmic Identities

$\log_e a = \ln a$ (natural logarithm)

$\log_{10} a = \log a$ (common logarithm)

$\log ab = \log a + \log b$

$\log \dfrac{a}{b} = \log a - \log b$

$\log a^n = n \log a$

B.11 Exponential Identities

$$e^x = 1 + x + \frac{x^2}{2!} + \frac{x^3}{3!} + \frac{x^4}{4!} + \cdots, \quad \text{where } e \simeq 2.7182$$

$e^x e^y = e^{x+y}$

$(e^x)^n = e^{nx}$

$\ln e^x = x$

B.12 Approximations for Small Quantities

If $|a| \ll 1$, then

$$\ln(1+a) \simeq a$$

$$e^a \simeq 1+a$$

$$\sin a \simeq a$$

$$\cos a \simeq 1$$

$$\tan a \simeq a$$

$$(1 \pm a)^n \simeq 1 \pm na$$

B.13 Matrix Notation and Operations

Finite element analysis procedures are most commonly described using matrix notation. These procedures eventually lead to solution of a large set of simultaneous equations. Therefore, we describe here the basics of matrix notation and matrix operations.

1. Matrices

    A *matrix* is a rectangular array of elements arranged in rows and columns. The array is commonly enclosed in brackets. Let a matrix A (expressed in boldface as $\mathbf{A}$ or in bracket as $[A]$) has m rows and n columns, then the matrix can be expressed by

$$\mathbf{A} = [A] = \begin{bmatrix} a_{11} & a_{12} & \cdot & \cdot & \cdot & a_{1j} & \cdot & \cdot & \cdot & a_{1n} \\ a_{21} & a_{22} & \cdot & \cdot & \cdot & a_{2j} & \cdot & \cdot & \cdot & a_{2n} \\ \cdot & \cdot & \cdot & \cdot & \cdot & \cdot & \cdot & \cdot & \cdot & \cdot \\ \cdot & \cdot & \cdot & \cdot & \cdot & \cdot & \cdot & \cdot & \cdot & \cdot \\ a_{i1} & a_{i2} & \cdot & \cdot & \cdot & a_{ij} & \cdot & \cdot & \cdot & a_{in} \\ \cdot & \cdot & \cdot & \cdot & \cdot & \cdot & \cdot & \cdot & \cdot & \cdot \\ \cdot & \cdot & \cdot & \cdot & \cdot & \cdot & \cdot & \cdot & \cdot & \cdot \\ \cdot & \cdot & \cdot & \cdot & \cdot & \cdot & \cdot & \cdot & \cdot & \cdot \\ a_{m1} & a_{m2} & \cdot & \cdot & \cdot & a_{mj} & \cdot & \cdot & \cdot & a_{mn} \end{bmatrix}$$

    where the element a_{ij} has two subscripts, of which the first denotes to the row ith and the second denotes to the column jth which the element locates in the matrix. A matrix with m rows and n columns, $[A]$, is defined as a matrix of order or size $m \times n$ (m by n), or a $m \times n$ matrix. A vector is a matrix that consists of only one row or one column.

    Location of an element in a matrix:

$$\text{Let } A = \begin{bmatrix} a_{11} & a_{12} & a_{13} & a_{14} \\ a_{21} & a_{22} & a_{23} & a_{24} \\ a_{31} & a_{32} & a_{33} & a_{34} \\ a_{41} & a_{42} & a_{43} & a_{44} \end{bmatrix} \text{ is matrix with size } 4 \times 4$$

    a_{11} is element a at row 1 and column 1.
    a_{12} is element a at row 1 and column 2.
    a_{32} is element a at row 3 and column 2.

2. Special common types of matrices

    a. If $m \neq n$, then the matrix $[A]$ is called *rectangular matrix.*

    b. If $m = n$, then the matrix $[A]$ is called *square matrix of order n.*

    c. If $m = 1$ *and* $n > 1$, then the matrix $[A]$ is called *row matrix or row vector.*

    d. If $m > 1$ *and* $n = 1$, then the matrix $[A]$ is called *column matrix or column vector.*

    e. If $m = 1$ *and* $n = 1$, then the matrix $[A]$ is called a scalar.

f. A *real matrix* is a matrix whose elements are all real.

g. A *complex matrix* is a matrix whose elements that may be complex.

h. A *null matrix* a matrix whose elements are all zero.

i. An *identity* (or *unit*) *matrix*, [*I*] or **I**, is a square matrix whose elements are equal to zero except those located on its *main diagonal* elements, which are unity (or one). *Main diagonal* elements have equal row and column subscripts. The main diagonal runs from the upper left corner to the lower right corner. If the elements of an identity matrix are denoted as e_{ij}, then

$$e_{ij} = \begin{cases} 1, & i = j \\ 0, & i \neq j \end{cases}$$

j. A *diagonal matrix* is a square matrix which has zero elements everywhere except on its main diagonal. That is, for diagonal matrix $a_{ij} = 0$ when $i \neq j$ and not all a_{ii} are zero.

k. A *symmetric matrix* is a square matrix whose elements satisfy the condition $a_{ij} = a_{ji}$ for $i \neq j$.

l. An *anti-symmetric (or skew symmetric) matrix* is a square matrix whose elements $a_{ij} = -a_{ji}$ for $i \neq j$, and $a_{ii} = 0$.

m. A *triangular matrix* is a square matrix whose all elements on one side of the diagonal are zero. There are two types of triangular matrices; first, an upper triangular **U** whose elements below the diagonal are zero, and second, a lower triangular **L**, whose elements above the diagonal are all zero.

n. A *partitioned matrix* is a matrix whose can be divided into smaller arrays (*submatrices*) by horizontal and vertical lines.

3. Matrix operations

a. Transpose of a matrix
The *transpose* of a matrix $\mathbf{A} = [a_{ij}]$ is donated as $\mathbf{A}^T = [a_{ji}]$ and is obtained by interchanging the rows and columns in matrix **A**. Thus, if a matrix **A** is of order $m \times n$, then $\mathbf{A}^T$ will be of order $n \times m$.

b. Addition and subtraction
Addition and subtraction can only be performed for matrices of the same size. The addition is accomplished by adding corresponding elements of each matrix. For addition, $\mathbf{C} = \mathbf{A} + \mathbf{B}$ implies that $c_{ij} = a_{ij} + b_{ij}$.

Now, the subtraction is accomplished by subtracting corresponding elements of each matrix. For subtraction, $\mathbf{C} = \mathbf{A} - \mathbf{B}$ implies that $c_{ij} = a_{ij} - b_{ij}$ where c_{ij}, a_{ij}, and b_{ij} are typical elements of the **C**, **A**, and **B** matrices, respectively. Both **A** and **B** matrices are in the same size $m \times n$. The resulting matrix C is also of size $m \times n$.

Matrix addition and subtraction are associative:

$$A + B + C = (A + B) + C = A + (B + C)$$

$$A + B - C = (A + B) - C = A + (B - C)$$

Matrix addition and subtraction are commutative:

$$\mathbf{A} + \mathbf{B} = \mathbf{B} + \mathbf{A}$$

$$\mathbf{A} - \mathbf{B} = -\mathbf{B} + \mathbf{A}$$

c. Multiplication by scalar
    A matrix is multiplied by a scalar by multiplying each element of the matrix by the scalar. The multiplication of a matrix A by a scalar c is defined as

$$c\mathbf{A} = \left[ca_{ij}\right]$$

The scalar multiplication is commutative.

d. Matrix multiplication
    The product of two matrices is $\mathbf{C} = \mathbf{AB}$ if and only if the number of columns in $\mathbf{A}$ is equal to the number of rows in B. The product of matrix $\mathbf{A}$ of size $m \times n$ and matrix $\mathbf{B}$ of size $n \times r$ resulting in matrix $\mathbf{C}$ of size $m \times r$. Then, $c_{ij} = \displaystyle\sum_{k=1}^{n} a_{ik}b_{kj}$. That is, the (ij)th component of matrix $\mathbf{C}$ is obtained by taking the dot product

$$c_{ij} = (i\text{th row of } \mathbf{A}) \cdot (j\text{th column of } \mathbf{B})$$

Matrix multiplication is associative:

$$\mathbf{ABC} = (\mathbf{AB})\mathbf{C} = \mathbf{A}(\mathbf{BC})$$

Matrix multiplication is distributive:

$$\mathbf{A}(\mathbf{B} + \mathbf{C}) = \mathbf{AB} + \mathbf{AC}$$

Matrix multiplication is not commutative:

$$\mathbf{AB} \neq \mathbf{BA}$$

e. Transpose of matrix multiplication
    Transpose of matrix multiplication, is usually denoted $(\mathbf{AB})^T$, and is defined as

$$(\mathbf{AB})^T = \mathbf{B}^T \mathbf{A}^T$$

f. Inverse of square matrix
    The inverse of a matrix A is denoted by $\mathbf{A}^{-1}$. The inverse matrix satisfies

$$\mathbf{AA}^{-1} = \mathbf{A}^{-1}\mathbf{A} = \mathbf{I}$$

A matrix that possesses an inverse is called *nonsingular matrix (or invertible matrix)*. A matrix without an inverse is called a *singular matrix*.

g. Differentiation of a matrix

Differentiation of a matrix is differentiation of every element of the matrix separately. To emphasize, if the elements of the matrix A are a function of t, then

$$\frac{d\mathbf{A}}{dt} = \left[\frac{da_{ij}}{dt}\right]$$

h. Integration of a matrix

Integration of a matrix is integration of every element of the matrix separately. To emphasize, if the elements of the matrix A are a function of t, then

$$\int \mathbf{A}\, dt = \left[\int a_{ij} dt\right]$$

i. Equality of matrices

Two matrices are equal if they have the same sizes and their corresponding elements are equal.

4. Determinant of a matrix

The determinant of a square matrix **A** is a scalar number denoted by $|\mathbf{A}|$ or det **A**. The value of a second-order determinant is calculated from

$$\det \begin{bmatrix} a_{11} & a_{12} \\ a_{21} & a_{22} \end{bmatrix} = \begin{vmatrix} a_{11} & a_{12} \\ a_{21} & a_{22} \end{vmatrix} = a_{11}a_{22} - a_{12}a_{21}$$

By using the sign rule of each term, the determinant is determined by the first row in the diagram $\begin{vmatrix} + & - & + \\ - & + & - \\ + & - & + \end{vmatrix}$.

The value of a third-order determinate is calculated in form

$$\det \begin{bmatrix} a_{11} & a_{12} & a_{13} \\ a_{21} & a_{22} & a_{23} \\ a_{31} & a_{32} & a_{33} \end{bmatrix} = \begin{vmatrix} a_{11} & a_{12} & a_{13} \\ a_{21} & a_{22} & a_{23} \\ a_{31} & a_{32} & a_{33} \end{vmatrix} =$$

$$a_{11}\begin{vmatrix} a_{22} & a_{23} \\ a_{32} & a_{33} \end{vmatrix} - a_{12}\begin{vmatrix} a_{21} & a_{23} \\ a_{31} & a_{33} \end{vmatrix} + a_{13}\begin{vmatrix} a_{21} & a_{22} \\ a_{31} & a_{32} \end{vmatrix}$$

B.14 Vectors

1. Vector derivatives
   a. Cartesian coordinates

Coordinates	(x,y,z)
Vector	$A = A_x\,a_x + A_y\,a_y + A_z\,a_z$
Gradient	$\nabla A = \dfrac{\partial A}{\partial x}\,a_x + \dfrac{\partial A}{\partial y}\,a_y + \dfrac{\partial A}{\partial z}\,a_z$
Divergence	$\nabla \cdot A = \dfrac{\partial A_x}{\partial x} + \dfrac{\partial A_y}{\partial y} + \dfrac{\partial A_z}{\partial z}$
Curl	$\nabla \times A = \begin{vmatrix} a_x & a_y & a_z \\ \dfrac{\partial}{\partial x} & \dfrac{\partial}{\partial y} & \dfrac{\partial}{\partial z} \\ A_x & A_y & A_z \end{vmatrix}$
	$= \left(\dfrac{\partial A_z}{\partial y} - \dfrac{\partial A_y}{\partial z} \right) a_x + \left(\dfrac{\partial A_x}{\partial z} - \dfrac{\partial A_z}{\partial x} \right) a_y + \left(\dfrac{\partial A_y}{\partial x} - \dfrac{\partial A_x}{\partial y} \right) a_z$
Laplacian	$\nabla^2 A = \dfrac{\partial^2 A}{\partial x^2} + \dfrac{\partial^2 A}{\partial y^2} + \dfrac{\partial^2 A}{\partial z^2}$

   b. Cylindrical coordinates

Coordinates	(ρ,ϕ,z)
Vector	$A = A_\rho a_\rho + A_\phi a_\phi + A_z a_z$
Gradient	$\nabla A = \dfrac{\partial A}{\partial \rho}\,a_\rho + \dfrac{1}{\rho}\dfrac{\partial A}{\partial \phi}\,a_\phi + \dfrac{\partial A}{\partial z}\,a_z$
Divergence	$\nabla \cdot A = \dfrac{1}{\rho}\dfrac{\partial}{\partial \rho}(\rho A_\rho) + \dfrac{\partial A_\phi}{\partial \phi} + \dfrac{\partial A_z}{\partial z}$
Curl	$\nabla \times A = \dfrac{1}{\rho}\begin{vmatrix} a_\rho & \rho a_\phi & a_z \\ \dfrac{\partial}{\partial \rho} & \dfrac{\partial}{\partial \phi} & \dfrac{\partial}{\partial z} \\ A_\rho & \rho A_\phi & A_z \end{vmatrix}$
	$= \left(\dfrac{1}{\rho}\dfrac{\partial A_z}{\partial \phi} - \dfrac{\partial A_\phi}{\partial z} \right) a_\rho + \left(\dfrac{\partial A_\rho}{\partial z} - \dfrac{\partial A_z}{\partial \rho} \right) a_\phi + \dfrac{1}{\rho}\left(\dfrac{\partial}{\partial x}(\rho A_\phi) - \dfrac{\partial A_\rho}{\partial \phi} \right) a_z$
Laplacian	$\nabla^2 A = \dfrac{1}{\rho}\dfrac{\partial}{\partial \rho}\left(\rho \dfrac{\partial A}{\partial \rho} \right) + \dfrac{1}{\rho^2}\dfrac{\partial^2 A}{\partial \phi^2} + \dfrac{\partial^2 A}{\partial z^2}$

c. Spherical coordinates

Coordinates	(r,θ,ϕ)
Vector	$A = A_r a_r + A_\theta a_\theta + A_\phi a_\phi$
Gradient	$\nabla A = \dfrac{\partial A}{\partial r} a_r + \dfrac{1}{r}\dfrac{\partial A}{\partial \theta} a_\theta + \dfrac{1}{r\sin\theta}\dfrac{\partial A}{\partial \phi} a_\phi$
Divergence	$\nabla \cdot A = \dfrac{1}{r^2}\dfrac{\partial}{\partial r}(r^2 A_r) + \dfrac{1}{r\sin\theta}\dfrac{\partial}{\partial \theta}(A_\theta \sin\theta) + \dfrac{1}{r\sin\theta}\dfrac{\partial A_\phi}{\partial \phi}$
Curl	$\nabla \times A = \dfrac{1}{r^2 \sin\theta} \begin{vmatrix} a_r & r a_\theta & (r\sin\theta)a_\phi \\ \dfrac{\partial}{\partial r} & \dfrac{\partial}{\partial \theta} & \dfrac{\partial}{\partial \phi} \\ A_r & r A_\theta & (r\sin\theta)A_\phi \end{vmatrix}$
	$= \dfrac{1}{r\sin\theta}\left(\dfrac{\partial}{\partial \theta}(A_\phi \sin\theta) - \dfrac{\partial A_\theta}{\partial \phi}\right)a_r + \dfrac{1}{r}\left(\dfrac{1}{\sin\theta}\dfrac{\partial A_r}{\partial \phi} - \dfrac{\partial}{\partial r}(rA_\phi)\right)a_\theta$
	$+ \dfrac{1}{r}\left(\dfrac{\partial}{\partial r}(rA_\theta) - \dfrac{\partial A_r}{\partial \theta}\right)a_\phi$
Laplacian	$\nabla^2 A = \dfrac{1}{r^2}\dfrac{\partial}{\partial r}\left(r^2\dfrac{\partial A}{\partial r}\right) + \dfrac{1}{r^2 \sin\theta}\dfrac{\partial}{\partial \theta}\left(\sin\theta\dfrac{\partial A}{\partial \theta}\right) + \dfrac{1}{r^2 \sin^2\theta}\dfrac{\partial^2 A}{\partial \phi^2}$

2. Vector identity
   a. Triple products

$$\mathbf{A}\cdot(\mathbf{B}\times\mathbf{C}) = \mathbf{B}\cdot(\mathbf{C}\times\mathbf{A}) = \mathbf{C}\cdot(\mathbf{A}\times\mathbf{B})$$

$$\mathbf{A}\times(\mathbf{B}\times\mathbf{C}) = \mathbf{B}(\mathbf{A}\cdot\mathbf{C}) - \mathbf{C}(\mathbf{A}\cdot\mathbf{B})$$

   b. Product rules

$$\nabla(fg) = f(\nabla g) + g(\nabla f)$$

$$\nabla(\mathbf{A}\cdot\mathbf{B}) = \mathbf{A}\times(\nabla\times\mathbf{B}) + \mathbf{B}\times(\nabla\times\mathbf{A}) + (\mathbf{A}\cdot\nabla)\mathbf{B} + (\mathbf{B}\times\nabla)\mathbf{A}$$

$$\nabla\cdot(f\mathbf{A}) = f(\nabla\cdot\mathbf{A}) + \mathbf{A}\cdot(\nabla f)$$

$$\nabla(\mathbf{A}\times\mathbf{B}) = \mathbf{B}\cdot(\nabla\times\mathbf{A}) - \mathbf{A}\cdot(\nabla\times\mathbf{B})$$

$$\nabla\times(f\mathbf{A}) = f(\nabla\times\mathbf{A}) - \mathbf{A}\times(\nabla f) = \nabla\times(f\mathbf{A}) = f(\nabla\times\mathbf{A}) + (\nabla f)\times\mathbf{A}$$

$$\nabla\times(\mathbf{A}\times\mathbf{B}) = (\mathbf{B}\cdot\nabla)\mathbf{A} - (\mathbf{A}\cdot\nabla)\mathbf{B} + \mathbf{A}(\nabla\cdot\mathbf{B}) - (\nabla\cdot\mathbf{A})$$

c. Second derivative

$$\nabla \cdot (\nabla \times \mathbf{A}) = 0$$

$$\nabla \times (\nabla f) = 0$$

$$\nabla \cdot (\nabla f) = \nabla^2 f$$

$$\nabla \times (\nabla \times \mathbf{A}) = \nabla(\nabla \cdot \mathbf{A}) - \nabla^2 \mathbf{A}$$

d. Addition, division, and power rules

$$\nabla(f + g) = \nabla f + \nabla g$$

$$\nabla \cdot (\mathbf{A} + \mathbf{B}) = \nabla \cdot \mathbf{A} + \nabla \cdot \mathbf{B}$$

$$\nabla \times (\mathbf{A} \times \mathbf{B}) = \nabla \times \mathbf{A} + \nabla \times \mathbf{B}$$

$$\nabla\left(\frac{f}{g}\right) = \frac{g(\nabla f) - f(\nabla g)}{g^2}$$

$$\nabla f^n = n f^{n-1} \nabla f \quad (n = \text{integer})$$

3. Fundamental theorems

a. Gradient theorem

$$\int_a^b (\nabla f) \cdot d\mathbf{l} = f(b) - f(a)$$

b. Divergence theorem

$$\int_{volume} (\nabla \cdot \mathbf{A}) dv = \oint_{surface} \mathbf{A} \cdot d\mathbf{s}$$

c. Curl (Stokes) theorem

$$\int_{surface} (\nabla \times \mathbf{A}) \cdot d\mathbf{s} = \oint_{line} \mathbf{A} \cdot d\mathbf{l}$$

d. $$\oint_{line} f d\mathbf{l} = - \oint_{surface} \nabla f \times d\mathbf{s}$$

e. $$\oint_{surface} f d\mathbf{s} = \int_{volume} \nabla f dv$$

f. $$\oint_{surface} \mathbf{A} \times d\mathbf{s} = - \int_{volume} \nabla \times \mathbf{A} dv$$

Index

A

Ab initio molecular dynamics
  Born–Oppenheimer approximation, 341
  description, 341
  DFT, 341–342
  Hartree–Fock theory, 341
ACP, *see* Amorphous calcium phosphate (ACP)
Adhesion and wear test methods, 385
AFEM, *see* Atomic-scale finite element
      modeling (AFEM)
Amorphous calcium phosphate (ACP), 375
Amplification method, 482, 489
Anisotropic constitutive law, 205–206
Approximations, small quantities, 599
Atomic-scale finite element modeling (AFEM)
  vs. continuum mechanics solutions
    material anisotropy effects, curvature
        radius, 229–231
    self-positioning nanostructures,
        226–228
    thickness effects, curvature
        radius, 228–229
  crystalline structures, 209–210
  defined, 206
  equation system, 206–207
  geometrically nonlinear problems, 208–209
  interactions potential function, 210, 212–213
  mesh free interpolation techniques, 230–235
  PCG algorithm, equation system, 214–217
  strain estimation methods, 217–223
  Tersoff-Nordlund potential validation, gaAs
      and inAs, 213–214

B

Base-centered cubic (BCC) lattices, 103
BCB, *see* Benzocyclobutene (BCB)
Bean
  description, 574
  lens effect, 575, 576
  velocity field, 574–575
BEM, *see* Boundary element method (BEM)
Benzocyclobutene (BCB)
  Cu inter-wafer vias, 37
  design parameters, 22, 38
  3D IC structures, 37
  von Mises stress, 30, 31

wafer-level 3D integrations, 21
Young's modulus, 29
Biological systems, applications
  deployment, stent, 283
  dirac delta function, 280–281
  3D multiscale technique, 282
  Eulerian and Lagrangian description, 280
  fluid and solid domain, 279
  fluid-structure interaction force, 280
  Lagrangian formulation, 281
  Newton–Raphson method, 281
  simulation, cell motility, 282
Boundary element method (BEM), 249, 274, 275
Built-in linear drives mechanism
  parallel micro-robot structures
    accessible zone, vertex C, 528, 529
    three built-in linear drives, 529–531
    two simultaneously operating linear
        drives, 529, 530
  spatial three-pod
    angular displacement, magnitude,
        533, 537
    characteristics, 533, 536
    description, 533, 535
    linear displacement, vertex C, 533, 537
Built-in rotary drives mechanism
  parallel micro-robot structures
    configurations and trace path,
        vertex C, 526
    three, rotary drives, 527, 528
    two, rotary drives, 527
  spatial three-pod
    angular displacement, magnitude,
        532, 535
    characteristics, 532, 533
    description, 532
    linear displacement, vertex C, 532, 534

C

Cachy–Born rule, 260
Carbon nanotubes (CNTs)
  axial loading and boundary
      conditions, 302
  chirality, 44
  CoNTub1.0 program, 43
  defined, 253

description, 291–292
FE model, 293
fibers, 264, 265
fracture toughness, 295
geometry, graphitic plane rolling, 44
glass fiber, 293
graphene sheet, *see* Graphene sheet
inclusions, 267
matrix condensation approach, 43
MWCNT and SWCNT, 301, 302
nano-composites, NEMS, 43
nanomaterial tester, 50–52
nanomotor, 52–53
NEMS resonator, 53
PP, 292
reaction forces
    axial loading, 301, 303
    transverse loading, 301, 304
reduced-order modeling, 47–48
rubber matrix, 292–293
RVE, 292
structural mechanics model, 44–47
sugarcube design and simulation, 50, 51
sugar design
    deflected models implementation, 48, 49
    3D MEMS simulation, 48
    reduced-order lumped model, 48–50
SWCNT, 295
synthesis, RVE, 260, 261
tensile behavior, 294
transverse loading and boundary
        conditions, 302, 303
Von Mises stresses
    axial loading, 298, 301, 304
    transverse loading, 298, 301, 305
Carreau–Yasuda and Ostwald law, 434, 435
Cauchy–Born hypothesis, 334
Cell chips
    bifurcations and pinched channel
            microflows
        flow, branched network, 452, 453
        flow focusing, 454–455
        PFF coupled, flow focusing device, 456–458
        pinched flow fractionation, 455–457
        trajectories, spherical particles, 452, 454
    culture
        bolus, concentration, 465, 466
        concentration map, 466
        convective microchannel, 464–465
        velocity field, microsystem, 465
    defined, 448–449
    DLD, *see* Deterministic lateral
            displacement (DLD)

single-phase flow focusing, 450–453
trapping chambers, 462–464
CGMD, *see* Coarse-grained molecular
        dynamics (CGMD)
CGMD *vs.* macro-model
    characteristics, 110
    comparison, 110, 111
    description, 109
    JKR contact model, 110
    local attraction and repulsion, 111
    system geometry, 109, 110
Chemical and physical bonds, MDFEM
    covalent bonding, 335–336
    description, 335
    hydrogen bonding, 338
    ionic bonding, 336–337
    metallic bond, 337
    Van der Waals bonding, 337–338
Chemical vapor deposition (CVD)
    coatings and films, 382
    flexibility, 382
    PVD, 382
Clamping devices
    aim, 510–511
    averaged radial displacement, 512, 516
    description, 510
    distribution, von Misses stress, 512, 513
    3D model, 511
    eligible solutions, 512
    loaded clamping module, 511, 512
    numerical experiments, *see* Numerical
            experiments, clamping devices
    shear modulus, 512
    spatial finite elements, 511
    vertex nodes, "proboscis" surface,
            512, 513
CM, *see* Continuum mechanics (CM)
CNTs, *see* Carbon nanotubes (CNTs)
CNTs and nanocomposites
    ANSYS, COMSOL and Nastran, 6
    approaches, 6
    buckminsterfullerene synthesis, 5
    definition, 5
    electronic properties, 5
    gap bridging, 6
    softwares, 6
    SWCNTs, 5
Coarse-grained molecular dynamics (CGMD)
    comparison, RDFs, 108
    electromechanical process, 107
    procedures, 108
    SC potential, 108
    trial and error process, 108

Coefficients of thermal expansion (CTEs)
    SiLK, BEOL processing, 26
    thermally induced stresses, 22
    Young's moduli, 27
Coherent precipitation
    composition, shape and size, 160
    description, 160
    energy processes, 160
    Eshelby construction, 160, 161
    FEM computations, 160–161
    isotropic conditions, 162
    misfit dislocations, 161–162
    parent *vs.* product phase, 160
    preexisting dislocations, 162
    spherical, interfacial misfit loop, 161
Complex variables, 592
Computational methods, nanotechnology
    atoms/molecules manipulation, 1–2
    CNTs and nanocomposites, *see* Carbon
        nanotubes (CNTs)
    complex systems, 2
    description, 4–5
    DNA and proteins, 2
    economic and social significance, 3
    electronic quantum dots, *see* Electronic
        quantum dots
    FE, microscale circuits, 8–9
    microscale circuits interconnection, *see*
        Finite element method (FEM)
    nanoscale structures
        description, 3–4
        elements, 3
        groups, 4
    nanostructures, 2
    novel and unique physical–chemical
        properties, 2
    quantum wires and nano-MOSFET
        devices, 8–9
    theoretical skills and computer power, 12
    transistors and electronic processors, 1
COMSOL
    defined, *cho2comsol,* 78
    with 3D orientations, 78, 79
    geometries, 79
    parameterizable shapes, 78
Continuum mechanics (CM)
    description, 94
    nanorobotic problem, FEM, *see*
        Nanorobotics, FEM
Corks
    capillary net, 575, 577
    gas pressure, 575, 576
Coulomb blockade/Pauli exclusion principle, 492

Coupled atomistic/dislocation dynamics
    (CADD) method, 98
Coupled thermo-electrical-mechanical finite
    element model
    description, 20
    equations, components, 20–21
Creeping Flow Reversibility, 428–429
Crystalline structures modeling, AFEM
    atomic, material axes orientations, 210
    GaAs and InAs, zincblende, 209, 210
    zincblende atom configurations, 210, 211
CVD, *see* Chemical vapor deposition (CVD)

D

2D approach, *see* Hele-shaw model, pseudo 3D
    microflow
DC, *see* Direct coupling (DC)
Dean effect, 447
Dean flows
    concentration, 448–449
    hydrodynamics, 447–448
    Reynolds number, 447
Density functional theory (DFT), 341–342
Derivatives, 592–594
Deterministic lateral displacement (DLD)
    analyses, 459
    cell lysis solution, 460, 461
    D_c/m *vs.* ε, 460
    numerical model, 461–462
    particle diameter, 459
    principle, 457, 458
DFT, *see* Density functional theory (DFT)
Diamond-like carbon (DLC) properties, 384
3D IC structures, FE method
    BCB bonding, 21–22
    brittle materials, 33
    computations, 22
    computed von Mises stresses, 25, 26, 29–31
    COMSOL multiphysics, 24
    contour and isosurface, 32–33
    Cu metallization, 26
    description, 27–28
    distributions, stress, 27, 28
    experimental data
        electrical failures, 23, 24
        low-k dielectrics, 22
        metal and vias levels, test structure, 23
        SEM image, SiLK, 23
    grain boundaries, 34
    Hall–Petch formula, 25
    materials properties, chain test
        simulations, 24–25, 29

mechanical stability, 21
micromachined Cu thin-film beams, 25
MLM layers, 21
parametric studies, 26–27
Poisson's ratio, Cu, 29
SiLK and SiCOH, 22
static analysis, 25
stress calculation, point *P,* 27
stress-free reference state, 25
symmetry, x-z and y-z planes, 28
temperature change, Cu vias, 31, 32
TEOS oxide and methyl silsesquioxane
    dielectrics, 34
tetrahedral FE mesh, barrier and capping
    layers, 25, 26
thermally induced stresses, 22
vertical displacement, 33
volume-averaged stresses, 33
wafer curvature, 22
wafer-level integration, 21
Young's moduli and CTEs, 27
Direct coupling (DC), 98
Dislocations and coherent nanostructures
advantages, 150
assumptions and quality, results, 180
boundary conditions, 152–153
coherent precipitation, *see* Coherent
    precipitation
definition, 150
domain choice, 153
edge dislocations, *see* Edge dislocations
eigenstrains, *see* Eigenstrains
FEM, *see* Finite element method (FEM)
heteroepitaxy and structures, *see*
    Heteroepitaxy
improvements, 180–181
material properties, simulations, 181
Mesh, 152
microstructure, 151
objectives, 150
problems and limitations, 151
simulation process and computation
    parameters, 150
simulations, FEM, 163
theoretical equations, 163
DLC, *see* Diamond-like carbon (DLC)
DLD, *see* Deterministic lateral displacement
    (DLD)
DNA recognition, heterogeneous reactions
adsorption, 466, 467
convection-diffusion-reaction problem,
    470–472
kinetic constants, 470

kinetics, adsorption and desorption, 468–469
surface concentration kinetics, 468

E

Edge dislocations
finite domains
    contour plot, σ_{xx} and τ_{xy} stress fields,
        155, 156
    cylindrical, 154–155
    description, 154
    image forces, 155–158
nanocrystals and hybrids
    compute model, harder material, 165, 166
    domain and boundary conditions, 165
    energy, edge dislocation and image force,
        165, 166
    glide component, image force, 165, 167
    image force calculation, 165
    isotropic moduli, 164
    mesh size, 164
    numerical model, 163, 164
    stress state, 164–165
    structure and energy, 164
    σ_x contours and deformations plot,
        165, 167
neutral equilibrium and stability
    deformed configurations, domain,
        167, 169
    "material" and "structural"
        characteristics, 168
    simulation model, thin free-standing Al
        plate, 168
Eigenstrains
heating, 153, 154
matrix evaluation, 154
"stress-free strains", 153
Electro magnetic (EM) waves, 269, 271, 272
Electronic quantum dots
definition, "nano-interferometer", 7
2D quantum dot plane, 7
GaAs/AlGaAs, 7
interplays, 8
magnetic field and arrays radiation, 7–8
properties, 6
QMC methods, 8
single-electron/reliable wave functions, 8
Electro-optical modeling
dark current *vs.* voltage, 488–490
device heating, 491, 492
optimization, device structure and doping
    levels, 488, 489
photocurrent and amplification, 488–489

potential distribution and electric field, 490–491

quantum efficiency, 489–490

Electrospinning

conical-coil process, 324–325

description, 311–312

edge-plate, 326

electric field

description, 314–315

fiber generators, 316

needle nozzle and tip, 315–316

PDEs, 315

and electrospun nanofibers, 312–314

parameters

different applied voltages, 323

fiber productivity, 324

rotary needleless spinnerets, *see* Rotary needleless spinnerets

Electrospun nanofibers

applications, 313

jet formation, 313–314

technology, 312

Electro-Thermo-Actuator model

dog-shaped polysilicon test specimen, 75

FEA, 75

MEMS nanomaterial tester, 76

thermal actuator displacement, 75

EM waves, *see* Electro magnetic waves

Eshelby method, 249

Established mathematical model, parallel micro-robot structures

design solution, implementation, 524–525

geometrical solution, spatial mechanism, 522, 523

nonlinear spatial mathematical system, 524

pseudo-2D four bar mechanism, 523

Exponential identities, 599

F

Fabrication technology, 187

Face-centered cubic (FCC) lattices, 103, 104

FDM, *see* Finite difference method (FDM)

FEA, *see* Finite element analysis (FEA)

FEM, *see* Finite element method (FEM)

Filtration

description, 560

fictitious soil, porosity

cube, volume, 566

diagonal cross sections, rhombohedron, 562

free space, 565

rhombohedron, 561

Slichter's formula, 564

sphere, 562–563

spheres dispositions, 561–562

viscosity of fluid, 561

Finite difference method (FDM), 106

Finite element analysis (FEA)

ACP, 375

bioceramics, 374

biomimetic processing, 376

bone mineral, 375

cardioelectrical circulation, 412–413

chrisotherapeutic gold compounds, 418

computer-based approach, 402

3D body, finite volume and surface, 245, 246

dental and medical implants, 381

dental filling material, 411

dentistry, 405–406

description, 373–374, 401–402

DLC, 384

electricity transfer, 404

engineering technique, 402–403

fever, red blood cells, 413

flow cytometry, 410–411

force equilibrium, 247

heat transfer, 404

hydrogen, nanosensor diagnostic property, 414–415

MRI, 416

nanocomposite, 376

nanocrystalline coatings

CVD, 382

description, 381–382

PVD, 382

Sol–gel process, 382–384

nanodrug, cancerous treatment, 417

nanofiber patch, 415–416

nanomedicine

clarification, 407

crestal bone loss, 380–381

FE meshes, 379

fracture mechanisms, 380

GNPs, 378–379

nanofiber and nanoparticle, 378

nanoscale, 406

nanostructured materials, 377

prediction, 407–408

quadratic remodeling formula, 377–378

silico technique, 406

nanostructured materials, 375

orthopedics, 404–405

strain-displacement relations, 247–248

thin film mechanical test, *see* Thin film mechanical test

tools, biomedical data
    ANSYS, 410
    COMSOL multiphysics, 410
    differential and integral equation, 409
    FEBio, 410
    GAGSim3D, 410
    graphical model, 409
    MATLAB®, 409
    nodes, 408–409
    parameters, 408
    PostView, 410
    PreView, 410
  vascular injury, 413–414
Finite element method (FEM)
  anisotropic constitutive law, 205–206
  anisotropic structures, 203–205
  approaches, 10
  complexity, 10
  vs. continuum mechanics solution
    bilayer system, 223
    curvature radius calculation, 225, 226
    material properties, 224
    self-positioning nanostructure
      shape, 224–225
    strain components, orientation angles,
      225, 226
  defined, 196
  electrostatic environment, 11
  equation system, 196–197
  ICs and MCMs, 10
  load vector, 199–200
  macro-, micro-, and nano-systems, *see*
    Macro-, micro-, and nano-systems,
    FEM
  manufacturing processes, 151
  micro and nano-systems, *see* Micro and
    nano-systems, biotechnology
  micro-/nanoscale silicon-IC process, 10
  microscale coupled interconnect line,
    Si–SiO_2 substrate, 11–13
  microscale single interconnect line, Si–SiO_2
    substrate, 11–12
  nanocomposites, *see* Nanocomposites, FEM
  nanoinjection detectors, *see* Nanoinjection
    detectors, FEM
  nanoscale problems, 152
  shape functions, 201
  stiffness matrix, 197–199
  stress update, 200–201
  structures, 151
  time integration, 202–203
First-order analysis, N/MEMSs
  boundary conditions, 72–73
  description, 71
  geometric design parameters, V-beams
    pair, 73
Flow focusing device, PFF
  separation, dispersed particulate
    suspension, 456, 458
  spherical particles trajectories, 456, 458
  streamlines, 456, 457
Forcefield potentials, MDFEM
  Coulomb/electrostatic interactions, 346
  hydrogen bondings, 345, 346
  parameter, 344
Fourier's law, 265

G

Gilbert equations, 274, 277
Gold nanoparticles (GNPs), 378–379
Graphene sheet
  boundary conditions, 297
  deformed and undeformed shapes
    first load case, 298
    second load case, 298, 299
  molecular representation, 296–297
  reaction forces
    first load case, 298, 299
    horizontal direction, 298, 300
  Von Mises stresses
    horizontal direction, 298, 301
    vertical direction, 298, 300

H

Halpin–Tsai method, 249
Hartree–Fock theory, 341
Hele–shaw model, pseudo 3D microflow
  application, microfluidic resonator, 442–443
  axial pressure profile, rectangular channel,
    441, 442
  flow velocity map, 440, 441
  friction, 440
  microchamber, 440, 441
  pressure distribution, microchamber,
    440, 441
  quadratic vertical profile, 440
Herringbone structures, microsystems
  shapes, 445, 446
  transport, particles, 447
  Vy-component, velocity, 445, 446
Heteroepitaxy
  Burgers vector, 177
  coherent to semicoherent transition, 177–178
  Cu-4wt.%Co solid solution, 176

description, 158
force balance approach, 159–160
growth modes, 158–159
InGaAs overlayer, 158
material properties, 177
in nanocrystals, 178–180
stress state (σ_{yy} plot), 177
stress state simulation, 176
thin films and critical thickness
advantages, finite element simulation,
172–173
determination, misfit edge dislocation,
172, 173
dislocation stress fields, 173
numerical model, 171
sapphire substrate, 170–171
stress state, epitaxial system, 171, 172
theoretical formulations, 173
thin substrates, stripes and islands
finite, 176
geometry, 173
models, 173–174
strain interaction effects, 176
stress states, 173, 175
Heterostructure field-effect transistors (HFETs)
description, 34
2D FE model, transistor, 34–35
electrical field distribution, 35–36
failure mechanisms, WBG devices, 34
modules, COMSOL multiphysics, 35
TEM image, AlGaN layer, 34, 35
temperature distribution, 36
von Mises stress distributions, 37
HFETs, *see* Heterostructure field-effect
transistors (HFETs)
Hooke's law, 189, 192, 205, 248
Hot particles, 487
Huggins law, 432
Hyperbolic functions, 591–592

I

Integrals, 594–596
International system, units (SI) prefixes, 589
Inverse problem of kinematics (IPK)
actuating devices (AD), parameters, 539
angular displacement, 538
control of micromotions, 541
description, 538
direct (DPK), 540–541
five-link mechanism, operating actuators,
542, 543
operational area, actuators, 540

quadratic equations, 539
two-crank closed five-link mechanism,
538, 539
variations of angles, 541–542
vector contour, kinematical chain, 543

K

Kinematics, spatial mechanisms
angular velocity and acceleration, 554
climate projections, angular velocity vector,
554, 555
coordinate system, Coupler's angular
velocity, 553–554
position and velocity characteristic, 552–553
SFL, 551, 552
Kriging interpolations
construction, coefficient vectors
calculation, 220
contour plots, AFEM, 233, 234
displacement, 220
GaAs atomic configuration, 233, 234
InAs atomic configuration, 233, 235
plane strain solution and AFEM, 233
quadratic polynomial basis, 221
strain components calculation, 222

L

Lagrangian formulation, 197, 200, 202, 235
Langmuir equation, 468, 469
Langmuir–Hinshelwood mechanism, 466–467
Langmuir law, 466
Laplace equation, 274, 275
"Local-global" approach, *see* Representative
volume element (RVE) modeling
Logarithmic identities, 599
Lumped modeling
CNTs
applications, 50–53
description, 43–44
design and simulation, sugarcube, 50
reduced-order, 47–48
structural mechanics, 44–47
sugar design, 48–50
MEMS, *see* Microelectromechanical systems
(MEMSs)

M

Macro-electromechanical systems (MEMS)
applications, 132
linear and nonlinear results, 137

nano-electrical generators, 139
nonlinear behavior, 144
Macro-field (MF) and nano-field (NF) coupling
    approaches, 121
    calculation, virtual hardness, 122
    definition, center of mass, 122
    FB and CB regions analysis, 121
    regions
        closed systems, 119–120
        open systems, 120–121
Macro-field modeling, FEM
    coordinate systems, 101
    description, 100
    element's geometry and nonisotropic
        characteristics, 101
    hysteresis model, 102
    hysteretic behavior, 101
    nonlinear piezoelectric equation, 102
    piezoelectric sensors and actuators, 100
    structural relationships, 102
Macro-mechatronic system, FEM
    angular kinematic parameters, 546, 548, 549
    CAD/CAE model, 544
    dead weight of links, 546, 549
    difference functions, trace paths of
        vertex D, 546
    graphs of functions, 546
    IPK, *see* Inverse problem of kinematics (IPK)
    kinematic parameters, motion of
        vertex D, 546–548
    piezo-actuators, 544
    rotary drives, 544
    spatial orientation, 546
    trace paths of vertex D, 544, 545
Macro-, micro-, and nano-systems, FEM
    benefits, cooperation, 507–508
    bioreactor systems, 509–510
    clamping device, *see* Clamping devices
    description, 506
    disadvantages, cooperation, 508
    integration, 506, 508
    kinematic analysis, hybrid robot, 509
    macro-mechatronic system, *see* Macro-
        mechatronic system, FEM
    mechanical bioreactors, *see* Mechanical
        bioreactors
    nano-motor and-manipulator, 506, 507
    parallel micro-robot structures, *see* Parallel
        micro-robot structures, FEM
    rotary/linear drive, 509
    three-finger clamp, 508–509
Macro–Micro-Coupling Model
    atomic process calculations, 117

crack propagation problems, 119
definitions, closed and open systems, 118
description, 116–117
fourth-order Runge–Kutta method, 117
MM methods, 116–117
multiscale coupling methods, 119
piezoelectric properties, 119
Macroscale nanorobots
    comparison, CGMD approach,
        127, 128
    hysteresis behavior
        micro-actuators, 132–133
        nanogenerators, *see* Nanogenerators
    limitations, 131
    material, substrate, 129, 130
    MF and NF parts, 126–127
    nanomanipulation methods, 91–92
    nanoscale forces, *see* Nanoscale forces
    NF *vs.* MF, open system, 127
    nonlinearities modeling, *see* Nonlinearities
        modeling
    notches, substrate, 129–131
    physical properties, 128
    roughness, substrate
        configurations, holes, 128–129
        nickel nanoparticle, travelled distance,
            128, 129
    scanner limitations
        CM-based model, 132
        coupling, MF and NF, 132
        creep and hysteresis model, 132
        ferromagnetism, 131–132
        piezoelectric actuators, 131
    SPMs, 130–131
    system configuration, 90, 91
Magnetic properties, nanocomposites
    Co nanoelement and finite element mesh,
        276, 277
    Gilbert equation, 274
    matrix vector multiplication, 275
    mixed finite element, 275
    NiFe elements, 276–278
    Poisson and Laplace equation, 274
    test, magnetostatic field calculations, 276
Material constants
    conductivity, 585
    dielectric constant and strength, 586
    relative permeability, 586
Matrix notation and operations
    definition, 600–603
    determinants, 603
    finite element analysis procedures, 600
    types, 600–601

Maxwell's equations, 271
MC methods, *see* Monte Carlo methods
MD simulation, *see* Molecular dynamics
    simulation
Mean free paths (MFPs), 96
Mechanical bioreactors
    bioreactor systems, 551
    cells, 547
    disadvantages, 550
    force fluid circulation, 550
    innovative solutions, 555–556
    microcarriers, 550
    natural fluid circulation, 550
    spatial mechanisms, kinematics, *see*
        Kinematics, spatial mechanisms
    tissue engineering, 550–551
Mechanical properties, nanocomposites
    contact stiffness, 253
    3D nanoindentation, 253, 254
    elastic modulus, 253
    indentation geometry, 253, 254
    load-displacement curve, 251, 252
    material modeling techniques,
        249, 250
    multiscale RVE modeling, 258–260
    nanomaterials computational modeling,
        249, 250
    object-oriented modeling, 256–258
    Oliver–Pharr method, 251–252
    power law, 253
    spherical/pyramidal indenter, 252
    strains, displacements, 251
    total elastic energy, 251
    unit cell modeling, 255–256
MEMSs, *see* Microelectromechanical systems
    (MEMSs)
Mesh free interpolation techniques
    Kriging interpolation, 233–235
    MLS approximation, 231–233
MFPs, *see* Mean free paths (MFPs)
Micro and nano-systems, biotechnology
    "artificial" isotropic diffusion, 444
    cell chips, *see* Cell chips
    coflows
        concentration contours, 445, 446
        concentration profile, diffusing species,
            444, 445
        miscible fluids, T-junction,
            444, 445
    convection-diffusion problem, 443–444
    creeping flow reversibility, 428–429
    Dean flows, 447–448
    determination, fluid flows, 425

DNA recognition, 465–471
herringbone structures, 445–447
laminarity, microfluidic flows, 429, 430
mass conservation and Navier-Stokes
    equations, 426–427
multi-physics approach, 424–425
non-Newtonian fluids, *see* Non-Newtonian
    fluids
numerical modeling, 425
pressure drop, 429–431
shallow channel/Hele–shaw model,
    440–443
Stokes equations, 427–428
Microelectromechanical systems (MEMSs)
    attributes, 54
    CAD, 54
    characteristics, 53
    components, 77
    COMSOL and MATLAB, 77–78
    2D/3D plots, 67
    design cycles, 76–77
    domains and numerical methods, 76
    efficiency and versatility, 77
    features, 67
    gyroscopes and microgyroscopes, 54
    iSugar framework, 78
    layout
        fabrication, 66, 67
        GDS-II format, 66
    librarian, 57–58
    MEMSolver, 54
    optimization
        algorithm, 63
        parameters, sugarcubes, 64, 65
        static and sinusoidal analysis, 64
    parameterization, 58
    practical design-rule limits, 54
    simulation, 58
    sinusoidal analysis, 62, 63
    static analysis
        strain gradient, ADXL accelerometer,
            58–59
        thermal actuator, 59–60
    steady-state analysis
        description, 60–61
        frequency response, gyroscope, 61
        parameterized frequency response, crab
            leg resonator, 62
    sugar integration, COMSOL, 78–79
    sugar to sugarcube, *see* Sugar to sugarcube
        modelling
    tools, 77
    transient analysis, 62–64

Microfluidic network, 438
Microscale coupled interconnect line, Si–SiO$_2$ substrate
  cross section, 11, 12
  potential distribution, 11, 13
Microscale single interconnect line, Si–SiO$_2$ substrate
  geometry, 11
  potential distribution, 11, 12
Molecular dynamic finite element method (MDFEM)
  ab initio molecular dynamics, 341–342
  carbon nanotube, 360–361
  Cauchy–Born hypothesis, 334
  chemical and physical bonds, *see* Chemical and physical bonds, MDFEM
  constitutive material models, 332
  elastomeric material
    loading step, 363–364
    relaxation step, 361–363
  inversion
    force field potentials, 358–359
    modeling, improper torsion, 359
    planar structures, 357–358
    spacial structures, 357
  matrices, 367–370
  Newton–Raphson method, *see* Newton–Raphson method
  requirement
    finite elements, rotational degrees of freedom, 348–349
    force fields, 344
    linear and nonlinear analysis, 347–348
    material parameters, 348
    multi-body force field potentials, 346–347
    natural lengths and angles, 347
    reference configuration, 351–352
    short-and long-range forcefield potentials, 344–346
    time integration schemes, 349–351
  requirements, 334
  Schrödinger wave equation, 339–341
  semiempirical models, *see* Semiempirical models
  simulation models, 332–333
  user elements
    four-node element, torsion, 355–356
    three-node element, bending, 354–355
    two-node element, bondstretch, 353–354
Molecular dynamics (MD) approach
  "automatic nanomanipulation" procedure, 111

CGMD, *see* Coarse-grained molecular dynamics (CGMD)
CGMD *vs.* macro-model, 109–111
classical Newtonian and Quantum mechanics, 103
continuum-based methods, 103
equations, motion, 106
example, tip damage, 112
FCC and BCC lattices, 103
FDM, 106
"interatomic potential energy", 104
mathematical models, 103
Monte Carlo method, 103
nanomanipulation procedure, 109
nanoparticle and desired position, 113
nanoparticle crushing, 112, 113
NVT ensemble, 107
parameters, 103
Rafii-Tabar–Sutton (RTS) potential, 104–105
SP parameter *vs.* ε_t and ε_p, 115, 117
Sutton–Chen (SC) interatomic potential, 104, 114–115
traveled distance, 112–113
Verlet algorithm, 106–107
Molecular dynamics (MD) simulation, 189
Monte Carlo (MC) methods, 242
Mori–Tanaka method, 249, 256
Moving least squares (MLS) approximation
  coefficients, 217–218
  quartic spline weight function, 218
  strains
    AFEMand plane solution, 231, 232
    contour plots, AFEM, 230, 232
    defined, 231
  vector, 218
MQ polynomials, *see* Multiquadratic polynomials
Multilayer self-positioning nanostructures
  AFEM, *see* Atomic-scale finite element modeling (AFEM)
  crystal lattice mismatching, 186, 187
  defined, 186
  fabrication procedure, rolled-up nanohinges, 186, 187
  FEM, *see* Finite element method (FEM)
  generalized plane strain solution
    anisotropic structures, 195
    bilayer structures, 195
    directions and axes, 192, 193
    force, 194
    Hooke's law, 192
    isotropic materials, 193

and ordinary, 196
    parameters, 195
MD simulation, 189
nanotubes, 187
strain energy minimization, 188
transformation, constitutive matrix
    crystalorientation, structures, 190, 191
    cubic crystals, 190
    Hooke's law, 189
    isotropic materials, 192
    stress and strain vector, 189
Multilevel metallization (MLM) layer
    circular Cu vias, 28
    Cu inter-wafer interconnection, 21
    material properties, 28
    planar ICs, 22
    thermal stresses, 22
Multi-physics approach, biotechnology,
    424–425
Multiquadratic (MQ) polynomials, 217, 219–220
Multiscale method
    coupling, MF and NF, *see* Macro-field (MF)
        and nano-field (NF) coupling
    damped dynamics, MF, 123–124
    damped dynamics, NF, 125–126
    iterative and static equilibrium, 122
    macro-field modeling, 100–102
    macro–micro-coupling model, 116–119
    nano-field modeling, *see* Molecular
        dynamics (MD) approach
    problems, 122–123
Multiwalled nanotube (MWNT)
    development, continuum model, 261, 262
    wire frame model, fiber composite inclusion,
        264, 265
MWNT, *see* Multiwalled nanotube (MWNT)

N

Nano-and microelectromechanical systems
    (N/MEMSs)
    CNT, *see* Carbon nanotubes (CNTs)
    description, 43
    electro-thermo-actuator model, 75–76
    FEA simulation, 68
    first-order analysis, thermal actuator, 71–73
    material-testing device, 67–68
    mechanisms, 42
    MEMS, *see* Microelectromechanical systems
        (MEMSs)
    nanowires and CNTs, 67
    reduced-order modeling, 42–43
    software tools, 42

structural mechanics-based lumped
    model, 68
sugar, CNT model, 70, 71
sugarcube model, *see* Sugarcube model
sugar integration, Simulink®, 79–82
thermo-mechanical response, device, 73–74
traditional CAE tools, 68
Nanocomposites, FEM
    analysis classification, 249
    application, biological systems, 279–283
    computation, thermal conductivity
        adiabatic boundary conditions, RVE,
            267–269
        CNTs, 264
        continuum model development, MWNT,
            261, 262
        expression, effective solid fiber, 262
        Fourier's law, 265–266
        homogenization, 260
        interfaces modeling, 265, 266
        Legendre polynomials, 263
        MWNT inclusion, 261
        RVE, 266–267
        single equivalent solid MWNT fiber
            composite inclusion, 264, 265
        temperature profile, RVE, 267
        volume fraction, notube phase, 263
        wire frame model, 264, 265
    damage computation, 278–279
    FEA, *see* Finite element analysis (FEA)
    hierarchy, multiscale modeling
        techniques, 242
    implementation, 245, 246
    magnetic properties, *see* Magnetic
        properties, nanocomposites
    mechanical properties, *see* Mechanical
        properties, nanocomposites
    nanomaterials and properties, 243–245
    optical properties, *see* Optical properties,
        nanocomposites
Nano-electromechanical systems (NEMS)
    applications, 132
    linear and nonlinear results, 137
    nonlinearity behavior, 139
Nano-field modeling, *see* Molecular dynamics
    (MD) approach
Nanogenerators
    double-layered plate reinforcement, 139, 140
    dynamic response, nanobar voltage, 140, 141
    FEM simulations, 139
    hysteresis loop construction, 140, 143
    linear and nonlinear generated voltage,
        140, 143

longitudinal force, nanobar, 140, 141
nanobeam voltage, 140, 142
nanotechnology research, 138
nondimensionalization, 139
nonlinearity behavior modeling, 139
phase diagram, nanobar voltage, 140, 142
piezoelectric materials, 138
ZnO nano-cantilevers, 139
ZnO nanostructures, 139
Nanoinjection detectors, FEM
  description, 478
  development, simulator
    Auger recombination, 485–486
    2-D cross modeling, cylindrical
      symmetry, 484–485
    hot electron effects, 487
    impact ionization, 486–487
    ionization, dopants, 485
    nonlinear mobility, 486
    self-consistent 3-D core physics, 484
    surface recombination effects, 487–488
    temperature effects, 488
    thermionic emission, 487
  electro-optical modeling, *see* Electro-optical
    modeling
  layer and band structure, 482
  measured performance
    current and optical gain–voltage, 492, 493
    imagers, 494–503
    noise *vs.* voltage, 493, 494
    passivated, 494, 495
    shot noise suppression/enhancement,
      Fano factor, 492
  photodetection and imaging, SWIR,
    479–480
  principles, 481
  SPD, *see* Single photon detector (SPD)
  theory of operation, 483
Nanoinjection imagers
  current and gain voltage, integration,
    496–497
  current voltage plot, 496
  hybridized test pixels
    bandwidth, 497
    noise performance, 497–498
    spatial sensitivity, 498
    temperature behavior, 499, 500
  microscope image, gap, 494, 496
  SEM image, detectors, 494, 495
  signal-to-noise analysis, 500–502
  signal-to-noise level, commercial camera
    measured sensitivity, detector, 499, 500
    short-wave infrared images, 499, 501

Nano level
  3D model
    and capillaries, 580
    current flows straight lines, 578, 579
    deformed current lines and equipotential
      surface, 578, 579
    matrix of coordinates, 576–577
    matrix of elements, 577
    principle, 575
    tetrahedrons, cube full, 578
  filtration, *see* Filtration
  fluid flow, capillaries
    input face, 580
    velocity field, 580–581
  geometry objects
    analytical approach, 567
    bean, 574–575
    capillaries, 571–572
    corks, 575
    current lines, 570
    matrix *TrP*, 567–568
    meshes, 567
    microwave sintering, 572–573
    triangles, square, 568–571
  granular material, 559–560
  heterogeneous materials, 559
  microwave heating
    net, light and thermal currents, 581, 582
    tetrahedrons surface, temperature, 581
Nanomanipulation strategy
  AFM, 99
  cluster deflection results, 100
  force feedback data, 99
  photodiode data, 100
  substrate *vs.* nanocluster, 99
Nano-/microelectronics, FEM
  coupled thermo-electrical-mechanical,
    20–21
  degradation, HFETs, *see* Heterostructure
    field-effect transistors (HFETs)
  description, 19
  thermal stress, 3D IC structures, *see* 3D IC
    structures, FE method
Nanorobotics, FEM
  Buckingham method, 95
  CADD, 98
  CM, nanorobotic science, *see* Continuum
    mechanics (CM)
  concurrent models, 97
  coupling, CM and MM, 144
  coupling models, 97–98
  DC, 98
  defined, nanomanipulation, 87

description, 86
dimensional analysis, 95
equation, motion, 94–95
flowchart, research, 86, 87
hierarchical models, 97
linear and nonlinear behavior, NMES, 144–145
macrodimensions modeling, 98
macroscale nanorobots, *see* Macroscale nanorobots
mass–spring–damper model, 96
MFPs, 96
micro-actuators/sensors, 96
micrometer and microsecond dimensions, 97
MM-based approaches, 96
multiscale method, *see* Multiscale method
nanomanipulation, CGMD, 144
nanorobots
    applications, 89–90
    types, 87–88
nondimensionalizing and scaling techniques, 95
nonlinear behavior, MEMS, 144
proposed nanomanipulation strategy, 99–100
QC method, 98
teleoperated systems, 86
Nanoscale forces
    CM-based approach, 94
    humidity and electrostatic charge, 92
    mechanical instability, 93
    photodiode data, 93–94
    tip, particle and substrate, electrostatic, 92–94
    van der Waals forces, 92–93
Navier–Stokes equations and mass conservation, 426–428
NEMS, *see* Nano-electromechanical systems (NEMS)
Neural networks (NNs) model
    calculation, 517, 518
    creation, 517
    description, 517
    "proboscis", 517
    use, 538
Newtonian fluids, 427
Newton–Raphson iteration algorithm, 202–203
Newton–Raphson method
    description, 365–366
    non-linear equations, 366–367
NNs model, *see* Neural networks (NNs) model
Nonlinearities modeling
    advantages and disadvantages, 133
    comparison, exact and present models, 134
    electrical excitation, 134, 136

electromechanical excitation, 134, 136
experimental work, 134, 135
hysteresis loops, composite configuration, 138
hysteretic behavior calculation, 133–134
mechanical excitation, 134, 135
MEMS and NEMS, 137
nanorobotic systems, 136
Piezo layers, 138
PI operator, 133
sinusoidal excitation, 136, 137
triangular excitation, 137
Non-Newtonian fluids
    defined, 430
    networks, 438–439
    pressure drop, 436–437
    square microchannel, 437–438
    viscosity
        concentration, 432
        shear rate, 433–435
        temperature, 433
Numerical experiments, clamping devices
    material characteristics, brass, 516
    modulus of elasticity, 513, 516
    NNs model, 517–519
    nodal displacements, 512, 514
    Poisson's ratio, 513, 516
    radial displacements, 512, 515
Numerical model, DLD
    streamlines, 461
    transport, concentration stream, 462
Numerical modeling, biotechnology, 425

O

Object-oriented modeling, mechanical properties computation
    3D microstructure-based model, 258, 259
    RVE, assumptions, 257
    SEM micrographs, 257, 258
Oliver–Pharr method, 251
Optical properties, nanocomposites
    dielectric constant, 269
    electrical conductivity, reciprocity theorem, 270
    EM waves, 272
    Lorentz–Lorenz model, 269–270
    Maxwell's equations, 271
    nanopores/nanowires, 273
    Poynting vectors, 272–273
    VAT, 270
    wave propagation, time-harmonic form, 271
Ostwald law, 433, 434

P

Parallel micro-robot structures, FEM
 advantages, 520
 aim, 519–520
 built-in linear drives mechanism, 528–531
 built-in rotary drives mechanism, 526–528
 3D structure, 520, 522
 established mathematical model, 522–525
 sixpod, 520, 521
 spatial three-pod mechanism, *see* Spatial
  three-pod mechanism, FEM
 three-arm delta robots, 520, 521
 UPU joint, 520, 521
Partial differential equations (PDEs), 315
PCG, *see* Preconditioned conjugate gradient
  (PCG)
PDEs, *see* Partial differential equations (PDEs)
Physical constants, 587
Physical vapor deposition (PVD), 382
Pinched flow fractionation (PFF), *see* Flow
  focusing device, PFF
Piola–Kirchhoff stress, 281
Plain polypropylene (PP), 292
Poisson's equation, 274
Power law, 253, 432, 433, 436
Preconditioned conjugate gradient (PCG)
 norm scaling algorithm, 216
 procedure, 214–216
Pressure drop
 3D calculation, COMSOL, 430, 431
 flow rate, 429
Probability distributions, 596–597

Q

Quantum wires and nano-MOSFET
  devices, 8–9

R

Radial basis function (RBF) interpolations
 coefficient vector, 219
 displacement, 219
 MQ, 219, 220
Radial distribution functions (RDFs)
 comparison, MD and CGMD models, 108
 SC potential, 108
 trial and error process, 108
Rafii-Tabar–Sutton (RTS) potential, 104–105
RBF interpolations, *see* Radial basis function
  interpolations
RDFs, *see* Radial distribution functions (RDFs)

Reduced-order modeling, CNTs
 6 degree-of-freedom nodes, 47–48
 description, 47
 interfacial links, 47
 lumped model, 48
 matrix condensation technique, 47
Representative volume element (RVE)
  modeling
 assumptions, 258
 CNTs synthesis, 260, 261
 tools, 260
Rotary needleless spinnerets
 ball, 318
 coils, 322, 323
 cylinder, 317–318
 disk, 317, 319–320
 electric field profiles, 317
 performance and spinneret shape, 322
 productivities, 320, 321
RVE modeling, *see* Representative volume
  element modeling

S

Scanning electron microscopy (SEM), 257
Scanning probe microscopes (SPMs)
 and AFM, 88
 and applications, 90
 identification and operation tasks, 90
 optical and electron microscopes, 88
 teleoperated systems, 86
 tube actuators, 131
 use, piezotube, 126
Schrödinger wave equation
 description, 339
 feature, 340–341
SEM, *see* Scanning electron microscopy (SEM)
Semiempirical models
 classical molecular dynamics
  Hamilton's equations, 343
  Lagrange's equation, 343
  Newton's equation, 342
 TDSCF, *see* Time-dependent self-consistent
   field (TDSCF)
SFL, *see* Spatial four-bar linkage (SFL)
Shallow channel, *see* Hele–shaw model, pseudo
  3D microflow
Shear thinning, 433
Short-wave infrared (SWIR) spectrum,
  479–480
Signal-to-noise ratio (SNR)
 and imager performance, 496
 measurements, 500

and photon count, 501, 502
quantum cryptography, 479
SPDs, 480
Simulink®
lumped analysis, 79–82
with spice, 80
Single-phase flow focusing
channel center, sheath flows, 450, 452
2D calculation, particles transportation, 450, 451
principle, 3-D, 450, 452
sheath flow rates, 450, 453
streamlines and concentration slices, 450, 451
Single photon detector (SPD)
defined, 480
properties, 480–481
quantum cryptography, 480
"Single photon detectors", 480
Single-walled CNT (SWCNT), 294
SNR, *see* Signal-to-noise ratio (SNR)
Sol–gel process
colloidal particles, 383
description, 382–383
nanohydroxyapatite, 383–384
Spatial four-bar linkage (SFL)
angle, 551, 552
kinematic analysis, 551, 552
planes of motion, 551
use, 555–556
Spatial three-pod mechanism, FEM
advantages, 535
aluminum alloy, 531–532
built-in linear drives, 533–537
built-in rotary drives, 532–535
hybridization, 537
kinematic tasks, 535
optimization problems, 538
tetrahedrons, 532
UPU kinematic couples, 530–531
SPD, *see* Single photon detector (SPD)
SPMs, *see* Scanning probe microscopes (SPMs)
Stiffness matrix
Cyclic rule, 199
3D analyses, 199
Lagrangian formulation, 197
tangent, 197–198
Stokes equations, *see* Navier–Stokes equations
Strain energy minimization, 188
Strain estimation methods, AFEM
FEM, 217
MLS approximation, 217–218
moving Kriging interpolation, 220–222

neighbor atom search, 222–223
RBF interpolations, 219–220
Structural mechanics model, CNTs
description, 44
elemental equilibrium equation, 45–46
flexures, 45
"original CNT model", 47
properties, 44
Sugarcube model
GDS-II layout, 69, 71
MEMSCAPMUMP design rules, 70
nanomaterial-testing device, 69, 70
ready-made M/NEMS, 69
static, modal and transient analysis, 69
Sugar to sugarcube modelling
built-in automation, 56
conversion, 54, 56
description, 54
display, electrothermal actuator, 54, 56
framework, 54, 57
micromirror and sugar representation, 54, 55
netlist, electrothermal actuator, 54, 56
Summations, series
finite element, terms, 597
infinite element, terms, 598
Sutton–Chen (SC) interatomic potential, 104, 114–115
SWIR spectrum, *see* Short-wave infrared spectrum
Symbols and formulas, 589–607

T

TDSCF, *see* Time-dependent self-consistent field (TDSCF)
Tersoff model, 212
Thermo-mechanical response, N/MEMSs
description, 73
free-body diagram, 73
specimens, 74
Thin film mechanical test
adhesion and wear test methods, *see* Adhesion and wear test methods
instrumented nanoindentation
coating adhesion, 386–387
load–displacement curve, 385, 386
micro-and nanocoatings, FEA
bone stress and strain distributions, 392
DLC, 391
elastic and plastic deformation, 394
substrate, 393
zirconia, hydroxyapatite and alumina, 393–394

micro-and nanoindentation
    ceramic structures, 390–391
    compressive and tensile tests, 387
    experimental tests and finite element
        simulations, 389–390
    loading and unloading, 387
    mechanical properties, 391
    mineralization, 387
    three-dimensional, 388
Time-dependent self-consistent field
    (TDSCF), 342
Trapping chambers
    particles, recirculation grooves, 462, 463
    recirculation, diamond-shaped
        microchamber, 463
    slices, concentration, 464
Trigonometric identities, 590–591

U

Unit cell modeling, 255–256

V

VAT, *see* Volume averaging theory (VAT)

Vectors
    derivatives, coordinates
        Cartesian, 604
        cylindrical, 604
        spherical, 605
    fundamental theorems
        curl (stokes), 606–607
        divergence, 606
        gradient, 606
    identity
        addition, division and power
            rules, 606
        product rules, 605
        second derivative, 606
        triple products, 605
Volume averaging theory (VAT), 270

Y

Young's modulus, 27, 29

Z

Zincblende atom configurations, 210, 211

9 781138 076884